LARGE-SCALE STRUCTURES AND PECULIAR MOTIONS IN THE UNIVERSE

A SERIES OF BOOKS ON RECENT DEVELOPMENTS IN ASTRONOMY AND ASTROPHYSICS

Printed by BookCrafters, Inc.

First published 1991

Library of Congress Catalog Card Number: 91-70003
ISBN 0-937707-34-1

D. Harold McNamara, Managing Editor of Conference Series
293 ESC Brigham Young University
Provo, UT 84602

ASTRONOMICAL SOCIETY OF THE PACIFIC

CONFERENCE SERIES

Volume 15

LARGE-SCALE STRUCTURES AND PECULIAR MOTIONS IN THE UNIVERSE

Edited By

David W. Latham and L. A. Nicolaci da Costa

A SERIES OF BOOKS ON RECENT DEVELOPMENTS IN ASTRONOMY AND ASTROPHYSICS

Vol. 1-Progress and Opportunities in Southern Hemisphere Optical Astronomy: The CTIO 25th Anniversary Symposium
ed. V. M. Blanco and M. M. Phillips ISBN 0-937707-18-X

Vol. 2-Proceedings of a Workshop on Optical Surveys for Quasars
ed. P. S. Osmer, A. C. Porter, R. F. Green, and C. B. Foltz ISBN 0-937707-19-8

Vol. 3-Fiber Optics in Astronomy
ed. S.C. Barden ISBN 0-937707-20-1

Vol. 4-The Extragalactic Distance Scale: Proceedings of the ASP 100th Anniversary Symposium
ed. S. van den Bergh and C. J. Pritchet ISBN 0-937707-21-X

Vol. 5-The Minnesota Lectures on Clusters of Galaxies and Large-Scale Structure
ed. J. M. Dickey ISBN 0-937707-22-8

Vol. 6-Synthesis Imaging in Radio Astronomy: A Collection of Lectures from the Third NRAO Synthesis Imaging Summer School
ed. R. A. Perley, F. R. Schwab, and A. H. Bridle ISBN 0-937707-23-6

Vol. 7-Properties of Hot Luminous Stars: Boulder-Munich Workshop
ed. C. D. Garmany ISBN 0-937707-24-4

Vol. 8-CCDs in Astronomy
ed. George H. Jacoby ISBN 0-937707-25-2

Vol. 9-Cool Stars, Stellar Systems, and the Sun
ed. G. Wallerstein ISBN 0-937707-27-9

Vol. 10-The Evolution of the Universe of Galaxies, The Edwin Hubble Centennial Symposium
ed. Richard G. Kron ISBN 0-937707-28-7

Vol. 11-Confrontation Between Stellar Pulsation and Evolution
ed. C. Cacciari and G. Clementini ISBN 0-937707-30-9

Vol. 12-The Evolution of the Interstellar Medium
ed. L. Blitz ISBN 0-937707-31-7

Vol. 13-The Formation and Evolution of Star Clusters
ed. K. Janes ISBN 0-937707-32-5

Vol. 14-Astrophysics with Infrared Arrays
ed. R. Elston ISBN 0-937707-33-3

Alonso, María Victoria
CNPq - Observatório Nacional
Rua General José Cristino 77
20921, São Cristóvão, RJ
Brasil

Andernach, Heinz
INPE - Inst. de Pesquisas Espaciais
Depart. Radioastronomia e Física Solar
C.P. 515
12201, São José dos Campos, SP
Brasil

Bergman, Thaisa Storchi
Inst. de Física - UFRGS
C.P. 15051
91500, Porto Alegre, RS
Brasil

Bertschinger, Edmund
Massachusetts Inst. of Technology
Dept. of Physics
Cambridge, MA 02139
USA

Bruno, Ruth
CNPq - Observatório Nacional
Rua General José Cristino 77
20921, São Cristóvão, RJ
Brasil

Burstein, David
Dept. of Physics
Arizona State Univ.
Tempe, AZ 85728
USA

Capelato, Hugo Vicente
INPE - Inst. de Pesquisas Espaciais
Av. dos Astronautas 1758
12201, São José dos Campos, SP
Brasil

Chan, Roberto
CNPq - Observatório Nacional
Rua General José Cristino 77
20921, São Cristóvão, RJ
Brasil

Chian, Abraham
INPE - Inst. de Pesquisas Espaciais
C.P. 515
12201, São José dos Campos, SP
Brasil

Chung, Kai Cheong
CNPq - Centro Brasileiro de Pesquisas Físicas
Rua Xavier Sigaud 150
22290, Urca, RJ
Brasil

Cunha, Kátia
CNPq - Observatório Nacional
Rua General José Cristino 77
20921, São Cristóvão, RJ
Brasil

da Costa, Luiz A. Nicolaci
CNPq - Observatório Nacional
Rua General José Cristino 77
20921, São Cristóvão, RJ
Brasil

da Silva, Jr., Waldemar M.
UFF - Inst. de Física
Outeiro de São João Batista s/n
24210, Niterói, RJ
Brasil

da Silva, Maria de Fátima A.
CNPq - Observatório Nacional
Rua General José Cristino 77
20921, São Cristóvão, RJ
Brasil

Davis, Marc
Dept. Astronomy
Univ. California
Berkeley, CA 94720
USA

de Souza, Carlos Alexandre
INPE - Inst. de Pesquisas Espaciais
Depart. de Astrofísica
C.P. 515
12201, São José dos Campos, SP
Brasil

Dekel, Avishai
The Hebrew Univ. of Jerusalem
Racah Inst. of Physics
Jerusalem, 91904
Israel

Dressler, Alan
The Observatories of the
Carnegie Inst. of Washington
813 Santa Barbara St.
Pasadena, CA 91101
USA

Fagundes, Hélio
UNESP - Inst. de Física Teorica
Rua Pamplona 145
01405, São Paulo, SP
Brasil

Fairall, Anthony P.
Dept. of Astronomy
Univ. of Cape Town
Rondebosch, 7700
South Africa

Feretti, Luigina
Inst. di Radioastronomia
Via Irnerio 46
I-40100, Bologna
Italy

Focardi, Paola
CNPq - Observatório Nacional
Rua General José Cristino 77
20921, São Cristóvão, RJ
Brasil

Gaia, Miguel Ayan
CNPq - Observatório Nacional
Rua General José Cristino 77
20921, São Cristóvão, RJ
Brasil

García-Lambas, Diego
Observatorio Astronómico
Univ. Nacional de Córdoba
Laprida 854, 5000 Córdoba
Argentina

Giovanelli, Riccardo
NAIC - Arecibo Observatory
P.O. Box 995
Arecibo, PR 00612
USA

Gott, III, J. Richard
Dept. of Astrophysical Sciences
Peyton Hall
Princeton Univ.
Princeton, NJ 08540
USA

Junqueira, Selma
CNPq - Observatório Nacional
Rua General José Cristino 77
20921, São Cristóvão, RJ
Brasil

Kaiser, Nicholas
Canadian Inst. Theoretical Astrophysics
Univ. of Toronto
60 Saint George St.
Toronto, Ontario M5S 1A1
Canada

Kashlinsky, A.
Astronomy Unit, School of Mathematics
Mile End Rd.
Queen Mary College
London E1 4NS
England

Kirhakos, Sofia Demetrios
INPE - Inst. de Pesquisas Espaciais
Av. dos Astronautas 1758
12201, São José dos Campos, SP
Brasil

Kofman, Lev
Tartu Astrophysical Observatory
Toravere 202444
Estonia
USSR

Kraan-Korteweg, Reneé
Astronomisches Inst. der Univ. Basel
Venusstr. 7
Ch - 4102 Binningen
Switzerland

Latham, David W.
Center for Astrophysics
60 Garden St.
Cambridge, MA 02138
USA

Lemos, José P. S.
CNPq - Observatório Nacional
Rua General José Cristino 77
20921, São Cristóvão, RJ
Brasil

Lucchin, Francesco
Dipart. di Fisica *Galileo Galilei*
Univ. di Padova
Via Marzolo 8
35131 Padova
Italy

Lucey, John
Dept. of Physics
Univ. of Durham
Science Lab., South Road
Durham, DH1 3LE
England

Lynden-Bell, D.
Inst. of Astronomy
Univ. of Cambridge
The Observatories, Madingley Rd.
Cambridge, CB3 0HA
England

Maddox, S. J.
Dept. of Astrophysics
Oxford Univ.
South Park Rd.
Oxford, OX1 3RQ
England

Maia, Márcio A. G.
CNPq - Observatório Nacional
Rua General José Cristino 77
20921, São Cristóvão, RJ
Brasil

Marques, Gil da Costa
USP - Inst. de Física
C.P. 20516
01498, São Paulo, SP
Brasil

Matarrese, Sabino
Dipart. di Fisica *Galileo Galilei*
Univ. di Padova
Via Marzolo 8
35131 Padova
Italy

Maurellis, Achilles
Univ. of Cape Town
Dept. of Applied Mathematics
Univ. of Cape Town
Rondebosch 7700
South Africa

Mosconi, Mirta Beatriz
Observatorio Astronómico
Univ. Nacional de Córdoba
Laprida 854, 5000 Córdoba
Argentina

Muriel, Hernán
Observatorio Astronómico
Univ. Nacional de Córdoba
Laprida 854, 5000 Córdoba
Argentina

Olowin, R. P.
Dept. of Mathematical Sciences
Saint Mary's College
Moraga, CA 94575
USA

Opher, Reuven
USP - Inst. Astronômico e Geofísico
Av. Miguel Stefano 4200
01000, São Paulo, SP
Brasil

Ortega, Vladimir Garrido
CNPq - Observatório Nacional
Rua General José Cristino 77
20921, São Cristóvão, RJ
Brasil

Pech, Gerson
CNPq - Cent. Brasil. de Pesq. Físicas
Rua Xavier Sigaud 150
22290, Urca, RJ
Brasil

Pellegrini, Paulo Sergio S.
CNPq - Observatório Nacional
Rua General José Cristino 77
20921, São Cristóvão, RJ
Brasil

Pereira, Flavio Irineu M.
CNPq - Observatório Nacional
Rua General José Cristino 77
20921, São Cristóvão, RJ
Brasil

Primack, Joel
Physics Dept.
Univ. California
Santa Cruz, CA 95064
USA

Rabaça, Carlos Roberto
USP - Inst. Astronômico e Geofísico
Av. Miguel Stefano 4200
01000, São Paulo, SP
Brasil

Rabaça, Duilia F. de Mello
USP - Inst. Astronômico e Geofísico
Av. Miguel Stefano 4200
01000, São Paulo, SP
Brasil

Ryden, Barbara
Center for Astrophysics
60 Garden St.
Cambridge, MA 02138
USA

Salim, José Martins
CNPq - Cent. Brasil. de Pesq. Físicas
Rua Xavier Sigaud 150
22290, Urca, RJ
Brasil

Santiago, Basilio Xavier
CNPq - Observatório Nacional
Rua General José Cristino 77
20921, São Cristóvão, RJ
Brasil

Santos, Nilton O.
CNPq - Observatório Nacional
Rua General José Cristino 77
20921, São Cristóvão, RJ
Brasil

Scaramella, Roberto
SISSA
Strada Costiera 11
34014 Trieste
Italy

Shandarin, Sergei
Inst. for Physical Problems
Kosygina 2
117334 Moscow
USSR

Shapiro, Paul
Dept. of Astronomy
Univ. of Texas
Austin, TX 78712
USA

Silveira, Tomás de Aquino
ICEX-UFMG, Depart. de Física
C.P. 1208
30161 Belo Horizonte, MG
Brasil

Smith, Rodney M.
Dept. of Physics
Univ. of Durham
Science Lab., South Rd.
Durham, DH1 3LE
England

Soares, Maria Cristina
ICEX-UFMG, Depart. de Física
C.P. 1208
30161, Belo Horizonte, MG
Brasil

Strauss, Michael
Astronomy Dept. 105-24
California Inst. Technology
Pasadena, CA 91125
USA

Sutherland, Will
Dept. of Astrophysics
Oxford Univ.
South Park Rd.
Oxford, OX1 3RQ
England

Szalay, Alex
Dept. of Physics and Astronomy
The Johns Hopkins Univ.
Baltimore, MD 21218
USA

Telles, José Eduardo
USP - Inst. Astronômico e Geofísico
Av. Miguel Stefano 4200
01000, São Paulo, SP
Brasil

Tomimura, Nazira A.
UFF - Inst. de Física
Outeiro de São João Batista s/n
24210, Niterói, RJ
Brasil

Triay, Roland
CNRS - Inst. Physique Teorique
13288, Marseille Cedex 9
France

Viegas-Aldrovandi, Sueli M.
USP - Inst. Astronômico e Geofísico
Av. Miguel Stefano 4200
01000, São Paulo, SP
Brasil

Villumsen, Jens Verner
Dept. of Astronomy
Ohio State Univ.
Volumbus, Ohio 43210
USA

Wegner, Gary
Dept. of Physics and Astronomy
Dartmouth College
Wilder Lab.
Hanover, NH 03755
USA

White, Simon
Steward Observatory
Univ. of Arizona
Tucson, AZ 85721
USA

Willmer, Christopher N. A.
CNPq - Observatório Nacional
Rua General José Cristino 77
20921, São Cristóvão, RJ
Brasil

FOREWORD

In recent years Cosmology has been one of the most exciting areas of research. New theoretical ideas have emerged, enormous progress has been made with *N*-body simulations, and the database of observed galaxy redshifts and distances has grown enormously. It has not been an easy task for workers in the field to keep pace with this rapid progress, and periodic meetings have played a major role in keeping the information flowing among different research groups. This motivated the organization of a Workshop on the *Large Scale Structures and Peculiar Motions in the Universe*, in 1989 (May 22-27) when no other major meeting on the subject was scheduled. The main goal of the workshop was to review the progress in observational and theoretical work and outline future research efforts. The idea of holding the meeting in Rio de Janeiro was very appealing for Brazilian researchers and students because it represented a rare opportunity to interact with some of the most active researchers in cosmology from around the world.

Besides M. Davis and A. Dekel, members of the Scientific Committee, many people helped in elaborating the idea of a meeting in Rio de Janeiro, giving suggestions about speakers and important hints about the organization. In particular we would like to acknowledge the help of S. Djorgovski, who unfortunately could not participate.

During the Workshop exciting new results were presented, giving promise that some of the fundamental questions concerning the structure of the Universe may be answered in the near future. In particular, enormous progress has been made in understanding peculiar velocity flows and in using these data to determine the underlying density field.

This book summarizes the contributions presented at the workshop. We have made a special attempt to reproduce faithfully the discussions that followed each presentation. We thank the participants for their collaboration and patience in this effort.

Judging from the comments at the end of the Workshop we believe that the meeting was a great success, and we thank all the participants for the high-quality of their contributions and the friendly atmosphere that prevailed during the meeting.

We are indebted to the members of the Local Organizing Committee, G. da Costa Marques, J. Danon and P. S. Pellegrini, for their help in raising the necessary funds for the meeting during a very difficult time. Our special gratitude to B. Santiago, V. Bertholini, L. Penalva and P. dos Santos and to all other members of the Observatório Nacional staff for their invaluable help in keeping everything running.

We would also like to express our appreciation to the many funding agencies of Brasil that made it possible the organization of this important meeting, in alphabetical order: CAPES, CNPq, FAPERJ, FAPESP and FINEP. We also acknowledge the partial sponsorship of the Hotel Copacabana Palace, Varig S. A. and Amsterdam Sauer.

Much of the labor in formatting the proceedings was carried out at the University of Canterbury, in Christchurch, New Zealand. DWL is grateful for the Erskine fellowship which made this possible.

Finally, our special thanks to Olavo and Betsy Monteiro de Carvalho, Aloisio Salles, Tereza Muniz and Rosario Nicolaci da Costa for a incredible dinner party, which will be fondly remembered as a ten-sigma event.

L. Nicolaci da Costa and David. W. Latham
Tuesday 27 November 1990

OPENING REMARKS

DAVID W. LATHAM
Harvard-Smithsonian Center for Astrophysics
60 Garden Street, Cambridge, MA 02138

Nine years ago, in March of 1980, Luiz da Costa came into my office at the Center for Astrophysics to talk about the possibilities for carrying out a redshift survey in the southern hemisphere, to extend the work of the CfA Redshift Survey that a group of us was doing in the north. Luiz had finished his graduate studies at Harvard in theoretical astrophysics, and had returned here to his native city of Rio to work at the Observatório Nacional and to root for Vasco, and he was looking for a good observational project that could take advantage of the Brazilian 1.5-m Boller and Chivens telescope located near Itajubá. Luiz had already talked to Marc Davis and was brain-washed to think that a redshift survey was the most important project to do in observational cosmology, and he wanted to talk to me about what kind of instrument that would be needed. That was the beginning of our collaboration and friendship, and of a long and complicated path that has eventually led to this Workshop on Large Scale Structures and Peculiar Motions in the Universe.

At CfA we figured out that our collaboration to build a duplicate of our Z-Machine was serious when Marcos Nunes, the chief engineer from the Observatório Nacional showed up at John Geary's laboratory one day, ready to get to work making a copy of our photon-counting Reticon detector system. For several years this system was used at Itajubá, where the weather was not very good and there was a lot of competing pressure for observing time. Nevertheless, by dogged determination one problem after another was overcome, and the redshifts began to pile up. The culmination of this first stage of the effort was the publication of the Southern Sky Redshift Survey, in collaboration with astronomers working at Las Campanas in Chile and at Sutherland in South Africa. Then the system was transferred to the 2.2-m reflector at El Leoncito in Argentina, where it is now being used to complete the redshift work on a deeper slice in the south.

More than two dozen papers on a variety of aspects of observational cosmology have emerged from this effort, with more on the way. In addition to the basic redshift survey publications, there have been results on groups, clusters, and superclusters; on velocity dispersions and distances, and on the deeper slice.

Part of Luiz's secret of success has been the loyal team that he has put together and led at the Observatório Nacional, people like Pellegrini, Marcos Nunes, Charles Rité, Darcy Nascimento, and several good students, four of whom have already earned (or will soon earn) their PhDs; Reinaldo de Carvalho, Márcio Maia, and Christopher Willmer in addition to the faithful companion Pelle.

It is an indication of Luiz's vision, and energy, and success that we are gathered here for this Workshop, in such a pleasant and exotic venue, for a program that combines both observationalists and theorists. Here is a golden opportunity for us not only to talk about what has been accomplished, but to discuss where we should go from here. My own view is that the observationalists have more to gain by listening to the theorists, rather than the other way around. I was reading the proceedings of the George Mason meeting on brown dwarfs last week, and I couldn't help but be amused by John Bahcall's advice in the symposium summary, that smart observers should ignore the theoreticians if they want to make progress. Maybe that is good advice for the field of brown dwarfs or quasars, but when it comes to the next steps in observational cosmology, I think the observers should look for some guidance from the theory when they map out their plans. The problem is that we have reached the point where vast amounts of telescope time and large sustained efforts will undoubtedly be needed to advance as far in the next decade as we have in the past ten years. So here is my plea to the theorists; don't be bashful, tell us what you think the observers should be doing next.

Another side of the same coin is that we should try to avoid too much duplication of effort as we push ahead with the observations. Competition can be a very powerful driving force for progress, but when the observing programs get so large and expensive, we really have an obligation to organize the efforts between different groups.

Now let me turn to some practical remarks about the publication of the proceedings of the workshop. The Astronomical Society of the Pacific has agreed to publish the proceedings of the workshop on *Large-Scale Structures and Peculiar Motions in the Universe* as a volume in the *A.S.P. Conference Series*, with D. W. Latham and L. N. da Costa as editors. The volume will be produced in a uniform manner using WordPerfect 5.1 on an IBM PC and will be printed in final form using a HP LaserJet II. To allow easy conversion into our system we request that manuscripts be submitted as ASCII files, either via bitnet to LATHAM@CFA3, or else on an MS-DOS compatible diskette. We also ask that you provide a printed copy, mailed to Dave Latham, so that we can see where you want your figures and tables placed, and so we can see how your equations and any special characters should look.

We will try to exercise some flexibility on the length of papers. As a guideline, invited authors should aim for about 12 pages and should try not to exceed 16 pages; contributed papers should aim for about 6 pages and should not exceed 8 pages.

For the discussion we will pass out sheets as each participant opens his or her mouth to make a significant comment. If you don't want your comment to appear in print, refuse to accept the sheet. If you do want to have your comment published, write down on the sheet what you meant to say and hand the sheet back to Dave Latham or one of his helpers. Then the sheet will be given to the responder for his or her reply. You will get a chance to review our version of your comments when we send out the final printed version for approval.

LIGHT AND GRAVITY

D. LYNDEN-BELL
Institute of Astronomy, University of Cambridge
The Observatories, Madingley Road, Cambridge CB3 0HA, England

ABSTRACT The Local Group and nearby galaxies are moving in the direction from which the extragalactic light flux comes. The value of the light flux is already quite large for a motion of 600 km s^{-1} and yields a value of $\Omega \sim 0.2$ if we take no biasing on the large scales involved ($\sim$3000 km s^{-1}). New major attractors or domain-wall theories of galaxy streaming are not wanted. They could only lower Ω still further or lead to large-scale biasing factors greater than 3. Raychaudhury's survey of the Hydra and Centaurus regions behind the Great Attractor shows that the agglomeration of clusters around Shapley 8 contributes only some 10 - 15% of the dipolar light flux and can only account for a similar fraction of the Local Group's motion.

INTRODUCTION

In the early 1920s V. M. Slipher showed that most extragalactic nebulae were moving away from us (Eddington 1922). Hubble's great struggle to determine extragalactic distances led him to the law governing this expansion $v = Hr$ (Hubble 1929). To use this law not only must the distance r be known, but the correct zero point from which the radial velocities should be measured must be determined. This may be done by rewriting the law $v_H + \mathbf{v}_\odot \cdot \hat{\mathbf{r}} = Hr$, where v_H is the heliocentric radial velocity, and choosing $\mathbf{v}_\odot$ to give best fit. In his pioneering paper Hubble got a $\mathbf{v}_\odot$ of 305 km s^{-1} towards $\ell = 71°$, $b = -14°$ or in components (96, 280, −74) km s^{-1}. This is not far from 300 km s^{-1} towards $\ell = 90°$, $b = 0°$ which was adopted by the IAU following extensive work by de Vaucouleurs. Although later determinations give some small transverse components too, de Vaucouleurs' prescription is within 50 km s^{-1} of the best zero point today.

Thus, when radiometers were used by Conklin and Bracewell (1967) and Conklin (1969) to find the motion of the Sun through the Cosmic Microwave Background (CMB), astronomers expected a motion of around 300 km s^{-1} towards a direction within 20° of $\ell = 90°$, $b = 0°$. They were wrong! The radio observations consistently gave velocities of around the present value (see review by Lubin and Villela 1986) 360 km s^{-1} but toward $\ell = 267°$, $b = 50°$, about 130° away from the direction predicted. This implies that the locally determined zero point found by de Vaucouleurs' prescription is in reality moving at 600 km s^{-1} towards $\ell = 268°$, $b = 27°$. Not only is this motion (hereafter referred to as the Local Group's motion) non-zero, it is even greater than the raw motion of the Sun relative to the CMB. The correction to the local zero point made things worse. Why?

What is found is not just a motion of the Local Group but a motion of all nearby galaxies over and above the general Hubble expansion, since it is these that are used to find the optical zero point. Current theory attributes motions of galaxies to gravitational perturbations due to density fluctuations in the universe which grow themselves by Jeans's gravitational instability. Although 600 km s^{-1} is quite large for a large-scale velocity perturbation, nevertheless it can be accommodated within current theories. There are three ways to test the hypothesis that the velocity has been caused by gravity: (1) to look for the density enhancements that cause the gravity; (2) to map the flow field of galaxies, determine the deviations from Hubble flow more generally and to relate them to the density distribution, using the galaxies as tracers; and (3) to determine the gravity field on the Local Group and so check whether it is parallel to the velocity generated and of sufficient magnitude.

Here we shall consider (1) and (2) rather briefly and concentrate on (3) where work with O. Lahav has made the most recent progress.

DENSITY ENHANCEMENTS

To get a picture of the density field, the best we can do is to map the density of luminous matter. The dark matter may be differently distributed, but it can not be observed at present. The theory of biased galaxy formation suggests that galaxies form more readily per unit mass in regions of higher density, so that the density of luminous matter will give a somewhat exaggerated picture of the total density variations. To parameterize this idea in the simplest possible way we shall take

$$\Delta\rho/\rho_0 = (1/b)\, \Delta\rho_L/\langle\rho_L\rangle \tag{1}$$

where ρ_0 and $\langle\rho_L\rangle$ are the current mean total density of the universe and the mean density of luminous matter, respectively, and b is the bias parameter, which is one if galaxies trace the total matter in an unbiased fashion. Unfortunately, distances to galaxies are not generally available, so the best one can hope for is a large number of galaxies with redshifts. In spite of sterling work by Huchra, Fairall, da Costa and others, complete redshift data is only available for a somewhat biased sample of rather bright galaxies, so in much of the sky the observed galaxy density is strongly affected by observational selection. To get around such problems I show here not the density of galaxies but their distribution over the sky. In Figures 1 and 2 the galaxies which give the greatest light on Earth are given the largest and brightest symbols. The figures are equal area projections of the distribution of galaxies with major diameters $\geq 1'$ in the northern and southern hemispheres. The northern data are from the Uppsala General Catalog (Nilson 1973); the southern from the ESO-Uppsala Catalog (Lauberts 1982) supplemented in the $-2.5° > \delta > -17.5°$ strip by the Morphological Catalog of Galaxies (Vorontsov-Velyaminov and Archipova 1963-74), which is less complete. In the northern sky the most prominent feature is the Virgo Cluster (v ~ 1000 km s^{-1}), through which passes the strong supergalactic band of bright galaxies. In these pictures the Milky Way region $|b| < 15°$ has been replaced by clones of neighboring equal-area strips of greater $|b|$. The strong density enhancement in Centaurus $\ell \sim 300°$, $b \sim 25°$ has been even further emphasized by this process, and the link to the Pavo-Indus-Telescopium band in the southern Galactic hemisphere is made more obvious. Fairall (1988) gives redshift evidence that this linking is real.

STREAMING MOTIONS

The Local Group's motion lies towards Hydra, and both Shaya (1984) and Tammann and Sandage (1985) suggest that the Hydra-Centaurus supercluster, with some help from Virgo, gives the gravity that caused the Local Group's motion. This was contested by the Seven Samurai (Burstein *et al.* 1986, Dressler *et al.* 1987, Lynden-Bell *et al.* 1988, Lynden-Bell

1988) in their studies of galaxy streaming, who found the Centaurus clusters to have a rapid excess motion. They therefore suggested that these were victims, not culprits, and the real cause of the motion might lie in the Great Attractor, which they identified with the concentration of galaxies in northern Centaurus at $\sim$4500 km s^{-1} redshift. Currently there are conflicting data on the excess outward motions of the Centaurus clusters (Lucey and Carter 1988, Aaronson *et al.* 1989, Faber *et al.* 1989), so these arguments must be considered uncertain. However, our most recent results do show strong excess motion outward for most galaxies in this region (Lucey *et al.* this volume). The main finding of the Seven Samurai, that a large-scale streaming of galaxies exists out to at least 3200 km s^{-1} with a value of around 600 km s^{-1}, but on average more towards Centaurus than Hydra, still stands. Much of this motion arises in the southern sky, and it is not parallel to the Local Group's motion but about 35° further around in Galactic longitude. There is no reason why these motions should be parallel – one gives some average velocity of galaxies over quite a large region, and the other gives a local mean velocity of very nearby galaxies.

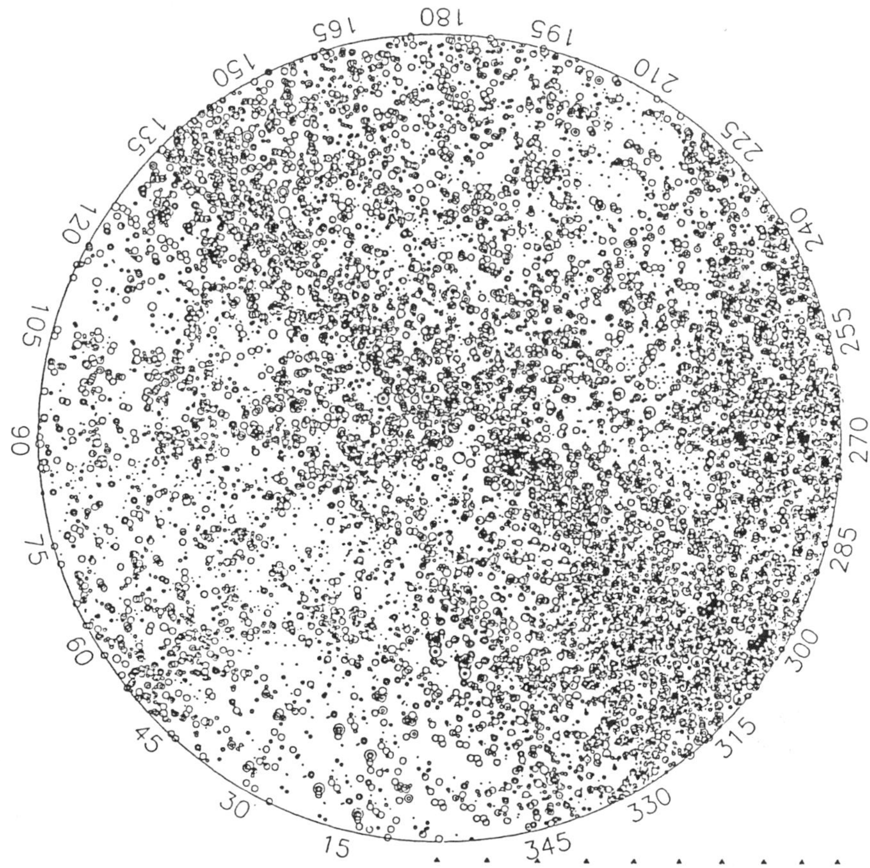

Fig. 1. Equal area projection of the north Galactic hemisphere showing galaxies of the MOCK SKY from which the dipoles has been calculated. In the MOCK SKY strips above and below the Galactic plane are cloned and the clones are then moved into the zone of avoidance.

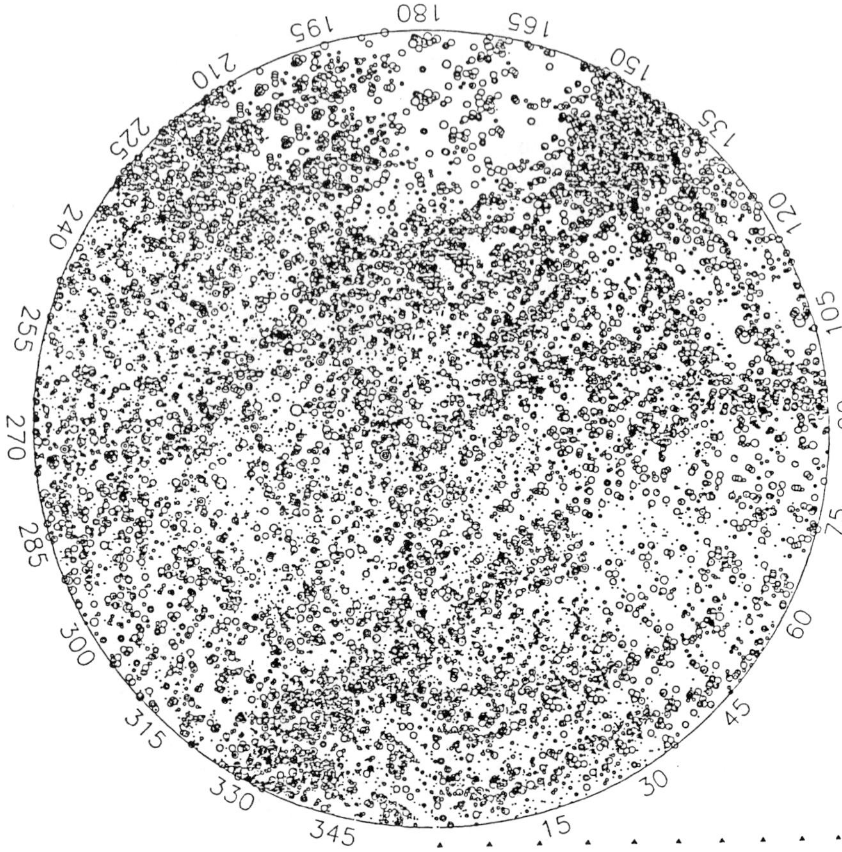

Fig. 2. As Figure 1 for the south Galactic hemisphere.

LOCAL GRAVITY

The light from a galaxy falls off as r^{-2}. The gravity field from a galaxy falls off as r^{-2}. Thus light fluxes and gravity fields remain proportional, however far from the galaxy we may be. Furthermore, another galaxy of twice the luminosity is not unlikely to have roughly twice the total mass. Thus, for statistical purposes we may probably assume there is some overall average mass-to-light ratio and that the gravity field from every galaxy is some multiple of the light flux received from it. Adding all such contributions vectorially, we see that the net gravity field on the Local Group should be proportional to the net directed flux of extragalactic light. Since the gravity field is thought to be responsible for the Local Group's velocity, it follows that this velocity should be aligned with the direction from which the net extragalactic light flux comes.

Are these vectors parallel? To find out we must evaluate the light flux. The available galaxy catalogs do not allow a direct evaluation, but they do give major angular diameters, θ, for the galaxies. Since θ^2 falls off like r^{-2}, and since luminosities are roughly proportional to θ^2, we use $\Sigma\theta^2\hat{\mathbf{r}}$ in place of the net light flux. I exclude details of our compensation for different catalogs, galactic absorption and incomplete data, *etc.* Those interested can find a detailed description in Lynden-Bell, Lahav and Burstein (1989).

The Net Directed Extragalactic Light Flux.

The surface brightness σ of the extragalactic sky after exclusion of Local Group galaxies may be expanded in spherical harmonics in the form

$$\sigma(\ell,b) = M + \mathbf{P}\cdot\hat{\mathbf{r}} + \hat{\mathbf{r}}\cdot\mathbf{Q}\cdot\hat{\mathbf{r}} + \ldots \qquad (2)$$

where $\hat{\mathbf{r}} = (\cos\ell\cos b, \sin\ell\cos b, \sin b)$. $\mathbf{P}$ is the dipole moment of the extragalactic brightness, $\mathbf{Q}$ is the (traceless) quadrupole moment, *etc.* $4\pi M$ is the total extragalactic light falling on earth. The total directed extragalactic light flux is

$$2\pi\int_{-1}^{+1} \mathbf{P}\mu^2 d\mu = (4\pi/3)\mathbf{P},$$

where μ is cos the angle measured from the direction of $\mathbf{P}$. In accord with the procedures of the last section, we replace luminosity by θ^2. The evaluation of the corresponding $\mathbf{P}$ is made by summing over galaxies,

$$\mathbf{P} = (3/4\pi)\sum \theta^2\hat{\mathbf{r}}. \qquad (3)$$

We may also define $\mathbf{P}(>\theta)$ by summing over only galaxies with corrected diameters $>\theta$ and $\mathbf{P}(<\mathrm{v})$ by summing only over galaxies with actual or estimated redshifts less than v. Similarly, we find the corresponding monopoles

$$M(>\theta) = (1/4\pi)\sum \theta^2 \quad \text{(for galaxies with diameter } > \theta). \qquad (4)$$

Our studies of $\mathbf{P}(>\theta)$ (Lahav *et al.* 1988, Lynden-Bell and Lahav 1988) and of $\mathbf{P}(<\mathrm{v})$ (Lynden-Bell *et al.* 1989) show that most contributions to $\mathbf{P}$ arise at velocities <4000 km s^{-1} and from galaxies with diameters $>2'$. However, for comparison with the observed velocity we need the total dipole, not just that from galaxies with diameters $>1'$. From our diameter function for galaxies, it is clear that more that 1/3 of all the light from aggregates of galaxies at 10,000 km s^{-1} will come from galaxies $>1'$ in diameter. We would like to correct the contributions of all the observed galaxies to allow for galaxies that form part of the same luminosity function as those observed, but which are smaller than $1'$. Beyond 10,000 km s^{-1} we have too few galaxies to allow us to perform such a correction with accuracy, but since most of the dipole arises from v $<$ 4000 km s^{-1}, the omission of galaxies beyond 10,000 km s^{-1} where the catalogs are becoming seriously incomplete, is not of great concern. Averaged over such large scales, it appears that the universe is too uniform to give a significant extra contribution to the dipole. From our diameter function for galaxies we find the function $L(>D)$, which gives that fraction of the total light (or rather, the total D^2) which arises from those galaxies with diameters $>D$. The function $L(D)$ is well fitted by

$$L(D) = (1+t/2)/(1+t/5)^4$$

where $t = (D/D_L)^2$ and $D_L = 6538\ '$ km s^{-1} (*i.e.* a galaxy with $D = D_L$ would subtend $1'$ if it were placed at a distance corresponding to 6538 km s^{-1} in the Hubble flow). A galaxy with recession velocity v will have diameter $D = \mathrm{v}\theta$. With a diameter limit of θ_m, the associated galaxies with $D \leq \mathrm{v}\theta_m$ will have been omitted. Thus we will have recorded only a fraction $L(\theta_m)$ of the total light.

To compensate for their omission we must weight each observed galaxy with a factor $1/L(v\theta_m)$. We have only $\sim 1/3$ of the redshifts of our galaxies, so when v is unknown we replace $1/L(v\theta_m)$ by its average taken over all those galaxies with the same angular diameter and with known redshifts. Thus our formula for the total **P** is

$$\mathbf{P} = (3/4\pi) \sum \theta^2 \hat{\mathbf{r}}/L \text{ (for galaxies with diameter } > 1') \qquad (5)$$

where $1/L$ is replaced by $\langle 1/L \rangle$ (a function of θ) when the velocity of the galaxy is unknown. The directions of the vectors $\mathbf{P}(<v)$ as v is varied from 500 km s^{-1} to >10000 km s^{-1} are shown in Figure 3. Evidently there is remarkable agreement between the direction of the total **P** and the motion of the Local Group through the Cosmic Microwave Background. This is a significant improvement on the first uses of the IRAS data for this purpose (Yahil, Walker and Rowan-Robinson 1986, Meiksin and Davis 1986) but the Meurs and Harmon (1988) IRAS sample gives similar agreement (Harmon, Lahav and Meurs, 1987).

It is interesting that the principal positive axis of the brightness quadrupole points straight at the Great Attractor.

Figure 4 was produced by Robin Harmon to demonstrate that although the optical catalogs have many more galaxies, nevertheless they show density contrast on the sky that is far greater than the IRAS galaxies. It is the higher contrast which leads to the high value of Ω determined in the next section from the optical catalogs. Thus the bias factor b determined over the large scales ~ 3000 km s^{-1} over which the dipole arises must be greater

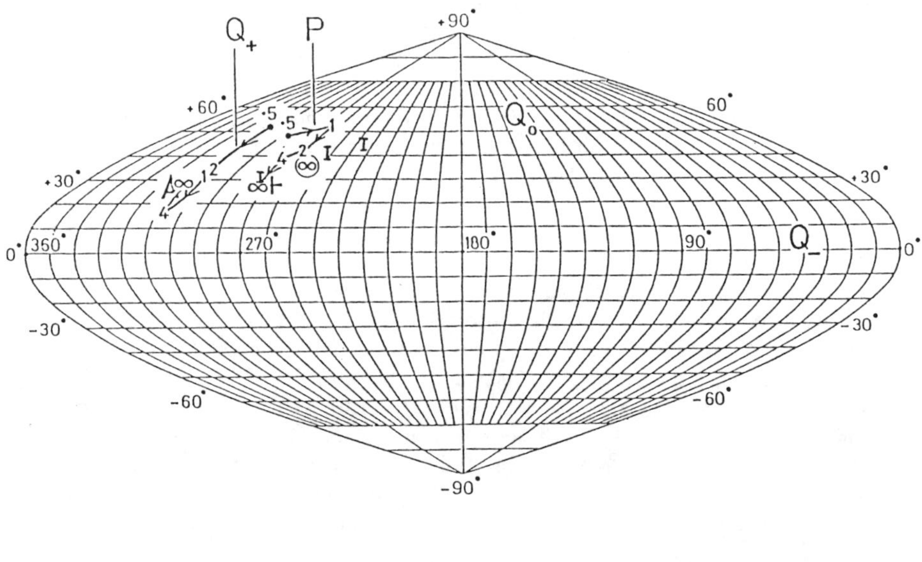

Fig. 3. The directions of the net optical dipole **P** arising from galaxies at <500, <1000, <2000 and <4000 km s^{-1} together with an estimate '∞' of the total dipole. ∞ is the equivalent direction in which a 'mean sky' estimate of the obscured regions has replaced the 'mock sky' estimate that interpolates from neighboring areas. Also shown are the principal axes of the light quadrupole **Q**. Q_- is close to the supergalactic pole, Q_+ points at the Great Attractor *A*. Q_0 is generally small, but it is larger than Q_+ at $v < 1000$ km s^{-1} where its direction (illustrated) lies towards the Ursa Major cloud of nearby galaxies.

for the optical catalogs than it is for the IRAS catalog. A contributor to the lack of contrast seen in Figure 4 is the broad spread of the IRAS luminosity function that tends to wash out all structures seen in projection by diluting them with more distant objects.

Recently there has been much interest in finding new objects that might be the cause of large-scale streaming motions. Hoffmann and Zurek (1988) consider loops of cosmic string. Scaramella *et al.* (1989), following the earlier work by Melnick and Moles, have suggested that the aggregation of clusters around Shapley 8 in northern Centaurus might be the primary cause of streaming. Here I emphasize that extra causes of streaming are not needed. Already we have a more than sufficient light flux to explain the Local Group's motion by the associated gravity. Thus, if loops of cosmic string provide the gravity, we

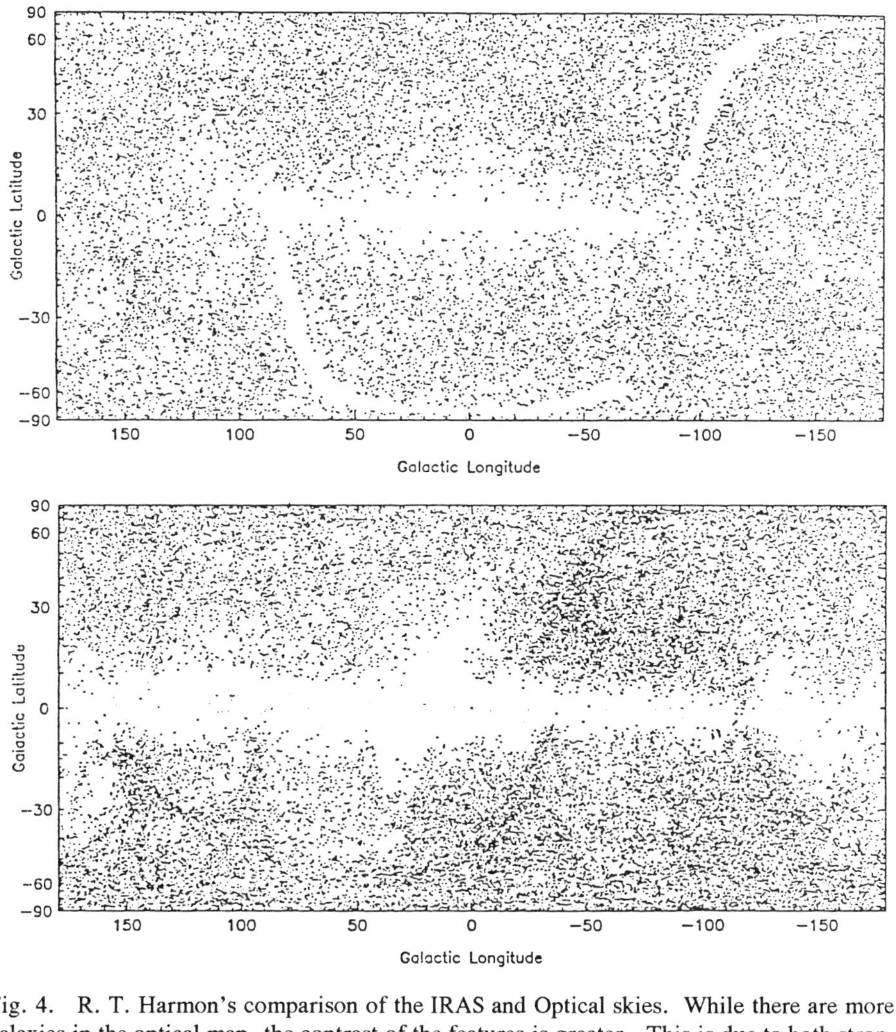

Fig. 4. R. T. Harmon's comparison of the IRAS and Optical skies. While there are more galaxies in the optical map, the contrast of the features is greater. This is due to both stronger representation of high density regions in the optical and to the broader luminosity function of IRAS galaxies diluting these features.

would have to associate a very low Ω with the clustering properties of the galaxies. A direct check on the Shapley 8 concentration is Raychaudhury's survey of 17,000 galaxies with diameters $\theta \geq 0.3'$ in Hydra and Centaurus. The APM scanning machine was used, but with every galaxy further identified and typed by eye. There is a very significant enhancement of galaxy surface density within 20° of Shapley 8 as seen in Figure 5. However, this large excess of galaxies, smaller than those already accounted for, produces only 10 - 15% of the total dipole observed at the Local Group. It is unlikely that this aggregate, centered around 14,000 km s^{-1}, produces more than a sixth of the Local Group's motion.

THE COSMOLOGICAL DENSITY Ω

Peebles shows that small perturbations in the density of the universe generate peculiar velocities **v** and gravity fields **g** related by

$$\mathbf{v} = (2/3)\Omega^{-0.4}\mathbf{g}/H \tag{6}$$

where Ω is the density parameter of the universe and H is Hubble's constant. For discussion of non-linear corrections to this formula see Lynden-Bell (1988). For small θ, where the universe is almost uniform, $\theta^2 dM/d\theta$ is given by

$$\theta^2 dM/d\theta = (\langle \rho_L \rangle /4\pi) \int_0^\infty (-dL/dD_0) D_0 \, dD_0 = \langle \rho_L \rangle \bar{D}_0/4\pi. \tag{7}$$

D_0 is the true diameter of a galaxy, which is related to the velocity times angular diameter, while $D = HD_0$.

Now, from the definitions of **P** and b

$$\mathbf{P} = (3/4\pi) \int \hat{\mathbf{r}} \Delta\rho_L (4\pi r^2)^{-1} d^3r = (3b/4\pi) \int \hat{\mathbf{r}} (\Delta\rho/\rho_0) \langle \rho_L \rangle (4\pi r^2)^{-1} d^3r$$

$$= (3b/16\pi^2)(\langle \rho_L \rangle / \rho_0)\mathbf{g} \tag{8}$$

Using equations (6) and (7) this may be re-written

$$\mathbf{P}/(\theta^2 dM/d\theta) = 3\Omega^{-0.6}(b/\bar{D})\mathbf{v} \tag{9}$$

where $8\pi G\rho_0/3 = \Omega H^2$ and $\bar{D} = H\bar{D}_0$. Since H is poorly known but the velocities of galaxies are known, we know $L(D)$, not $L(D_0)$, so $H\bar{D}_0 = \bar{D}$ is far more accurately known than $\bar{D}_0$. From our catalogs data we evaluate **P** and $M(>\theta)$, while $\bar{D}$ is evaluated from our determination of $L(D)$. We give two evaluations of $\Omega^{0.6}$/b using different compensations for missing regions of the sky, called Mock Sky and Average Sky. Since the main problems occur in the Milky Way, we also evaluate $\Omega^{0.6}$/b from the z components of equation 9, which are hardly affected by the missing regions behind the Milky Way.

TABLE I Compensation for Missing Sky Regions

	Mock Sky	Average Sky
$\Omega^{-0.6}$/b	0.31	0.49
z component	0.32	0.37

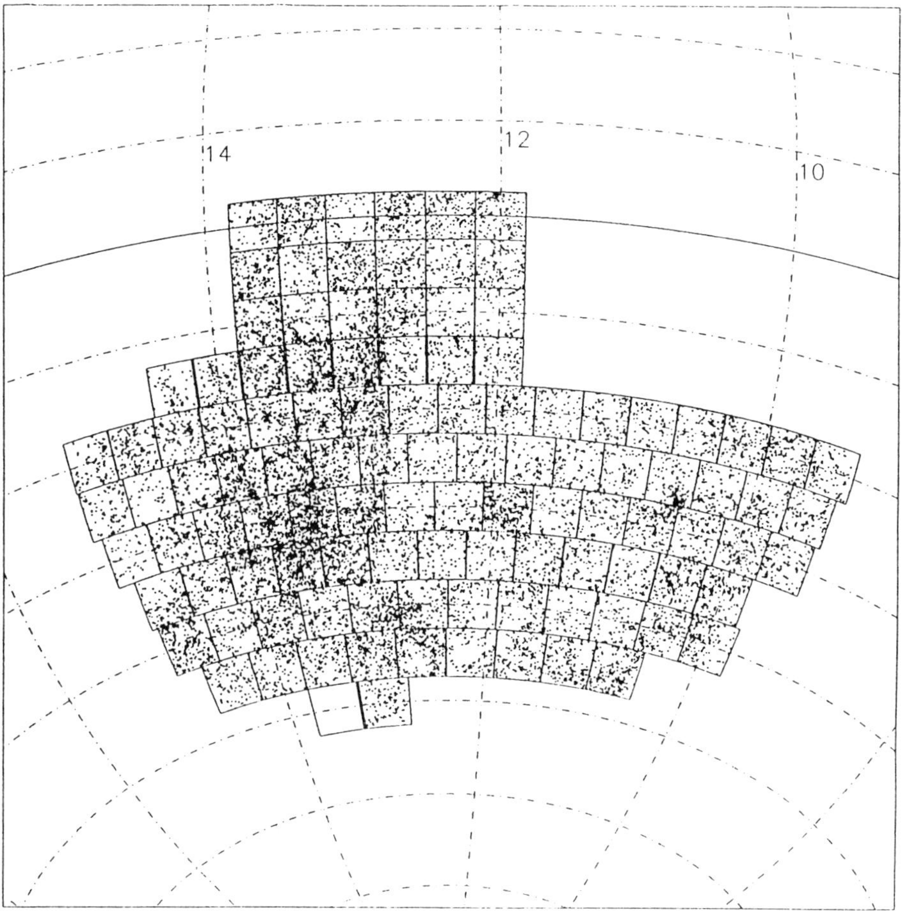

Fig. 5. S. Raychaudhury's map of galaxies with angular diameters greater than 0.3′ in over 100 Schmidt fields covering Centaurus and Hydra. The major features are centered around Shapley 8 in northern Centaurus.

We can determine Ω only if we know b. The unbiased assumption (b = 1) yields $\Omega = 0.17$ for $\Omega^{0.6}/b = 0.347$, but for that same data the very acceptable b = 2.9 yields $\Omega \sim 1$. Thus, provided galaxies give a three times greater density contrast than the mean of all matter, the universe will be closed and finite.

A further reason for considering $\Omega \leq 1$ to be less physical than $\Omega > 1$ may be found from the fact that when $\Omega \leq 1$, the Big Bang did not start from one point but from a line or a volume. In the latter cases the initial conditions look very contrived (see Lynden-Bell, Katz and Redmount 1989).

REFERENCES

Aaronson, M., Bothun, G. D., Cornell, M. E., Dawe, J. A., Dickens, R. J., Hall, P. J., Sheng, H. M., Huchra, J. P., Lucey, J. R., Mould, J. R., Murray, J. D., Schommer, R. A., and Wright, A. E. 1989, *Ap. J.*, **338**, 654.

Burstein, D., Davies, R. L., Dressler, A., Faber, S. M., Lynden-Bell, D., Terlevich, R. J., and Wegner, G. 1986, in *Galaxy Distances and Deviations from Universal Expansion*, ed. R. B. Tully and B. F. Madore (Dordrecht: Reidel), p. 123.

Burstein, D., and Lebofski, M. J. 1986, *Ap. J.*, **301**, 683.

Conklin, E. K., and Bracewell, R. N. 1967, *Nature*, **216**, 777.

Conklin, E. K. 1969, *Nature*, **222**, 971.

Eddington, A. S. 1922, *The Mathematical Theory of Relativity*, (Cambridge: Cambridge University Press).

Faber, S. M., Wegner, G., Burstein, D., Davies, R. L., Dressler, A., Lynden-Bell, D., and Terlevich, R. J. 1989, *Ap. J. Suppl.*, **69**, 763.

Fairall, A. P. 1988, *M.N.R.A.S.*, **230**, 69.

Harmon, R. T., Lahav, O., and Meurs, E. J. A. 1987, *M.N.R.A.S.*, **228**, 5P.

Hoffmann, Y., and Zurek, W.H. 1988, *Nature*, **333**, 46.

Hubble, E. P. 1929, *Proc. Nat. Acad. Sci.*, **15**, 158.

Lahav, O., Rowan-Robinson, M., and Lynden-Bell, D. 1988, *M.N.R.A.S.*, **234**, 677.

Lauberts, A. 1982, *The ESO/Uppsala Survey of the ESO(B) Atlas*, (Garching: European Southern Observatory).

Lubin, P. M., and Villela, T. 1986, in *Galaxy Distances and Deviations from Universal Expansion*, ed. B. F. Madore and R. B. Tully, (Dordrecht: Reidel) p. 169.

Lucey, J. R., and Carter, D. 1988, *M.N.R.A.S.*, **235**, 1177.

Lucey, J. R., Lahav, O., Lynden-Bell, D., Melnick, J., and Terlevich, R. J. this volume, p. 31.

Lynden-Bell, D. 1987, *Quart. J.R.A.S.*, **28**, 187.

Lynden-Bell, D. 1988, *Observatory*, **108**, 145.

Lynden-Bell, D., Faber, S. M., Burstein, D., Davies, R. L., Dressler, A., Terlevich, R. J., and Wegner, G. 1988, *Ap. J.*, **326**, 19.

Lynden-Bell, D., Katz, J., and Redmount, I. H. 1989, *M.N.R.A.S.*, **239**, 201.

Lynden-Bell, D., and Lahav, O. 1988, in *Large-Scale Motions in the Universe: A Vatican Study Week*, ed. V. C. Rubin and G. V. Coyne, S. J., (Princeton: Princeton University), p. 199.

Lynden-Bell, D., Lahav, O., and Burstein, D. 1989, *M.N.R.A.S.*, **241**, 325.

Meiksin, A., and Davis, M. 1986, *A.J.*, **91**, 191.

Melnick, J., and Moles, M. 1987, *Rev. Mex. Astr. Ap.*, **14**, 72.

Meurs, E. J. A., and Harmon, R. T. 1988, *Astr. Ap.*, **206**, 53.

Nilson, P. 1973, *Uppsala General Catalogue of Galaxies*, (Uppsala: University of Uppsala).

Peebles, P. J. E. 1980, *The Large-Scale Structure of the Universe* (Princeton: Princeton University).

Scaramella, R., Baiesi-Pillastrini, G., Chincarini, G. L., Vettolani, G., and Zamorani, G. 1989, *Nature*, **338**, 562.

Shaya, E. J. 1984, *Ap. J.*, **280**, 470.

Tammann, G. A., and Sandage, A. 1985, *Ap. J.*, **294**, 81.

Vorontsov-Velyaminov, B., and Archipova, V. 1963-74, *Morphological Catalog of Galaxies*, Parts 2 to 5, (Moscow: Moscow University).

Yahil, A., Walker, D., and Rowan-Robinson, M. 1986, *Ap. J. (Letters)*, **301**, L1.

DISCUSSION

M. Strauss: In what sense is there no 'new attractor'? Do you mean there is no unseen attractor?

D. Lynden-Bell: The light actually seen, supplemented by the simplest interpolations through the Galactic Plane, gives good directional agreement with the velocity of the Local Group relative to the Cosmic Microwave Background. Extra light in the direction of the motion could only lower Ω still further or increase the large-scale bias factor still further, both of which would make greater difficulties. Significant unseen attractors are likely to upset the directional agreement seen so far.

S. White: This type of analysis is founded on two general assumptions, namely that the apparent motion with respect to the microwave background is gravitationally induced by relatively local inhomogeneities, and that inhomogeneities in light are directly related to inhomogeneities in mass (although not, perhaps, by simple proportionality). To what extent do you think these assumptions are justified by the data?

D. Lynden-Bell: I believe the directional agreement is quite significant and is hard to explain if light and mass are not well correlated on the 1000 to 2000 km s^{-1} scale. The significance of the magnitudes is not as great since one could happily accept different values which give a larger value of $\Omega^{0.6}/b$. Furthermore, different allowances for the missing zone do give significantly different magnitudes to the dipole even though the directions agree quite well.

E. Bertschinger: I have one remark and one question. First, a more detailed test of Simon White's questions whether mass traces light and whether gravitationally induced motions produce the dipole comes from comparing the predicted gravity field and the observed velocity field *throughout space* rather than only at the local group. My question is, have you tried to estimate the 'shot noise' uncertainty in the dipole arising from the declining selection function at large distance and small galactic latitude? Although your estimated $\Omega^{0.6}/b$ changes little with your treatment of the zone of obscuration, the dipole direction seems to be uncertain by up to $\sim 20°$, with corresponding change in velocity vector of ~ 200 km s^{-1}.

D. Lynden-Bell: First, I totally agree that the best test is to compare gravity and light flux at lots of places. We can only get all three components at the Local Group, and peculiar velocities at other places are subject to greater errors. But, there are very many places where the comparison could and should be made.

On the shot noise arising from the declining selection function, this is only large around 8000 km s^{-1}. At 4500 km s^{-1} the incompleteness corrections are factors of 1.5 or so; leaving them out altogether does not lead to a significantly different dipole. We did try leaving out the incompleteness correction.

N. Kaiser: Regarding the question raised by Simon White concerning the extent to which these observations support the assumption that gravity causes the motions we see: I know that if Jerry Ostriker were here he would argue that all we are really testing is the equation of continuity - if the structure grew by galaxies moving to overdense regions (by whatever cause) then much of what we see can be accounted for. I'm not sure if I fully accept this - important additional ingredients are the assumptions that the flows are curl free and that structure grows everywhere at the same rate - but the point certainly deserves further debate.

D. Lynden-Bell: I believe it is not impossible to get the local group moving in the opposite direction to the light dipole if the continuity equation is all I have to go on.

M. Davis: Michael Strauss and I have compared the optical dipole of Lynden-Bell, Lahav, and Burstein with the IRAS dipole. Somewhat to our surprise, we found a significant difference only occurred in the Galactic **x** direction (towards the Galactic center). Only for velocities less than 1500 km s^{-1} was there any substantial difference, where both catalogs should be reliable. At present we do not understand *why* these dipole vectors disagree.

D. Lynden-Bell: It would be very interesting to track down this difference. Could it be due to the diameter-squared weighting that we use and you don't? There is a larger local shot noise in the IRAS data due to the broader luminosity/mass distribution.

A. Dressler: However, if both the optical and IRAS surveys were tracing the same distribution, wouldn't the derived values of $\Omega^{0.6}/b$ be closer than the factor of two difference you found?

D. Lynden-Bell: They can't both be, because the contrast is so different between the IRAS and optical pictures of the sky. I think the bias factor on large scales must be bigger for optical galaxies, or the errors in the magnitudes of $\Omega^{0.6}/b$ are bigger than we think in both cases. In the optical the monopole evaluation of the mean luminosity density is far from perfect.

A. Dekel: On Simon White's second question, what do you mean by allowing the galaxy density to be a general function of the mass density? You could allow an arbitrary 'biasing field' b(**x**), but wouldn't this affect the resultant effective $\Omega^{0.6}/b$ as well as the direction of the dipole?

D. Lynden-Bell: Given that the dipole and velocity are parallel then one can define an effective bias factor in terms of their ratio. This effective b is a weighted average of the b in the scales over which the dipole arises, *i.e.* the 100 to 4500 km s^{-1} radius scales.

S. White: The thrust of my question was somewhat more general. I wondered how well the data support the assumption that the density of galaxies is a unique monotonic function of the mass density. For example, could there be dark massive enhancements of density with few associated galaxies?

POSSIBLE RELEVANCE OF LARGE SCALES ON PECULIAR VELOCITIES: DO WE SEE GIANT ATTRACTORS?

R. SCARAMELLA
Osservatorio Astronomico di Roma, via Frascati 5, 00040 Monteporzio Catone, Italy, and International School for Advanced Studies (SISSA), Strada Costiera 11, 34014 Trieste, Italy

G. BAIESI-PILLASTRINI
Dipartimento di Astronomia, Università di Bologna, via Zamboni 33, 40126 Bologna, Italy

G. CHINCARINI
Università di Milano and Osservatorio Astronomico di Brera, via Brera 28, 20121 Milano, Italy

G. VETTOLANI
Instituto di Radioastronomia CNR, via Irnerio 46, 40126 Bologna, Italy

G. ZAMORANI
Osservatorio Astronomico, via Tiepolo 11, 34131 Trieste, Italy, and Instituto di Radioastronomia CNR, via Irnerio 46, 40126 Bologna, Italy

ABSTRACT We report on work in progress to assess the possible dynamical significance of large-scale structures, as traced by rich clusters, on the observed peculiar motions. If clusters of galaxies are mostly responsible for the observed peculiar velocities, these are dominated by the contribution from the nearest cluster aggregate that we have recently identified in the southern hemisphere.

We reported from the analysis of the southern extension of the Abell catalog (Abell *et al.* 1989, hereafter ACO) a unique, very rich concentration of clusters beyond Centaurus, with distance $D \approx 140\ h^{-1}$Mpc (Scaramella *et al.* 1989). This concentration of clusters is composed of 28 clusters within 200 h^{-1}Mpc which have measured or estimated redshift through the magnitude of the tenth ranked galaxy, m_{10}, with a typical error of 25% for small z.

The radial distribution appears to be double peaked with four clusters at $\sim 40\ h^{-1}$Mpc and the others in the range $\approx$ 100 - 200 h^{-1}Mpc, with the majority at $\approx 140\ h^{-1}$Mpc. From a uniform distribution at these galactic latitudes, one would expect only ≤ 3 clusters in the whole region: we find therefore a number overdensity in clusters of ≤ 10. The sky barycenter of these clusters lies only $\sim 10°$ above the latest reported direction for the 'Great Attractor' model (Faber and Burstein 1988) and less than 30° away from the inferred direction of the peculiar motion of the Local Group (LG) with respect to the Cosmic Microwave Background (CMB) rest frame (Lubin and Villela 1986).

Because the pull of the two aligned peaks combines very effectively, we try to determine to what extent the peculiar acceleration due to this particular feature is predominant compared to other nearby cluster aggregates. Also, to consider the net sum of various acceleration vectors induced on us from nearby clusters can give information on the relevance of large structures on peculiar velocities. To this end, we have jointly considered the Abell and the ACO cluster catalogs.

The present sample consists of Abell and ACO clusters of galaxies with: $R \geq 0$, $z \leq 0.1$, $|\mathrm{b}| \geq 20°$ (see Figure 1). From this sample we compute the total acceleration on the LG as:

$$\mathbf{A} = \langle M \rangle \, \Sigma_i \, (M_i/\langle M \rangle)(\hat{\mathbf{r}}_i/r_i^2) \equiv \langle M \rangle H_0^2 \gamma$$

with $M_i/\langle M \rangle \propto \text{Richness}_i$.

The analysis is still at a preliminary stage, due to many difficulties, of which the most important are those due to systematic differences between Abell and ACO: different number densities, depths and sky coverage, stronger incompleteness for Abell's $R = 0$ class, fewer z measured in ACO and unsurveyed areas.

From Figure 2, which shows the absolute value of the total acceleration vector per unit mass as a function of the sample depth, one can *preliminarily* deduce that: 1) a very good convergence of $|\mathbf{A}|$ is obtained well inside the sample depth ($\approx$75% of the sample is

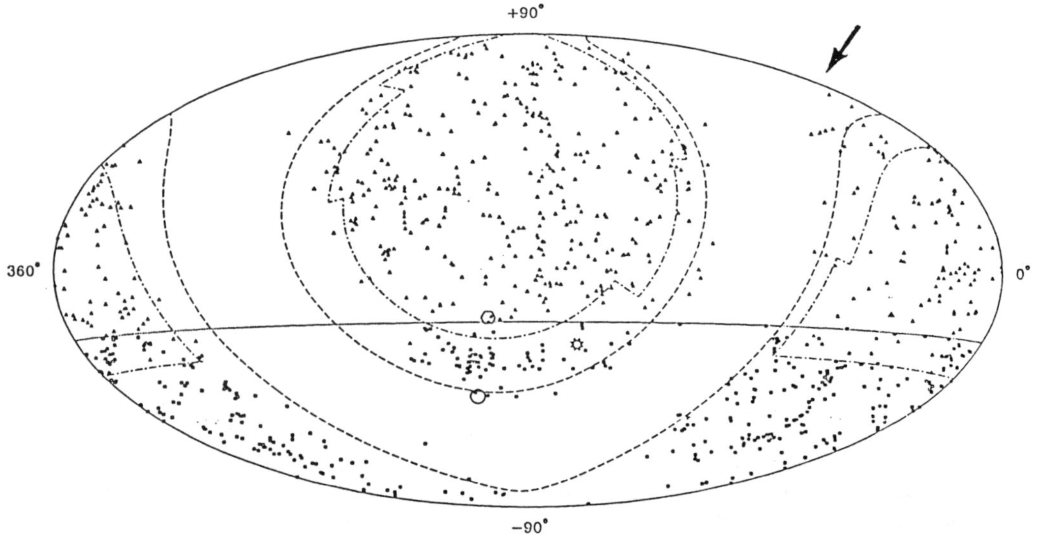

Fig. 1. Equi-area projection in Right Ascension and Declination of all the rich clusters with measured or estimated redshift smaller than 0.1. Triangles and squares represent Abell and ACO clusters, respectively. The dashed lines indicate the avoidance zone due to the Galaxy plane ($|\mathrm{b}| \leq 20°$). The dotted line shows the ACO boundary ($\delta = -17°$). The dashed-dotted lines represent the regions of the Abell statistical sample. The star shows the direction of the cosmic microwave dipole, the empty circle shows the latest direction of the Great Attractor (Faber and Burstein 1988), and the empty hexagon shows the direction of the peculiar flow reported by Collins *et al.* (1986). The arrow denotes the line of sight of Figure 3.

complete in redshift within the convergence length $\approx$180 h^{-1}Mpc), and 2) a plateau in the behavior of $|\mathbf{A}|$ for $50 < D < 100$ h^{-1}Mpc resembles the 'convergency' detected from the *IRAS* galaxies in the same range of distances (Strauss and Davis 1988, Yahil 1988). The direction of **A**, in the convergency region, is within 10° - 20° from the direction of the CMB dipole, but corrections are still to be included, mainly Virgo infall and contribution from unsurveyed areas.

In Figure 3 we show the 3-D acceleration field due to the considered clusters, as seen from top right corner of Figure 1. The highest peak is due to the supercluster beyond Centaurus at $D \approx 140$ h^{-1}Mpc, which, together with the Hydra-Centaurus supercluster, has a relatively dominant effect.

If $\Omega_0 = 0.2$ and the mass related to a typical Richness class $R = 2$ cluster, *e.g.* Coma, is $M \sim 2\text{-}3 \times 10^{15}$ $M_\odot$, one can explain a peculiar velocity of 600 km s^{-1}.

Work is in progress to acquire more data on both this specific region and the overall ACO catalog in order to reduce the uncertainties due to distance estimates, as well as to measure the galaxy overdensity associated to this region.

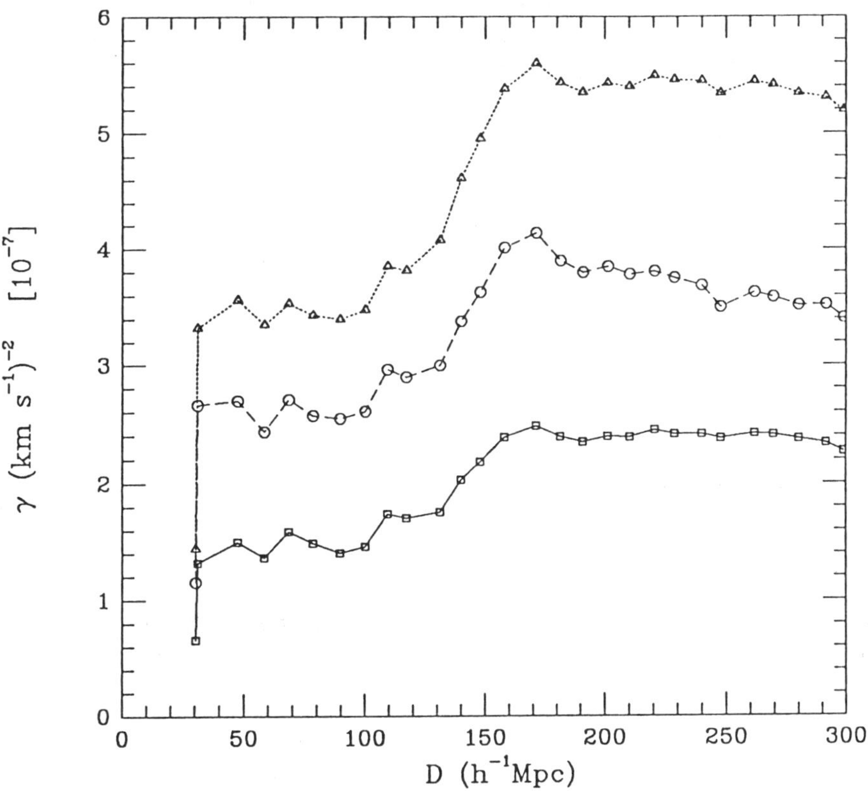

Fig. 2. Absolute value of the acceleration as a function of the distance. The upper curve is obtained by weighting each cluster by its richness and applying to the data a correction for incompleteness as a function of the galactic latitude. The middle curve is obtained by adding another correction which makes the mass densities of the two catalogs equal. The lower curve is obtained by weighting each cluster by its richness only, with no correction for incompleteness with the galactic latitude.

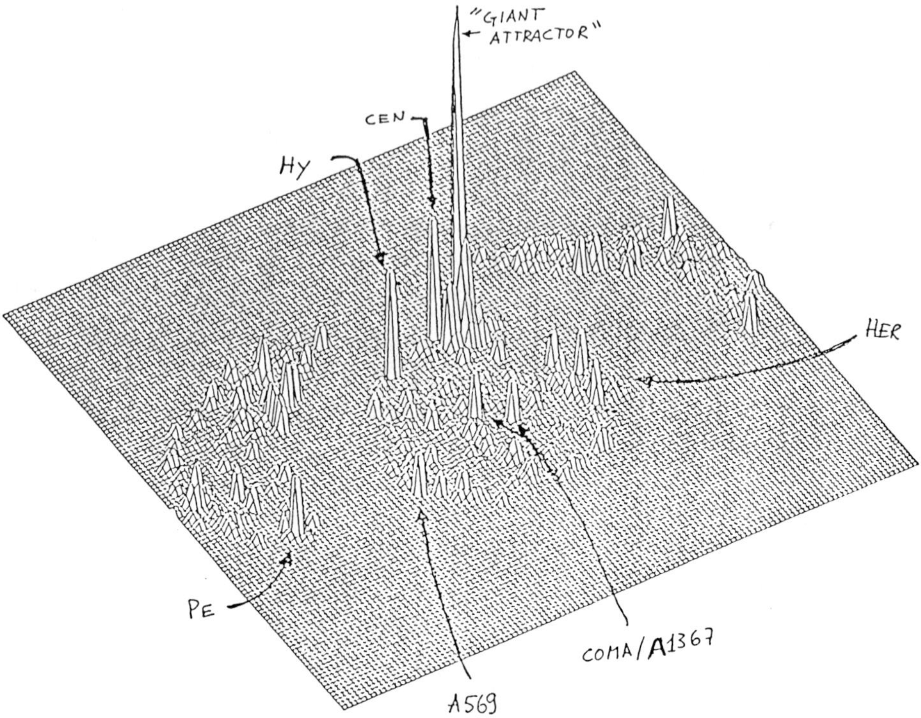

Fig. 3. Three-dimensional view of the smoothed acceleration field due to clusters shown in Figure 1 as seen by looking along the arrow in Figure 1.

REFERENCES

Collins, C. A., Joseph, R. D., and Robertson, N. A., 1986, *Nature*, **320**, 506.

Faber, S. M., and Burstein, D., 1988, in *Large Scale Motions in the Universe: A Vatican Study Week*, ed. V. C. Rubin and G. V. Coyne, S. J. (Princeton: Princeton University), p. 115.

Lubin, P., and Villela, T., 1986, in *Galaxy Distances and Deviations from Universal Expansion*}, ed. B. F. Madore and R. B. Tully, (Dordrecht: Reidel) p. 169.

Scaramella, R., Baiesi-Pillastrini, G., Chincarini, G., Vettolani, G., and Zamorani, G., 1989, *Nature*, **338**, 562.

Strauss, M. A., and Davis, M. 1988, in *Large Scale Motions in the Universe: A Vatican Study Week*, ed. V. C. Rubin and G. V. Coyne, S. J. (Princeton: Princeton University), p. 256.

Yahil, A. 1988, in *Large Scale Motions in the Universe: A Vatican Study Week*, ed. V. C. Rubin and G. V. Coyne, S. J. (Princeton: Princeton University), p. 219.

DISCUSSION

D. Burstein: To what extent are your results on the growth of the dipole affected by the fact that the cluster distribution is incomplete, due to the zone of avoidance and declination limits?

R. Scaramella: Work is in progress to fully deal with the angular incompletenesses. However, the effect of the major correction (cosec$|b|$ selection function) on the modulus of the acceleration has been shown in Figure 2: you get an increase in final amplitude, but the qualitative behavior as well as the convergence depth changes little, although you have a large change in the final direction. I expect that further refinements to correct for Abell's boundaries (see Figure 1) down to $|b| \approx 20°$ won't change much the amplitude. To try to account for the zone of avoidance can instead have quite an effect, which will depend strongly on the method used as was shown in the literature, but this has still to be worked out.

D. Burstein: Your cluster distribution is much more structured than the galaxy distribution. In the limit there are only two clusters in the sky. If one cluster is located in the zone of avoidance, you would get a very different answer than if you could see both clusters.

R. Scaramella: I think that a rich nearby cluster, completely hidden by the Galactic plane, would be a serious problem for everyone! On the other hand, I fully agree with the fact that to have too few objects gives you undersampling of structures and greater noise, but here roughly 300 clusters are responsible for the signal, whose direction happens - for the moment - to be in an interesting area. This fact is lost when you consider only very rich clusters or x-ray selected ones (Lahav *et al.*), which evidently are too rare objects.

J. Primack: If mass is distributed uniformly on very large scales, then peaks in mass on smaller scales must be compensated by valleys - voids. And voids 'repel' just as peaks attract. I'm worried that your assumption that rich clusters trace the mass is implausible both observationally and theoretically. Two more technical points: the increasing incompleteness in the cluster surveys near the zone of avoidance is a big problem for your analysis, especially since you claim the interesting structure is there. Also, since there is much structure in the Centaurus direction, there is a real danger that redshifts will be those of projected galaxies - so it's important to get at least four redshift measurements per cluster.

R. Scaramella: When you have departures from homogeneity, underdense regions show less tracers with respect to the average number. Therefore 'voids' give a smaller 'pull' than the expected one, and when you subtract the signal expected from the homogeneous background - zero if you consider the whole sky - this becomes a 'push'. This is a common behavior for three-dimensional samples. Here we are *testing* the hypothesis that masses traced by clusters (do not forget $\xi_{gc} \neq 0$) can give you some net acceleration because of the presence of medium-distant structures: whether or not this hypothesis will be shown to be true, we will learn a lot in both cases, and the numbers we are getting here do not look so crazy to not warrant further investigation.

As remarked during the talk, while most other catalogs have to worry mainly for undersampling in radial distance and not too much for direction, here we have the opposite situation, in that the cluster catalogs we examine are radially but not angularly complete for the volume of interest. On the other hand, at low latitudes ($|b| \sim 20$ to $40°$) the selection function for clusters is roughly $\Phi \approx ½$, not bad if you consider that other samples have

$\Phi << 1$ and large uncertainties from shot noise at interesting depths (*e.g.* see IRAS).

Confusion by foreground galaxies is a well-known problem which, however, becomes crucial at depths of $z \sim 0.1$ for Abell (see Postman *et al.* 1987) and deeper for ACO, while here we are speaking mostly of clusters at $z \sim 0.05$. Indeed we just took two weeks ago an average of four spectra for most of the clusters that had estimated redshifts among the 28 of the 'α region' (see our *Nature* paper.) The data will be reduced soon.

N. Kaiser: I wonder if there is a conflict between the idea proposed here and the results of Raychaudhury (referred to by Donald Lynden-Bell) which indicate that the contribution to the optical dipole from distant galaxies in this region is quite small (less than about 10%)?

R. Scaramella: I did not see the work you are referring to, so I can only point out that this preliminary result indicates that ~30% of this signal is gained between 100 and 200 h^{-1}Mpc. However, here we start with nonzero signal from ~30 h^{-1}Mpc, in that rich clusters of course do not trace our immediate surroundings and Virgo in particular, which I think contributes for a large fraction of the optical dipole. To correct for this zero-point difference would probably lessen the discrepancy you are mentioning. Moreover, to make a full, very interesting comparison, the optical dipole should come from a galaxy sample which covers the volume we are considering here: $|b| > 20°$ and complete at least up to 200 h^{-1}Mpc.

R. Gott: Lynden-Bell's picture of the optical galaxies shows them to be more clustered on the sky than the IRAS galaxies. One effect that can cause this is that galaxies may get the dust knocked out of them if they are in tight groups or clusters. Ed Turner and I once separated galaxies in the Zwicky Catalog in to 'associated' or 'single' galaxy categories depending on whether they had a near neighbor on the sky within 0.75°. The 'single' galaxies had $\xi(r) = -1$ out to 0.75° as expected but, surprisingly, had $\xi(r) = 0$ at all larger scales, a result that was later explained by Fall *et al.* as a natural result of the three-point correlation function. So, if we select a sample of galaxies which excludes ones with near neighbors, we can produce from a clustering hierarchy a subset which is even sub-biased, *i.e.* $b < 1$.

LARGE-SCALE STRUCTURE AND THE GREAT ATTRACTOR

ALAN DRESSLER
The Observatories of the Carnegie Institution of Washington
813 Santa Barbara Street, Pasadena, CA 91101

ABSTRACT Observations of large-scale structure are reviewed, including a brief summary of the distribution of galaxies compared to the CDM model of structure formation. The importance of mapping the Hubble flow, in order to determine the distribution of *mass* for comparison with the galaxy distribution, is emphasized. New data taken by Sandra Faber and the author are used to support the model proposed by the 'Seven Samurai' that a Great Attractor dominates the mass distribution of the local universe.

INTRODUCTION

The physics of the early universe responsible for the formation of structure left its imprint in the stars, galaxies, and clusters we see today. Unfortunately, gaseous dissipation in the collapse of these structures and the enormous energy released in star formation has largely erased evidence of initial conditions. On the other hand, superclusters and voids, the largest structures in the universe, are thought to have remained unaltered by such processes. As such, they are directly descended from the seeds of primal structure. Furthermore, because density enhancements of such structures are still small, their growth is fairly simple to model as a primarily linear amplification of the initial perturbation. Thus, it should be possible to recover the original power spectrum from today's distribution. With relatively few free parameters, the 'cold-dark-matter' (CDM) model (Davis *et al.* 1985, Frenk *et al.* 1985) has successfully accounted for the statistics of small-scale clustering and some structural properties of galaxies. Hot-dark-matter, string models, explosions, and various combinations of these, are still contending mechanisms of structure formation, but are more-or-less problematical and are less well constrained. I will, then, continue the tradition of hurling data at CDM to see how it holds up as a standard bearer for these contemporary ideas about the birth of structure in the early universe.

THE DISTRIBUTION OF GALAXIES

The most straightforward observation of large-scale structure in the universe is, of course, the distribution of galaxies. The extensive mapping of large volumes of space by the acquisition of thousands of galaxy redshifts has fundamentally altered our picture of the universe. Begun as the CFA redshift survey (Huchra *et al.* 1983), there is now worldwide activity in this important program, including valuable contributions by da Costa and collaborators here at the Observatório Nacional in Brazil.

Although the original CfA Survey and its extension into the southern sky, the SSRS (da Costa *et al.* 1988), produced thousands of galaxy redshifts and gave us our first three-dimension picture of the galaxy distribution, the sampling in these surveys was too sparse to uncover the lacy structure which emerged from the first 'slice' published by de Lapparent, Geller and Huchra (1986). The description of a bubble-like topology was soon ascribed to the remarkable patterns, which resemble a cut through a bath of suds. This in turn led to much rediscussion of the Ostriker and Cowie (1981) suggestion that structure in the universe was formed by explosive release of energy as, for instance, in the birth of the first generation of stars. However, the largest 'bubbles' strain the power requirements even of exploding cosmic strings.

Now, two additional slices adjacent to the original survey confirm that these remarkable structures are coherent from one cut to the next, but they have not added much support for the belief that voids are spherical and, thus, that the structure is bubble-like. An equally good interpretation is that the voids are tube-like. Gott, Weinberg, and Melott (1987) have pointed out that structures which grow from the gravitational instability of small-amplitude, random-phase fluctuations would not be expected to produce either a bubble-like topology, in which case the voids are isolated, or its opposite, a meatball topology, where the clusters are isolated. Rather, one expects a symmetric distribution like a sponge where both high- and low-density regions are interconnected through a complex network. Gott *et al.* have devised an ingenious test to quantify the topology in these terms, and early analysis of some catalogs favor the sponge interpretation over the meatball or bubble pictures. Another clever statistical measure, the contour crossing statistic (Ryden *et al.* 1989), has yielded similar results.

The new CfA slices have not yet been made available to these statistical analyses, although we can hope that they soon will be. My subjective judgement is that the new CfA data are just as consistent with the slicing of a sponge as a slice through soapsuds, but the sampled volume still appears too small for a definitive test. This points up the need for continuation of large-scale surveys of the galaxy distribution, and for clever tests to quantify the topology of structures. For example, the long-standing debate over whether there are filamentary structures in the galaxy distribution remains unresolved (*cf.* Sutherland *et al.*, this volume), and Geller comments that as surveys have covered larger and larger regions, they consistently find what appears to be coherent structure as large as the volume surveyed! On the other hand, the isotropy of radio source counts and the smoothness of the Cosmic Background Radiation (CBR) reassures us that at some level the universe can be considered homogeneous and isotropic.

How does the CDM model fare in reproducing these slices of the galaxy distribution? At first it was suggested that the strength of these features might present a problem for CDM, but one must remember that the models by Davis and collaborators have a high 'biasing parameter' that results in a much higher contrast for the galaxy distribution than for the underlying mass distribution. In fact, the group has sliced through their *N*-body simulations and produced pictures that are qualitatively similar to the observed distributions (White *et al.* 1987). It was my impression three years ago that the actual data had more rounded contours than the CDM simulations, in which the structures seem more chopped-up and linear. However, the new observations of adjacent slices by the CfA group appear to me more similar to the CDM simulations. The real data include other effects, like the 'fingers-of-God' due to regions of high velocity dispersion, and the likelihood of coherent motions that confuse a simple translation of a redshift into a distance. This being the case, it is hard to argue that the CDM simulations have failed to reproduce the observed structure.

A bit more troubling for CDM, however, is the fact that these structures are seen equally well in dwarf galaxies as in luminous galaxies. The strong biasing needed in the CDM models to assure a critical density of $\Omega = 1$ further implies a stronger clustering for galaxies with deep potential wells compared to less-tightly-bound systems. Yet the first studies to compare the distribution of dwarf galaxies to that of luminous galaxies (Bothun *et al.* 1986, Thuan *et al.* 1987, Eder *et al.* 1988) show no obvious differences. Just how embarrassing this is awaits further quantitative analysis, but it seems likely that there are data for too few dwarf galaxies to allow a decisive test. Also, Nick Kaiser (1989) has suggested a substantially lower biasing parameter for CDM, which could reduce the expected difference to a level that would be difficult to confirm observationally.

Finally, there is the challenge made to CDM based on the cluster-cluster correlation function (Bahcall and Soneira 1983; Bahcall, Batuski and Olowin 1988). The strength of this correlation and the positive correlation found for large scales $D > 20\ h^{-1}$ Mpc is not predicted in the standard CDM model. However, others have doubted that statistical biases in the Abell catalog are sufficiently well understood for it to be used in this way (Sutherland 1988, Dekel *et al.* 1989). As we will hear from Primack at this workshop, this remains a contentious issue. We eagerly await the completion of computer-made cluster catalogs by Maddox, Sutherland, and Efstathiou for the APM Survey, and by Picard and Mould at CalTech. Even though these catalogs will suffer from similar selection effects, the well-understood procedure will allow an identical treatment of model universes, with the result that a definitive comparison of the models and the real universe can be made.

Analysis of the galaxy-galaxy correlation function from the APM Survey, and its important implications for CDM, are discussed here by Sutherland *et al.* (this volume).

COHERENT MOTIONS OF GALAXIES AND THE GREAT ATTRACTOR

The crucial importance of studying the galaxy distribution is obvious, but it is also clear that progress is frustrated by the fact that models describe the mass distribution, not the galaxy distribution. The link between the two is heavily laced with messy astronomical matters like how one turns some baryons plus dark matter into a galaxy of stars. It will probably be some time before we can describe such a process with any confidence, and, in fact, we would prefer to use a comparison of the matter and galaxy distributions to help us understand the process of galaxy formation, rather than the other way around. Fortunately, measurements of the peculiar motions of galaxies provide a measure of the distribution of all matter, seen and unseen. In the simple gravitational instability model for the growth of structure, any non-uniformity in the mass distribution produces departures from a uniform Hubble flow that can, in principle, be measured.

This was first convincingly demonstrated by Aaronson *et al.* (1982), who showed that the expansion in the Local Supercluster is slowed by its higher-than-average density. This non-Hubble motion, in which our Galaxy and its Local Group neighbors share, amounts to about 200-300 km s^{-1} toward the approximate direction of the Virgo cluster. This motion is referred to as the 'infall to Virgo,' although Burstein (this volume) argues that it is, in fact, a more general contraction of the Local Supercluster. Regardless, this $\sim$250 km s^{-1} infall falls far short of the $\sim$600 km s^{-1} motion of the Local Group relative to the distant, presumably more fundamental reference frame of the CBR. This disagreement prompted a search to find the scale over which the larger peculiar velocity arises. A prime candidate for the necessary gravitational pull was the Hydra-Centaurus supercluster, about three times more distant than the center of the Local Supercluster. Burstein, Davies, Faber, Lynden-Bell, Terlevich, Wegner, and I, now known collectively as the 7 Samurai (7S), obtained redshifts, velocity dispersions, and photometry for $\sim$400 elliptical galaxies, data we used to predict distances and thus map the Hubble flow. Our sample extended far enough from the Local Group to allow us to test this hypothesis by looking for the expected signature of a retarded expansion field in the Hydra-Centaurus region.

Our surprising and not particularly popular conclusion was that the peculiar motion of the Local Group is shared by galaxies over a large volume of space, probably extending beyond Hydra-Centaurus (Dressler *et al.* 1987). It was immediately recognized that the amplitude of this flow, reaching ~ 1000 km s^{-1} and extending at least 30 h^{-1} Mpc, was a challenge for the CDM model. A simple model, described by Lynden-Bell *et al.* (1988), attributed most of the Local Group's motion to an enormous distributed mass at a distance of 42 h^{-1} Mpc, which I dubbed 'the Great Attractor' (GA). Such a mass concentration would be exceedingly rare if not nonexistent in a CDM universe with a highly biased distribution of baryonic matter to dark matter, the 'standard model' adopted by Davis and collaborators. However, although the data were best fit by this simple GA model, they did not strongly constrain it. It was apparent that further studies of the region were crucial.

Later studies only partially supported our findings and conclusions. Two of these have centered on the motions of clusters in the Centaurus region. Aaronson *et al.* (1989) have confirmed a large peculiar motion for some clusters in the region, the largest they have yet observed anywhere in the sky, but at an amplitude only about half that found by the 7S. Lucey and Carter (1988) claim that Cen45 has an even larger velocity than found by the 7S, but that Cen30 is nearly at rest. The average, however, is little different from the 7S result, suggesting the interpretation that both clusters are participating in the flow but have significant infall towards each other. More about the complexity of the Centaurus region can be found in Burstein (this volume) and Lucey *et al.* (this volume).

The GA model made three predictions beyond the scope of data then available. The first was that a large overdensity would be found in the distribution of galaxies in the direction specified by the model. Second, as predicted by measurements of the quadrupole moment in the peculiar velocity field, the centroid of this overdensity would be found at approximately 4000 km s^{-1}. Third, the subsequent measurements of peculiar velocity should show a decline in peculiar velocity as one approached the mass centroid, and a reversal as one crossed through to the far side of the GA.

The first prediction was confirmed in a picture of the galaxy distribution made by Ofer Lahav (Figure 8 of Lynden-Bell *et al.*). A broadening of the supergalactic plane in nearly the precise direction specified by the model proved to be the most prominent concentration of galaxies in this hemisphere of the sky. In order to check the second prediction, *i.e.*, to see if this was a coherent structure at a distance of ~ 40 h^{-1} Mpc, I conducted a redshift survey of galaxies in a ~ 1 steradian area in this direction, which I refer to as the SPS survey (Dressler 1988). This survey, which now completed includes ~ 1300 redshifts, shows peaks in the galaxy distribution at 2500 km s^{-1} and 4500 km s^{-1}, with a centroid at ~ 4000 km s^{-1}, in good agreement with the GA model.

To test the final prediction, Sandra Faber and I undertook the task of obtaining new measurements of velocity dispersion and photometry for new distance determinations. Our observations at Las Campanas Observatory during runs in March, 1988 and April, 1989 resulted in high-quality data for 136 E and S0 galaxies in the SPS sample. These new data are the ingredients for determining peculiar motions, and are an improvement over our original 7S data because the sample is much more extensive and has been observed in a uniform way.

I will summarize our results, presented in more detail in Dressler and Faber (1990). We find that the new data dramatically confirm the peculiar velocity measurements made by the 7S, and move toward validation of the GA model. To show this, we have returned to a simple form of displaying the data which uses Hubble diagrams that plot distance against observed velocity. The data are as close to 'raw' as possible. Only small and non-controversial corrections for cosmology and aperture affects have been made. The galaxies have neither been collected into groups and clusters (to lower the noise due to inherent velocity dispersion) nor have any corrections for the Malmquist-like bias been applied. The original 7S data for a magnitude-limited sample of E galaxies outside the approximately 1 steradian region covered in the SPS survey is shown in Figure 1a. Note that even in its raw form, the Hubble diagram for most of the sky is close to its simple form: a straight line of slope 1. A trend of increasingly too-large observed velocity for a given distance is the expected signature of a Malmquist-like bias. Figure 1b shows the data for ellipticals in the SPS survey region. The signature of a large outflow, which was the principal result found by the 7S, shows up dramatically as a too-high velocity at a given distance, approximately 1000 km s^{-1} all the way to the Centaurus mass concentration.

Figure 2a shows the new data for 136 E and S0 galaxies. The principal 7S result is

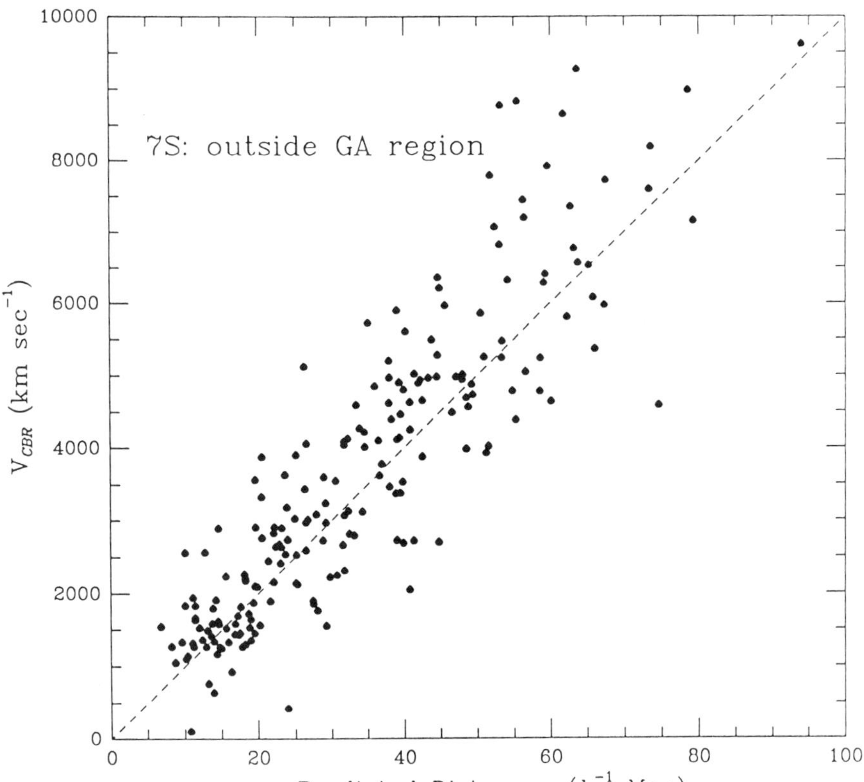

Fig. 1a Hubble diagram of predicted distance *versus* observed velocity for galaxies in the 7S sample outside of the GA region, where only moderate flow velocities were found. Distances have been determined using the D_n - σ relation. Observed velocities are with respect to the CBR. Only small, non-controversial corrections for cosmology and aperture size have been applied. Galaxies have not been collected into groups, nor has the Malmquist-like bias, responsible for the increasing systematic drift of the points above the pure Hubble line, been removed. This diagram shows that, on average, the technique produces a uniform Hubble flow for most of the sky.

confirmed. Galaxies in this direction out to 35 h^{-1} Mpc show large, positive peculiar velocities. It is clear that the flow is coherent, that is, the earlier work did not, by misfortune, merely undersample a series of flows moving both toward and away from the Local Group. Furthermore, there is new evidence that the source of the gravitational pull is at about 40 h^{-1} Mpc, since the 36 galaxies with $D > 40$ h^{-1} Mpc do not share in the flow but appear to be back at rest with the CBR. This decline in the peculiar motions is primary evidence that the center of attraction has been reached, rather than being further out, as suggested by Scaramella *et al.* (1989).

What is not immediately apparent is any infall from the far side, which would be expected if the region is populated with E and S0 galaxies and is as strong a concentration as we have suggested. However, this conclusion is strongly influenced by the effect of Malmquist-type bias on the plot. Figure 2b shows our new data compared to 'Hubble lines' (the locus of points at rest with respect to the CBR) appropriate for different models of Malmquist-type bias. One is an empirical fit to the data of Figure 1a, assumed to represent a uniform distribution since it covers so much of the sky. The other is based on the Lynden-Bell *et al.* prescription for removing the bias in the case of uniform density. In the case of non-uniform density, for example, the GA model by Faber and Burstein (1988), a

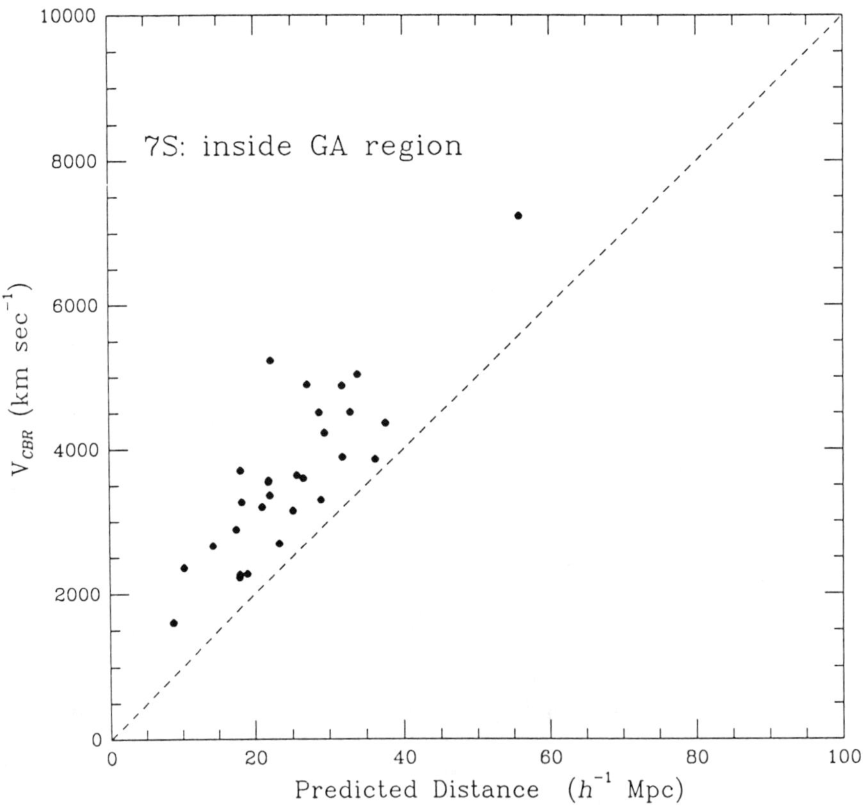

Fig. 1b The principal result of 7S of a strong outflow velocity in the GA region, the ~1 steradian area of the SPS survey.

third possible Hubble line would rise above the Lynden-Bell *et al.* solution by ~200 km s^{-1} above for D ~ 20 h^{-1} Mpc and fall below by ~500 km s^{-1} at D ~ 60 h^{-1} Mpc. The differences between these Hubble lines are small, indicating that our conclusions are not strongly influenced by uncertainties in the density distribution and the appropriate correction for Malmquist-like bias.

Viewed in reference to these Hubble lines, the new data do show evidence for backside infall, as most of the points beyond 40 h^{-1} Mpc reside below the Hubble lines. However, as discussed by Dressler and Faber (1990), we cannot eliminate the possibility that the region $D > 50$ h^{-1} Mpc is empty of elliptical and S0 galaxies, in which case many of the points below the Hubble line would be galaxies at D ~ 40 h^{-1} Mpc whose distances had been overestimated by ~20%, the typical accuracy of the D_n - σ method. In order to rule out this extreme case of bias, a deeper sample of E and S0 galaxies is needed. However, at this point the data are completely consistent with the prediction from the simple GA model of a

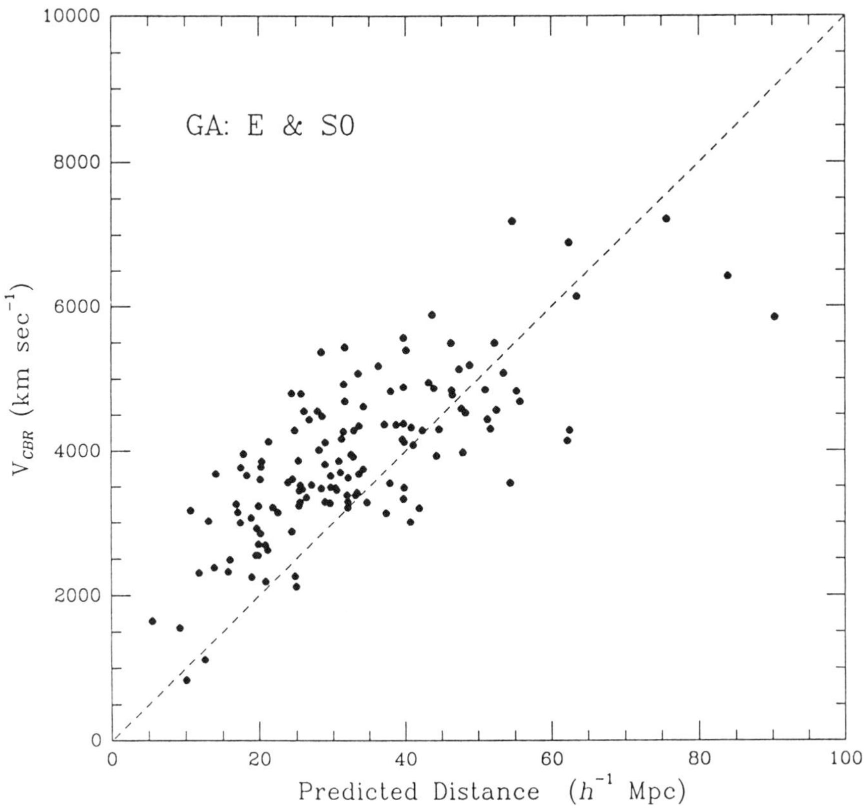

Fig. 2a. The new data of the present study in the GA region. Distances to 136 E and S0 galaxies following 7S. The original finding of a large outflow out to 30 h^{-1} Mpc is confirmed. The new data, which extend beyond 42 h^{-1} Mpc, the putative center of the GA, show that the peculiar velocities decrease beyond 30 h^{-1} Mpc. This trend rules out the possibility that systematic errors, like Galactic reddening or Malmquist-like bias, were responsible for the original 7S result.

full S-wave in the peculiar velocities, as demonstrated in Figure 2b where we compare the predicted velocity as a function of distance for Faber and Burstein's GA model. I emphasize that our results on the large infall velocity on the near side of the GA and the zero-crossing at $D \sim 40\ h^{-1}$ Mpc are not subject to this uncertainty in the Malmquist-like bias.

An analysis of all the published distance determinations for galaxies in the GA region, including the new Dressler-Faber data, can be found in Burstein, Faber and Dressler (1990). In my opinion, these studies provide a significant confirmation of the original measurements and conclusions of 7S. The new data do not sample far or extensively enough to fully confirm the simple GA model, but it is clear that they are moving the discussion in that direction. In particular, the curving back toward the Hubble line proves that systematic

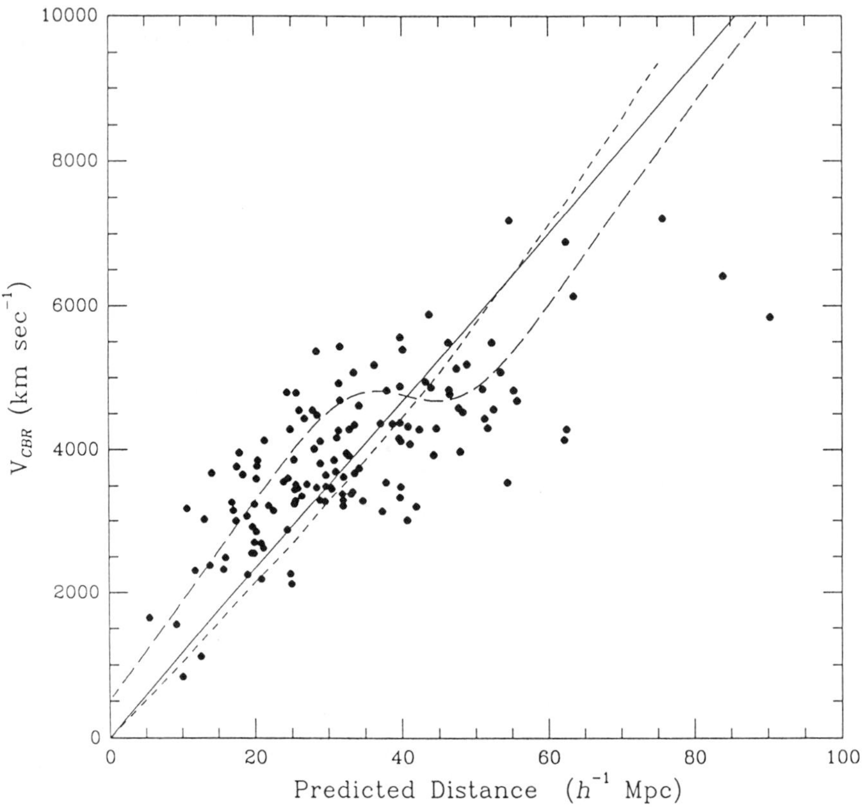

Fig. 2b. The new data compared to Hubble lines appropriate for different forms of the Malmquist-like bias, as described in the text. The short-dash curve is the Hubble line appropriate for bias determined empirically by median fitting the data in Figure 1a in bins of 1000 km s^{-1} in observed velocity. The solid line is appropriate for Malmquist-like bias following Lynden-Bell *et al.* for the case of uniform density. Assuming that these curves are representative Hubble lines, the data can be interpreted as showing evidence for backside infall beyond the centroid of the GA. An alternate possibility is described in the text. The long-dash curve line is a representation of the prediction of slowed Hubble expansion in an overdense region of space for the Faber and Burstein (1988) GA model.

errors like Galactic extinction and the Malmquist-like bias were not responsible for the original result, and that the high velocity (~ 1000 km s^{-1}) flow is unlikely to extend over a much larger region than our GA model predicted. As discussed in Burstein, Faber, and Dressler, this is additional evidence that the CBR is a proper rest frame for measurements of peculiar motion.

To the extent that the original GA model has held up, it continues to challenge CDM as a model for the formation of structure, particularly if a high biasing of baryonic matter relative to dark matter is needed to match other observations. These new data should help in comparing the observed peculiar velocity field to the predictions of *N*-body models, thus sharpening our ideas about the composition of the universe and its early physics.

Many projects are now underway to increase the sample in the GA region, including efforts by Lucey and collaborators to obtain more data for E galaxies, and programs by Schommer, Bothun and Mould, and by Faber and myself, to obtain distances to spirals. These, combined with new surveys around the sky, including a massive effort by Giovanelli and collaborators to measure distances to several thousand spirals with the infrared - HI Tully-Fisher technique, the program by Burstein and collaborators to map the Hercules-A2199 supercluster, and the Faber-Courteau study of the Rubin-Ford effect, are greatly increasing the data base for mapping the Hubble flow. By the mid-1990's our knowledge of the local mass distribution should be much improved.

REFERENCES

Aaronson, M., Huchra, J. P., Mould, J. R., Schechter, P. L., and Tully, R. B. 1982, *Ap. J.*, **258**, 64.

Aaronson, M., Bothun, G. D., Cornell, M. E., Dawe, J. A., Dickens, R. J., Hall, P. J., Sheng, H. M., Huchra, J. P., Lucey, J. R., Mould, J. R., Murray, J. D., Schommer, R. A., and Wright, A. E. 1989, *Ap. J.*, **338**, 654.

Bahcall, N. A., and Soneira, R. M. 1983, *Ap. J.*, **270**, 20.

Bahcall, N. A., Batuski, D. J., and Olowin, R. 1988, *Ap. J. (Letters)*, **333**, L13.

Bothun, G. D., Beers, T. C., Mould, J. R., and Huchra, J. P. 1986, *Ap. J.*, **308**, 510.

Burstein, D. this volume, p. 39.

Burstein, D., Faber, S. M., and Dressler, A. 1990, *Ap. J.*, **354**, 18.

da Costa, L. N., Pellegrini, P. S., Sargent, W. L. W., Tonry, J., Davis, M., Meiksin, A., Latham, D. W., Menzies, J. W., and Coulson, I. A. 1988, *Ap. J.*, **327**, 544.

Davis, M., Efstathiou, G., Frenk, C. S., and White, S. D. M. 1985, *Ap. J.*, **292**, 371.

de Lapparent, V., Geller, M. J., and Huchra, J. P. 1986, *Ap. J. (Letters)*, **302**, L1.

Dekel, A., Blumenthal, G. R., Primack, J. R., and Olivier, S. 1989, *Ap. J. (Letters)*, **338**, L5.

Dressler, A. 1988, *Ap. J.*, **329**, 519.

Dressler, A., and Faber, S. M. 1990, *Ap. J.*, **354**, 13.

Dressler, A., Faber, S. M., Burstein, D., Davies, R. L., Lynden-Bell, D., Terlevich, R. J., and Wegner, G. 1987, *Ap. J. (Letters)*, **313**, L37.

Eder, J., Schombert, J. M., Dekel, A., Oemler, A. Jr. 1989, *Ap. J.*, **340**, 29.

Faber, S. M., and Burstein, D. 1988, in *Large-Scale Motions in the Universe*, ed. V. C. Rubin and G. V. Coyne, (Princeton: Princeton University), p. 115.

Frenk, C. S., White, S. D. M., Efstathiou, G., and Davis, M. 1985, *Nature*, **317**, 595.

Gott, J. R., III, Weinberg, D. H., Melott, A. L. 1987, *Ap. J.*, **319**, 1.

Huchra, J. P., Davis, M., Latham, D. W., and Tonry, J. 1983, *Ap. J. Suppl.*, **52**, 89.

Kaiser, N. 1988, in *Large-Scale Structures and Motions in the Universe*, ICTP Trieste, in

press.
Lucey, J. R., Lahav, O., Lynden-Bell, D., Melnick, J., and Terlevich, R. J. this volume, p. 31.
Lucey, J. R., and Carter, D. 1988, *M.N.R.A.S.*, **235**, 1177.
Lynden-Bell, D., Faber, S. M., Burstein, D., Davies, R. L., Dressler, A., Terlevich, R. J., and Wegner G. 1988, *Ap. J.*, **326**, 19.
Ostriker, J. and Cowie L. L. 1981, *Ap. J. (Letters)*, **243**, L127.
Ryden, B. S., Melott, A. L., Craig, D. A., Gott, J. R., III, Weinberg, D. H., Scherrer, R. J., Bhavsar, S. P., and Miller, J. M. 1989, *Ap. J.*, **340**, 647.
Scaramella, R., Baiesi-Pillastrini, G., Chincarini, G., Vettolani, G., and Zamorani, G. 1989, *Nature*, **338**, 562.
Sutherland, W. 1988, *M.N.R.A.S.*, **234**, 159.
Sutherland, W., Maddox, S. J., and Efstathiou, G. this volume, p. 239.
Thuan, T. X., Gott J. R., III, Schneider, S. E. 1987, *Ap. J. (Letters)*, **315**, L93.
White, S. D. M., Frenk, C. S., Davis, M., Efstathiou, G. 1987, *Ap. J.*, **313**, 505.

DISCUSSION

M. Davis: My recollection of the IRAS maps is that the six points showing backflow in your new sample occur in regions where IRAS predicts backflow. Three of them are in the Pavo-Indus region.

A. Dressler: You would know this better than I. I want to emphasize, however, that it is the *amplitude* of the infall pattern, not its existence, that is in disagreement, at least with your technique of normalizing to the Local Group's motion with respect to the Microwave Background. The IRAS prediction you have derived is a factor of two lower in velocity than the new observational data for 136 elliptical and S0 galaxies in the Great Attractor region.

J. Villumsen: The Great Attractor is a strong potential well. Where is the x-ray gas?

A. Dressler: The potential well is not deep. That is why the Faber-Burstein model includes a large core radius (about 1400 km s^{-1}). As the Lahav map shows, there is no central concentration, hence no strong x-ray emission is expected, and none is observed.

E. Bertschinger: You mentioned that the Malmquist correction can take either sign, and noted that it could even produce an S-shape distortion in the Hubble diagram. I agree with you that this is not likely to be a serious problem in the Great Attractor region - the Monte Carlo simulations Avashi Dekel, Sandy Faber, and I have been doing to estimate errors in our three-dimensional reconstructions show evidence for very minor Malmquist errors in this region. Moreover, I want to emphasize that the Malmquist correction depends on $r^2 n(\mathbf{r})$ along the line of sight (n is the galaxy density) and consequently is sensitive to large density gradients as may exist in the foreground and background of the Great Attractor. Ultimately it should be possible to correct density-dependent bias, provided that the selection function is known well enough so that the sample itself can be used to calculate $n(\mathbf{r})$.

A. Dressler: I agree.

N. Kaiser: Why are there so few ellipticals on the rear side of the Great Attractor?

A. Dressler: Actually, the new data we got this last year filled the hole somewhat. I'm less sure that the drop off is significant, but if so, it's probably due to the large number of high-density groups on the front side in the Centaurus region, compared to the far side, that is smoother with fewer peaks.

J. Primack: You mentioned that there is a lot more data in the pipeline, especially on spirals. Could you please summarize these observational projects?

A. Dressler: Riccardo Giovanelli will talk about an all-sky Tully-Fisher study later this week. Lucey and coworkers are observing elliptical and S0 galaxies in northern and southern clusters, as he will report. Schommer, Bothun, and coworkers are observing Tully-Fisher with Fabry-Perot rotation curves, and Faber and I are using long-slit Hα measurements. Both these studies are in the Great Attractor region. Courteau, Faber, and I are also measuring spirals at greater distances to check the Rubin-Ford result. Burstein and collaborators are studying the Hercules-A2199 Supercluster with the D_n - σ relation for ellipticals. There are a lot of new data on the way!

P. Shapiro: I would like to address a question to Ed Bertschinger. It is my impression from the Gorski *et al.* (1989) figure of yours which Alan just showed us, that the CDM model prediction of the large-scale velocity field, even for small values of the bias parameter (*i.e.* b = 1), is in substantial disagreement with the data. Can we infer from this that the CDM model is now in conflict with the data for *all* reasonable choices of b? The so-called 'standard' CDM models, at least, with b = 2.5, which would produce a curve even lower than the b = 2 curves shown on the figure, appears to be in trouble.

E. Bertschinger: First of all, I must urge some caution against over-interpreting bulk-flow modelling, since the observed flow is more complex, and our theoretical calculations only approximate the sample selection effects, and they neglect possible biases in the distance estimators. However, unless there is some terrible error in the distances, the b = 2 CDM model seems to be excluded with a statistically high level of confidence. Even with the b = 1 model we predict bulk velocity only half as large as measured. Some statistics, such as those used by Nick Kaiser, indicate better agreement with CDM, while others (*e.g.* see Bertschinger and Juszkiewicz 1988) yield even larger discrepancies. Until we understand better the possible systematic errors in the peculiar velocities, however, we must not assume Gaussian statistics for 2σ or 4σ events.

DISTANCES AND PECULIAR VELOCITIES OF CLUSTERS IN CENTAURUS

J. R. LUCEY
Department of Physics, University of Durham, Science Laboratories, South Road Durham, DH1 3LE, England

O. LAHAV AND D. LYNDEN-BELL
Institute of Astronomy, University of Cambridge
The Observatories, Madingley Road, Cambridge, CB3 0HA, England

R. J. TERLEVICH
Royal Greenwich Observatory, Herstmonceaux Castle, Hailsham, East Sussex BN27 1RP, England

J. MELNICK
European Southern Observatory, Casilla 567, La Serena, Chile

ABSTRACT A new study of cluster distances and peculiar velocities is described. The first results for the Centaurus (1246-41) and IC4329 (1346-30, Klemola 27) clusters are presented. The Centaurus cluster is found to be even more complicated than previously thought, possessing at least a rapidly streaming component and a component that lies approximately at rest with respect to the Cosmic Microwave Background (CMB). The IC4329 cluster is a relatively poor aggregate that is part of the candidate Great Attractor system at $\sim$4650 km s^{-1}. We find this cluster has a peculiar velocity of $\sim 800 \pm 300$ km s^{-1}.

INTRODUCTION

The measurement of relative distances to galaxies and galaxy clusters in the local universe allows distortions in the nearby Hubble flow to be determined. The derived peculiar velocities allow the major mass concentrations in the nearby universe to be identified as well as providing a powerful test for theories of large-scale structure. Recently several studies (*e.g.* Aaronson *et al.* 1986, Lynden-Bell *et al.* 1988 (7S), Faber and Burstein 1988, Staveley-Smith and Davies 1989) have provided a clearer picture of the distortions in the local Hubble flow. An unexpected result was the discovery of large-scale streaming in the Centaurus region which led to the concept of an extremely large mass concentration (the Great Attractor, GA) located at $\ell = 307°$ $b = 9°$, $\langle R \rangle = 4350 \pm 350$ km s^{-1} (Lynden-Bell *et al.* 1988). New independent data is needed to fully assess the GA model and to provide a reliable measure of the *rms* peculiar velocity of galaxy clusters.

We are undertaking a program to measure the relative distances for a set of nearby galaxy clusters. Our derived distances are based on an application of the diameter *versus* velocity dispersion (D - σ) relationship to early-type galaxies in each cluster (Dressler *et al.* 1987, Dressler 1987). Diameters have been derived from CCD frames and velocity dispersions with the Anglo Australian Telescope's (AAT) fiber system (FOCAP, Gray 1986).

Results for 5 southern clusters using this technique have recently been reported by Lucey and Carter (1988). Because of the special interest in the GA region we chose to observe many of the most prominent clusters that lie in this general direction (*i.e.* 1038-46, Centaurus, Shapley 8, IC4296, IC4329, 1415-27, 1610-60). The rest of our cluster sample is more uniformly distributed over the sky accessible from the AAT.

CURRENT STATUS

We currently have reliable velocity dispersions for ~250 early-type galaxies in 15 clusters in the redshift range 3000 to 14,000 km s^{-1}. CCD frames from a variety of telescopes have been used to measure diameters at an integrated V-band surface brightness (corrected for extinction and cosmological effects) of 19.80 mag $arcsec^{-2}$. This value has been chosen for close compatibility with the system of Burstein and collaborators (Burstein *et al.* 1987). We have yet to fully analyze all our CCD data; however, we have reduced data for two important clusters near the GA position and the results for these are presented here.

THE CENTAURUS CLUSTER (1246-41)

The Centaurus cluster possesses a number of intriguing characteristics including being close (~8°) to the GA position. On the sky the cluster appears very substructured with several prominent groupings of galaxies. The cluster has a bimodal velocity distribution with peaks at ~3000 and ~4500 km s^{-1} (Lucey *et al.* 1980). Lucey and collaborators (Dickens *et al.* 1986, Lucey *et al.* 1986a, 1986b) found good evidence that the majority of the higher velocity component (Cen45) was located at the same spatial location as the lower velocity component (Cen30) and they argued that Cen45 mostly arose from a subgroup infalling into the main Centaurus cluster (Cen30) at ~1500 km s^{-1}. However, they suggested that a minority of the Cen45 component arose from a true background system.

Recently several groups have obtained independent distance estimates to the Cen30 and Cen45 components.

TABLE I Derived distances for the Centaurus Components

	Derived distances in km s^{-1}		
	Cen30	Cen45	
Lucey and Carter (1988)	3115 ± 277	2672 ± 436	E&S0s
Faber *et al.* (1989, 7S)	2221 ± 208	3177 ± 334	Es
Aaronson *et al.* (1989)	2835 ± 253	4259 ± 453	Ss
mean redshift in CMB	3300	4900	

For Cen30, Lucey and Carter (1988) found a small peculiar velocity while Faber *et al.* (1989, 7S) found a large peculiar velocity. Aaronson *et al.*'s (1989) IRTF distance is midway between these two estimates. For Cen45, while Lucey and Carter (1988) and Faber *et al.* (1989) both measured a large peculiar velocity, the IRTF work failed to detect any significant motion. This may be due to the early-type and spiral galaxies being located in separate systems. The GA model has been designed to account for the high peculiar velocities of isolated ellipticals and clusters in the Hydra-Centaurus region including Faber *et al.*'s result for the two components of Centaurus. Improving the data for Centaurus is thus particularly desirable and in our new work it was given a high priority.

Our new data for Centaurus consists of (a) a repeat fiber observation of the Lucey-Carter field which was centered near NGC4696, (b) 2 independent observations of a second fiber field in the cluster core 30′ east of NGC4696 and (c) new CCD photometry for the galaxies in Centaurus with either our fibre σ's or σ's from Davies *et al.* (1987). Our new Centaurus sample (both Cen30 and Cen45) consists of 37 early-type galaxies.

Compared to other well-studied clusters, Centaurus' D - σ relationship (Figure 1) has a larger scatter, 0.14 in log(D) as against 0.08 found for early-type galaxies in Virgo and Coma. Either the D - σ relationship is non-universal or there are real depth effects present in the Centaurus cluster. We will not discuss the former possibility here, and in the following discussion we will assume that various components along the line-of-sight in Centaurus have broadened the D - σ relationship. Although there is a distinct band on the right hand side of the distribution, this does not correspond to either of the two velocity components of Centaurus.

In our analysis of cluster distances we use $D \propto \sigma^x$ with x = 1.2 or 1.5. While Lynden-Bell *et al.* (7S) use a slope of 1.2 for a combined sample of early-type galaxies in well-studied clusters, a slope of 1.5 is, in general, a better fit. For the preliminary analysis presented here the Malmquist bias correction has been ignored. In the derived distance versus redshift diagrams (Figures 2a and 2b) several subgroups are apparent. Possible subgroups are:

- A foreground group with $\langle R \rangle \simeq 2000$ km s^{-1}, $cz_{CMB} \simeq 3500$ km s^{-1} and hence with $v_{pec} \simeq 1500$ km s^{-1}. This group contains mainly brighter galaxies and corresponds to the rapidly streaming component of Cen30 previously identified by the 7S.

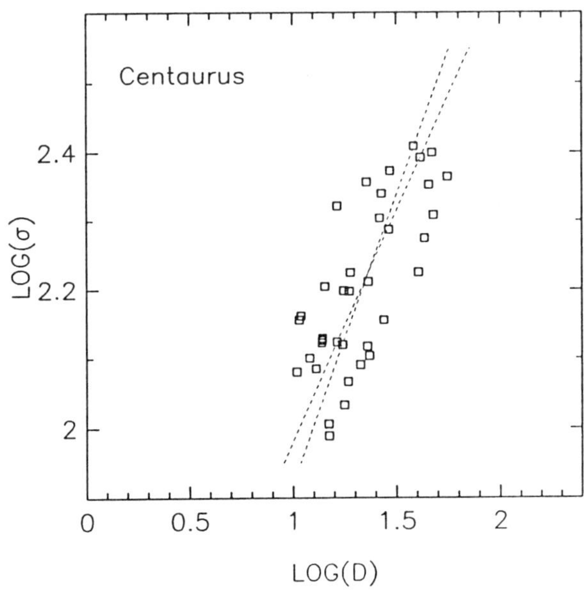

Fig. 1. The log(D) versus log(σ) plot for the 37 early-type galaxies in the Centaurus cluster. The lines have slopes of 1.2 and 1.5.

- A group with $\langle R \rangle \simeq 2800$ km s^{-1}, $cz_{CMB} \simeq 4800$ km s^{-1} and hence with $v_{pec} \simeq 2000$ km s^{-1}. This group contains mainly the Cen45 early-type galaxies in the cluster core. It is a compact group centered on the large elliptical NGC4709 and corresponds to the rapidly streaming Cen45 system first identified by Lucey *et al.* (1986) and subsequently confirmed by Lucey and Carter (1988) and the 7S.

- A 'main' group with $\langle R \rangle \simeq 3500$ km s^{-1}, $cz_{CMB} \simeq 3500$ km s^{-1} and hence with $v_{pec} \simeq 0$ km s^{-1}. There is a large overlap between this group and Lucey and Carter's Cen30 sample, with the exception that the largest streaming galaxies with distances of ~2000 km s^{-1} have now been placed in a separate subgroup, *i.e.* the first of this list.

The dominant central elliptical in the Centaurus cluster, NGC4696, lies between the foreground and 'main' subgroups. Our bias would be to associate it with the 'main' subgroup. As noted above, in this partition of the data the previously identified Cen30 velocity component is now subdivided into two subgroups. Although many galaxies in the

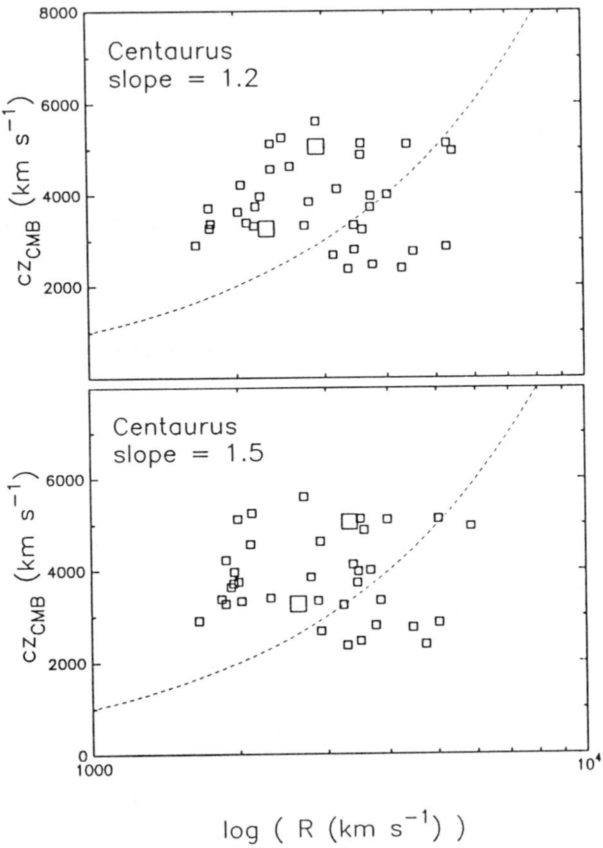

Fig. 2. Log(derived distance) *versus* redshift plots for the early-type galaxies in Centaurus. The plots show the distances derived from adopting slopes for the $D - \sigma$ relation of 1.2 and 1.5. The two large central ellipticals in Centaurus are shown as larger squares: NGC4696 at $cz_{CMB} \simeq 3300$ km s^{-1} and NGC4709 at $cz_{CMB} \simeq 5000$ km s^{-1}. The Hubble lines are shown.

Centaurus region have redshifts in the Cen30 component, their distances are found to stretch from 2000 to nearly 4000 km s^{-1}.

Other partitions of the data set are possible. For example, there may be a subgroup at $\langle R \rangle$ = 3800 km s^{-1}, cz_{CMB} = 2500 km s^{-1}, which could represent a system rapidly streaming towards the observer. We believe that most partitions of the data set will identify both a rapidly streaming foreground component and a component that is approximately at rest with respect to the CMB. In addition, to account for the 4500 km s^{-1} redshift peak, a separate Cen45 subgroup is probably required. Derived distances suggest that this subgroup has a large peculiar velocity which may be caused by infall to the 'main' subgroup.

Thus the Centaurus cluster appears to be composed of two possibly interacting components (the 'main' and Cen45 subgroups) and the chance alignment of a foreground system. The Centaurus cluster lies near the edge of the Centaurus Concentration noted by Lynden-Bell *et al.* as a possible candidate for the GA. This large aggregate has a diameter of ~25° or ~1500 km s^{-1} across the line-of-sight. Most of the galaxies with measured distances in the Centaurus cluster lie within the cluster's core, diameter ~2°. At a distance of 3500 km s^{-1} this core corresponds to only 120 km s^{-1} across the line-of-sight. The distance between the foreground system and the 'main' subgroups down the line-of-sight is 1500 km s^{-1}, *i.e.* the size of the Centaurus Concentration. It is tempting to suggest that an observer residing, say, in the Hydra cluster and studying a rich cluster in the Concentration might be misled by a similar chance alignment. Overall it is disappointing (or perhaps more interesting) that the Centaurus cluster has turned out to be so complicated.

THE IC4329 CLUSTER (1346-30, KLEMOLA 27)

The IC4329 cluster (Shapley 1936, Ritcher 1984) is a relatively poor aggregate in the Hydra-Centaurus supercluster. The cluster has a mean $cz_{\text{CMB}} \simeq 4650 \pm 100$ km s^{-1}. It is part of the GA candidate at $cz_{\text{CMB}} \simeq 4650$ km s^{-1}. Thus knowledge of its peculiar velocity is particularly desirable. From four galaxies Faber *et al.* (1989) derived $\langle R \rangle = 3895 \pm 473$ km s^{-1}, *i.e.* a measured peculiar velocity of $\sim 750 \pm 500$ km s^{-1}.

Our independent sample of 6 early-type galaxies yields $\langle R \rangle = 3950 \pm 350$ km s^{-1} for an adopted slope of 1.2 and 3750 ± 250 km s^{-1} for a slope of 1.5. Thus we find that the IC4239 cluster has a peculiar velocity of $\sim 800 \pm 300$ km s^{-1}.

IMPLICATIONS FOR THE GREAT ATTRACTOR

The 7S study of the motions of nearby elliptical galaxies argued strongly for a large mass concentration located at $\ell = 307°$, $b = 9°$, $\langle R \rangle = 4350 \pm 350$ km s^{-1}. The Hydra-Centaurus supercluster lies in this general direction with its members having redshifts ranging from 3000 to 4800 km s^{-1}. The 7S argued that the large positive peculiar velocities they found for the ellipticals in the Hydra-Centaurus region contradicted the idea that this supercluster was the source of the rapid streaming. However, Lucey and Carter's (1988) study of the Centaurus and Hydra clusters found that these aggregates possessed, at most, only small peculiar motions and hence argued that Hydra-Centaurus was indeed the major cause of streaming, with the 7S's rapidly streaming ellipticals lying foreground to this supercluster.

Dressler's (1988) supergalactic plane redshift survey of the GA region found velocity peaks at 3000 and 4500 km s^{-1}, with the lower velocity peak being associated with the Cen30 cluster and its halo. After correcting for solar motion relative to the CMB, adopting the 7S

1000 km s^{-1} peculiar velocity for Cen30 and assuming that the 4500 km s^{-1} system was at rest with respect to the CMB, he argued that the 'true' velocity distances of the peaks were ~2500 km s^{-1} and ~4500 km s^{-1}. So, although the lower velocity peak was the most prominent, he argued that his diameter-limited redshift survey greatly undersampled the 4500 km s^{-1} system and hence suggested that the higher velocity system possessed ~4 times the mass of the lower velocity system. He associated the centroid of the velocity distribution at 4000 km s^{-1} with the GA and hence argued that the GA comprised both the Hydra-Centaurus supercluster and the more distant, more massive, 4500 km s^{-1} concentration. This nomenclature is misleading because the IC4329 cluster, at a redshift of 4650 km s^{-1}, is part of the Hydra-Centaurus supercluster (Chincarini and Rood 1979, Hopp and Materne 1984).

Our new work on two lines-of-sight through the GA region yields an unexpected result. We have measured peculiar velocities for two independent systems of galaxies with redshifts of ~4500 km s^{-1} (those in the Cen45 velocity component of the Centaurus cluster and those in the IC4329 cluster). The new measurement for Cen45 confirms that this system has a distance considerably nearer than its redshift. For the IC4329 cluster we find a similar result. Therefore these two aggregates that contribute to Dressler's 4500 km s^{-1} velocity system both possess sizeable peculiar velocities. This puts into doubt the claim that this system is the major component of the GA. In addition, we have found that a Centaurus subgroup, at a distance of ~3500 km s^{-1}, lies at rest with respect to the CMB. This suggests that the source of rapid streaming in the Centaurus direction probably lies at a distance less than 4000 km s^{-1}. Clearly there is still much to learn about this complex region.

REFERENCES

Aaronson, M., Bothun, G. D., Cornell, M. E., Dawe, J. A., Dickens, R. J., Hall, P. J., Sheng, H. M., Huchra, J. P., Lucey J. R., Mould, J. R., Murray, J. D., Schommer, R. A., and Wright, A. E. 1989, *Ap. J.*, **338**, 654.

Aaronson, M., Bothun, G. D., Mould, J. R., Huchra, J. P., Schommer, R. A. and Cornell, M. E. 1986, *Ap. J.*, **302**, 536.

Burstein, D., Davies, R. L., Dressler, A., Faber, S. M., Stone, R. I. S., Lynden-Bell, D., Terlevich, R. J. and Wegner, G. 1987, *Ap. J. Suppl.*, **64**, 601.

Chincarini, G. L., and Rood, H. 1979, *Ap. J.*, **230**, 648.

Davies, R. L., Burstein, D., Dressler, A., Faber, S. M., Lynden-Bell, D., Terlevich, R. J. and Wegner, G. 1987, *Ap. J. Suppl.*, **64**, 581.

Dickens, R. J., Currie, M. J. and Lucey, J. R. 1986, *M.N.R.A.S.*, **220**, 679.

Dressler, A. 1987, *Ap. J.*, **317**, 1.

Dressler, A. 1988, *Ap. J.*, **329**, 519.

Dressler, A., Lynden-Bell, D., Burstein, D., Davies, R. L., Faber, S. M., Terlevich, R. J., and Wegner, G. 1987, *Ap. J.*, **313**, 42.

Faber, S. M., and Burstein, D. 1988, in *Large-Scale Motions in the Universe*, ed. V. C. Rubin and G. V. Coyne, S. J. (Princeton: Princeton University), p. 115.

Faber, S. M., Wegner, G., Burstein, D., Davies, R. L., Dressler, A., Lynden-Bell, D. and Terlevich, R.J. 1989, *Ap. J. Suppl.*, **69**, 763.

Gray, P. M. 1986, *Proc. S.P.I.E.*, **627**, 96.

Hopp, U., and Materne, J. 1985, *Astr. Ap. Suppl.*, **61**, 93.

Lucey, J. R. and Carter, D. 1988, *M.N.R.A.S.*, **235**, 1177.

Lucey, J. R., Currie, M. J. and Dickens, R. J. 1986a, *M.N.R.A.S.*, **221**, 453.

Lucey, J. R., Currie, M. J. and Dickens, R. J. 1986b, *M.N.R.A.S.*, **222**, 427.

Lucey, J. R., Dickens, R. J. and Dawe, J. D. 1980, *Nature*, **285**, 305.

Lynden-Bell, D., Faber, S. M., Burstein, D., Davies, R. L., Dressler, A., Terlevich, R. and Wegner, G. 1988, *Ap. J.*, **326**, 19.
Ritcher, O-G. 1984, *Astr. Ap. Suppl.*, **58**, 131.
Shapley, H. 1936, *Harvard Coll. Obs. Bull.*, **903**.
Staveley-Smith, L., and Davies, R. D. 1989, *M.N.R.A.S.*, **241**, 787.

DISCUSSION

S. White: One firm conclusion from your data would seem to be that early-type galaxies in the Centaurus region are not confined within a single volume of space but must be spread out substantially along the line of sight. Thus your original model with Bob Dickens and Malcolm Currie cannot be the whole story.

J. Lucey: Yes, the new data suggest that our original view was not correct, although some of the cluster components do appear to be at the same distance.

E. Bertschinger: I was alarmed when I first saw that you could fit the log D_n - log σ relation with a slope of 1.5 better than with the slope of 1.2 used by Lynden-Bell *et al.* I was reassured to see that this change causes the inferred distance of the IC4329 group to change by only 200 km s^{-1}. Still, I do not understand the difference in slopes. Are you performing a regression of log D_n on log σ or *vice versa*? Caution should be applied since your sample is presumably diameter-limited.

J. Lucey: Our initial reduction indicates that a slope of 1.5 is a better fit than 1.2 for a cluster sample of early-type galaxies.

A. Dressler: I believe that the slope of 1.5 you find for the D_n - σ relation in some clusters is to be compared with the average value of 1.33 in Dressler *et al.* (1986). The slope of 1.2 used in Lynden-Bell *et al.* (1988) is not the best fit to the data, but the appropriate slope to use in the prediction of distances.

J. Lucey: I agree that the choice of the best slope for distance estimation needs careful consideration.

D. Burstein: Sandy Faber, Alan Dressler, and I have also looked at the peculiar velocities of the Centaurus clusters, using Seven-Samurai data, your 1988 data taken with Dave Carter, and the unpublished Dressler-Faber data. It may come as a shock, but we agree that it will be very difficult, if not impossible, to disentangle the peculiar motions of these galaxies. Ironically, we can measure distance, but what radial velocity should we use? The intrinsic velocity dispersion of the clusters further confuses a confused region (creating 'errors' in velocity as well as distance). We would need ten times more galaxies to make sense of the particular motions, and that number of galaxies for which we can get distances is probably not much more than we have now.

J. Lucey: I'm glad to learn that you agree with our conclusion that the Centaurus Cluster is very complex, and I agree that it will be very difficult to disentangle.

A. Dekel: What is the explanation for the difference in the measured σ by your method and by the Seven-Samurai method, and how do you take it into account?

J. Lucey: Dave Carter and I found that our FOCAP velocity dispersions were between 2 and 5% larger than the Seven-Samurai values. We thought that the difference might be due to the smaller aperture size of FOCAP work. However, at this workshop Alan Dressler has reported that his new velocity dispersions for ellipticals in the Great Attractor region are offset from the Seven-Samurai values by a similar amount. Thus any offset for the FOCAP data appears to be small. In my work with Dave Carter, our velocity dispersions were adjusted by a constant to match the Seven-Samurai system.

M. Strauss: Could you comment on the fact that the redshift histogram of the Centaurus cluster looks very similar to Dressler's survey of a much larger region?

J. Lucey: The main component of the Centaurus Cluster has a mean redshift of ~ 3000 km s^{-1}, and it is the dominant nearby structure in the Great Attractor region. Thus it is not surprising that both studies show a peak at 3000 km s^{-1}. The origin of the 4500 km s^{-1} peak in the two studies is not exactly the same. In our early study of Centaurus we attributed this peak to a combination of 1) a component infalling to the main Centaurus Cluster and 2) a more widespread background component. Dressler's SPS sampled the background component and found this structure extended over a considerably larger area.

THE COMPONENTS OF THE CBR MOTION OF THE LOCAL GROUP, AS INTERPRETED FROM THE PECULIAR MOTIONS OF GALAXIES

DAVID BURSTEIN
Department of Physics and Astronomy, Arizona State University
Tempe, AZ 85287-1504

ABSTRACT The peculiar motions of 898 spiral and elliptical galaxies with radial velocities less than 5000 km s^{-1} yield evidence that the 600 km s^{-1} Cosmic Background Radiation (CBR) motion of the Local Group has *at least* four components: 1) An amplitude of ~500 km s^{-1} as part of a global flow shared by all galaxies out to a distance of ~4000 km s^{-1}; 2) ~150 km s^{-1} as part of a general collapse of the Local Supercluster; 3) ~350 km s^{-1} in a direction perpendicular to the Supergalactic plane and of unknown origin; and 4) ~150 km s^{-1} generated by the distribution of mass closer than ~500 km s^{-1}. Additional components of the CBR motion, at levels of ~150 km s^{-1}, cannot be ruled out by present data (*e.g.*, from the more distant mass concentration proposed by Scaramella *et al.* this volume)

INTRODUCTION

The dipole anisotropy of the temperature of the Cosmic Background Radiation (CBR), most easily interpreted as produced by the absolute space motion of the Local Group, is currently measured to be 614 km s^{-1} towards $\ell = 269°$, $b = 28°$ (Lubin and Villela 1986, Boughn *et al.* 1989). Although the large amplitude of this anisotropy has led some to question its origin in a motion of the Local Group (*e.g.* Gunn 1989), the bulk of current evidence is in favor of this straightforward interpretation (Burstein 1990, Burstein *et al.* 1990).

The question of the origin of the Local Group's motion has been of central interest since the discovery of the CBR anisotropy. Is the CBR motion plausibly gravitational in origin? If so, on what size scale(s) is it generated? The present paper attempts to address the latter question, without intentionally prejudicing the answer to the first question, using only the measured peculiar motions of galaxies. In what follows, it is instructive to express the CBR motion of the Local Group in terms of two coordinate systems: 1) Supergalactic (SG) coordinates X, Y and Z, with $v_X = 331$ km s^{-1}, $v_Y = 277$ km s^{-1}, $v_Z = -355$ km s^{-1}; and 2) coordinates centered on the Virgo cluster and within the plane of the Local Supercluster: $v_V = 421$ km s^{-1}, $v_y = -361$ km s^{-1}, and $v_z = -371$ km s^{-1}, where direction V is towards Virgo, y is perpendicular to v but within the SG plane, and z is perpendicular to the SG plane (Aaronson *et al.* 1982a).

THE DATA SAMPLE

The distance estimates from which peculiar motions are derived come from the efforts of many workers over the past decade. Those data considered here include: 1) infrared Tully-Fisher (1977) distances for spiral galaxies derived by Aaronson *et al.* (1982b, 1986, 1989;

Bothun *et al.* 1984); 2) distances to elliptical galaxies determined by the D_n - σ method of the Seven Samurai (Faber *et al.* 1989, Lucey and Carter 1988, Dressler and Faber 1989); and 3) blue-magnitude Tully-Fisher distances for spiral galaxies from de Vaucouleurs and Peters (1984; limited here to galaxies with radial velocities less than 1400 km s^{-1}).

The number of galaxies in each data set, listed in order of accuracy, are: the 'good' subset of the Aaronson *et al.* sample (281 galaxies, 16% error in distance), the D_n - σ distances (544 galaxies, 21%), the 'fair' Aaronson *et al.* data (287 galaxies, 23%), and the de Vaucouleurs-Peters data (76 galaxies, 26%) (see Faber and Burstein 1988 for details). Of these galaxies 898 have Local Group radial velocities, v_{LG} less than 5000 km s^{-1}.

The calculation of peculiar motion requires both a radial velocity and a distance. Radial velocities can be defined in either the CBR reference frame, v_{CBR}, or the Local Group reference frame, v_{LG}. Following Lynden-Bell *et al.* (1988), distances (R) are expressed in units of km s^{-1}, with the distance of the Coma Cluster (here taken to be 7380 km s^{-1}) determining the overall scale. Thus, the CBR peculiar motion of a galaxy is expressed as ($v_{CBR} - R$) km s^{-1}; the Local Group peculiar motion as ($v_{LG} - R$). Also following Lynden-Bell *et al.*, distances are corrected for Malmquist bias assuming a homogeneous distribution of galaxies in space.

METHOD OF ANALYSIS

Previous studies (*e.g.* Faber and Burstein 1988) have shown that the motions of galaxies within a distance of 5000 km s^{-1} are fairly complex, varying coherently over a large range of scale and with a large range in amplitude. If one's eventual goal is to unravel the motion of the Local Group, it is therefore prudent to determine first the largest scale over which the Local Group shares a common motion with other galaxies. If this largest-scale motion can be modelled, motions shared by the Local Group on smaller size-scales can then be more easily found. This is the approach taken by the present analysis.

The principal advantage of this approach is that it can be done graphically and is little dependent on complicated mathematical modelling. However, if the CBR motion of the Local Group arises from a cacophony of many motions on many size scales, this method will tend to underestimate the real number of such motions. As such, the reader should take the number of components derived here as a lower limit to the actual number of discrete components of the CBR motion of the Local Group.

THE LARGEST-SCALE COMPONENT: A GLOBAL FLOW

The current evidence is that galaxies in the vicinity of the Local Group, out to distances of ~4000 km s^{-1}, share a common motion (Lynden-Bell *et al.* 1988, Faber and Burstein 1988, Burstein 1990). One way to demonstrate this is presented in Figure 1 (upper panel), in which the peculiar motions ($v_{CBR} - R$) km s^{-1} of 898 galaxies with $v_{LG} < 5000$ km s^{-1} are plotted against the cosine of the angle θ of the galaxy to the direction $\ell = 309°$, $b = +18°$. The sources of data, field galaxies, and clusters are identified by plotting symbol.

Essentially all galaxies out to 4000 km s^{-1} share a common motion towards this direction. This 'global' flow is not a linear function of cos θ, but rather shows a net acceleration to cos $\theta = 1.0$. Of course, the direction for this motion was not chosen arbitrarily, but lies at the center of a large region given the name the 'Great Attractor'. A current discussion of the motions of galaxies in the Great Attractor region is given by Dressler (this volume), Dressler and Faber (1990) and Burstein *et al.* (1990), and will not be elaborated on here.

Rather, the reader is asked to note four things about the global flow exhibited in Figure 1 (upper): 1) The net flow is smooth and continuous over its whole length, with the Local Group situated in the slowly-rising part of the flow. 2) At any one position of cos θ, the scatter in peculiar motions relative to the global flow is as large as the amplitude of the net flow itself. Faber and Burstein (1988) and Burstein (1990) show that this scatter is primarily due to real, additional motions of galaxies, many of which are generated separately from the global flow (see also below). Thus, the global flow is evident not because it is the dominant motion of every galaxy in this volume of space, but because it is the only motion *shared* by almost all galaxies. 3) Removal of galaxies with cos $\theta > 0.7$ from the graph does not affect the overall coherency of motion. The evidence for a global flow does not depend, *per se*, on the observation of motions in the Great Attractor region, or the observed motion of any one particular cluster or group of galaxies. 4) The strong acceleration evident for cos $\theta > 0.7$, combined with a more detailed analysis (Dressler this volume), indicates that, if the

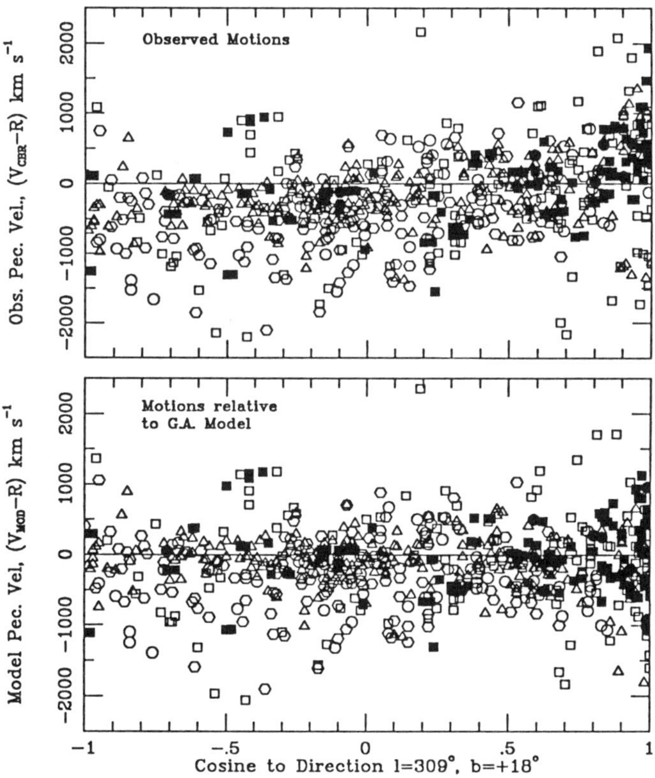

Fig. 1. Cosine diagrams of peculiar motions for 898 E's, S0's and spirals having Local Group radial velocities ≤5000 km s^{-1}, taken relative to the center of the Great Attractor region. Upper panel: as observed in the CBR reference frame. Lower panel: relative to the Great Attractor model of Faber and Burstein (1988). Symbols - open squares: E/S0's, single galaxies; closed squares: E/S0's, in clusters; triangles: Aaronson *et al.* 'good' spirals; hexagons: Aaronson *et al.* 'fair' spirals; closed circles: Aaronson *et al.* clusters; open circles: de Vaucouleurs and Peters spirals.

global flow is gravitational in origin, the source of this flow is at a distance of 4000 - 5000 km s^{-1} in this direction.

As first shown by Lynden-Bell *et al.* (1988), and further by Faber and Burstein (1988), this global flow can be fit by a simple, spherical velocity field model, with a core radius of ~ 1400 km s^{-1}, centered at a distance of 4200 km s^{-1} in the given direction, and having a radial power-law fall-off of peculiar velocity of slope -1.7. The component of motion of the Local Group predicted from this Great Attractor model is 535 ± 50 km s^{-1}. This model, with just five free parameters (including direction) fits the observed global flow remarkably well, as shown in Figure 1 (lower panel). The goodness of this fit, and the lack of a strong residual flow, implies that the more distant gravitational source proposed by Scaramella *et al.* (this volume) could not produce the strong acceleration that is seen. The peculiar motions indicate that this more distant 'attractor' is unlikely to contribute to the CBR motion of the Local Group at a level greater than 100-200 km s^{-1}.

In sum, the peculiar motions of galaxies, out to distances of ~ 4000 km s^{-1} (taking a conservative estimate of the volume enclosed) are dominated by a global flow towards the direction $\ell = 309°$ $b = +18°$ and centered at a distance of ~ 4000 - 5000 km s^{-1}. This global flow is of amplitude ~ 500 km s^{-1} at the position of the Local Group, making it the component of the Local Group's CBR motion that is both shared on the largest scale and of the largest amplitude. Most available evidence indicates that this global flow, now commonly termed the Great Attractor flow, is of gravitational origin (Dressler 1988; see also discussions by other papers in this volume).

THE COLLAPSE OF THE LOCAL SUPERCLUSTER

The Local Group lies on the outskirts of a Supercluster that is dominated in its center by the Virgo and Ursa Major clusters (*e.g.* de Vaucouleurs 1953). These are the nearest large clusters to the Local Group, both at distances of ~ 1350 km s^{-1}. This fact, combined with the fact that the CBR motion of the Local Group has a component of 420 km s^{-1} towards the Virgo Cluster (see above), has led many investigations of large-scale motions to concentrate on the 'Virgocentric Infall' velocity of the Local Group (Aaronson *et al.* 1982a, Davis and Peebles 1983, Pierce and Tully 1988, Huchra 1989).

The motions of galaxies towards these two clusters are shown as observed (Figure 2) and as predicted relative to the spherical Great Attractor velocity field model (Figure 3). Only the more accurate data sets are used in these diagrams, to emphasize the form of motion towards these clusters. These diagrams are also constructed in a manner that emphasizes form as it would be observed from a position outside the flow: distance is calculated for each galaxy projected along a line joining the Local Group to the center of each cluster. Projected distances are positive in the direction of the cluster, negative in the opposite direction. When distance is negative, the sign of peculiar motion is reversed. The motions of galaxies are used for these diagrams only if 1) the galaxies lie within a 30° radius cone of the center of the cluster and 2) we can observe at least half of their motion towards the cluster center. These latter two constraints ensure that only those galaxies are included for which a reliable measure can be obtained of motion relative to the center of the cluster.

Comparison of Figure 2 with Figure 3 shows that the Great Attractor model removes much of the apparent scatter in the observed motions, revealing net infall patterns towards each cluster that are remarkably similar. These patterns of infall are *not* of the form predicted by the 'standard' model of Virgocentric infall: The Virgocentric infall model of Aaronson *et al.* (1982a) predicts peculiar motion to *decrease* with radius relative to the center of the Virgo cluster. The observed peculiar motions become *larger* with distance r

from the center of both the Virgo cluster and the Ursa Major cluster, with $(v_{CBR} - R) \sim r^{+1}$. The observed form of infall motion is one of compression, or overall collapse, instead of motion towards a central mass concentration.

The similarity of form of motion towards each cluster, once the differential velocity field of the Great Attractor is modelled, suggests that the observed infall pattern is not specific to either of these clusters, but may be a more general phenomenon of the Local Supercluster as a whole. This possibility is explored in Figure 4 (upper panel), which plots peculiar motions relative to the Great Attractor model for galaxies that lie within 60° of a direction midway between the Virgo and Ursa Major clusters (ℓ = 225°, b = 80°) and within 25° of the plane of the Local Supercluster (the Supergalactic plane). Given the wide angle of viewing, the distance used for Figure 4 is simply R. To complete this picture, the motions of galaxies within 60° of ℓ = 225°, b = 80°, lying >25° outside the Supergalactic plane are plotted in Figure 4 (lower panel).

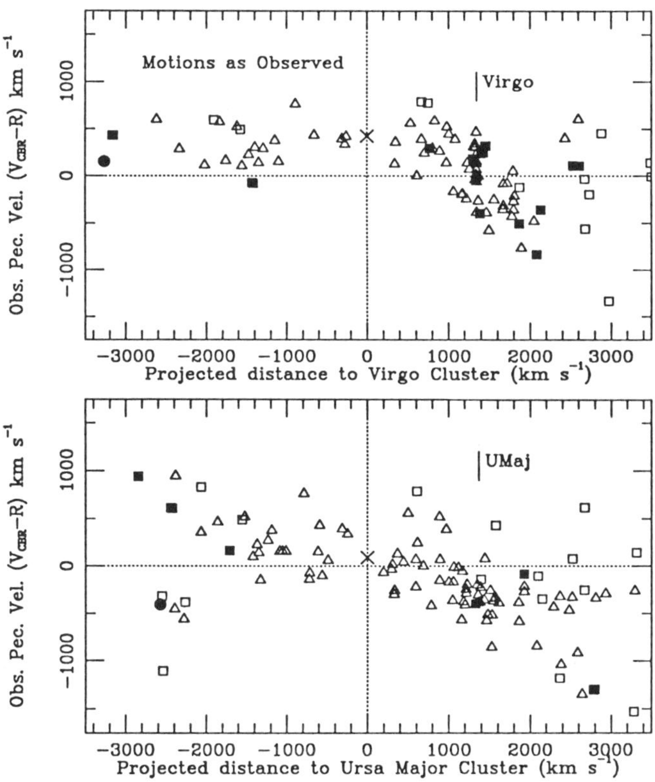

Fig. 2. Observed peculiar motions as a function of distance towards the Virgo cluster (ℓ = 284°, b = 74°) and the Ursa Major cluster (ℓ = 140°, b = 62°), upper and lower panels, respectively. Symbols are as in Figure 1; only the most accurate data are used. Galaxies are included if they lie within a 30° cone towards the cluster, and have less than half of their motion directed towards it. Positive distance is from Local Group towards cluster. Negative distances are for directions opposite the Local Group from the cluster, with the sign of peculiar motion also reversed (*i.e.*, as seen by an 'outside' observer). X marks the component of the Local Group's CBR motion in each direction.

Motions of galaxies relative to a velocity field model that combines both the Great Attractor model and the standard model of Virgocentric Infall are shown in Figure 5. Comparison of Figure 4 and Figure 5 shows that the 'standard' Virgocentric model matches some of the peculiar motions of galaxies, but does not model the more general r^{+1} infall pattern within the Local Supercluster. The standard Virgocentric infall model has little impact on the motions of galaxies perpendicular to the Supergalactic plane (lower panel in Figures 4 and 5).

Figures 4 and 5 (upper) show that the r^{+1} infall pattern evident towards the Virgo and Ursa Major clusters in Figure 3 is shared by most galaxies within the Local Supercluster. Moreover, the scatter in this infall pattern is little more than that seen towards the Virgo cluster alone, despite the much wider viewing angle. This supports the interpretation that this pattern of collapse is not directed towards a particular point, *per se*, but rather is a generalized collapse (*i.e.*, shrinking) of the plane of the Supercluster along lines parallel to the central region. Given the possible confusion in terminology, it should be emphasized that

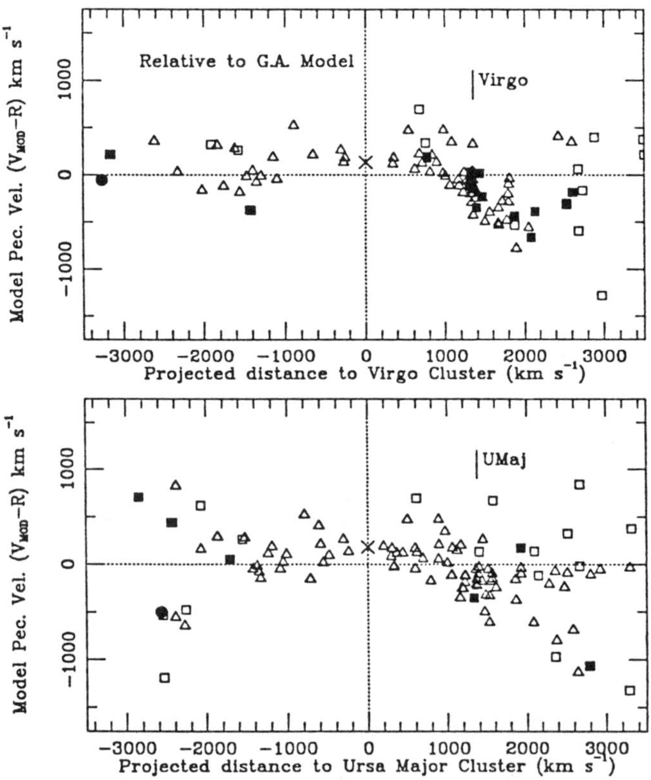

Fig. 3. As in Figure 2, but now using peculiar motions relative to the Great Attractor model of Faber and Burstein (1988; *cf.* bottom panel of Figure 1). The motions of galaxies are similar in the directions of both clusters when the differential motion across the whole volume is removed by the Great Attractor model. The peculiar motions around each cluster increase in amplitude *away* from the cluster, not towards the cluster; *i.e.*, *not* as would be predicted for gravitational infall to a centralized mass.

the infall pattern seen in Figures 4 and 5 (upper) is for galaxies *within* the Supergalactic plane, directed *within* the plane.

The r^{+1} compression-like nature of the collapse cannot be produced by a strongly-centralized mass. Such an infall pattern can be produced, however, by a sphere of homogeneous mass density, with $M \sim r^{+3}$, $\mathbf{F}_g \sim r^{-2}$, acceleration $\sim r^{+1}$. This is not to say that the mass distribution of the Local Group is predicted to be smooth and homogeneous, but simply to point out that a gravitational model of this r^{+1} infall pattern probably will require a mass distribution within the plane of the Local Supercluster that is *not* centrally-*dominated*.

The motions of galaxies *towards* the Supergalactic plane from *outside* the plane are shown in the lower panels of Figures 4 and 5, respectively. Uncovering evidence of infall towards the Supergalactic plane is difficult due to the majority of galaxies lying close to the Supergalactic plane. This means that the relatively few galaxies for which we could observe

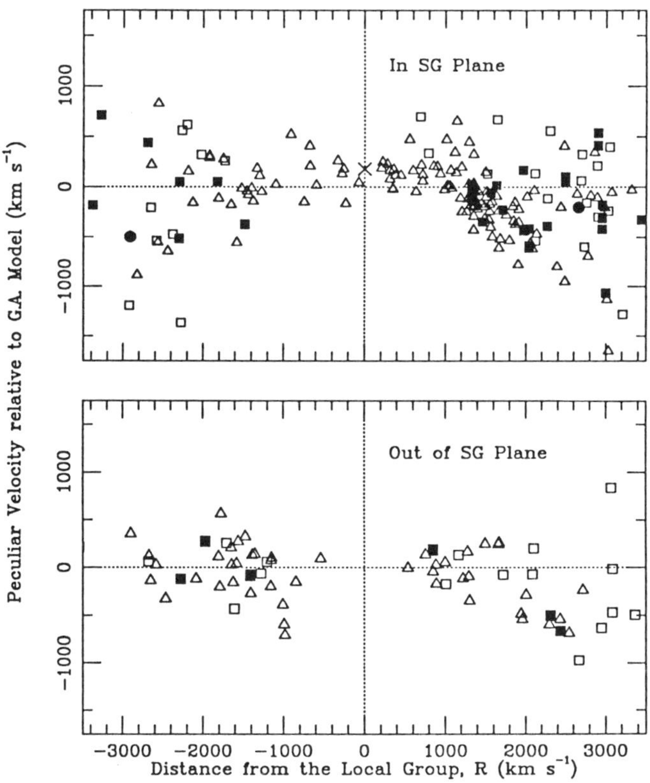

Fig. 4. Motions of galaxies relative to the Great Attractor model, as in Figure 3, but here for the whole Supergalactic plane, using distance R from the Local Group. Upper panel: galaxies within $\pm25°$ of the Supergalactic plane and within $\pm60°$ of $\ell = 225°$, $b = 80°$. Lower panel: galaxies $>25°$ outside of the Supergalactic plane and within $\pm60°$ of $\ell = 225°$, $b = 80°$. Symbols as in Figure 1; only the best data are used. Motions within the whole Local Supercluster (to $R \sim 2000$ km s^{-1}) are similar to those towards Virgo and Ursa Major in Figure 3. Motions of galaxies outside the Supergalactic plane and in the direction of the Local Supercluster tend to be inward, towards the Supergalactic plane.

motion towards the Supergalactic plane will tend to lie at large distances from the Local Group. Nonetheless, one notes that most galaxies and clusters outside the Supergalactic plane and in the direction of the Local Supercluster tend to have motions towards the plane. A net motion towards the Supergalactic plane is not evident in the opposite direction, where the Local Supercluster does not extend.

In sum, the Local Group has a motion of ~ 130 km s^{-1} relative to the CBR that is part of a more general collapse of the Local Supercluster. This collapse is occurring *within* and *along* the plane of the Local Supercluster but, in absence of a realistic velocity field model, the direction of the Local Group's motion due to this component cannot be reliably determined. This collapse motion is shared by almost all galaxies in the plane of the Local Supercluster, and is in the sense of shrinking this plane. The motion of the Local Group relative to the center of the Local Supercluster is about 300 km s^{-1}; the smaller CBR motion of the Local Group is due to the fact that the Supercluster itself has a net motion relative to the CBR of ~ -170 km s^{-1}.

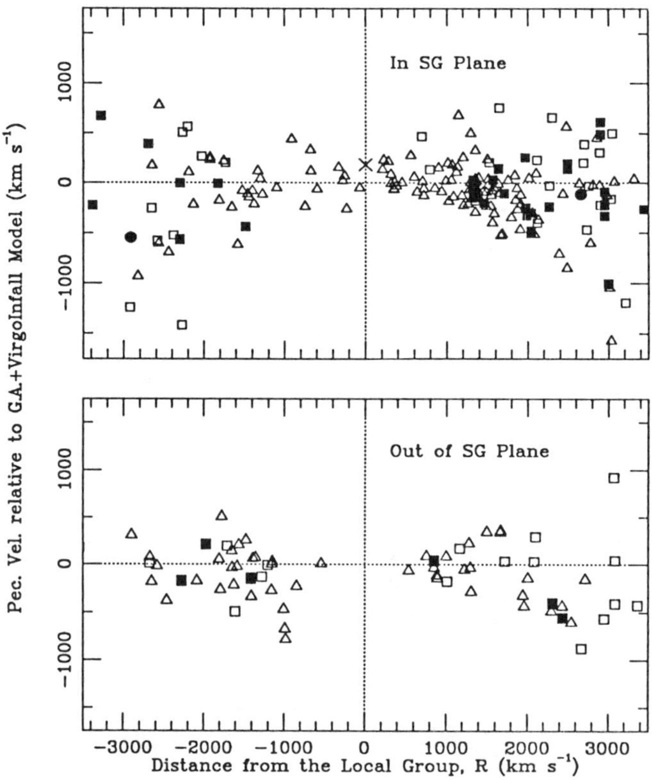

Fig. 5. Motions of galaxies relative to the combined Great Attractor and the standard Virgocentric infall model, plotted in the same format as Figure 4. Although the addition of the Virgo infall model removes some of the motions within the Local Supercluster, compression-like infall motions of ± 300 km s^{-1} remain, especially near the Local Group (upper panel). The standard Virgo infall model does not fit infall motions toward the Supergalactic plane by galaxies outside the Supergalactic plane, as evidenced by comparing the lower panel here with that of Figure 4.

THE MOTION OF THE LOCAL GROUP WITHIN THE COMA-SCULPTOR CLOUD

The distribution of galaxies within 750 km s^{-1} of the Local Group is that of a thin, elongated structure that points in the direction $\ell = 202°$, $b = 65°$ (Raychaudhury 1989). The long axis of this thin 'sausage' is parallel to, and within the Supergalactic plane. As described by Tully and Fisher (1987), the Local Group is part of the Coma-Sculptor cloud of galaxies which dominates this thin 'sausage'.

As shown by both Peebles (1988) and Faber and Burstein (1988), the galaxies within 750 km s^{-1} of the Local Group predominantly share the CBR motion of the Local Group. This fact is shown in Figure 6, taken from Burstein (1990). The upper panel in Figure 6 plots the Local Group peculiar motions, ($v_{LG} - R$) km s^{-1} versus distance R km s^{-1} for all galaxies having $v_{LG} \leq 750$ km s^{-1} and measured distances. This sample is primarily made

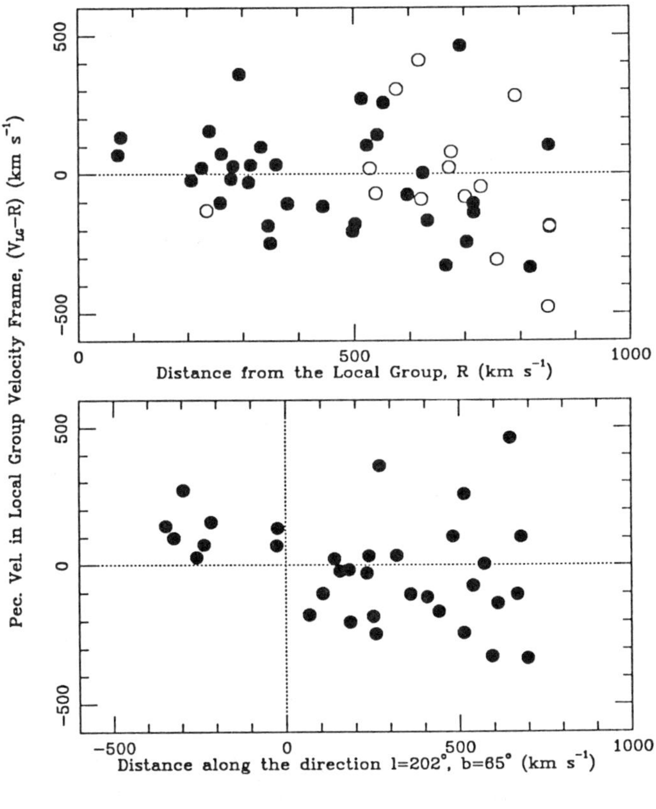

Fig. 6. The motions of those galaxies nearest the Local Group ($v_{LG} \leq 850$ km s^{-1}), as measured relative to v_{LG} plotted as a function of distance (upper panel), and as a function of distance along the major axis of the Coma-Sculptor cloud (lower panel). The plotting convention in the lower panel is as in Figures 2-3. Closed circles represent galaxies identified as Coma-Sculptor members by Tully-Fisher (1987). Almost all Coma-Sculptor galaxies share the Local Group's CBR motion. The ~100 km s^{-1} dispersion of peculiar motions within the cloud implies ~140 km s^{-1} of the Local Group's motion is generated by the net gravity of the nearest galaxies/mass.

of spiral galaxies, with only three ellipticals having $v_{LG} < 750$ km s^{-1}. ($v_{LG} - R$) peculiar motions are used here to emphasize the fact that most of these galaxies share the CBR motion of the Local Group.

The motion of the Local Group relative to galaxies within the Coma-Sculptor cloud is better shown in Figure 6 (lower), which plots the peculiar motions of only the Coma-Sculptor galaxies as a function of distance along the long axis of the Coma-Sculptor cloud, in the style of Figures 2-3. When viewed in this manner, it is seen that the Coma-Sculptor cloud extends about 700 km s^{-1} in the direction $\ell = 202°$, $b = 65°$, and about 350 km s^{-1} in the opposite direction. The apparent infall within the Coma-Sculptor cloud seen in Figure 6 (lower) can be adequately modelled by a combination of the Great Attractor model velocity field and the overall collapse of the Local Supercluster (Burstein 1990), as opposed to a collapse of the Coma-Sculptor cloud alone (as suggested by Tully 1988).

The apparent velocity dispersion of galaxies within the Coma-Sculptor cloud is ~ 100 km s^{-1} (Faber and Burstein 1988). From this fact, one would infer that the Local Group should also have a net motion relative to the most nearby galaxies of amplitude $\sim 100 \times 3^{1/3}$ ~ 140 km s^{-1}. Unfortunately, the linear, coplanar nature of the Coma-Sculptor cloud makes the Z motion (*i.e.*, motion perpendicular to the Supergalactic plane) of the Local Group and the other Coma-Sculptor galaxies difficult to measure. Nonetheless, it is clear that the CBR motion of the Local Group should include a component ~ 100 km s^{-1} in amplitude, generated by the net gravity field of the nearest galaxies, with direction uncertain.

THE MOTION PERPENDICULAR TO THE SUPERGALACTIC PLANE

The second largest component of the CBR motion of the Local Group is both the most straightforward to identify and the most difficult to interpret. From the introduction, the Local Group has a motion of ~ 350 km s^{-1} perpendicular to the Supergalactic plane. This cannot be due to infall within the Supergalactic plane, nor motion towards the Great Attractor region, which lies very close to the Supergalactic equator.

It is plausible to think that this 350 km s^{-1} motion is shared by most of the galaxies in the Coma-Sculptor cloud. However, the high degree of coplanarity of the galaxies in the Coma-Sculptor cloud yields little direct evidence that this is so. The interpretation of this motion is made even more difficult by the fact that the direction of this Z motion is essentially within the Galactic plane: $\ell \sim 200°$, $b \sim 0°$.

It has been suggested that the Z component of the Local Group's motion could be due to a 'push' from the Local Void (Tully and Fisher 1987) that lies in approximately the opposite direction (Lynden-Bell and Lahav 1989, Faber and Burstein 1988). It was also suggested at the present meeting by several people that it could be a pancake-like collapse motion, with the Local Group and the Coma-Sculptor cloud sharing a net residual motions towards the Supergalactic plane.

In sum, although the direction ($\ell \sim 200°$, $b \sim 0°$) and amplitude (~ 350 km s^{-1}) of the Z component of the Local Group's motion are known, little else about it is currently known.

DISCUSSION

Of the four components of the CBR motion of the Local Group that can be discerned from a study of peculiar motions, the origin of three can be plausibly identified: The global flow ascribed to the Great Attractor, the general collapse of the Local Supercluster and the net gravity of those galaxies and/or mass nearest the Local Group. The origin of the fourth component, that perpendicular to the Supergalactic plane, cannot be determined from the present data.

In terms of amplitude two components are dominant: the motion of the Local Group as part of the global flow ($\sim$500 km s^{-1}) and the motion perpendicular to the Supergalactic plane ($\sim$350 km s^{-1}). The directions for both of these components are reasonably well determined. The amplitude of the other two components are considerably smaller, of comparable size to each other, $\sim$100 - 150 km s^{-1}, and have directions much more poorly defined. The low amplitude of the component of motion due to the collapse of the Local Supercluster is partly due to the fact that the Local Supercluster itself has a net motion relative to the CBR reference frame.

As stated at the outset, these four components are the minimum that are required to describe the CBR motion of the Local Group. There obviously can be more, but none that are likely to be of amplitude over $\sim$150 km s^{-1} at the position of the Local Group (*e.g.*, from the clusters discussed by Scaramella *et al.* in this volume).

Although there is nothing in the above analysis that requires these components of motions to arise from gravity alone, the fact that three of these components are directly correlated with known structures of mass near the Local Group certainly supports that interpretation. This issue should, in principle, be settled in the future by combining the present kind of analysis with separate analyses of the distribution of matter and mathematical modelling of the observed motions (*cf.* other papers in this volume, as well as those in Rubin and Coyne 1988).

Finally, it should be stated that this writer has freely distributed, upon request, the peculiar velocities that have gone into the present analysis. A first release was made starting in December 1987, and has been the source of the theoretical analyses of several workers at the present meeting. It is anticipated that a second release will be made by the end of 1989, incorporating the more recent data included in this paper. This list is typically distributed over electronic mail, although exceptions can obviously be made. Anyone wishing to be on the distribution list for the second release should send an electronic message to BURSTEIN@ASUCPS.BITNET), or write directly to the author.

ACKNOWLEDGEMENTS

Parts of this work can be traced back to the original Seven Samurai study, but the present study has benefitted principally from discussions with Alan Dressler, Sandra Faber, Donald Lynden-Bell and Ofer Lahav. The generosity of AD and SMF in permitting the use of their data before publication is very much appreciated. The hospitality and support of the Institute of Astronomy, Cambridge, England, where much of the research for this paper was begun, is gratefully acknowledged. This research has also been supported by a Faculty Grant-in-Aid from Arizona State University.

REFERENCES

Aaronson, M., Bothun, G. D., Mould, J. R., Huchra, J. P., Schommer, R. A., and Cornell, M. 1986, *Ap. J.*, **302**, 536.

Aaronson, M., Bothun, G. D., Cornell, M. E., Dawe, J. A., Dickens, R. J., Hall, P. J., Sheng, H. M., Huchra, J. P., Lucey J. R., Mould, J. R., Murray, J. D., Schommer, R. A., and Wright, A. E. 1989, *Ap. J.*, **338**, 654.

Aaronson, M., Huchra, J. P., Mould, J., Schechter, P. L. and Tully, R. B. 1982a, *Ap. J.*, **258**, 64.

Aaronson, M., Huchra, J. P., Mould, J. R., Tully, R. B., Fisher, J. R., van Woerden, H., Goss, W. M., Chamaraux, P., Mebold, U., Siegman, B., Berriman, G., and Persson, S. E. 1982, *Ap. J. Suppl.*, **50**, 241.

Bothun, G. D., Aaronson, M., Schommer, R. A., Huchra, J. P., and Mould, J. R. 1984, *Ap. J.*, **278**, 475.

Boughn, S., Cottingham, D., Fixsen, D. and Cheng, E. 1989, private communication.

Burstein, D. 1990, *Rep. Prog. Phys*, **53**, 421

Burstein, D., Faber, S. M., and Dressler 1990, *Ap. J.*, **354**, 18.

Davis, M. and Peebles, P. J. E. 1983, *Ann. Rev. Astr. Ap.*, **21**, 109.

de Vaucouleurs, G. 1953, *A. J.*, **58**, 30.

de Vaucouleurs, G. and Peters, W. L. 1984, *Ap. J.*, **287**, 1.

Dressler, A. 1988, *Ap. J.*, **329**, 519.

Dressler, A. this volume, p. 19.

Dressler, A., and Faber, S. M. 1990, *Ap. J.*, **354**, 13.

Faber, S. M. and Burstein, D. 1988, in *Large-Scale Motions in the Universe: A Vatican Study Week*, ed. V. C. Rubin and G. V. Coyne, S. J. (Princeton: Princeton University), p. 115.

Faber, S. M., Wegner, G., Burstein, D., Davies, R. L., Dressler, A., Lynden-Bell, D. and Terlevich, R. J. 1989, *Ap. J. Suppl.*, **69**, 763.

Gunn, J. 1989, in *The Extragalactic Distance Scale*, ed. S. van den Bergh and C. J. Pritchet, *A.S.P. Conf. Ser.*, **4**, (Provo: Astronomical Society of the Pacific), p. 344.

Huchra, J. P. 1989, in *The Extragalactic Distance Scale*, ed. S. van den Bergh and C. J. Pritchet, *A.S.P. Conf. Ser.*, **4**, (Provo: Astronomical Society of the Pacific), p. 257.

Lubin, P. and Villela, T. 1986, in *Galaxy Distances and Deviations from Universal Expansion*, ed. B. F. Madore and R. B. Tully (Reidel: Boston), p. 169.

Lucey, J. R. and Carter, D. 1988, *M.N.R.A.S.*, **235**, 1177.

Lynden-Bell, D., Faber, S. M., Burstein, D., Davies, R. L., Dressler, A., Terlevich, R. J. and Wegner, G. 1988, *Ap. J.*, **326**, 19.

Lynden-Bell, D. and Lahav, O. 1988, in *Large-Scale Motions in the Universe: A Vatican Study Week*, ed. V. C. Rubin and G. V. Coyne, S. J. (Princeton: Princeton Univ. Press), p. 199.

Peebles, P. J. E. 1988, *Ap. J.*, **332**, 17.

Pierce, M. J. and Tully, R. B. 1988, *Ap. J.*, **330**, 579.

Raychaudhury, S. 1989, *Nature*, **342**, 251.

Rubin, V. C. and Coyne, G. V., S. J. 1988, *Large-Scale Motions in the Universe: A Vatican Study Week*, (Princeton: Princeton University).

Scaramella, R., Baiesi-Pillastrini, G., Chincarini, G., Vettolani, G., and Zamorani, G., this volume, p. 13.

Tully, R. B. 1988, *Nature*, **334**, 209.

Tully, R. B. and Fisher, J. R. 1977, *Astr. Ap.*, **54**, 661.

Tully, R. B. and Fisher, J. R. 1987, *Nearby Galaxies Atlas*, (Cambridge: Cambridge University).

DISCUSSION

A. Dekel: I think we are confusing two things: mapping the velocity field using toy spherical flow models (Virgo, the Great Attractor, etc.) is useful, but it is probably hopeless to try to track down the gravitational 'sources' for these velocities in terms of simple spherical attractors and repellors.

D. Burstein: I would agree that the use of the term 'sources' is misleading. Perhaps the term 'direction' conveys a better meaning. As my written remarks will reflect, the term directions will be used for the four components of the Cosmic Microwave Background vector deduced from peculiar motions.

S. White: Once the dipole anisotropy of the Cosmic Microwave Background had been established in the 1970s, it became clear that it could not be interpreted with a spherical model centered on the Virgo cluster. Thus there was strong motivation to consider nonspherical models. The simplest such model considers a homogeneous but nonspherical perturbation. About ten years ago Joe Silk and I generalized earlier work on such models (mainly by Vincent Icke) to try to explain the apparent microwave motion in detail. These models predict a retarded expansion field in the Supergalactic Plane with the form you have exhibited and which has long been advocated by de Vaucouleurs. A large velocity perpendicular to the plane can be present as a residual of a collapse to a sheet or a 'pancake'. Silk and I concluded that the observed velocity vector could (just) be accommodated by this kind of model, but we worried that the *galaxy* distribution appeared much more concentrated to Virgo than the model would suggest (biased galaxy formation was not in vogue at the time). I think this kind of model can account at least qualitatively for at least three of your scales within a single framework which is rather similar to Zel'dovitch's old ideas about pancake collapse.

D. Burstein: I would agree. A uniform mass distribution ($\rho(r)$ independent of r) would go a long way to explaining the fact that $v(r)$ varies roughly as r. However, if one concludes that $\rho(r)$ is a constant, I think one would also have to conclude that light does not trace mass within the supercluster. Are we prepared to do that yet (see also my question to Marc Davis)?

THE PECULIAR VELOCITY FIELD PREDICTED FROM THE DISTRIBUTION OF *IRAS* GALAXIES

MICHAEL A. STRAUSS AND MARC DAVIS
Astronomy Department, University of California, Berkeley, CA 94720

ABSTRACT We present recent results from our full-sky redshift survey of *IRAS* galaxies. After briefly describing our sample selection and observations, we use the linear-theory relation between acceleration and peculiar velocity to make predictions of the peculiar-velocity field in the local universe. We compare our predictions with directly measured peculiar velocities from the Local Supercluster spiral-galaxy sample of Aaronson *et al.* (1982) and the elliptical-galaxy sample of Lynden-Bell *et al.* (1988). In a reference frame at rest with respect to the Cosmic Microwave Background, there is a systematic bulk flow in the residuals between observations and predictions. The Local Group itself takes part in this motion, so the bulk flow disappears in a frame at rest with respect to the Local Group. We show with the use of *N*-body techniques that this type of systematic effect is at least partly due to small-number statistics in the tracing of the density field. We conclude with a discussion of the challenges facing us in the future.

THE *IRAS* SAMPLE

Peebles (1976) describes the velocity field expected at late times in linear theory; the dipole anisotropy of the galaxy distribution around a given galaxy is directly proportional to the peculiar velocity of that galaxy:

$$\mathbf{v} = (H_0\Omega^{0.6}/4\pi) \int \hat{\mathbf{x}}\, \delta(\mathbf{x})/x^3\, d^3x \tag{1}$$

where $\mathbf{v}$ is the peculiar velocity of the galaxy in question, Ω is the cosmological density parameter, and $\delta(\mathbf{x})$ is the fractional density perturbation at position $\mathbf{x}$. The following year, a dipole anisotropy was discovered in the Cosmic Microwave Background (CMB) (*e.g.*, Lubin and Villela 1986), which was interpreted as due to a peculiar motion of the Local Group (LG) with respect to the rest frame of the CMB of 600 km s^{-1}. It was immediately realized that a measurement of the distribution of nearby galaxies could be used in concert with equation (1), and the measured LG peculiar velocity, to make an estimate of Ω. This has been tried by many authors, usually with the simplifying assumption that the Virgo cluster is the only gravitating mass point (see, *e.g.*, Davis and Peebles 1983; Huchra 1988). To do better requires galaxy samples that cover a large fraction of the sky. Pioneering work using optically selected samples of galaxies has been done by Lahav and collaborators (Lynden-Bell, Lahav, and Burstein 1989, and references therein). In this paper we describe a survey of galaxies observed by the Infrared Astronomical Satellite (*IRAS*). We have discussed our work before in two other conference proceedings (Strauss and Davis 1988a, 1988b; Yahil 1988), and in this paper will concentrate on new results obtained since the writing of those articles. A much more detailed account can be found in Strauss (1989), and several papers in preparation.

The *IRAS* satellite surveyed the entire sky in four broad bands in the far-infrared, with a resolution of approximately 1′ at 60μm. Because of the full-sky coverage and the lack of Galactic extinction in the far-infrared, the *IRAS* database offers a unique opportunity to study the large-scale distribution of galaxies. We have selected galaxy candidates from the *IRAS* Point Source Catalog (1985) using the following criteria:

1) $f_{60}/f_{12} > 3$, where f_λ is the flux density at wavelength λ. This discriminates effectively between stars and galaxies;

2) $|b| > 5°$, in a region not affected by confusion at 60μm. With these restrictions, our sample covers 87.6% of the sky;

3) $f_{60} > 1.936$ Jy.

We have obtained optical identifications and redshifts of the approximately 4500 objects meeting these criteria, thus obtaining a sample of roughly 2550 galaxies. The vast majority of the non-galaxies in the sample are associated with stars and infrared cirrus at $|b| < 10°$ (see Yahil 1988 for a description of how optical identifications were made at these low latitudes). In a section below we discuss some biases of the sample selection and our attempts to correct for them. The distribution of all galaxies in the sample is shown in Galactic coordinates in Figure 1. Note the strong continuity of structures across the Galactic plane, in particular, the overdensities associated with Centaurus and the Pavo-Indus-Telescopium region ($\ell \sim 330°$) appear to be physically associated. The Pisces-Perseus filament ($\ell \sim 135°$) and the Hydra region ($\ell \sim 280°$) also show continuity across the plane, as do the underdense regions centered at $\ell = 90°$ and $\ell = 220°$. Compare this figure with the corresponding figure 1a in Yahil (1988); note how much stronger this sense of continuity is now that we have identified galaxies in the two strips $5° < |b| < 10°$.

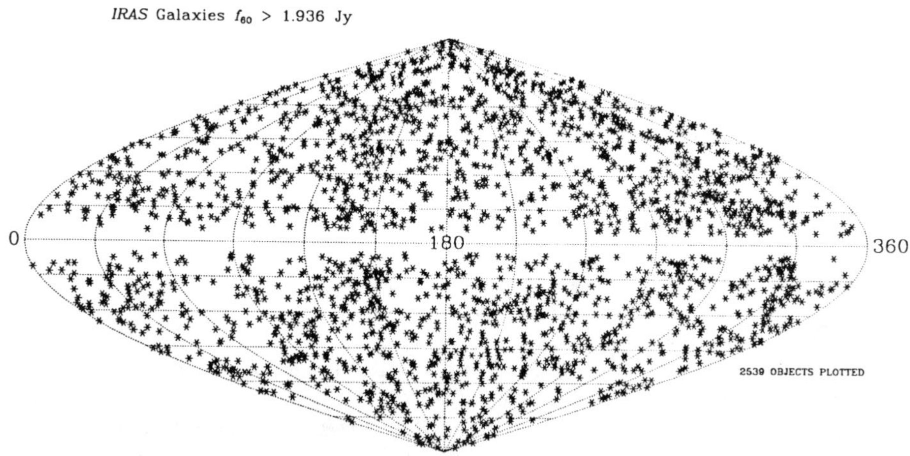

Fig. 1. The distribution of galaxies in the survey, plotted in Galactic coordinates. The sample does not include galaxies at $|b| < 5°$, and there are other regions at high Galactic latitudes which are not covered due to confusion or incompleteness of the *IRAS* survey.

THE PREDICTED PECULIAR VELOCITY FIELD

As described in detail in Strauss and Davis (1988b) and Yahil (1988), we have used equation (1) to predict the velocity flow field within 8000 km s^{-1} of the Local Group. The procedure we follow is slightly modified from that of Yahil (1988). We first fill the missing Galactic plane strip with random 'galaxies' with number density given by an interpolation of the density in strips with $5° < |b| < 20°$, under the assumption that the continuity across the plane described in the last section is real. A selection function is derived using a maximum-likelihood estimator (Yahil 1988), which is used to weight the galaxies to compensate for the magnitude limit of the sample. A value of Ω is derived by requiring that equation (1) reproduce the 600 km s^{-1} peculiar velocity of the Local Group inferred from the CMB dipole. For an adopted value of $\Omega = 0.7$, the predicted peculiar velocity of the Local Group is 650 km s^{-1} towards $\ell = 250°$, $b = +45°$, some 22° from the CMB dipole direction of $\ell = 270°$, $b = +30°$ (Lubin and Villela 1986). Strauss and Davis (1988b) show that the majority of this acceleration is due to material within 4000 km s^{-1} of the Local Group (we will express distances in terms of radial velocities in order to bypass the necessity of specifying the Hubble Constant).

We can go further, by using equation (1) to make estimates of the peculiar velocity of every galaxy in the sample. We restrict ourselves to objects within 8000 km s^{-1} of the Local Group, because the sample becomes prohibitively sparse at greater distances. In practice, our estimate of the peculiar velocity of a galaxy at position **x** is given by:

$$\mathbf{v}(\mathbf{x}) = \frac{H_0\Omega^{0.6}}{4\pi n_1} \sum_i \frac{S(|\mathbf{x}-\mathbf{x}_i|)(\mathbf{x}-\mathbf{x}_i)}{\phi(x_i)|\mathbf{x}-\mathbf{x}_i|^3} - H_0\Omega^{0.6}\,\mathbf{x}/3 + W(S) \tag{2}$$

Here n_1 is the measured galaxy density, $\phi(x_i)$ is the value of the selection function at distance x_i, and S is a smoothing on small scales, to be described further below. The second term corrects for the fact that the sum is not spherically symmetric for a point displaced from the origin, and $W(S)$ is an additional correction applied when the source is within one smoothing length of the edge of the sample.

The measured redshift, cz, of a galaxy may be expressed as:

$$cz = \text{distance} + (\mathbf{v} - \mathbf{v}_{LG}) \cdot \hat{\mathbf{x}} \tag{3}$$

where $\mathbf{v}$ and $\mathbf{v}_{LG}$ are the peculiar velocity vectors of the galaxy and the Local Group, respectively, and $\hat{\mathbf{x}}$ is the unit vector pointing from the Local Group to the galaxy. We separate peculiar velocities and distances using an iterative procedure: galaxies are initially assumed to be at their redshift distances ($\mathbf{v} \equiv 0$), and equation (2) is used to make a first estimate of the peculiar velocity field. Distances are then updated using equation (3) and the process is repeated until convergence. Ω is kept fixed throughout this procedure, and the value of $\mathbf{v}_{LG}$ used is that *derived from the IRAS sample itself* on each iteration. A 'buffer zone' between 8000 km s^{-1} and 10,000 km s^{-1} prevents galaxies from being lost from the sample if their peculiar velocities take them beyond 8000 km s^{-1}. The resulting peculiar velocity field converges with a standard deviation of residuals less than 10 km s^{-1} after eight iterations.

There are several subtleties associated with this process. In regions that are virialized, or even overdense by a factor of several, equation (1) no longer holds, and this iterative procedure may cause chaos. Thus we will never be able to properly model the 'Finger of God' effect seen in the cores of clusters. We have thus collapsed the galaxies in seven cluster cores (Virgo, Ursa Major, Fornax, Eridanus, Centaurus, Hydra, and Perseus) to a common redshift, and have suppressed the mutual gravitational attraction of galaxies within each cluster. Another related problem is that equation (2) diverges at small separations. Note that because the selection function drops sharply with distance, a pair of galaxies separated by a small amount at a large distance from us can give each other a tremendous kick. In order to suppress this, we adopt a smoothing law, $S(r) = (r/r_s)^3$ for $r < r_s$, where r_s, the smoothing length, is taken to be the mean interparticle spacing at that distance, or 500 km s^{-1}, whichever is larger.

The resulting velocity field is similar to that shown in Figure 10 of Yahil (1988) and will not be reproduced here. As those figures show, the predicted velocity field is dominated by two large mass concentrations in Hydra-Centaurus and Perseus-Pisces. We will compare these predictions with the observed velocity field in the next section.

COMPARISON WITH THE OBSERVED PECULIAR VELOCITY DATA

Burstein (1990) reviews the current status of direct observations of the peculiar velocity flow field; see the contributions by him, Dressler, Lucey, Bertschinger, and Dekel in this volume for interpretations of these observations. In this paper, we will make comparisons of our

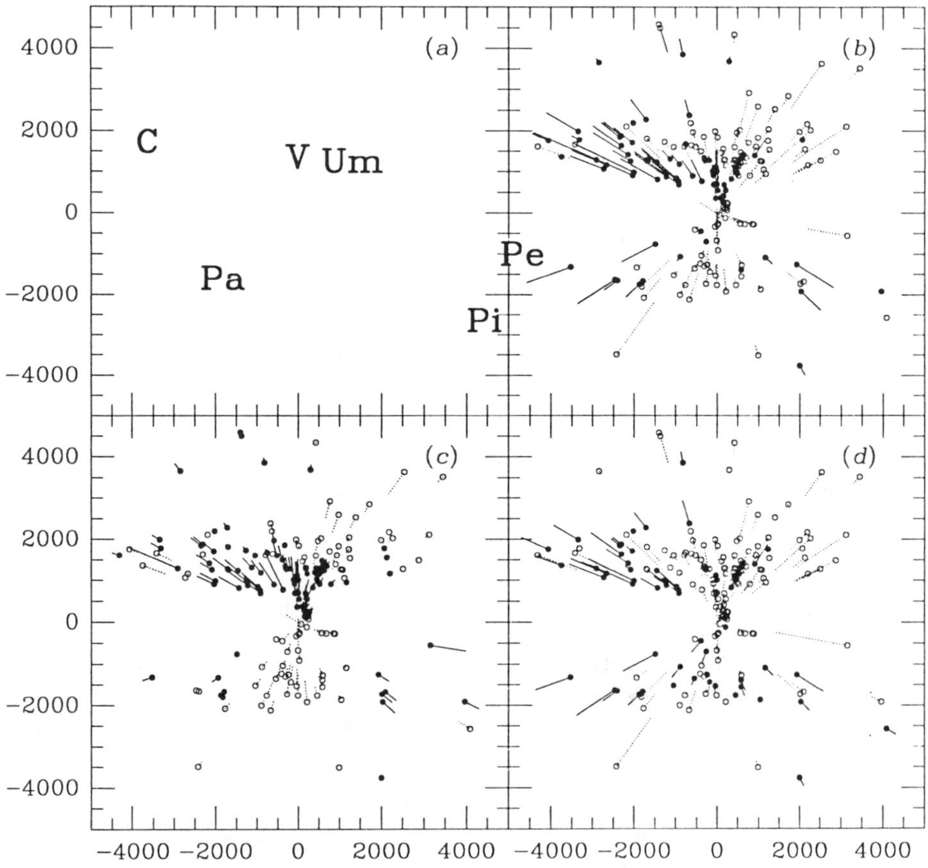

Fig. 2. The velocity field of the Aaronson *et al.* (1982) and Faber *et al.* (1989) galaxy samples in a slice through the Local Supercluster plane, in the CMB frame: (a) The positions of the major clusters falling in the slice; V = Virgo, Um = Ursa Major, C = Centaurus, Pa = Pavo, Pe = Perseus, and Pi = Pisces; (b) The observed velocity field; (c) The *IRAS* predicted velocity field; (d) The difference between (b) and (c).

predictions with two data sets: the subset of the Aaronson *et al.* (1982) Local Supercluster spiral galaxies with high-quality observations defined by Faber and Burstein (1988), and the independent peculiar velocities of individual galaxies, groups, and clusters of the Seven Samurai elliptical galaxy sample, listed in Table 4 of Faber *et al.* (1989). Figure 2 shows the distribution of peculiar velocities in a slice through the plane of the Local Supercluster, using the plotting technique introduced by Lynden-Bell *et al.* (1988). All galaxies in the two samples within 22.5° of the plane are plotted, and the length of the line attached to each point is equal to the radial peculiar velocity. Solid symbols are used for positive peculiar velocities, and open symbols are used for negative. In the first panel of this figure, we show the positions of the well-known clusters that fall in this projection; the symbols are indicated in the Figure caption. In panel (b), the *observed* peculiar velocities are plotted, panel (c) shows the *IRAS predictions*, and panel (d) shows the difference of (b) and (c). It is indeed a remarkable fact that *IRAS* is able to reproduce the qualitative nature of the velocity flow field, perhaps the first *direct* indication that peculiar velocities are generated by gravity, and that galaxies at least approximately trace the mass distribution of the underlying dark matter. However, there are some systematic differences between observations and predictions. Note for instance that whereas the observed peculiar velocities (panel b) to the upper right of the

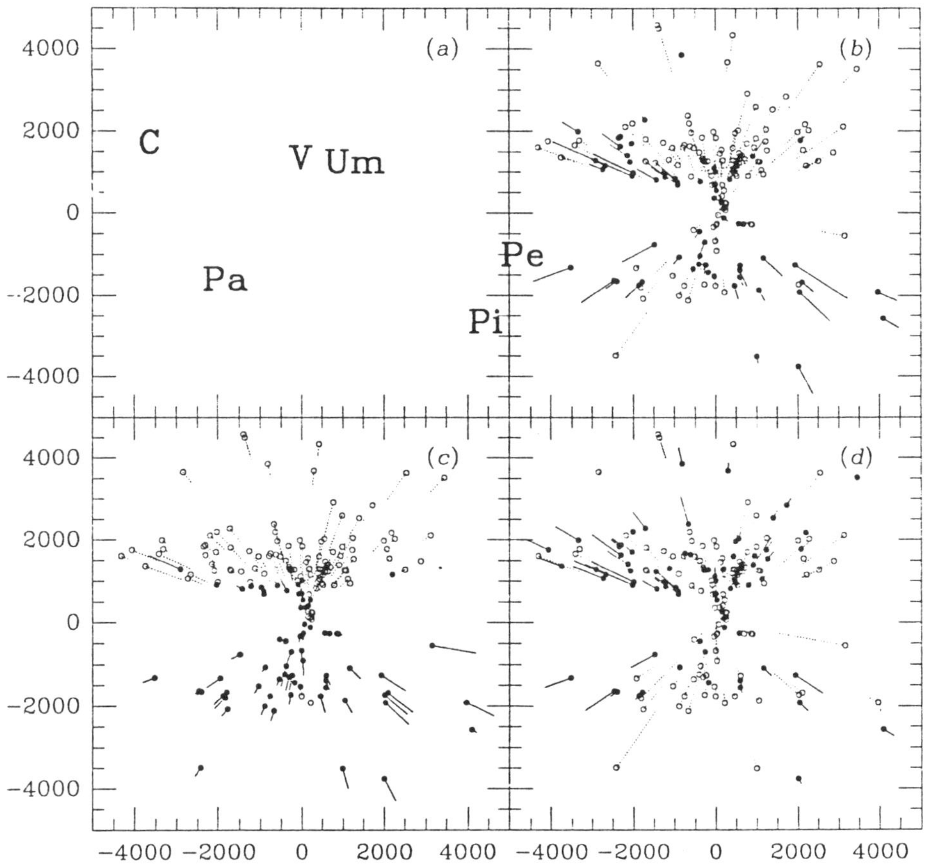

Fig. 3. As in Figure 2, but plotted in the rest frame of the Local Group.

Ursa Major region are large and negative, the predicted motions (panel c) are very quiet; *IRAS* predicts that the pulls of the Pisces-Perseus and Hydra-Centaurus superclusters approximately balance in this region. Note also the very large observed peculiar velocities towards Centaurus; *IRAS* predicts motion of the same sign, but of smaller amplitude. In panel (d), there is a systematic motion seen in the difference between predicted and observed peculiar velocities, as if there is a bulk flow component to the motion that *IRAS* does not see. We find consistent solutions for the bulk flow using the Aaronson *et al.* and Faber *et al.* galaxies separately as well as together: 250 km s^{-1} pointing towards $\ell = 320°$, $b = -35°$.

In Figure 3, the same comparison of observed and predicted peculiar velocities is made, this time in a frame at rest with respect to the Local Group. This is not a trivial change, because in order to transform the *IRAS* predicted motions, we use the Local Group motion *as predicted by IRAS*, while the *observed* LG frame peculiar velocities are merely the difference between their redshift and observed distance. Note now in panel (d) that the sense of an overall bulk flow to the residuals between observed and predicted peculiar velocities has vanished, and we are left with mostly small-scale motions that we have not modelled correctly. Thus the Local Group itself takes part in the bulk flow seen in the residuals in Figure 2(d). The only significant systematic effect remaining in the LG frame is a strong infall in the direction of Hydra-Centaurus that *IRAS* is unable to reproduce. This plot should be compared with Figure 7b, panel 2, of Faber and Burstein (1988), which shows residuals of observed peculiar velocities after subtracting away the predictions of their Great Attractor model.

This comparison shows that there is a bulk motion within a sphere of $\sim$4000 km s^{-1} radius, which *IRAS* is not properly modelling. Moreover, the Local Group also takes part in this motion; most of the deviation of the *IRAS* calculated peculiar velocity for the Local Group from the value measured from the CMB dipole can be attributed to this motion. We show in the next section this may be explained as a purely statistical effect caused by dilute sampling of the density field.

COMPARISON WITH *N*-BODY MODELS

Our iteration scheme can be tested with *N*-body models for which we know the true distances of every point. We used *N*-body models consisting of 262,144 particles in a box 360 Mpc on a side ($H_0 = 50$ km s^{-1} Mpc^{-1}) in an $\Omega = 1$ universe, with an initial power spectrum of perturbations appropriate for universe dominated by Cold Dark Matter, and evolved until the *rms* density contrast in a sphere of radius 16 Mpc is $b^{-1} = 0.5$. The velocity field was smoothed with a Gaussian of width 100 km s^{-1}. Particles with peculiar velocities of 600 km s^{-1}, lying in regions of local overdensity δ with $-0.2 < \delta < 1.0$, and with smooth local velocity field, were chosen as Local Group candidates. '*IRAS*' flux-limited catalogs were constructed around these Local Group candidates by selecting objects according to the number density and selection function of the real data; no biasing was applied in the galaxy selection, so these objects do trace the mass of the simulation. For more details, see Górski *et al.* (1989).

We then ran our iteration program on the *N*-body '*IRAS*' catalog. That is, we gave the program only the information on the redshift of each particle as observed in the rest frame of the LG particle, and let it run to predict the true distances and the peculiar velocity field of the particles in the simulation. Ω was set equal to unity for these simulations. We emphasize that the motivation is not so much to test the Cold Dark Matter cosmogony, but rather to test our iteration scheme of peculiar velocities in the presence of non-linear effects and shot noise. A total of 18 realizations were run: nine with different 'Local Groups', and nine more with a single 'Local Group', but different random numbers selecting the galaxies in the survey. In Figure 4 we show the direction for the predicted peculiar velocity of the Local Group in each of these models. They all have been rotated so that the true direction of the Local Group particle in each case is $\ell = 270°$, $b = +30°$. Notice that the scatter is of the order of 15-20°, comparable to the difference between the true and predicted motions of the Local Group.

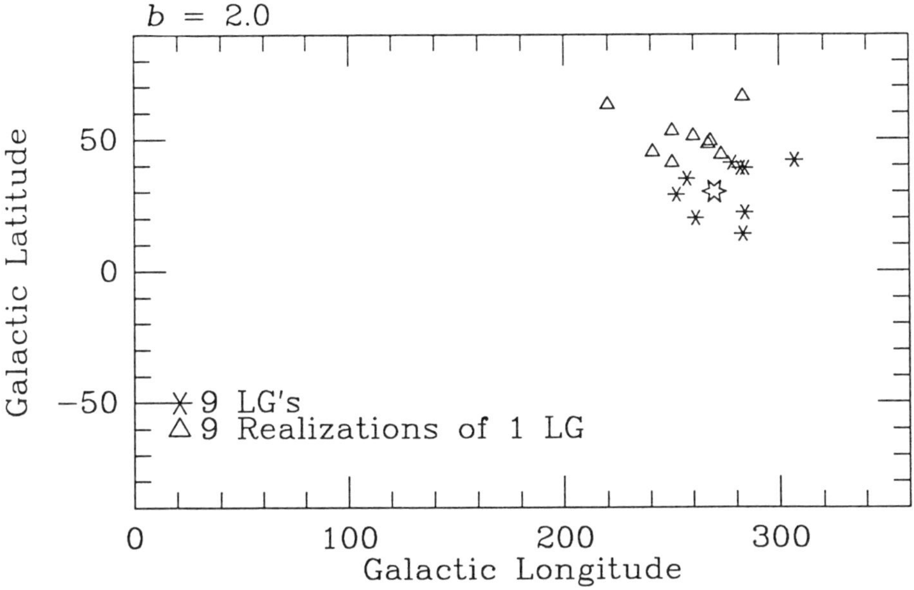

Fig. 4. The distribution of predicted directions of the 'Local Groups' in eighteen *N*-body realizations of the *IRAS* sample. The large star is placed at $\ell = 270°$, $b = 30°$.

In one of these models, we selected a sample of points to match closely the distribution of galaxies with observed peculiar velocities in the samples used in the previous section. With this sample we can make a direct comparison of true and predicted radial peculiar velocities and see how these results compare with the real universe described in the previous section. Unlike the real observations, we assume here that the distances are measured exactly. In Figure 5 we show the results. Panel (a) shows the distribution of true peculiar velocities in the CMB frame, in exact analogy with Figure 2(b). Panel (b) is then the predicted peculiar velocity as given by the iteration procedure, and (c) is the difference of the two. Finally, (d) is the difference of 'observed' and predicted, now in the LG frame. Note the very strong bulk flow seen in the residuals in panel (c), with an amplitude of ~200 km s^{-1}. Unlike the real universe, however, there is still a significant bulk flow in the LG frame residuals, with an amplitude of ~100 km s^{-1} (panel d), which simply says that the deviation of the predicted LG motion from its true direction is at least partly due to small-scale effects in this model.

We have repeated this experiment, but have assigned 15% errors to the true distances of the galaxies to more closely simulate the observations; we find that this seriously degrades the good agreement between 'measured' and predicted peculiar velocity.

Why is it that there seems to be a missing component of gravity in the CDM model, where the points are explicitly chosen to trace the mass? In these models, there is very little power on scales larger than the sample (8000 km s^{-1}) to cause the whole sample to translate (Juszkiewicz *et al.* 1990). Our interpretation is that this effect is caused by shot noise. Imagine a toy model in which the bulk of the motion of the region within 3000 km s^{-1} is caused by a large structure at 6000 km s^{-1}. Because of the extremely dilute sampling of the catalog, the overdensity of this structure may be traced by only a small number of points, and this number will be subject to Poisson statistics. Thus the inferred gravity from this structure can be in error, affecting the predicted peculiar velocities in a systematic way. Detailed calculations of the noise in the acceleration of galaxies relative to the Local Group is consistent with the observed amplitude of the effect, which must be at least partly responsible for the missing bulk motions seen in the predicted velocity field of the real *IRAS* galaxies.

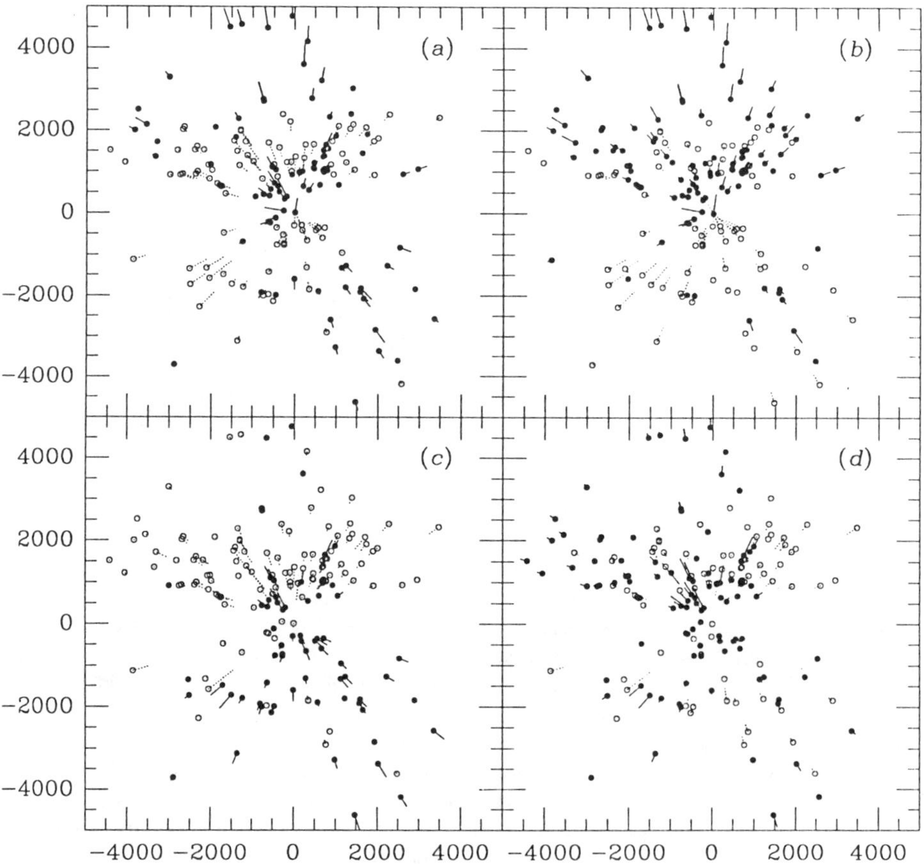

Fig. 5. The results of an *N*-body experiment simulating the comparison between observed and predicted peculiar velocities: (a) the true peculiar velocity field in the CMB frame; (b) the predicted peculiar velocities for the same galaxies; (c) the difference between (a) and (b); (d) the difference between observed and predicted in the LG frame.

UNRESOLVED ISSUES, AND THE WORK AHEAD

In the previous section, we have identified shot noise as a stumbling block in the interpretation of the comparison of our predictions with observations. Thus we have started a deeper survey of *IRAS* galaxies, flux-limited to 1.2 Jy, to rectify this situation. When the survey is completed, roughly one year after the publication of this volume, we will have redshifts for ~5500 *IRAS* galaxies selected uniformly over the sky. Figure 6 shows the distribution on the sky of the additional objects. Because this sample goes deeper, it emphasizes structures at distances of 3000 - 5000 km s^{-1}; note in particular the strong enhancements in the Perseus-Pisces region and the Hydra-Centaurus region.

We have identified an unfortunate systematic bias in our sample in its current form. We did not include sources of moderate flux quality at 60μm (*IRAS* Explanatory Supplement 1985) in the sample, because the fluxes have large uncertainties. Unfortunately, these sources are far from uniformly distributed over the sky, and including them will add an appreciable number of Local Supercluster objects. We hope to have complete redshift information for all such objects by the time of publication of this proceedings.

We are in the process of using the *IRAS* predicted distances for the Aaronson *et al.* galaxies to construct Tully-Fisher diagrams. We hope to make a more robust estimate of Ω by finding that value that minimizes the scatter in the TF diagram. The work of Dekel and Bertschinger described in these proceedings provides another powerful way of making comparison between *IRAS* predictions and observed peculiar velocities; we will soon be able to place constraints on the relative overdensities of galaxies and dark matter in different regions of space.

Finally, we realize that our iteration scheme is far from ideal, because it has no provision for recognizing the existence of triple-valued zones, the regions around clusters where there is a fundamental ambiguity of the distance of an object given its redshift. Direct evidence that our iteration scheme can be improved comes from our *N*-body experiments: if galaxies are placed at their true distances, the predicted peculiar velocities are in much

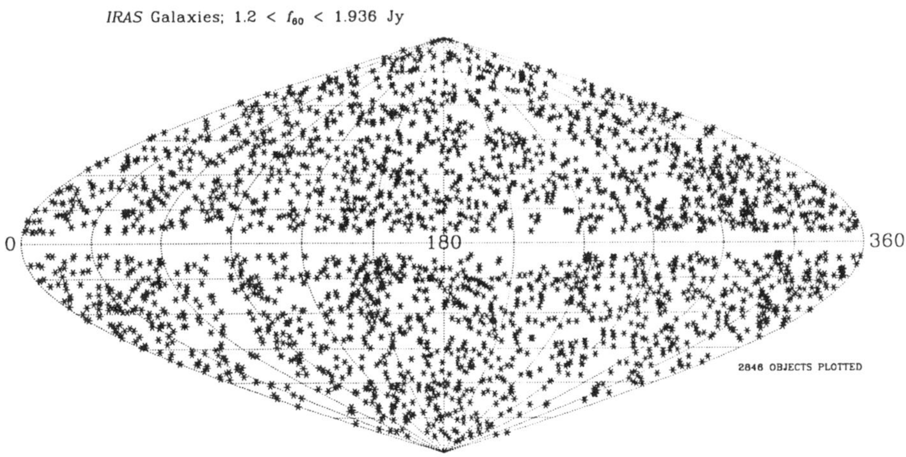

Fig. 6. The distribution of galaxies in Galactic coordinates of objects in a deeper *IRAS* redshift survey, in preparation. Note the appreciable overdensities associated with the Perseus-Pisces (ℓ ~ 120° to 150°, b ~ −30° to +30°) and Hydra-Centaurus (ℓ ~ 280° to 330°, b ~ 0° to +45°) regions.

better agreement with reality using equation (2) than when the iteration procedure is used. We are in the process of testing other iteration algorithms, with the hope that better prediction of peculiar velocities in high density regions will give improved comparison with observation.

ACKNOWLEDGEMENTS

We thank our collaborators in this project, Amos Yahil and John Huchra, for allowing us to use unpublished data, and for many helpful discussions. We are grateful to Dave Burstein for sending us machine-readable versions of the observed peculiar velocity data. MAS acknowledges the support of an NSF Graduate Fellowship.

REFERENCES

Aaronson, M., Huchra, J. P., Mould, J. R., Tully, R. B., Fisher, J. R., van Woerden, H., Goss, W. M., Chamaraux, P., Mebold, U., Siegman, B., Berriman, G., and Persson, S. E. 1982, *Ap. J. Suppl.*, **50**, 241.

Burstein, D. 1990, *Rep. Prog. Phys.*, **53**, 421

Davis, M. and Peebles, P. J. E. 1983, *Ann. Rev. Astr. Ap.*, **21**, 109.

Faber, S. M., and Burstein, D. 1988, in *Large Scale Motions in the Universe: A Vatican Study Week*, ed. V. C. Rubin and G. V. Coyne, S. J. (Princeton: Princeton University), p. 115.

Faber, S. M., Wegner, G., Burstein, D., Davies, R. L., Dressler, A., Lynden-Bell, D., and Terlevich, R. J. 1989, *Ap. J. Suppl.*, **69**, 763.

Górski, K., Davis, M., Strauss, M. A., White, S. D. M., and Yahil, A. 1989, *Ap. J.*, **344**, 1.

Huchra, J. P. 1989, in *The Extragalactic Distance Scale*, ed. S. van den Bergh and C. J. Pritchet, *A.S.P. Conf. Ser.*, **4**, (Provo: Astronomical Society of the Pacific), p. 257.

IRAS Catalogs and Atlases, Explanatory Supplement 1985, ed. C. A. Beichman, G. Neugebauer, H. J. Habing, P. E. Clegg, and T. J. Chester (Washington DC: U.S. Government Printing Office).

IRAS Point Source Catalog 1985, Joint *IRAS* Science Working Group (Washington DC: U.S. Government Printing Office)

Juszkiewicz, R., Vittorio, N., and Wyse, R. J. 1990, *Ap. J.*, **349**, 408.

Lubin, P. M. and Villela, T. 1986, in *Galaxy Distances and Deviations from Universal Expansion,* ed. B. F. Madore and R. B. Tully (Dordrecht: Reidel), p. 169.

Lynden-Bell, D., Faber, S. M., Burstein, D., Davies, R. L., Dressler, A., Terlevich, R. J., and Wegner, G. 1988, Ap. J., **326**, 19.

Lynden-Bell, D., Lahav, O., and Burstein, D. 1989, *M.N.R.A.S.*, **241**, 325.

Peebles, P. J. E. 1976, *Astr. Ap.*, **53**, 131.

Peebles, P. J. E. 1980, *The Large Scale Structure of the Universe*, (Princeton: Princeton University).

Strauss, M. A. 1989, PhD. thesis, University of California, Berkeley.

Strauss, M. A. and Davis, M. 1988a, in *IAU Symposium 130, Large Scale Structures of the Universe*, ed. J. Audouze, M.-C. Pelletan and A. Szalay (Dordrecht: Reidel), p. 191.

Strauss, M. A. and Davis, M. 1988b, in *Large Scale Motions in the Universe: A Vatican Study Week*, ed. V. C. Rubin and G.V. Coyne, S. J. (Princeton: Princeton University), p. 255.

Yahil, A. 1988, in *Large Scale Motions in the Universe: A Vatican Study Week*, ed. V. C. Rubin and G. V. Coyne, S. J. (Princeton: Princeton University), p. 219.

DISCUSSION

A. Dressler: From what you have concluded, could it be the case that, due to shot noise, you have accidentally undersampled the Great Attractor and/or oversampled Perseus-Pisces. Could this be the resolution of the apparent difference between the IRAS prediction of the velocity field and the actual observed velocity field?

M. Strauss: I doubt this is the full answer, although we have not explicitly tested for that. The sense of the disagreement in Centaurus is very large between IRAS and the Seven-Samurai observations. Besides, the largest peculiar velocities predicted anywhere in the IRAS model (with $\Omega = 0.7$) are ~ 700 km s^{-1}, which is appreciably smaller than the ~ 1500 km s^{-1} measured by the Seven Samurai for some of their points.

A. Dekel: Isn't $\langle \mathbf{v} \cdot \mathbf{v} \rangle$ at smaller r dependent on the smoothing of the velocity field?

M. Davis: Our smoothing of the *IRAS* velocity field does indeed reduce $\langle \mathbf{v} \cdot \mathbf{v} \rangle$ on small scales, but this obviously does not affect $\langle \mathbf{v} \cdot \mathbf{v} \rangle$ derived from the Aaronson *et al.* and Seven Samurai samples.

A. Dekel: Your collaborator Yahil argues that in CDM a significant fraction of the dipole in a sphere like the IRAS sphere comes from outside that sphere. Why do you argue otherwise?

M. Davis: I disagree with the assessment, and I think you have misunderstood Amos. Large scale power in a CDM model will contribute only about 100 km s^{-1} to our dipole from distances larger than 8000 km s^{-1}.

S. Shandarin: What spectrum was taken in the Peebles Baryon Isocurvature model?

M. Davis: $n = -1$ with full ionization as described by Peebles.

A. Szalay: How large a volume do we need to reach a fair sample from the N-body simulations?

M. Davis: A fair sample volume for velocity correlation studies would need to be at least 10,000 km s^{-1} in radius for a CDM universe. In a Peebles Baryon Isocurvature universe, a fair sample would be larger than 700 Mpc radius! This is obviously out of the question.

A. Szalay: Have you tested the isotropy of the reconstructed velocity field? Does this check the treatment of the triple-valued regions?

M. Davis: Since the velocity is derived from the gradient of a potential, it will inevitably be isotropic in the mean. Our treatment of the triple-valued regions is presently inadequate, and I hope to address this issue as soon as I get home.

A. Dressler: You showed a diagram indicating that contributions to the Local Group dipole converge by 4000 km s^{-1}. Amos Yahil has argued that this is an artifact of the cancellation (in the IRAS sample) of the Perseus-Pisces Supercluster and the Great Attractor. He argued that the dipole could, in fact, have large contribution from and beyond 4000 km s^{-1}, and his CDM simulations showed that it should.

In view of all this, isn't it misleading to say that the Local Group motion arises within 4000 km s^{-1}, and isn't it unwise to use the dipole to normalize the IRAS map of predicted peculiar velocities to the observed velocities?

M. Strauss: It is true that we seem to sit in a special place, *i.e.*, where the contributions from the Pisces-Perseus and Hydra-Centaurus Superclusters cancel to first order. This does not obviate the empirical fact that our acceleration does seem to converge at 4000 km s^{-1}. Amos Yahil has shown that in a $b = 1$ CDM universe, one does expect substantial power on scales larger than the sample, while for $b > 2$, there is negligible power on larger scales (see also Juszkiewicz, Vittorio, and Wyse, 1989).

With no *a priori* estimate of the power on scales larger than the sample, I agree it is a little foolhardy to use the dipole to set the value of Ω; far better would be to compare with all the observed peculiar velocities by means of, *e.g.*, minimizing the scatter in the Tully-Fisher diagram. This is work currently in progress.

D. Lynden-Bell: One of the surprises of our study was that the big quadrupole component came from deeper than the dipole. Have you looked at the generation of quadrupoles in your simulations and the depths from which they come?

M. Strauss: No, but we hope to do that soon.

N. Kaiser: You have used the density field to predict the velocity field which you compare with observed velocities. An alternative is to go the other way (à la Dekel and Bertschinger). I would like to open the debate on the relative merits of these techniques. It seems to me that three disadvantages of your approach are 1) the density in some region does not determine the velocity field in that region, but one can go the other way, 2) you have to go through an iterative procedure to get $\rho(r)$ from redshift space. If one reconstructs $\rho(r)$ from velocities, one can easily transform to redshift space. 3) You end up comparing vector fields rather than scalar fields (which are easier on the eye).

M. Davis: I agree that the method is imperfect and alternative methods must be explored. In response to your three points: 1) triple-valued zones are presently a real problem for us, 2) the iterations are not a problem, as convergence is rapid, and 3) I prefer to work as close to the observables as possible, and the basic observable in the velocity business is a scalar field, the radial component of the peculiar velocity.

D. Burstein: At what point do we (all of us) bite the bullet and decide to what extent light traces mass? There are hints in the comparisons among IRAS-optical peculiar motions along these lines (which way, I'll let the reader of this volume decide). What is your opinion?

M. Davis: The question of whether mass traces light on large scales is obviously a fundamental issue. In due course, as the data for the velocity field improves, methods such as described at this workshop by Bertschinger, Dekel, and Kaiser will allow us to construct estimates of the underlying mass field. These can then be compared to the galaxy density field, and we will be able to infer the bias field directly.

J. Primack: Following up on Nick Kaiser's point - it's obviously desirable to compare scalars rather than vectors, but perhaps the best scalars to compare are potentials rather than densities, either the velocity potential or gravitational potential, which are related (as Kofman and Shandarin have discussed). The potential is most directly calculated from peculiar velocity (as Bertschinger and Dekel are doing), and also theoretically easier to analyze since, for example, it is smoother on small scales.

M. Strauss: I could not agree more; Bertschinger and Dekel's work, reported at this workshop, is an attempt to do exactly what you suggest.

A. Dekel: There are pros and cons in using the potential for comparison. It is the integral quantity which is determined by the large-scale, linear components of the fluctuation spectrum, and is therefore our least noisy result. Unfortunately, it might smooth over interesting features on intermediate scales, which might be the distinguishing features between theories.

A. Dressler: Could I ask Marc Davis and Donald Lynden-Bell to comment on the relative 'strengths' of the Perseus-Pisces Supercluster and the Great Attractor as traced by IRAS and optical catalogs?

M. Davis: Centaurus is much closer than Perseus and indeed has a substantial influence on our motion, according to the IRAS procedure. But, Centaurus is also a big part of the Great Attractor. If Nigel could produce a map of galaxies with $v > 4000$ km s^{-1}, then we would expect roughly equal counts from the Great Attractor and Perseus regions.

D. Lynden-Bell: The clump one sees in Centaurus is composed of things between 2000 and 5000 km s^{-1}. What meets the eye are the density peaks. The net gravity field feels larger agglomerations (and voids negatively). The large-scale area around Centaurus is peculiarly rich.

MAPPING LARGE-SCALE FLOWS IN THREE DIMENSIONS: METHOD

EDMUND BERTSCHINGER
Department of Physics, Massachusetts Institute of Technology
Cambridge, MA 02139

AVISHAI DEKEL
Racah Institute of Physics, The Hebrew University of Jerusalem
Jerusalem, 91904 Israel

ABSTRACT We describe and test a method for recovering the three-dimensional potential, velocity and density fields from large-scale redshift-distance samples. Galaxies are taken as tracers of the velocity field - not of the mass. To reconstruct the three-dimensional velocity field from its radial component we assume that the smoothed velocity field is a potential flow. We motivate this assumption using gravitational instability theory, and demonstrate its validity by applying the method to a nonlinear cosmological *N*-body simulation of cold dark matter. A key element of the method is the construction of a smooth radial velocity field from observations. We present an analysis of our tensor window function smoothing technique addressing the biases and variance arising from distance errors and sparse and nonuniform sampling. The complete method is tested with a series of Monte Carlo simulations with errors added to the *N*-body simulation.

INTRODUCTION

According to gravitational instability theory, large-scale structure and peculiar motions arose from small-amplitude fluctuations in the matter distribution generated prior to recombination. Amplified by gravity, these fluctuations are manifested in the large-scale density and peculiar velocity fields. A single scalar field - the initial density or gravitational potential fluctuations of linear theory - contains all the information (aside from Ω) needed to specify fully the nonlinear gravitational evolution of structure in the universe. Consequently, the initial conditions on large scales can be recovered - leaving aside, for the moment, questions of nonlinearity and biased galaxy formation - from either the density field or the peculiar velocity field, and one can be used to predict the other.

Galaxy redshift surveys are the standard method for measuring the large-scale density distribution. They have resulted in the discovery of voids (Gregory and Thompson 1978; Kirshner *et al.* 1981), bubbles (de Lapparent, Geller, and Huchra 1986), and massive superclusters (Haynes and Giovanelli 1986; Tully 1987; Huchra and Geller this volume). Recently, peculiar velocities have been predicted from redshift surveys using either optical catalogs (Lahav, Rowan-Robinson, and Lynden-Bell 1988; Lynden-Bell this volume) or the *IRAS* redshift survey (Strauss and Davis 1988, this volume; Yahil 1988). Redshift surveys have the advantage of providing a three-dimensional representation of the (galaxy) density distribution, with redshifts providing an estimated distance either directly or after correction for estimated peculiar motions.

Peculiar velocities can also be used to map large-scale structure. The main advantage of this approach is that no assumptions are necessary concerning how mass traces light. The observational requirements, of course, are more difficult than they are for redshift surveys, as distances must be measured independently of the Hubble law. Fortunately, only relative distance measurements are needed, so that the uncertain calibration of H_0 does not spoil the measurement. The distance barrier seems to have been broken finally with the advent of the infrared Tully-Fisher (Tully and Fisher 1977; Aaronson, Huchra, and Mould 1979) and D_n - σ (Dressler *et al.* 1987; Djorgovski and Davis 1987) techniques. The main limitation now is that only the radial component of peculiar velocity is measurable: $u \equiv \hat{\mathbf{r}} \cdot \mathbf{v} = cz - r$. (Away with the Hubble constant! Both distance and velocity are measured in units of km s^{-1}).

Without all three components of velocity, the only way to recover the initial conditions and to predict the corresponding mass density field is by fitting a model to $u(\mathbf{r})$, such as a Virgocentric flow model (*e.g.*, Yahil, Sandage, and Tammann 1980) or a Great Attractor model (Lynden-Bell *et al.* 1988). This limitation has made analysis of peculiar velocities cumbersome compared with the direct picture provided by redshift surveys. It would be preferable to avoid the model by determining the three-dimensional fields directly. A simple counting argument suggests that this ought to be possible: the initial conditions are specified by one scalar field, while the radial peculiar velocity provides one constraint at each sampled point. Can this argument be converted into a practical method for recovering the three-dimensional velocity field (and subsequently the density and gravitational potential fields) directly from redshift and distance measurements?

If the gravitational instability theory is correct, then it *is* possible - at least on large scales - to recover all three components of peculiar velocity from the radial component. Away from regions of large-scale orbit-crossing, the theory predicts that the peculiar velocity field is irrotational, $\nabla \times \mathbf{v} = 0$. This condition imposes two constraints on the peculiar velocity, allowing it to be derived from a single scalar velocity potential:

$$\mathbf{v} = -\nabla\Phi. \tag{1}$$

If potential flow prevails on the large scales of interest, then the tangential components of the peculiar velocity field can be obtained using equation (1), with the potential determined by integrating the radial component along radial rays:

$$\Phi(r,\theta,\phi) = -\int_0^r u(r',\theta,\phi)\, dr' \ . \tag{2}$$

The data must first be interpolated and smoothed to provide an estimate of $u(\mathbf{r})$ throughout the volume of interest. This three-step procedure of smoothing, radial integration, and tangential differentiation, was introduced by Bertschinger and Dekel (1989; hereafter BD), who demonstrated its success by applying it to a cosmological *N*-body simulation of large-scale structure in an 'unbiased' cold dark matter universe.

Figure 1 shows an updated version of the *N*-body velocity- field reconstruction of BD (Dekel, Bertschinger, and Faber 1990; hereafter DBF). The *rms* and maximum errors in the three-dimensional velocity recovered on the spherical grid are 25 and 95 km s^{-1}, compared with the *rms* and maximum smoothed three-dimensional velocities of the simulation, 400 and 800 km s^{-1}, respectively. With perfect 'data', the method, which we have dubbed POTENT, works extremely well.

The velocity potential method shows great promise as a tool for studying large-scale structure. First, from the velocity and its potential, one may recover the *total* mass density field $\rho(\mathbf{r})$. Since the galaxies are being used only as tracers of the gravitational potential, the recovered density field does not distinguish between luminous and dark matter. Second, the method can allow cosmological theories for the initial perturbations to be tested more directly than before. Three-dimensional fields are clearly superior in this regard to the radial peculiar velocity field, and POTENT requires no assumptions about biased galaxy formation, such as those needed in the interpretation of galaxy redshift surveys. Third, as a consequence of the above, POTENT can be used, in conjunction with redshift surveys, to test whether mass traces light. In fact, it is possible to measure the large-scale 'bias' field (luminosity density divided by mass density), which will be useful in testing theories of biased galaxy formation. Fourth, one might hope that Ω itself can be determined from independent measurements of the peculiar velocity and density fields. Unfortunately, this measurement, like all dynamical measurements of Ω based on local cosmological linear perturbation theory, depends on assumptions about how mass traces light.

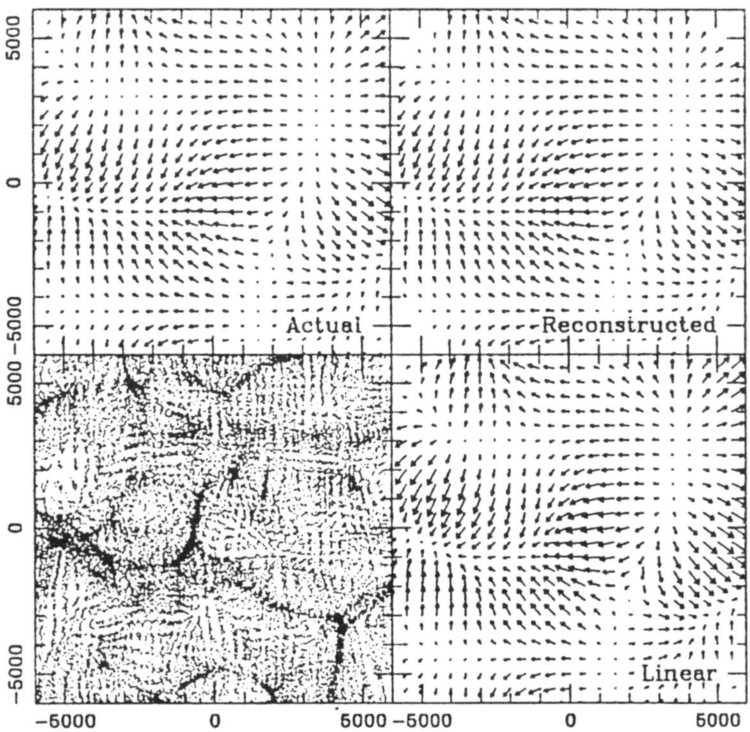

Fig. 1. The two-dimensional velocity field (*vector plots*) in a plane through an unbiased cold dark matter N-body simulation (*points plot*). The units of both distance and velocity are km s^{-1} and the spacing between vectors is 500 km s^{-1}. The panel labeled Actual shows the exact velocity field of the N-body simulation, convolved with a spherical Gaussian of radius 1000 km s^{-1}, while the panel labeled Reconstructed shows the result recovered using the velocity potential method with only the radial component of the actual field. The panel labeled Linear shows the result of extrapolating the initial Lagrangian velocity field of the N-body simulation to the present using linear perturbation theory. The particle plot (*lower left*) is a vertical projection of the final particle positions from the N-body simulation, in a Lagrangian slice of thickness 2000 km s^{-1} centered on the plane of the velocity field slices.

Along with the promise of the velocity potential method come several difficult challenges. Real data do not uniformly sample a large volume, nor are peculiar velocity errors negligible. Present redshift-distance samples contain ~ 1000 galaxies within 8000 km s^{-1} distance from the Local Group, with relative distance errors of $\sim 20\%$ per galaxy. Beyond ~ 3000 km s^{-1} distance, the standard deviation of the measured radial peculiar velocity for an individual galaxy exceeds the typical large-scale flow velocity of 600 km s^{-1}, and the mean separation of sampled objects exceeds 1000 km s^{-1}. Given sparse and noisy data, does POTENT live up to its name?

This paper summarizes the velocity-potential method and the smoothing procedure we have developed for use with sparse and noisy data. We present Monte Carlo simulations for testing the method and assessing errors and biases. In an associated paper, Dekel and Bertschinger (this volume) show the first results of applying POTENT to presently-available redshift-distance samples. Full discussion of the method and its first results are presented by DBF and by Bertschinger, Dekel, Faber, Dressler, and Burstein (1990).

POTENTIAL FLOW THEORY

Here we examine the justification for assuming a potential flow, and we describe two alternative methods (Eulerian and Lagrangian) for reconstructing the potential, velocity, and density fields. A word about nomenclature: Eulerian coordinates give the instantaneous comoving position $\mathbf{x} = \mathbf{r}/a$ of a particle or fluid element, while Lagrangian coordinates refer to the initial comoving position $\mathbf{q}$ of a particular particle or fluid element at large redshift. A standard cosmological *N*-body code is Lagrangian, since it uses particles of fixed Lagrangian position, while a standard finite-difference hydrodynamics code is Eulerian, because it represents the fluid variables at fixed Eulerian positions.

The Eulerian fluid equations in comoving coordinates for pressureless dust (*e.g.*, cold dark matter) are well known (*e.g.*, Peebles 1980):

$$\partial\delta/\partial t + \nabla\cdot[(1+\delta)\mathbf{v}] = 0,$$

$$(\partial\mathbf{v}/\partial t) + 2(\dot{a}/a)\mathbf{v} + (\mathbf{v}\cdot\nabla)\mathbf{v} = -(1/a^2)\nabla\phi,$$

$$\nabla^2\phi = 4\pi G\langle\rho\rangle a^2\delta = (3/2)\,\Omega H^2a^2\delta. \tag{3}$$

Here $a(t)\mathbf{v}$ is the proper peculiar velocity (a is the expansion factor normalized to $a = 1$ today) and $\delta = (\rho/\langle\rho\rangle) - 1$. By assumption, $\delta^2 << 1$ at large redshift. Solving the linearized equations gives the usual growing and decaying solutions for δ:

$$\delta(\mathbf{x},t) = \delta_1(\mathbf{x})D_1(t) + \delta_2(\mathbf{x})D_2(t),\ \ \delta^2 << 1. \tag{4}$$

The linear growth factor $D_1(t)$ equals $a(t)$ if $\Omega = 1$; for $\Omega < 1$, $d\ln D_1/d\ln a \approx \Omega^{0.6}$ is a good approximation. The decaying mode falls inversely with time for $\Omega = 1$ and $\Omega << 1$, so is assumed to be negligible because the perturbations were small at recombination ($a^{-1} \approx 1300$). The linear growing mode density perturbation, $\delta_1(\mathbf{x})$, is commonly assumed to be a Gaussian random process with primordial scale-invariant (Harrison-Zel'dovich) power spectrum.

The peculiar velocity field in linear theory has two terms associated with the growing and decaying modes of the density perturbation and satisfying the linearized continuity equation. These terms are irrotational. However, there is also the possibility of a rotational component that does not produce a density perturbation and, hence, does not couple to gravity:

$$\mathbf{v}(\mathbf{x},t) = \mathbf{v}_1(\mathbf{x})\dot{D}_1(t) + \mathbf{v}_2(\mathbf{x})\,\dot{D}_2(t) + a^{-2}(t)\,\nabla\times\mathbf{A}(\mathbf{x}). \tag{5}$$

The density perturbations correspond to the scalar perturbations of general relativity while the vorticity perturbation is a vector perturbation. It is reasonable to assume that vector perturbations were generated in the early universe just as were scalar perturbations. However, we see that the vorticity decays with expansion, while one mode of the density perturbation grows through gravitational instability. The proper rotational velocity, $a^{-1}\,\nabla\times\mathbf{A}$, decays inversely with expansion because of conservation of angular momentum (or circulation). Consequently, by the end of the linear regime the velocity field should be, to a good approximation, a potential flow, with the velocity potential proportional to the gravitational potential:

$$\mathbf{v}(\mathbf{x},t) = -a^{-2}\,\nabla\Phi, \quad \Phi = (2/3)\,(\Omega^{0.6}/\Omega H)\,\phi. \tag{6}$$

The peculiar gravitational potential ϕ is constant in the linear regime for $\Omega = 1$. For the linear growing mode, the velocity and density are related in the standard way by the linearized continuity equation:

$$\nabla\cdot\mathbf{v} = -H\,\Omega^{0.6}\,\delta. \tag{7}$$

Note that peculiar velocities measure $\Omega^{0.6}\,\delta$, not δ. Conversely, measurements of the density perturbations from redshift surveys allow the prediction of $\Omega^{0.6}\,\mathbf{v}$. It has long been recognized (*e.g.*, Silk 1974, Peebles 1976) that a comparison of the density perturbation and peculiar velocity fields yields an estimate of Ω. The new ingredient is that we can now hope to make the comparison *globally* rather than at only one place.

If the flow was irrotational in the linear regime, it remains so as long as the Kelvin circulation theorem applies. This theorem states that for barotropic flow the proper fluid circulation around a closed contour fixed in the fluid is conserved. The circulation vanishes for potential flow, which therefore remains irrotational. This is true even in the nonlinear regime. The conditions of the circulation theorem are violated only when dissipation occurs. For a collisionless fluid of dark matter this happens when fluid trajectories cross, if the fluid trajectory is defined subsequently as an average over the different velocity streams at a given Eulerian position. So, vorticity can be generated in collapsed regions where particle orbits mix (*e.g.*, galaxies and clusters). However, one expects the vorticity so generated to have negligible amplitude on scales that are still in the linear regime, so that when the Eulerian velocity field is smoothed to remove small-scale nonlinearity, potential flow will be a good approximation. The success of POTENT in reconstructing the smoothed velocity field of our highly nonlinear *N*-body simulation (*cf.* Figure 1) confirms this expectation.

If potential flow is a good approximation away from virialized regions, then the nonlinear Euler equation (equation 3) may be integrated to give the Bernoulli equation (Kofman, 1989):

$$\phi(\mathbf{x},\mathrm{t}) = (\partial\Phi/\partial t) - 1/2a^2\mathrm{v}^2. \tag{8}$$

This equation shows that, in the nonlinear regime, even if the flow remains irrotational, the gravitational and velocity potentials deviate from each other. If the time derivative is approximated using linear theory as in equation (6), equation (8) yields a quasi-linear estimate for the mass density, with δ computed from ϕ using the Poisson equation. We have not tried this method because the two numerical differentiations would severely increase the noise, but it looks interesting.

Our implementation of the Eulerian version of POTENT is straightforward. We simply integrate the radial peculiar velocity u along radial paths according to equation (2), using a spherical grid of 10 radial shells uniformly spaced within 7000 km s^{-1}, 16 latitude circles, and 16 longitude circles. This number of grid points is adequate given the $\geq$1000 km s^{-1} smoothing length we use prior to integration. We integrate cubic splines fit along each radial ray and then differentiate a cubic spline fit along the tangential directions to get the tangential velocity components from equation (1). To present the results on a regular Cartesian grid we use cloud-in-cell interpolation from the spherical mesh. No iteration is required, in contrast to the Lagrangian method described in BD. The reconstruction of Figure 1 was performed using the Eulerian version described here.

Alternatively, assuming that the flow was indeed irrotational in the linear regime, one can appeal to the irrotationality of the *initial* flow in Lagrangian space, as we did originally in BD. As we argued there, if the flow in the linear regime was derivable from a potential - indeed, we have seen that it follows from the gravitational potential - then we can derive this potential and the final three-dimensional velocity field from the radial velocity component. To accomplish this requires a mapping between Lagrangian (initial) comoving position $\mathbf{q}$ and the Eulerian (final) comoving position $\mathbf{x}$. In principle a cosmological N-body code could be used for this purpose if the initial conditions (or the final conditions) could be determined sufficiently accurately. In practice this is likely to be a bad idea because nonlinear gravitational clustering is chaotic. In regions where orbit mixing has occurred, one cannot trace the Eulerian positions back to a unique Lagrangian position. It is therefore necessary to smooth over scales larger than those on which orbit mixing occurs. After such smoothing, the *Eulerian* velocity field is essentially irrotational, as discussed above.

Nevertheless, we continue to use the Lagrangian procedure for reconstructing the density field. We use the Zel'dovich (1970) approximation to relate the present (Eulerian) comoving position $\mathbf{x}$ to the initial (Lagrangian) comoving position $\mathbf{q}$:

$$\mathbf{x}(\mathbf{q},t) \approx \mathbf{q} + D_1(t)\,\psi(\mathbf{q}). \tag{9}$$

The comoving displacement field $D_1\psi$ is computed in the linear approximation, neglecting the decaying compressional and rotational modes. The Lagrangian velocity, potential, and density fields follow at once:

$$\mathbf{v}(\mathbf{q},t) = \partial\mathbf{x}/\partial t = H\,\Omega^{0.6}\,D_1\,\psi(\mathbf{q}) = -(1/a^2)\,(\partial\Phi/\partial\mathbf{q}),$$

$$\delta(\mathbf{q},\mathrm{t}) = \left\| \partial\mathbf{x}/\partial\mathbf{q} \right\|^{-1} - 1. \tag{10}$$

Double vertical bars indicate the Jacobian determinant.

The Lagrangian version of POTENT computes the potential and velocity on a spherical grid in Lagrangian coordinates. The Eulerian position of a particle is displaced from the Lagrangian position by an amount proportional to the peculiar velocity (equation 10). Since this factor depends on Ω, we must assume a value for Ω and then divide the peculiar velocity by $\Omega^{0.6}$ to give the comoving displacement. In the N-body simulation we know $\Omega = 1$, but in the real universe we will have to assume a value for Ω *a priori*.

Given an estimate for the Lagrangian potential $\Phi(\mathbf{q})$, we compute the velocity field by cubic splines differentiation as in the Eulerian case, displace the grid points using equation (9), and then compare the resulting Eulerian radial velocity $\hat{\mathbf{r}} \cdot \mathbf{v}$ with the radial velocity

$u(\mathbf{r})$ deduced from the observations (or from an N-body simulation). From the differences we correct our estimate of the potential and iterate to convergence. Several correction schemes are described in BD. None of these schemes guarantees convergence to a solution, and, with noisy data we have to choose the grid spacing and smoothing carefully to prevent divergence. In the future we hope to improve significantly the Lagrangian implementation.

With the Lagrangian displacement field computed as described above, we recover the density perturbation field using equation (9) plus the second of equations (10). To evaluate the Jacobian determinant we simply interpolate the displacement field to a uniform Cartesian grid of equal-mass particles, displace the particles, and then sum over particles at any desired Eulerian position using a spherical Gaussian window function. This quasi-linear procedure gives us a more accurate density field than would be obtained directly from equation (7).

Figure 2 shows our reconstructed linear and evolved density fields, compared with the actual density field of the N-body simulation and the density field extrapolated from the initial conditions using linear theory. To reconstruct the linear field, we simply evaluated

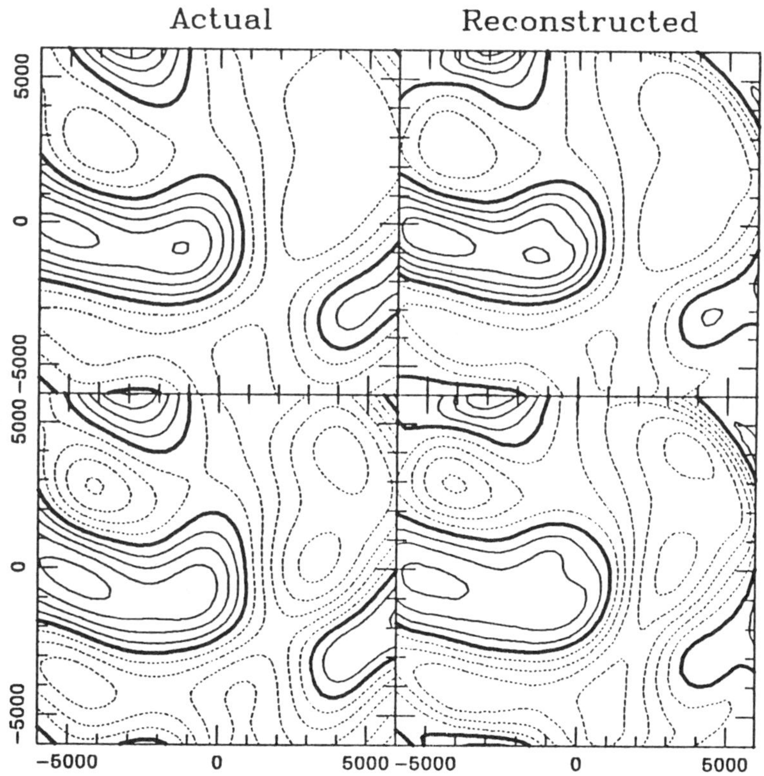

Fig. 2. Density perturbation fields from the N-body simulation. The contour interval is $\delta\rho/\rho = 0.1$ with the zero contour drawn heavy and negative contours dotted. The upper two panels show the nonlinear evolved density field, while the lower two panels show the density field at large redshift extrapolated to the present using linear theory. The Actual fields (*left panels*) have been computed using a Gaussian smoothing radius of 1300 km s^{-1} rather than 1000 km s^{-1} to account for the extra interpolations required in the reconstruction (*right panels*).

the Lagrangian displacement field at a large redshift ($D_1 = 0.03$) and then scaled the density field to the present using linear theory.

SMOOTHING SPARSE AND NOISY DATA

The greatest source of difficulty for POTENT comes not from nonlinearity, but from random distance errors and sparse, nonuniform sampling in real data sets. Random distance errors arise from the finite width of the D_n - σ and L - ΔV correlations used to estimate distances. Because the errors increase linearly with distance, the signal-to-noise ratio per galaxy falls below unity beyond about 3000 km s^{-1} distance from the Local Group. The nonuniform sampling of space creates serious additional problems, especially at large distances and low Galactic latitude.

These problems force us to smooth heavily to beat down the errors. Our estimate for the radial velocity smoothed with a window function W is (DBF);

$$u_e(\mathbf{r}) = \textstyle\sum_i u_i\, W(\mathbf{r},\mathbf{r}_i)\, [\hat{\mathbf{r}}_i \cdot \mathbf{A}^{-1}(\mathbf{r}) \cdot \hat{\mathbf{r}}], \qquad \mathbf{A}(\mathbf{r}) = \textstyle\sum_i W(\mathbf{r},\mathbf{r}_i)\, \hat{\mathbf{r}}_i\, \hat{\mathbf{r}}_i. \tag{11}$$

Tensor smoothing is required because we are averaging radial velocities from different directions $\hat{\mathbf{r}}_i$. This particular estimator was inspired by the maximum-likelihood bulk flow solution of Lynden-Bell *et al.* (1988). In the present case, the bulk flow solution is not global - we wish to reconstruct the large-scale velocity field rather than to model it by a bulk flow. Rather, we obtain a *local* estimate smoothed on a scale determined by W.

Because the random distance errors introduce a large variance into our estimated radial velocity field (see equation 17 below), we weight the window function inversely by the estimated variance σ_i^2 of the distance to each object. This weighting provides a minimum-variance, maximum-likelihood estimator. Our initial choice of window function was thus

$$W_B(\mathbf{r},\mathbf{r}_i) = (1/\sigma_i^2)\, \exp[-1/2\ |\mathbf{r} - \mathbf{r}_i|^2 / R_w^2]. \tag{12}$$

The Gaussian smoothing radius R_w is a constant.

As we quickly discovered, this window function suffers from biases. The most serious bias, which we term *sampling gradient bias*, arises from strong gradients in the selection function used to define the sample. It exists even if the distances and peculiar velocities are known exactly. Essentially, the radial velocity from densely sampled regions is propagated artificially into sparsely sampled regions. In DBF we demonstrate this bias analytically by averaging a given velocity field over an ensemble of (discrete) sampling positions with a selection function $\phi_s(\mathbf{r})$. For a galaxy density field $n(\mathbf{r})$ and a true peculiar velocity field $\mathbf{v}(\mathbf{r})$, the dominant term in the resulting mean estimated radial component is

$$\langle u_e(\mathbf{r})\rangle = \int d^3r_1\, \phi_s(\mathbf{r}_1)\, n(\mathbf{r}_1)\, \hat{\mathbf{r}}_1 \cdot \mathbf{v}(\mathbf{r}_1)\, W(\mathbf{r},\mathbf{r}_1)\, [\hat{\mathbf{r}}_1 \cdot \mathbf{A}^{-1}(\mathbf{r}) \cdot \hat{\mathbf{r}}]. \tag{13}$$

The source of bias is clear: the velocity field is weighted by an unwanted selection function in addition to the window function W. Where selection function gradients are large, equation (13) with window function W_B provides a poor estimate of the true radial velocity field smoothed with uniform sampling. This dependence on the sample selection makes comparison with other observations and theoretical models difficult.

A partial solution to the sampling gradient bias problem is to weight additionally by a volume factor approximating the inverse of the selection function. Since we want a robust method that can be applied to heterogeneous samples without well-defined selection functions, we approximate this factor by the inverse of the local density of objects in the sample, computed from the specific volume around each object:

$$W_F(\mathbf{r},\mathbf{r}_i) = [R_4^3(\mathbf{r}_i) / \sigma_i^2] \exp[-1/2|\mathbf{r} - \mathbf{r}_i|^2 / R_w^2], \tag{14}$$

where $R_4(\mathbf{r}_i)$ is the distance from object i to the fourth nearest neighboring object. The number 4 was chosen by trial-and-error based on minimum variance in the Monte Carlo simulations. By using the R_4^3 weights we accept slightly higher variance for a significant reduction in the bias of our estimator.

Given distance errors and sparse, nonuniform sampling, it is impossible to simultaneously eliminate the bias and variance of any estimated peculiar velocity. The window function W_F essentially minimizes the bias. Alternatively, we might try to control the variance by allowing the window radius R_w to expand or contract with galaxy density so as to maintain a constant number of objects in the window:

$$W_V(\mathbf{r},\mathbf{r}_i) = [R_4^3(\mathbf{r}_i) / \sigma_i^2] \exp[-1/2|\mathbf{r} - \mathbf{r}_i|^2 / R_5^2(\mathbf{r})]. \tag{15}$$

where $R_5(\mathbf{r})$ is the distance from the sampling position $\mathbf{r}$ to the fifth nearest neighboring object in the sample.

The sampling gradient bias arises from the nonuniform sampling of the peculiar velocity field, independently of the distance errors. Random distance errors themselves lead to a bias in our estimated radial velocity field - the *Malmquist bias*. This bias depends on the intrinsic scatter of the distance indicator employed and on the run of density along the line of sight. As shown by DBF, the measured distances are unbiased by selection effects but are biased by radial density gradients. If a uniform density Malmquist correction is applied (*e.g.* see Lynden-Bell *et al.* 1988), the residual bias is

$$\text{Bias}\,[u_e(\mathbf{r})] \approx \int d^3r_1\, \phi_s(\mathbf{r}_1)\, n(\mathbf{r}_1)\, W(\mathbf{r},\mathbf{r}_i)\, [\hat{\mathbf{r}}_1 \cdot \mathbf{A}^{-1}(\mathbf{r}) \cdot \hat{\mathbf{r}}]\, (\partial \ln n / \partial \ln r_1)\, \sigma_1^2 / r. \tag{16}$$

POTENT itself does not add any Malmquist bias: all of this bias arises even before we smooth the data. However, when we perform Monte Carlo experiments with the data, adding random distance errors to the distances in order to estimate the resulting errors of the three-dimensional velocity field, our 'noise' simulations have an extra artificial bias because we do not apply the observed selection function to our 'scattered' distances, nor do we correct these distances for the Malmquist effect. This *Monte Carlo bias*, described in more detail by DBF, is of no concern provided that we estimate our random errors using the second central moment of the noise simulations, with the bias removed.

It is sometimes erroneously thought that the only serious problem caused by distance errors is Malmquist bias. In reality, random distance errors create a scatter about the mean of any estimator that is generally much larger than the bias of that mean. With our tensor window function smoothing, the variance of the estimated radial velocity is (*cf.* DBF)

$$\text{Var}\,[u_e(\mathbf{r})] \approx \int d^3r_1\, \phi_s(\mathbf{r}_1)\, n(\mathbf{r}_1)\, W^2(\mathbf{r},\mathbf{r}_1)\, [\hat{\mathbf{r}}_1 \cdot \mathbf{A}^{-1}(\mathbf{r}) \cdot \hat{\mathbf{r}}]^2 \times$$
$$\{\sigma^2(\mathbf{r}_1) + [\hat{\mathbf{r}}_1 \cdot \mathbf{v}(\mathbf{r}_1) - \langle u_e(\mathbf{r})\rangle]^2\}. \tag{17}$$

The first term in the braces is due to measurement errors and is the variance counterpart of the Malmquist bias. The second term arises from Poisson sampling in the presence of velocity gradients within the window. This term is sometimes called *shot noise.* It can arise either from fluctuations in the discrete sampling of a large-scale velocity gradient across the window or from small-scale velocity dispersion in the galaxy distribution. The net variance of u_e is effectively $(\Delta u)^2/N_{eff}$, where Δu is the *rms* radial velocity error per galaxy, including contributions from measurement errors and shot noise, and N_{eff} is the effective number of galaxies, appropriately weighted, in the window function. The variance due to measurement errors generally exceeds the shot noise by a large factor. The standard deviation is also generally larger than the Malmquist bias of the mean, since the bias is smaller by a factor of Δ and the number of objects in our window is only a few beyond 3000 km s^{-1}.

MONTE CARLO SIMULATIONS OF POTENT

The real test of POTENT comes in applying it to a known velocity field with sparse sampling and distance errors added in. We have conducted a series of such Monte Carlo simulations using the same *N*-body simulation adopted above.

The Monte Carlo procedure consists of four parts. First, we obtain the exact three-dimensional velocity field of the *N*-body simulation, evaluated on a grid of spacing 500 km s^{-1} using cloud-in-cell interpolation. The interpolation removes 'fingers of god' in virialized regions and is a useful approximation to the grouping used in the observed samples. Second, we evaluate the radial component of the peculiar velocity *at the positions of observed galaxies*, with weight given by the estimated variance of the observed distance. The observed sample consists of spirals with infrared Tully-Fisher distances (Aaronson *et al.* 1982, 1986, 1989) and ellipticals with D_n - σ distances (Lynden-Bell *et al.* 1988; Dressler and Faber 1990). Our purpose for viewing the *N*-body simulation through the 'mask' of observed positions and errors is to study the effects of sparse and nonuniform sampling on the recovered fields, so that we can optimize our methods and can better understand our errors when POTENT is applied to real data. For this analysis it is desirable to apply the same sampling present in the observations to a known velocity field. Of course, the structures present in the simulation do not correspond to those in observed samples, so with this procedure we cannot study, *e.g.*, the effect of biased galaxy formation on our ability to reconstruct three-dimensional fields. Third, we add random distance errors, fixing the redshift of each object and scattering the distance with a Gaussian random variable of zero mean and variance equaling the estimated variance of the observed distance. We have not experimented with correlated or non-Gaussian (i.e., systematic) errors as might arise, *e.g.*, from residual Malmquist bias in the distance estimates. Finally, we apply POTENT, first constructing a smoothed radial velocity field $u_e(\mathbf{r})$, then reconstructing the three-dimensional fields from the potential.

Figure 3 shows how distance errors and sparse sampling affect the radial velocity field even before the potential-flow requirement is imposed. With a dirty input, we cannot expect POTENT to provide a clean output. The dipole pattern at the center of each panel arises from the coherent motion relative to the comoving frame (see Figure 1). A Gaussian window radius of R_w = 1200 km s^{-1} has been used for the exact field with no masking (panel labeled 'Actual') and for the fields obtained using the mask with tensor window functions *B* and *F*. Window function *V* uses the mask with a variable window radius, with R_w constrained to lie between 300 and 2000 km s^{-1}. The random distance errors have created large errors in the smoothed radial velocity field even before we apply the potential reconstruction algorithm. In well-sampled regions the errors are <300 km s^{-1}, but errors exceed 1000 km s^{-1} in some regions. The sampling gradient bias of window function *B* is evident: the velocity field is propagated outward from the densely sampled region within 3000 km s^{-1} of the center. The variable window function has much less coherence because the window shrinks where the sampling density is high.

Figure 4 shows the result of applying the Eulerian potential flow reconstruction algorithm to the N-body simulation with sparse sampling and noise. For the panels marked F and V, the radial velocities of the simulation were sampled using the observed 'mask' as in Figure 3, without the addition of random distance errors. We discarded W_B as useless because of the large sampling gradient bias. To evaluate the potential and the three-dimensional velocity fields we use the same Eulerian grid as in Figure 1. The error in the three-dimensional velocity reconstructed without distance errors added is less than 200 km s^{-1} at most of our spherical grid points, showing that nonlinearity and sparse and nonuniform sampling together are not significant problems for POTENT, although the reconstruction naturally is better with complete and uniform sampling. The results are consistent with what we learned from looking at the radial-velocity field alone: where the sampling is dense, we recover the velocity field reasonably well, but distance errors pose problems where the sampling is sparse. The error in the reconstructed three-dimensional velocity is generally less than twice the error in the radial component alone, showing that our procedure for computing the transverse components assuming potential flow is not a significant additional source of error. In the central region, which is densely sampled by field spirals, the errors in the fixed-window noise simulations are typically less than 300 km s^{-1}.

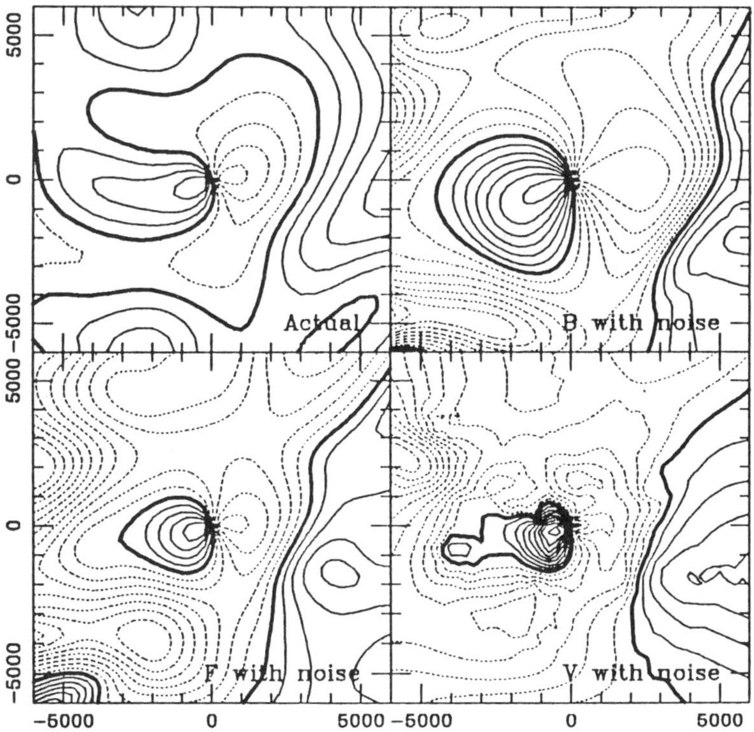

Fig. 3. Contours of the radial peculiar velocity field of the N-body simulation. The *Actual* field is smoothed with a Gaussian window of radius 1200 km s^{-1}. The other three panels show the radial velocity field obtained by sampling the N-body simulation at the positions of 526 elliptical and spiral galaxies and clusters, using various tensor window functions. Random distance errors have been added to the sampled peculiar velocities. The contour spacing is 100 km s^{-1}

Interpreting Figure 4 with the assistance of equations (13), (16), and (17), we conclude that with the present galaxy sample and with velocity fields of amplitude and coherence similar to the *N*-body simulation, the dominant source of uncertainty in the reconstructed velocity field is provided by *random distance errors* rather than by velocity-gradient shot noise or by systematic biases due to non-uniform sampling or the Malmquist effect.

We have applied the same Monte Carlo simulation techniques to the potential and density fields reconstructed by POTENT. The results appear in DBF. In summary, we find that the potential is our most stable recovered field, while the density is the least robust in the presence of noise. With a smoothing scale of 1200 km s^{-1}, the signal-to-noise ratio in the reconstructed potential is close to 10 over most of the volume, while that of the density is barely unity over much of the volume. These results are understandable: the potential is obtained by integrating a noisy velocity field, while the density effectively differentiates the velocity field. If the velocity errors are Poisson, the power spectrum of the noise component of the velocity is $P_V(k) \propto k^0$, while, for a Harrison-Zel'dovich power spectrum, the power spectrum of the signal is $P_V(k) \propto k^{-1}$. The expected signal dominates on large scales, where the potential is the most sensitive.

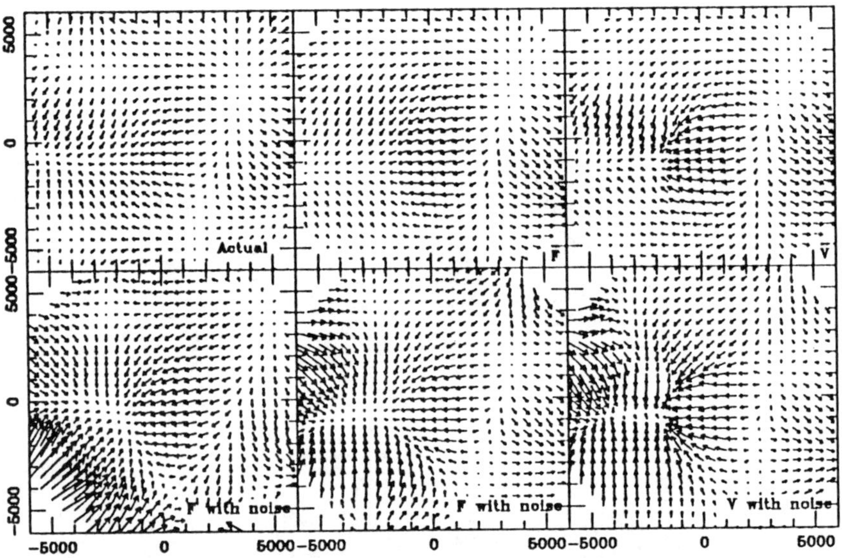

Fig. 4. Velocity field in the principal plane of the *N*-body simulation. The top three panels are vector versions of the actual and zero-noise reconstructed fields. The lower right two panels correspond to the reconstructed noise simulations of Figure 3. A second realization of noise with the fixed window function appears in the lower left panel.

CONCLUSIONS

Careful analysis of both the theoretical ideas of potential flow and of the effect of statistical errors in real data have led to the refinement of the potential-reconstruction program initiated by BD. We have developed a technique that shows much promise for the analysis of large-scale structure. Even with the relatively small redshift-distance samples available now, we are able to gain considerable insight into the dynamics of the local universe (*e.g.*, Dekel and Bertschinger this volume). We are optimistic that POTENT will prove to be a useful analysis tool as large ongoing redshift-distance surveys are completed in the next several years.

The tests we have described in this paper and elsewhere (BD, DBF) show, first of all, that nonlinear evolution is not a problem for a potential-flow analysis on large scales. After smoothing a highly nonlinear *N*-body simulation at two mass correlation lengths, we are able to recover the velocity, potential, and density-perturbation fields with an accuracy better than 10%. We could decrease the smoothing scale markedly if not for the sparseness of present galaxy surveys. However, we should caution that there is no law against cosmic vorticity, so that POTENT might be fooled if non-gravitational processes created large-scale circulation.

We have demonstrated that sparse and inhomogeneous sampling alone are not a serious problem, provided that we correct for the sampling gradient bias as mentioned above. This result means that we can combine heterogeneous data samples without well-defined selection functions. The only assumption is that the sampled objects should fall freely in the large-scale gravitational field.

On the other hand, we have found random distance errors to be a serious problem, given the current level of uncertainty in distance measurements, the sizes of the present samples, and the sparseness of the sampling. Fortunately, systematic biases such as the Malmquist bias are not serious in POTENT itself, although we can say little at present about possible systematic errors in the distance measurements themselves. What is needed from *future data*, for higher resolution results with smaller errors, besides improving the distance measurement techniques as much as possible, is *denser sampling*. For the best statistical analysis and constraints on theoretical models, dense uniform sampling is desirable, although uniformity is not strictly required (because we are not using the galaxies as mass tracers).

We have used Monte Carlo simulations to test POTENT with a known velocity field obtained from an *N*-body simulation. When we apply POTENT to real data, we can use the same techniques to estimate statistical errors in our reconstructed fields. The resultant errors are correlated and non-Gaussian, but that causes no problems. For every three-dimensional map we produce, we can produce a three-dimensional map of the standard deviation or any other statistic. The procedure is simple: add noise to the input data, follow it entirely through the POTENT processing, and then look at an ensemble of noisy outputs.

A *map* of the three-dimensional velocity, potential, or density field, with no built-in assumptions save gravity, is preferable to a *model* or a one-dimensional representation. We believe that the quality and quantity of peculiar velocity data being accumulated now will significantly improve our understanding of large-scale structure. POTENT provides tools useful for deriving the strongest conclusions from these data.

ACKNOWLEDGEMENTS

We wish to thank our collaborator Sandy Faber for allowing us to present results of our work here. Thanks also to Alan Dressler for permission to use data before publication, and to Dave Burstein for his help compiling the various data sets. This research was supported by the NSF and NASA and by the US-Israel Binational Science Foundation.

REFERENCES

Aaronson, M., Bothun, G. D., Cornell, M. E., Dawe, J. A., Dickens, R. J., Hall, P. J., Sheng, H. M., Huchra, J. P., Lucey, J. R., Mould, J. R., Murray, J. D., Schommer, R. A. and Wright, A. E. 1989, *Ap. J.*, **338**, 654.

Aaronson, M., Bothun, G. D., Mould, J. R., Huchra, J. P., Schommer, R. A., and Cornell, M. E. 1986, *Ap. J.*, **302**, 536.

Aaronson, M., Huchra, J. P., and Mould, J. R. 1979, *Ap. J.*, **229**, 1.

Aaronson, M., Huchra, J. P., Mould, J. R., Tully, R. B., Fisher, J. R., van Woerden, H., Goss, W. M., Chamaraux, P., Mebold, U., Siegman, B., Berriman, G. and Persson, S. E. 1982, *Ap. J. Suppl.*, **50**, 241.

Bertschinger, E. and Dekel, A. 1989, *Ap. J. (Letters)*, **336**, L5 (BD).

Bertschinger, E., Dekel, A., Faber, S. M., Dressler, A., and Burstein, D. 1990, *Ap. J.*, **364**, 370.

de Lapparent, V. Geller, M. J. and Huchra, J. P. 1986, *Ap. J. (Letters)*, **202**, L1.

Dekel, A. and Bertschinger, E. this volume, p. 83.

Dekel, A., Bertschinger, E., and Faber, S. M. 1990, *Ap. J.*, **364**, 349 (DBF).

Djorgovski, S. and Davis, M. 1987, *Ap. J.*, **313**, 59.

Dressler, A., and Faber, S. M. 1990, *Ap. J.*, **354**, 13.

Dressler, A., Lynden-Bell, D., Burstein, D., Davies, R. L., Faber, S. M., Terlevich, R. J., and Wegner, G. 1987, *Ap. J.*, **313**, 42.

Gregory, S. A. and Thompson, L. A. 1978, *Ap. J.*, **222**, 784.

Haynes, M. P. and Giovanelli, R. 1986, *Ap. J. (Letters)*, **306**, L55.

Huchra, J. P., and Geller, M. J. this volume, p. 143.

Kirshner, R. P., Oemler, A., Schechter, P. L. and Shectman, S. A. 1981, *Ap. J. (Letters)*, **248**, L57.

Kofman, L. A. 1989, preprint.

Lahav, O., Rowan-Robinson, M. and Lynden-Bell, D. 1988, *M.N.R.A.S.*, **234**, 677.

Lynden-Bell, D. this volume, p. 1.

Lynden-Bell, D., Faber, S. M., Burstein, D., Davies, R. L., Dressler, A., Terlevich, R. J. and Wegner, G. 1988, *Ap. J.*, **326**, 19.

Peebles, P. J. E. 1976, *Ap. J.*, **205**, 318.

Peebles, P. J. E. 1980, *The Large-Scale Structure of the Universe*, (Princeton: Princeton University).

Silk, J. 1974, *Ap. J.*, **193**, 525.

Strauss, M. A., and Davis, M. 1988, in *Large-Scale Motions in the Universe: A Vatican Study Week*, ed. V. C. Rubin and G. V. Coyne, S. J., (Princeton: Princeton University), p. 255.

Strauss, M. A., and Davis, M. this volume, p. 53.

Tully, R. B. 1987, *Ap. J.*, **323**, 1.

Tully, R. B. and Fisher, J. R. 1977, *Astr. Ap.*, **54**, 661.

Yahil, A. 1988, in *Large-Scale Motions in the Universe: A Vatican Study Week*, ed. V. C. Rubin and G. V. Coyne, S. J., (Princeton: Princeton University), p. 219.

Yahil, A., Sandage, A. and Tammann, G. A. 1980, *Ap. J.*, **242**, 448.

Zel'dovich, Ya. B. 1970, *Astr. Ap.*, **5**, 84.

DISCUSSION

S. White: I am glad to hear that you think that the uncertainties introduced by Malmquist bias are relatively small in practice compared to sampling noise. Although we can understand noise effects (with some effort) in your simulations because we know the characteristics of the errors you put in, the same is not true for real data. The amplitude (and even the sign!) of the Malmquist bias depends on a complex interaction between the observational quantities used to define the observational sample (total magnitude, isophotal radius, . . .) and the quantities used as distance indicators. I believe that the relationship between these quantities is still far from being understood at the required level.

E. Bertschinger: I agree with you. Our simulations assume a Gaussian distribution of D_n at fixed σ and neglect the possible dependence of these quantities on other variables. Clearly, more work needs to be done before we can be confident that residual Malmquist errors are entirely negligible. However, I have looked at some of these other dependences and have performed Monte Carlo simulations (unpublished) to assess their significance, and so far I have seen no residual bias comparable with the statistical errors due solely to the finite width of the D_n - σ relation.

R. Gott: I would like to compliment the authors on a fine piece of work. I think that this is a very important new technique, and I agree that it is important to begin making maps in three dimensions of the velocity flows, density, and potential fields. I wonder if galaxies were in some biased distribution so that they lived only on caustics, for example where velocity fields are crossing, if this could introduce any systematic effects. In other words, in the real data one never gets to see the velocity fields in the voids; could this cause any problems? An appropriate simulation could answer this.

E. Bertschinger: This is a good question that we have not yet addressed. Up to now we have been concerned primarily with sparse sampling and distance errors; we have sampled the *N*-body simulation at the positions of galaxies in the *real* universe but not at the positions of high-density peaks in the simulation. The sparse sampling alone seems not to cause problems, but it is possible that our method is biased by the lack of tracers in voids. We will test this in the future with *N*-body simulations.

A. Dekel: Note that the 'Great Attractor' that POTENT recovers in the simulations happens to be in the zone of avoidance - the region where the 'mask' of galaxies we use as tracers is empty! We can do that due to our smoothing procedure. It would have been harder with smaller smoothing length.

P. Shapiro: If galaxy formation is 'biased' in a way which is a function of the *velocity* field, rather than the density field, then are you still confident that your approach can produce an accurate map of the actual three-dimensional velocity and density fields from the observed velocity data, *without* knowing this bias? For example, suppose galaxies formed in pancakes where large, gravitationally-induced motions are *dissipated* by gas dynamical shocks and radiative cooling. In this case, there would be a tendency for galaxies to populate regions of low peculiar velocity, but which actually correspond to regions where the matter once had large gravitational acceleration.

E. Bertschinger: I am sure that it is possible for us to be fooled by a velocity-dependent bias. We will need to perform *N*-body simulations with such a bias built in before we can say how problematical this might be. However, in your scenario, the pancake is itself accelerated by the large-scale gravitational field, and so would be the galaxies in the pancake. I suspect that on large scales gravity will not hide itself, but we should test this quantitatively with simulations.

MAPPING LARGE-SCALE FLOWS IN THREE DIMENSIONS: APPLICATION

AVISHAI DEKEL
Racah Institute of Physics, The Hebrew University of Jerusalem
Jerusalem, 91904 Israel

EDMUND BERTSCHINGER
Department of Physics, Massachusetts Institute of Technology
Cambridge, MA 02139

ABSTRACT We present the first results of applying POTENT to real data, and preliminary results concerning the comparisons with other observations and with theory. The method for recovering the full large-scale potential, velocity and density fields from sparse and noisy redshift-distance samples, assuming gravity, is described in the preceding paper. The results are shown as maps of these fields, revealing the three-dimensional dynamical structure in a sphere of radius ~ 6000 km s^{-1} about us. The errors are estimated using Monte Carlo simulations. The main recovered features, largely confirming Lynden-Bell *et al.* (1988), are a dominant large-scale flow toward a Great Attractor (GA), away from a large void in the opposite side of the sky, and a weak signal of infall toward Pisces behind that void. The GA is roughly at rest relative to the microwave background, with infall seen in its far side as well, but its center is poorly determined at this stage. It can be described as a broad density peak centered near the Hydra-Centaurus Supercluster, extending toward another peak at the Pavo-Indus-Telescopium complex, and also toward the Local Supercluster – Virgo and Ursa Major. The flow in the Local Supercluster can be described as cylindrical; a minor circular infall toward the axis connecting Ursa Major, Virgo and the GA, modulated on a flow toward the GA along this axis. Preliminary comparison with the *IRAS* results show an agreement in the GA half of the sky, with $\Omega^{0.6}/b_I \simeq 1$, and a puzzling discrepancy in the Perseus-Pisces region. The mass distribution within 3000 km s^{-1} seems to qualitatively follow the distribution of optical galaxies. We briefly discuss preliminary applications of statistics to the data in comparison with theory.

INTRODUCTION

Mapping the large-scale structure has, until recently, been done based on galaxy redshift surveys, with the redshifts being used as crude distance indicators. By this method we can explore the galaxy distribution in large volumes, but the comparison with theory is not straightforward because galaxies are not necessarily faithful tracers of the underlying mass distribution – their formation might be biased toward certain regions due to uncertain physical processes (see Dekel and Rees 1987 for a review). For direct comparison with theory, we need measurements of dynamical quantities, such as the peculiar velocities and the fluctuations in the mass density, which are related to each other *via* gravity. Since most of the mass is unseen, our best hope is to measure peculiar velocities. This is achieved in samples of galaxies with measured redshifts and independently determined distances, which became available recently as a result of improvements in distance measurement techniques, and continue to improve rapidly. The radial peculiar velocity of a galaxy is obtained by subtracting the Hubble velocity at the measured distance of the galaxy from its total velocity as given by its redshift.

The first attempts to interpret such data involved fitting toy models of three-dimensional flows with a few free parameters. The best-fit parameters can be determined by minimizing the residuals between the radial velocities of the model and those observed. Our local neighborhood was traditionally modeled by a *spherical* infall of the *Local Supercluster* (LSC) toward the Virgo Cluster, lying at a distance of ≈1300 km s^{-1} from us, at (B = −2°, L = 103°), *i.e.* roughly in the direction of the North Galactic Pole (we measure distances in km s^{-1} to keep independent of the value of the Hubble constant, and use Supergalactic coordinates (B,L) or (X,Y,Z).] For example, by using a sample of 306 spiral galaxies with Tully-Fisher distances out to a distance of about 3000 km s^{-1}, Aaronson *et al.* (1982b) obtained a best fit with a systematic infall velocity at the radius of the Local Group of 250 ±64 km s^{-1}, a total LG velocity toward Virgo of 331 ±41 km s^{-1}, and rotation of 180 ±58 km s^{-1}.

The field has been revolutionized by the analysis of an extended full-sky sample of elliptical galaxies (Dressler *et al.* 1987, Lynden-Bell *et al.* 1988), which spans about 6000 km s^{-1} in radius. The peculiar velocity field was first modeled by a bulk flow, roughly parallel to the dipole in the Cosmic Microwave Background (CMB) pointing at (B = −37°, L = 137°). Then it developed into a multi-parameter model that is dominated by spherical infall into a '*Great Attractor*', centered at a distance of ≈4200 km s^{-1} and (B = −5°.5, L = 161°), *i.e.* just north of the Galactic zone-of-avoidance at the vicinity of the Hydra-Centaurus Supercluster [Centaurus at (−11°,156°), Hydra at (−37°,139°), N3557 at (−29°,150°)]. The GA infall component at the LG is 535 km s^{-1} in this model. This flow is modulated by spherical Virgo infall, of 100 km s^{-1} at the LG, plus a local bulk anomaly – the residual motion of the LG relative to this model (see Burstein this volume and references therein).

These models are very helpful in conveying an intuitive qualitative picture. However, it is difficult to asses the theoretical implications of such *a posteriori*, multi-parameter models (*e.g.* Bertschinger and Juszkiewicz 1989). For comparison with theory one wishes to apply simple, *a priori* statistics, such as the velocity correlation function, or the Cosmic Mach Number (Ostriker and Suto 1990), which relate to the desired power spectrum of the initial density fluctuations. Certain attempts have been made to compute such statistics based on the measured radial velocities (Górski *et al.* 1989; Groth, Juszkiewicz, and Ostriker 1989; Górski, Hoffman, and Bertschinger 1989), but these attempts involve certain assumptions concerning global isotropy, for example. It would have been better to use a detailed three-dimensional velocity field – if only one could extract it from the data. The recovery of the missing tangential components is the first goal of our effort.

If gravity is responsible for the growth of large-scale structure, then the velocity field contains, in principle, all the information required to recover the initial conditions. In particular, the mass density field can be extracted using the continuity and Poisson equations, and when the fluctuations are smoothed over large scales they can be traced back to the initial fluctuations using a theory for their evolution under gravity. One only needs to assume that the galaxies and clusters can serve as test bodies, which, when smoothed over large enough scales, faithfully trace the large-scale small amplitude component of the *velocity* field, independently of their particular type. Another advantage of dealing with the velocity field is that, when smoothed on a given scale, it is determined mostly by density fluctuations on *similar*, or *larger*, scales, so it can be approximated by quasi-linear theory better than the density field on the same scale.

Besides the fundamental problem of being able to observe only the *radial* component of the velocities, there is another serious problem facing us when trying to extract the large-scale velocity field – the fact that the data are *sparse* and *noisy*. Current samples contain only about 500 objects with estimated distances out to ~ 6000 km s^{-1}, and these are spread very non-uniformly throughout the volume. The better-sampled region, which extends only to ~ 2500 km s^{-1}, is populated by about 300 field spirals, but even here the Galactic zone-of-avoidance obscures a significant portion. Even more problematic is the fact that each distance estimate carries an error of order 20%. This error is reduced for galaxies grouped in clusters, but for single galaxies it is still a large ± 1200 km s^{-1} at the edge of the volume.

The method for dealing with these problems, which we named POTENT, is described in detail in the preceding paper (Bertschinger and Dekel this volume, Paper I), based on our major method paper (Dekel, Bertschinger, and Faber 1990, DBF), and on an earlier *letter* (Bertschinger and Dekel 1990, BD). The key ideas can be summarized as follows.

Given the radial velocities u_i of N sparsely sampled galaxies at measured positions $\mathbf{r}_i$, with their estimated measurement errors σ_i, we first use a window function to interpolate the data onto a smoothed radial velocity field, $u(\mathbf{r})$, at any desired point inside a spherical volume. Using a tensor window function we find at each point the maximum-likelihood solution of a local bulk flow inside a spherical Gaussian window centered on that point.

Then, in order to obtain the tangential components of the velocity field $\mathbf{v}(\mathbf{r})$, of which $u(\mathbf{r})$ is the radial component, we make the crucial assumption that the velocity field is derived from a scalar *potential*,

$$\mathbf{v}(\mathbf{r}) = -\nabla \Phi(\mathbf{r}), \tag{1}$$

i.e. it has zero vorticity. This *velocity* potential is proportional to the *gravitational* potential in the linear regime, and is related to it *via* the Bernoulli equation in the nonlinear regime (DBF). The velocity potential at each point on a spherical grid is then calculated by integrating the radial component along radial rays,

$$\Phi(\mathbf{r}) = -\int_0^{r} u(r',\theta,\phi)\, dr'. \tag{2}$$

Differentiation of this potential in the transverse directions finally recovers the two missing components of the velocity. We first demonstrated in BD that a potential analysis can be applied successfully to an idealized, smoothed cosmological N-body simulation, where the problem of sparse and noisy data was ignored. Then, in DBF (and Paper I), we showed how the method deals with non-uniform spatial sampling and measurement errors, by applying it to an N-body simulation using the actual 'mask' of observed objects. The errors where analyzed carefully both analytically, and using Monte Carlo noise simulations.

This paper is mostly based on a paper by Bertschinger, Dekel, Faber, Dressler, and Burstein (1989, BDFDB), where we presented the first results of applying POTENT to the real universe. Our intention here is to convey a visual impression of the three-dimensional dynamical cosmography in our neighborhood. We present maps of the recovered potential, velocity, and density fields projected onto the Supergalactic plane, along with a map of estimated errors in the velocity field. We also comment here on preliminary attempts to compare our results with other data and with theory.

In the next section we describe the data used. Then we describe the smoothing procedure; show the recovered potential, velocity and density fields, and discuss the errors; try to interpret the maps; discuss preliminary attempts to compare our results with other data and theory; and comment on ongoing work. Our conclusions are summarized in the final section.

THE DATA

The radial component of the velocity is given by $u_i = cz_i - r_i$, where cz_i is the observed velocity, corrected to the CMB frame, and r_i is the measured distance. Accuracy in u_i thus depends directly on the accuracy of r_i. The distance estimators used here have typical accuracies of 15 - 25%. Thus, at the edge of the volume surveyed (~6000 km s^{-1}), errors in u are 900 – 1500 km s^{-1}, which is large compared to typical velocity flows of ~600 km s^{-1}. Much of our effort has consequently been directed to understanding the errors and minimizing their effects.

Several studies have yielded large numbers of galaxies with measured distances. The distance indicators in these studies are based on galaxy scaling laws. For spirals, the infrared Tully-Fisher law is used, which relates rotation velocity to absolute H magnitude (*e.g.*, Aaronson, Huchra and Mould 1979). For elliptical and S0 galaxies, the relevant law is the D_n - σ relation, which relates velocity dispersion to a photometrically measured linear diameter (Dressler *et al.* 1987, Dressler 1988, Faber *et al.* 1989). The principles used to select galaxies are quite different in the different sources, and only two surveys (Lynden-Bell *et al.* 1988; Dressler and Faber 1990 [DF]) can claim to be magnitude limited in their respective areas of sky. Even with a magnitude-limited survey, sampling efficiency declines drastically with distance, and the patchiness of the galaxy distribution in space leaves large volumes empty of galaxies; this is a major problem.

Besides having large random statistical errors, distance measurements are subject to systematic errors, such as the so-called 'Malmquist bias' (Lynden-Bell *et al.* 1988). This results from the coupling of random measurement errors with the run of spatial density of objects as a function of distance. For uniform density, the true distance on average is higher than the estimated distance r_{ie} because the volume element grows as r^2; more galaxies are therefore scattered into each measurement volume from distant volumes than nearer ones. The correction for uniform density is

$$r_i = r_{ie}\,(1 + 3.5\Delta^2),$$

where $\sigma_i = \Delta r_i$ is the *rms* measurement error per galaxy (in natural logarithms), $\Delta = 0.21$ for ellipticals and $\Delta = 0.16$ for spirals. In our analysis, we routinely apply this uniform density correction to the input data.

The Malmquist bias for more clumpy distributions is more complicated and can be positive or negative, depending on whether a clump is in front of or behind the measurement volume. This bias is highly systematic and is not removed by smoothing, unless the smoothing scale is considerably larger than the typical errors. Since this bias is difficult to remove analytically, we resorted to Monte Carlo models to distinguish real velocities from bias artifacts (DBF and BDFDB).

To minimize Malmquist bias, we have grouped objects into clusters by adopting the group assignments of the original authors. The error per point is then $\sigma N^{-1/2}$, and the bias correction (for uniform density) is reduced by N^{-1}, where N is the number of objects per cluster.

The data set has been compiled by Dave Burstein from several different sources. The backbone of the nearby data is the all-sky field spiral survey of Aaronson *et al.* (1982a, A82), from which we used 208 galaxies with 'good' distances (Faber and Burstein 1988). This sample gives a fairly dense and accurate sampling of the velocity field within ~2500 km s^{-1} (Figure 1), although even within this volume large regions are empty. The backbone of the large-scale data is the all-sky elliptical survey by Lynden-Bell *et al.* (1988, 7S, Faber *et al.* 1989), containing a total of 87 individual galaxies and 115 groups. To this we add 10 spiral cluster distances obtained by Aaronson *et al.* (1986; A86; Virgo, Fornax, and Ursa Major are subtracted, as they are already contained in A82), plus two recent surveys of the Great Attractor region. These include seven new spiral clusters by Aaronson *et al.* (1988, A88) and the first results (98 galaxies) of a new magnitude-limited field elliptical survey by DF. Duplicate observations of ellipticals and S0s have been averaged with equal weights. The net result for the large-scale sample is a rather skeletal sampling, highly biased to certain high density regions, of the whole volume out to ~5000 km s^{-1}, plus a very densely sampled cone toward the Great Attractor (see Figure 1).

The combined sample of spirals, ellipticals, S0s and clusters (call it ESm), which contains 526 objects, is the basic sample for the current analysis. This sample was used accordingly in Paper I as the 'mask' through which the *N*-body simulation was 'observed' and analyzed. [In DBF and BDFDB, we use a somewhat improved data set, where the DF sample is merged and grouped with 7S and with other observations of E-S0s toward the GA by Lucey and Carter (1988), together with additional field spirals from Bothun *et al.* (1984).]

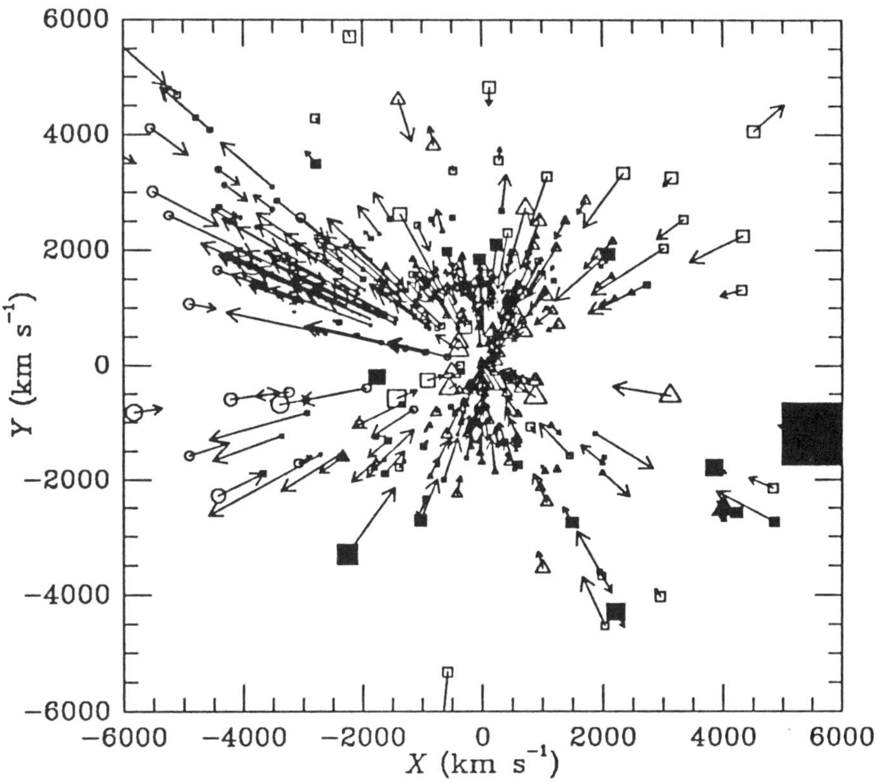

Fig. 1. The observed radial peculiar velocities of objects in a slice ±1500 km s^{-1} thick about the Supergalactic plane. The velocities are shown at the estimated distances of the objects, projected onto the Supergalactic plane. The symbols denote the type of object, where open and filled symbols mark field galaxies and groups respectively. Squares represent ellipticals from 7S, circles are ellipticals and S0s from DF, and triangles are spirals from A82, A86 and A89. The area of each symbol is proportional to the weight given to that data point in the smoothing procedure, $R_4^3(\mathbf{r}_i)\ /\sigma_i^2$ (see below).

Figure 1 shows the distribution of objects at their measured positions and the observed radial peculiar velocities in the Supergalactic plane. We use the data inside a whole sphere, and in BDFDB we plot the fields in nine different planes, but here we display only the Supergalactic plane, where most of the interesting activity occurs.

Note how sparse the spatial distribution is even in this relatively high density slice. The Galactic zone-of-avoidance is apparent as the empty angle about the Y = 0 axis. The GA region of Hydra-Centaurus (X < 0, Y > 0) is heavily populated, partly because there are several clusters there, but mostly because of the new ellipticals of DF which were selected intentionally in that region. The number density (*i.e.* completeness) drops steeply with radius in all directions. In the populated regions in the vicinity of Virgo (X = −300 km s^{-1}, Y = 1300 km s^{-1}), and toward Hydra-Centaurus, the mean separation between neighbors is less than 300 km s^{-1}, while within the sphere of 6000 km s^{-1} there are empty regions where the separation between neighbors exceeds 2000 km s^{-1}.

SMOOTHING

The first step of the analysis is interpolating the observed radial velocities u_i, at the galaxy positions $\mathbf{r}_i$, into a *smoothed radial* velocity field $u(\mathbf{r})$ at every desired point, using a tensor window function (DBF, Paper I). The smoothing serves to average over regions where vorticity might be present and to reduce the nonlinearity by averaging over the small-scale components. The major goal of the smoothing procedure, however, is to obtain an optimal mean velocity, with minimum bias due to the non-uniform spatial coverage and minimum variance arising from the measurement errors and the sparseness of the galaxy distribution.

Our estimate for the radial velocity $u_e(\mathbf{r})$, at a position $\mathbf{r}$, smoothed with a window function W, is

$$u_e(\mathbf{r}) = \sum_i \tilde{W}(\mathbf{r},\mathbf{r}_i)\, u_i, \tag{3a}$$

$$\tilde{W}(\mathbf{r},\mathbf{r}_i) \equiv [\hat{\mathbf{r}} \cdot \mathbf{A}^{-1}(\mathbf{r}) \cdot \hat{\mathbf{r}}_i]\, W(\mathbf{r},\mathbf{r}_i), \quad \mathbf{A} \equiv \sum_i W(\mathbf{r},\mathbf{r}_i)\, \hat{\mathbf{r}}_i\, \hat{\mathbf{r}}_i \tag{3b}$$

In effect, we are computing the maximum-likelihood bulk flow solution in the vicinity of $\mathbf{r}$, within a window defined by W. Tensor smoothing is required because we are averaging radial velocities from different directions $\mathbf{r}_i$. The tensor projection is absorbed into $\tilde{W}$, a normalized scalar. It is worth mentioning that a tensor smoothing similar to equation (3) can also provide, formally, an estimate for the tangential components of the velocity field. However, when the window size is smaller than r, except very close to the origin, this estimate carries a much larger error than the estimate of the radial component. Therefore, only the smoothed *radial* velocity of equation (3) is used as input for the potential flow analysis below.

Based on our error analysis, and after some trial-and-error effort, we have adopted the following two alternative window functions (DBF):

$$W_F(\mathbf{r},\mathbf{r}_i) = [R_4^3(\mathbf{r}_i)/\sigma_i^2] \exp[-|\mathbf{r}-\mathbf{r}_i|^2/(2R_w^2)], \tag{4a}$$

$$W_V(\mathbf{r},\mathbf{r}_i) = [R_4^3(\mathbf{r}_i)/\sigma_i^2] \exp[-|\mathbf{r}-\mathbf{r}_i|^2/(2R_5^2)]. \tag{4b}$$

$R_4(\mathbf{r}_i)$ is the distance from object i to its 4th nearest neighbor. By $R_4^3(\mathbf{r}_i)$ we thus weight the contribution from u_i to u_e by the inverse of the local density of objects in the vicinity of $\mathbf{r}_i$. This weighting corrects the *bias* of u_e relative to a uniform volume average, due to *sampling* gradients, which causes the information from well-sampled regions with small errors to propagate into regions with sparse and noisy data. The weighting by σ_i^{-2} minimizes the *variance* in u_e due to *measurement* errors. In this paper we took σ_i to be a given fraction ($\Delta = 0.21$ and 0.16 for ellipticals/S0s and spirals respectively) of the distance as estimated by the total redshift velocity of the galaxy in the Local Group frame. [In DBF and BDFDB, we take σ_i to be a constant fraction of the redshift-independent estimated distance, instead. This eliminates a Malmquist bias – see DBF.]

The radius of the Gaussian, R_w, defines the spatial extent of the smoothing window. The first window, W_F, has a fixed smoothing radius, which should keep the sampling gradient bias under control, but would allow large variance in poorly sampled regions, both due to measurement errors and *Poisson* sampling fluctuations. The second window, W_V, has a spatially varying smoothing radius, R_5, the distance from **r** to its 5th nearest neighboring object. This should control the variance by keeping fixed the effective number of objects summed over (~ 5), at the expense of large sampling gradient bias in the poorly sampled regions. An extended window should also reduce the Malmquist bias that is introduced by our smoothing procedure in clumpy regions. R_5 was chosen, again, by trial-and-error with the *N*-body simulation. The varying window takes advantage of the good quality of the data in some regions, while not asking too much from regions where the data are poor. However, for the purpose of quantitative comparison with other data and with theory, we would prefer to use the more uniform fixed window.

Figure 2 shows the smoothed radial velocity field, $u_e(\mathbf{r})$. Note the clear infall pattern, from the front and the back, into a point (the Great Attractor) in the upper left quadrant. Most of the errors of the resultant velocity field are present already at this stage of the analysis.

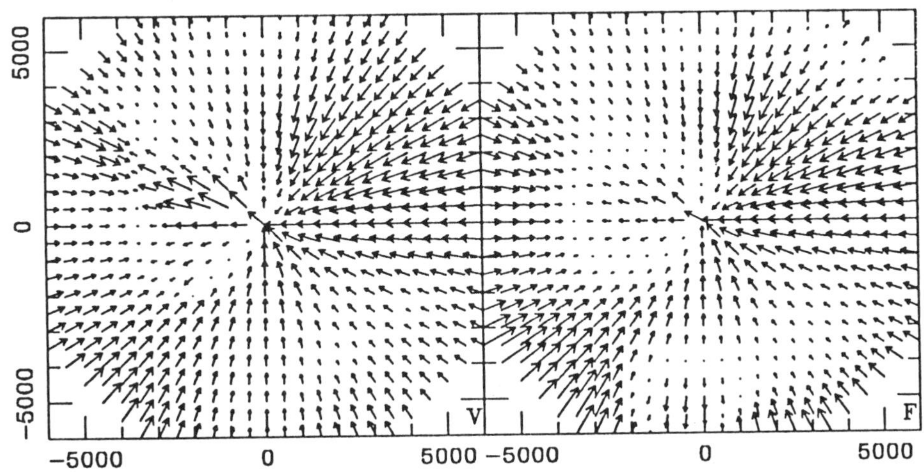

Fig. 2. The smoothed radial velocity field at the points of a cubic grid in the Supergalactic plane, projected onto the plane. *Left*: The result of smoothing with W_V, where the varying smoothing radius is given by R_5, except that it is limited to the range 300 to 2000 km s^{-1}. *Right*: After smoothing with W_F, with R_w = 1200 km s^{-1}.

RECOVERED POTENTIAL, VELOCITY AND DENSITY FIELDS

We now apply our Eulerian POTENT to the smoothed radial velocity field of the combined data set, with the windows W_V and W_F alternatively, and recover the potential and three-dimensional velocity fields. With W_V smoothing ($R_w = R_5$ between 300 and 2000 km s^{-1}), we used for the potential computation a spherical grid covering a sphere of radius 7000 km s^{-1}, with 10 equal radial bins, 24 equal latitude bins, and 48 equal longitude bins. With W_F smoothing (R_w = 1200 km s^{-1}), we used 10, 16, 32 bins respectively. Note that the resultant potential at a given radius depends only on the smoothed data in inner radii. The resultant fields are then evaluated at the points of a cubic grid, with 500 km s^{-1} spacing, by cloud-in-cell interpolation from the spherical grid. In the linear approximation, the fluctuations in the mass density field could be obtained from the velocity field using the linearized continuity equation, $\delta \simeq -\nabla \cdot \mathbf{v} /(H\,\Omega^{0.6})$. However, in order to recover structure on scales of order 1000 km s^{-1} or smaller, where δ might be on the order of a few, we would rather use an approximation that is more appropriate for moderately non-linear evolution. We cannot ignore peculiar displacements of matter relative to the comoving frame, as the linear approximation does (*i.e.* we must take into account the fact that the gravitational potential is not simply proportional to the velocity potential).

To do that, we use the iterative Lagrangian POTENT procedure described in DBF. It assumes a one-to-one correspondence between initial (*i.e.* Lagrangian) and final (*i.e.* Eulerian) position of particles. Imaginary particles are placed, at some early time, in points of a spherical grid. After making a first guess of the initial radial velocities, the tangential components are deduced by imposing the requirement of potential flow. The algorithm then moves the particles to their present positions using the *Zel'dovich* approximation (1970), assuming for example $\Omega = 1$, and corrects the initial guess iteratively until the present velocities agree with the radial velocity field $u(\mathbf{r})$ that we smoothed from the observations. After the iterations converge, the mass distribution is represented by a set of equal-mass particles that lie originally on the points of a cubic grid, and their current positions are determined by cloud-in-cell interpolation from the solution on the spherical grid. The density for the contour plots is computed by counting particles in a spherical Gaussian window of radius 500 km s^{-1} about each grid point, spaced by 250 km s^{-1}.

Figure 3 shows the resultant potential, velocity, and density fields in the Supergalactic plane. The difference between the windows is noticeable; in particular, the varying resolution of W_V *versus* the uniform resolution of W_F. As expected, the resultant potential field, $\Phi(\mathbf{r})$, is quiet and robust; it carries relatively small errors, but it hides small-scale features, being affected mostly by the large-scale, linear fluctuations. The velocity field, $\mathbf{v}(\mathbf{r})$, resolves better intermediate-scale features but with larger uncertainties. The density field, $\rho(\mathbf{r})$, is more uncertain and is sensitive to small-scale, non-linear structures. This is hardly surprising, recalling that the corresponding power spectra are related by

$$P_\rho(k) \propto k^2 P_v(k) \propto k^4 P_\phi(k). \qquad (5)$$

Before we proceed with the interpretation of the physical features in these maps, we must digress again to the issue of uncertainties. Based on the error analysis in BDF (and Paper I), we know that our main sources of uncertainty are the distance measurement errors. In order to evaluate the resultant errors better, we ran here Monte Carlo noise simulations, in which the measured distances, and therefore the peculiar velocities, were perturbed by random Gaussian increments of standard deviation σ_i, the total redshift being held unperturbed. From these simulations we extracted the *rms* scatter (standard deviation) of the final value at each grid point. This was done for each of the three fields. To illustrate the effect

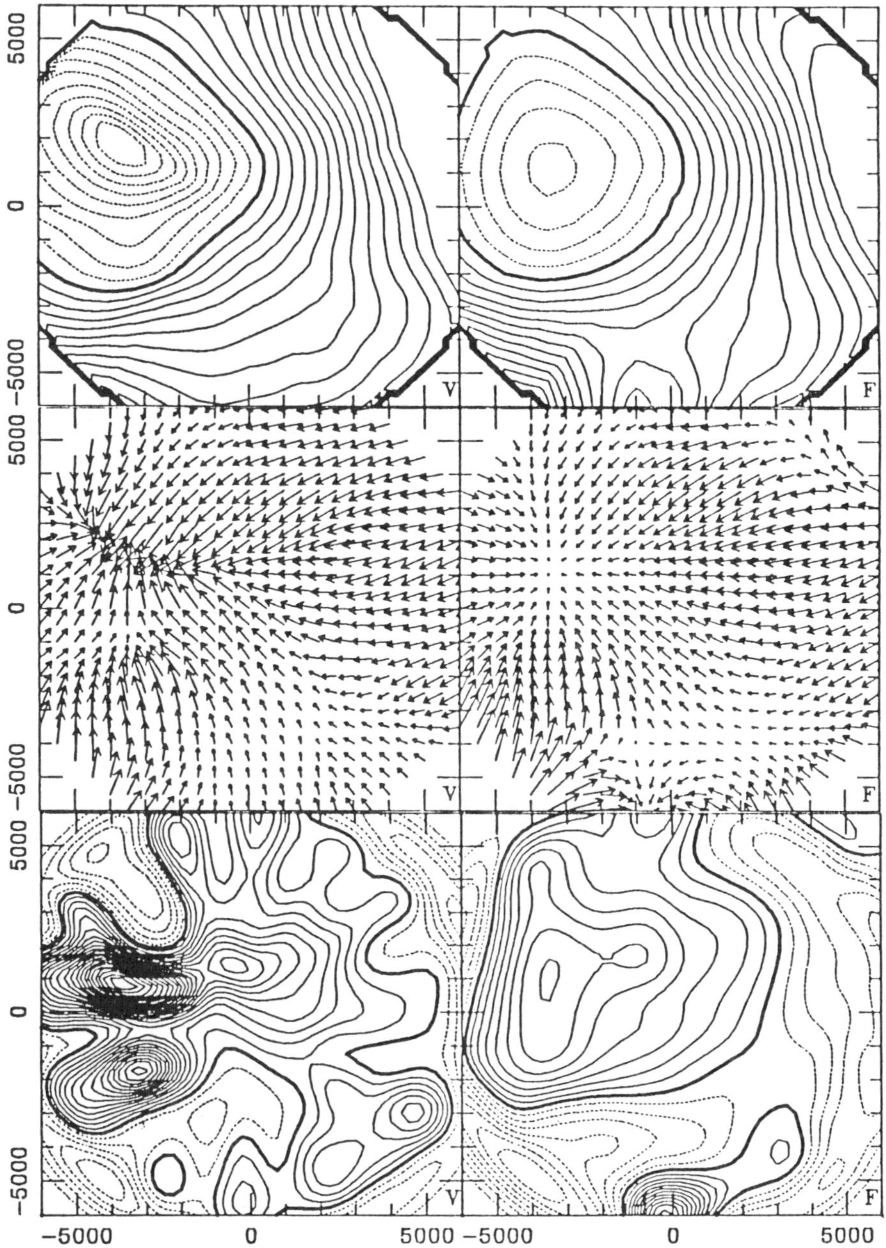

Fig. 3. Recovered potential, velocity, and density fields. *Left*: Window W_V. *Right*: W_F. *Top*: Potential contours. Spacing is 2.5×10^5 (km s^{-1})^2. The value at the Local Group, at the center of each plot, is arbitrarily set to zero (heavy contour), and negative contours are dotted. *Middle*: Velocity vector field on a cubic grid. *Bottom*: Contours of density contrast. Spacing is 0.1. The heavy contour marks $\delta\rho/\rho = 0$; the dotted contours are negative.

of noise, we show in Figure 4 the velocity field resulting from POTENT with the fixed window, out of one, arbitrary Monte Carlo realization. Figure 5 shows contours for the *rms* velocity scatter in 100 Monte Carlo simulations.

With the varying window, the scatter is relatively small even in most of the outer regions of the 7000 km s^{-1} sphere. The scatter is large in the better sampled regions because the window gets quite small there; the high resolution is achieved at the expense of larger Poisson sampling error. With the fixed window, the uncertainty is low in the inner regions ($r < 4000$ km s^{-1}) and in the well-sampled regions (*e.g.* toward the GA), but it becomes quite severe in most of the outer regions.

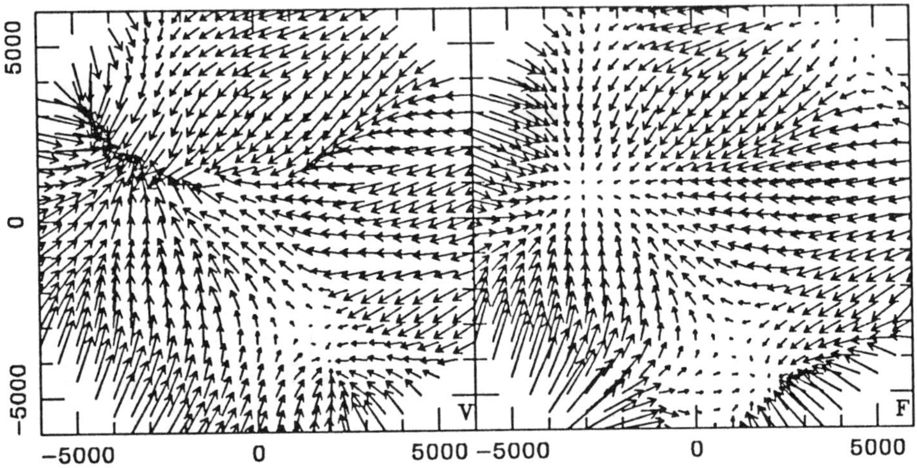

Fig. 4. The velocity field in one Monte Carlo noisy realization.

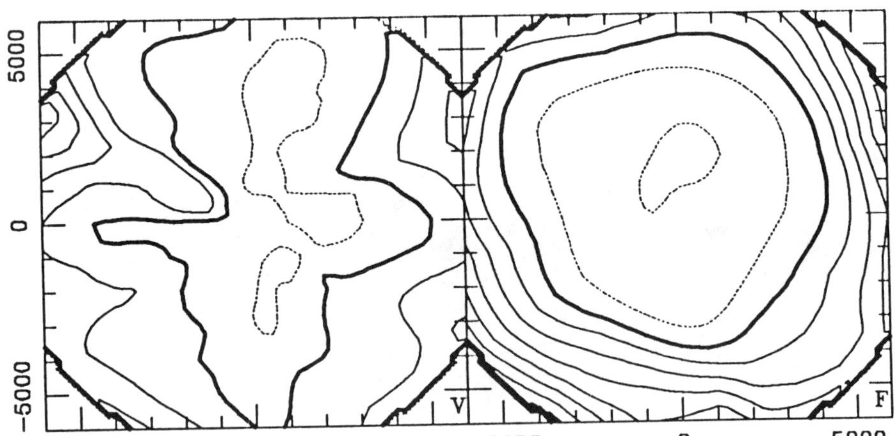

Fig. 5. Contours of standard deviation in 100 Monte Carlo noise simulations. Spacing is 100 km s^{-1}. The heavy contour marks a scatter of 300 km s^{-1}, and lower values are dotted.

The bias due to gradients in the selection function and the Malmquist bias are harder to estimate from the data itself. However, we were able to reduce the sampling gradient bias significantly using the volume weighting procedure, and we learned from the *N*-body simulation that the biases left are, in most regions, much less important than the scatter due to distance errors. There are, however, some very empty regions where the sampling bias might be severe and might possibly generate artificial coherent flows of no physical reality. Based on the *N*-body simulation, these voids are largely confined to regions away from the Supergalactic plane. We conclude that an appropriate choice of smoothing length (*e.g.* 1200 km s^{-1}) provides a reasonably large volume where the random errors are acceptable and the systematic biases are estimated to be small, so that the results are robust and meaningful. The results with the fixed window in these small error regions can be used for further statistical analysis.

INTERPRETATION OF THE MAPS

For better visual perspective, we show in Figure 6 a surface plot of the over-density distribution in the Supergalactic plane. This plot is the outcome of applying the varying window to the slightly modified data set used in BDFDB, but it is not very different from the density map of Figure 3.

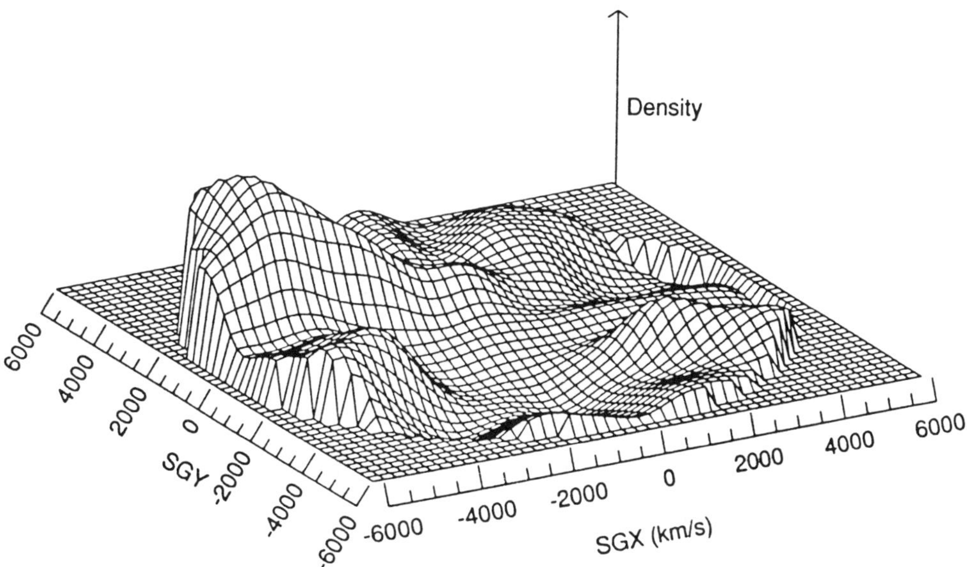

Fig. 6. The reconstructed mass density field in the Supergalactic plane, with W_V, from BDFDB. The Local group is at the center. The large peak is the Great Attractor. Virgo is the minor local hill along the extension to its right – this is the elongated Local Super-cluster. The peak in front of the GA is PIT. A large low-density region lies in front of the GA and the LSC extension. A trace of Perseus-Pisces (PP) is poorly recovered in the right corner.

What do the maps tell us about the local cosmography? The most robust feature, in all maps, is certainly the general phenomenon termed *The Great Attractor* (GA), confirming the original claims of its discoverers based on the radial velocities alone. In the potential maps it shows up as an extended, roughly spherical and somewhat featureless potential well, which overwhelms the entire sampled volume. Its central part, near the Centaurus Cluster(s), covers a region near the Supergalactic plane ($Z \simeq 0$), about $Y \sim 1000$ to 1500 km s^{-1} north of the galactic plane, and at a radial distance of 3300 to 5000 km s^{-1} from the Local Group. The velocity field, in accordance, shows a general flow over most of the volume towards the central region of this well; for example, the bulk motion in the CMB frame, inside a sphere of radius 4000 km s^{-1} around us, is about 400 km s^{-1}. The velocity and density fields reveal features on smaller scales, but the positive overdensity corresponding to the GA extends over about one half of the volume within 6000 km s^{-1}. With 1200 km s^{-1} fixed smoothing, the central region of the GA is a $\delta\rho/\rho \simeq 0.7$ peak.

This general GA flow is not affected by the controversial data in the complex Centaurus region itself (Lucey and Carter 1988); we have verified that there is no noticeable change in the results when the two Centaurus Clusters (Cen 30 and Cen 45) are removed from the data altogether.

The data, with the varying window, allows a higher resolution view of the central regions of the GA. It seems to reveal two centers: a high peak near Centaurus and a somewhat lower peak near the Pavo-Indus-Telescopium (PIT) Supercluster on the opposite side of the galactic plane, at $Y \simeq -1800$ km s^{-1} [Pavo is at ($B = 2.3°$, $L = 215°$)]. The saddle between those peaks might be largely an artifact of the poor sampling in and about the zone-of-avoidance (note that $\delta\rho$ depends on $\nabla \cdot \mathbf{v}$, and $\mathbf{v}$ is artificially smoothed in the zone-of-avoidance, so that $\delta\rho$ might be underestimated there.) It is therefore possible, as indicated by the density recovered with the fixed window, that the superclusters of Hydra-Centaurus and PIT should be regarded as one big superstructure.

With part of the new Dressler-Faber data added, POTENT reveals a clear, stable signal of infall into the GA from its back side. Recall, however, that because the density distribution falls off behind the GA, the Malmquist bias discussed in DBF might tend to produce a similar backflow (*cf.* Burstein, Faber, and Dressler 1990). Therefore, the reality of this backflow should await confirmation based on a deeper data set. A sample of spiral galaxies in that region, in preparation by Faber and colleagues, promises to provide the desired depth.

The direction to the center of the infall (or the minimum of the potential) is determined to be in the range $L = 151°$ to $162°$ (with a tendency toward lower L values when a smaller smoothing scale is used), in pleasant agreement with the original best fit of $L = 161°$ by Lynden-Bell *et al.* The distance to this center, however, varies between 3300 and 5000 km s^{-1} as we vary the smoothing scale or the σ_i weights in the window function. Because this is a broad minimum and the distribution is clumpy, an uncertainty of about 200 km s^{-1} in the velocities due to Malmquist bias translates to an uncertainty on the order 1000 km s^{-1} in the position of the center! We therefore do not attempt to determine accurately the distance to the center of the infall at this stage.

Note that in the North Galactic Pole, already at about 5000 km s^{-1} there is an indication for the periphery of the Coma Supercluster, with a lower density saddle at $r \sim 5000$ km s^{-1} between the GA and the Coma complex, but also a hint for a possible bridge between the two giants.

The present analysis allows a closer look at our veteran – the Local Supercluster – centered on the Virgo Cluster. It shows as an elongated extension of the GA, extending along the line from Centaurus, through Virgo towards Ursa Major (the line $Y \simeq 1000$ km s^{-1}, $Z \simeq 0$). Virgo is a minor local peak along this elongated extension. The flow in the LSC is cylindrical, as noticed by Faber and Burstein (1988): a circular infall into the line connecting Ursa Major and Virgo and a dominant flow along this line toward the GA. Virgo, smoothed over a sphere of radius 1200 km s^{-1}, is moving with a velocity of 475 km s^{-1} almost precisely toward the center of the GA as defined by Lynden-Bell *et al.*

Our Local Group lies at the outskirts of the extended ramp termed the GA, on the density contour $\delta\rho/\rho = 0.3$ (1200 km s^{-1} smoothing). Averaged over a 1200 km s^{-1} sphere, we find the velocity at the LG relative to the CMB to be 465 km s^{-1} in the direction (B = −17°, L = 157°). This can be compared with the actual, non-smoothed, LG motion (the CMB dipole) of 614 km s^{-1} toward (B = −37°, L = 137°); the difference can be attributed to the motion of the LG relative to the center-of-mass of the local sphere of radius 1200 km s^{-1} around us.

On the opposite side of the sky, in the south Galactic hemisphere, lying between the GA and the Perseus-Pisces Supercluster, at 2000 to 3000 km s^{-1} from us, is a large, elongated low-density region, which we might term 'The Great Repeller'. It is associated with known 'voids' of galaxies, where only a few galaxies have been observed.

But the most intriguing question is: "where is Perseus-Pisces?" Being the largest visible feature in many optical catalogs, one could have expected a density peak at roughly the same distance and in the opposite side of the sky from the GA [*i.e.* in the ($X > 0$, $Y < 0$) quadrant of the Supergalactic plane]. The *IRAS* maps also show it to be a large peak (see Figure 7 below). Our potential maps in the Supergalactic plane indicate only an extended valley of potential which leads from the Pisces region, through the LG, toward the GA deep potential well. In accordance, there is an indication for a cylindrical flow toward Pisces and along this valley towards GA. The density peak that could be identified with PP in the Supergalactic plane is minor, and it lies south of Pisces ($r \sim$ 4800 km s^{-1}, B = 5°.5, L = 328°) – more toward Pegasus ($r \sim$ 3600 km s^{-1}, B = 25°, L = 298°) than toward Perseus ($r \sim$ 5800 km s^{-1}, B = −15°, L = 348°).

The PP region, no doubt, is severely undersampled. Our maps there are very uncertain. So this must be the major target for more distance measurements, and indeed there are very promising efforts going on to measure a large sample of spiral galaxies in that region (R. Giovanelli and M. Haynes, private communication; J. Willick, private communication). The density maps do show, however, a strong 'mystery' peak at about Supergalactic coordinates (X,Y,Z) = (2500,−2500,−3000) (BDFDB), which does not seem to coincide with any known visible supercluster. It results from reliable data but of only 3 groups in that region, which is otherwise practically empty. This region deserves close attention in future surveys.

FURTHER APPLICATIONS

As a matter of fact, the maps discussed in this paper are just a by-product of the POTENT analysis, which aims at an *a priori* statistical analysis of the recovered dynamical fields, in order to allow a more quantitative comparison with the distribution of galaxies and with theoretical models. The statistical analysis of POTENT's output has just started. Here we briefly discuss some of this ongoing work, and quote very preliminary results, which should be taken with a grain of salt – until confirmed and properly published.

One ongoing attempt in this direction is a comparison with the output of the analysis of the *IRAS* redshift catalog of galaxies (see Strauss and Davis this volume), which we pursue in collaboration with M. Davis, M. Strauss and A. Yahil. Figure 7 shows a preliminary density map based on the *IRAS* catalog, in comparison with our density map.

Qualitative, visual comparison of the density maps (and the velocity maps – see Yahil 1988) reveals an encouraging agreement in the GA half of the sky and a puzzling discrepancy in the PP region. There are three possible explanations for this discrepancy: (a) Our results in the PP region are unreliable because of the poor current sampling in that region or because of unknown systematic errors (*e.g.* in the distance estimates, see Silk 1989; Djorgovski, de Carvalho, and Han 1989). (b) The *IRAS* analysis suffers from severe shot noise problems in that region. (c) The distribution of *IRAS* galaxies is not a simple function of the underlying density field. Possibly all three are true. The data available for POTENT analysis will certainly improve in the near future with the inclusion of many spiral galaxies in that region, which will hopefully allow us to reduce our uncertainties.

The main goal of a *quantitative* comparison is to address the value of the cosmological density parameter Ω and the relation between the distribution of *IRAS* galaxies and the underlying mass density field. First, we assume that the latter are related by a constant proportionality biasing factor, b_I, *i.e.* $\delta n_i/n_i = b_I(\delta\rho/\rho)$, and ask whether the *IRAS* data is consistent with being a noisy version of the ES data, or *vice versa*, given the measurement uncertainties in both, and given some reasonable statistics, *e.g.* a correlation coefficient, to compare them with. For this we run Monte Carlo noise simulations of both data sets, and cross-correlate the resultant fields, evaluating the distribution functions of the correlation coefficient and of the best fit value of the bias parameter, $b_I/\Omega^{0.6}$. (In the linear regime, unfortunately, this combination of Ω and b_I is inseparable; $\Omega^{0.6}$ coming from the proportionality factor relating $\delta\rho/\rho$ to the peculiar velocities.) The comparison could be done between either the potential, or the velocity, or the density fields. Unfortunately, the most error free quantity in the *IRAS* analysis is the density field, while for us it is the potential field. We therefore make the first comparison in the middle ground – the velocity field – despite the fact that this involves comparing vector, rather than scalar, fields. We compare the velocity fields in the LG frame; this is because the *IRAS* predicted velocities depend on the local sources of gravity, and are therefore not expected to recover properly the large-scale dipole flow of the local neighborhood relative to the CMB. Ω = 1 has been assumed in both

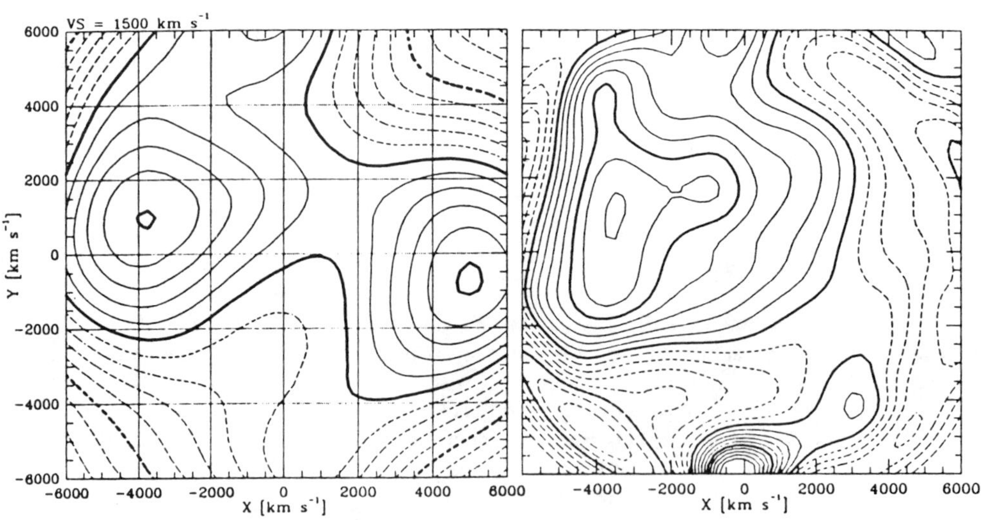

Fig. 7. *Left*: *IRAS* galaxy density contrast (courtesy of A. Yahil). *Right*: The underlying mass density contrast reconstructed by POTENT from the ESm peculiar velocities (same as in Figure 3). The effective smoothing in both cases is 1500 km s^{-1}. Spacing between contours is 0.1.

procedures, and the *IRAS* analysis assumed no biasing for the *IRAS* galaxies, $b_I = 1$. Our *very* preliminary tests of this sort indicate, in agreement with the visual impression, that the two data sets are consistent with each other in the GA half of the volume, with $b_I/\Omega^{0.6} \simeq 1$, but they seem to be inconsistent in the other half.

Next we can map the 'biasing' field, $b_I(\mathbf{r})$, up to a factor of $\Omega^{0.6}$, by comparing directly the density of *IRAS* galaxies, $n_i(\mathbf{r})$, with our reconstructed mass density field. A similar analysis is being done concerning the distribution of optical galaxies in redshift surveys, in collaboration with O. Lahav. Lynden-Bell (this volume) and his colleagues are developing a method for estimating the spatial number density of galaxies from a combined sample of galaxies with and without measured redshifts, and this would provide an appropriate data set for such a comparison. A very preliminary comparison of high resolution (R_w = 500 km s^{-1}) mass density maps recovered by POTENT with the groups and clusters in the Nearby Galaxies Atlas (Tully and Fisher 1987) shows an encouraging correlation between mass and light.

With 500 km s^{-1} fixed smoothing (see BDFDB), the velocity at the LG relative to the CMB is 565 $\pm$125 km s^{-1} in the direction (B = −23°, L = 147°). The component of this velocity in the direction of the GA as defined by Lynden-Bell *et al.* (B = −5.5°, L = 161°) is 522 km s^{-1}, namely, the velocity at the LG is mostly directed toward the GA. The velocity at Virgo is 658 $\pm$121 km s^{-1} almost precisely toward the GA. The residual velocity at the LG, perpendicular to the line-of-sight to the GA, is 215 km s^{-1}. This residual actually points at about 45° away from Virgo towards Ursa Major, and about 45° below the Supergalactic plane; its component toward Virgo is only 123 km s^{-1}. These velocities, in the linear approximation, can tell about the relative gravitational effects of Virgo and the GA on the Local Group.

The velocity at the LG, smoothed inside a sphere of a given radius, can be compared to the dipole of optical flux due to galaxies outside of this sphere. If light traces mass, the latter should provide an approximation for the gravitational acceleration of the center-of-mass of that sphere, and in the linear regime the velocity is proportional to the acceleration. We get from POTENT for the velocity at the LG, with Gaussian smoothing of radius 500 and 1200 km s^{-1} respectively, (v,B,L) = (565,−23°,147°) and (465,−17°,157°). Lynden-Bell, Lahav and Burstein (1989) obtain for the optical dipole, with top-hat smoothing of radius 500, 1000, and 2000 km s^{-1} respectively, (*D*,B,L) = (600,−37°,147°), (499,−36°,162°) and (300,−36°,174°). (The normalization of the optical dipole, *D*, has been arbitrarily determined at the 500 km s^{-1} sphere). The approximate agreement between these numbers seems to indicate a similarity between the light distribution and the mass distribution within the ~2000 km s^{-1} vicinity of the LG. We are currently studying the convergence with distance of the acceleration vector at the LG due to the mass density recovered by POTENT, for a better comparison with the optical flux dipole and the *IRAS* flux dipole.

Other statistics that we are applying to the POTENT output in three dimensions include bulk velocities relative to the CMB in spheres of varying radii around us, velocity-velocity correlation functions, the corresponding power spectrum of the initial density fluctuations and the potential structure function. The ultimate goal of the statistical tests is a comparison with theory, which we intend to pursue using detailed *N*-body simulations of the competing scenarios, to which we will apply the POTENT analysis.

One useful statistic for comparison with theory is the cosmic Mach number, recently proposed by Ostriker and Suto (1990). It is defined in a given volume by $M \equiv V/\sigma$, where V is the vector average bulk velocity in that volume, and σ is the standard deviation from that average within the same volume. This dimensionless quantity measures the relative contributions of large-scale fluctuations and small-scale ones, so it is sensitive to the shape of the power spectrum, but it is independent of the normalization of the initial power spectrum, *i.e.* independent of linear biasing. We compute the Mach number in the potent output, with 1200 km s^{-1} smoothing. Together with Y. Hoffman and T. Kolatt, we also calculate the distribution function of this quantity in several competing theoretical models, taking into account the correlation between V and σ in each given volume. There is a very preliminary indication for agreement between the Cold Dark Matter (CDM) power spectrum and the data, but there is a need for more careful analysis before this qualitative statement is confirmed.

Very preliminary velocity-velocity correlation functions seem to reveal a large coherence length, in agreement with $\Omega = 1$ CDM but with no biasing – a model which suffers from difficulties matching other observations.

CONCLUSIONS

We have processed the local universe through a 'cosmic CAT scan', converting the essentially one-dimensional peculiar velocity information into a three-dimensional picture, using a potent feature of gravity – that it produces potential flows. Even though the data consist of velocities only, the structure is recovered dynamically assuming Newton's laws. The output of this process is therefore a useful input for direct comparison with cosmological theories of the formation and evolution of large-scale density fluctuations. The comparison with the galaxy distribution, on the other hand, can give us some hints for how galaxies form, *via* the biasing field. A crucial feature of the POTENT analysis which will allow meaningful conclusions is the 'quality control' testing and error analysis developed for it, using analytic formalisms and Monte Carlo simulations with an N-body simulation and with the data itself, which we describe in detail here and elsewhere (DBF, BDFDB).

In the two papers we have presented here, we first described our technique and then presented first results in the form of maps of the potential, velocity, and density fields. In this paper we focused on the dynamical activity in the Supergalactic plane.

A visual inspection of the maps reveals the following main features:

• The dynamics in our 6000 km s^{-1} neighborhood is dominated by an extended potential well, the GA, which induces a coherent large-scale flow across the region. With a smoothing radius of 1200 km s^{-1}, the velocities at the LG and at Virgo are 465 km s^{-1} and 475 km s^{-1} respectively, both directed almost precisely toward the GA as defined by Lynden-Bell *et al.* (B = −5.5°, L = 161°). Under such severe smoothing, the 'infall' at the LG toward Virgo is merely a geometric effect.

• The center of this potential well is at about 4200 km s^{-1} from the LG. The dense data now available in the GA direction reveal infall also from the back side of the GA (see Dressler this volume). However, because of possible Malmquist bias, and because the minimum is quite broad, we cannot determine the distance to the center of the infall to an accuracy of less than ± 800 km s^{-1}.

• The Local Supercluster, with Virgo and Ursa Major, is an elongated extension of the GA.

• The Perseus-Pisces Supercluster does not seem to live up to one's expectations, dynamically speaking, based on its impressive appearance in galaxy catalogs. Its trace shows up in our maps as a lower density hill near Pisces. However, this conclusion might certainly be premature because of the currently poor sampling in this region. This will improve soon.

• There is an elongated, low density region, lying between GA/LSC and PP and between GA/LSC and Coma, which essentially half surrounds the GA overdensity. At least parts of it coincide with known voids in the galaxy distribution.

The global qualitative impression is of much similarity between the mass density distribution and the way galaxies are distributed, despite uncertain discrepancies in some regions such as PP. This seems to indicate that the galaxy density is a monotonic function of the mass density, as predicted by several scenarios of galaxy formation (*e.g.* Kaiser 1984, Davis *et al.* 1985), or, perhaps, the galaxy distribution is not biased at all, as expected in $\Omega \sim 0.2$ open universe models (*e.g.* Peebles 1987; Blumenthal, Dekel, and Primack 1988).

Is the GA telling the whole dynamical story? It is certainly possible that when more extended data accumulates the GA will turn to be an extension of an even grander attractor ('Attratorre di tuti Attratorri' – see Dekel 1988), much like the Local Supercluster turning into an extension of the GA. First hints in this direction are provided by the detection of an X-ray supercluster at $\sim 15{,}000$ km s^{-1} behind the GA (Lahav *et al.* 1989), and a supercluster of Abell clusters there (Scaramella *et al.* 1989.)

Our results essentially confirm the basic picture of Lynden-Bell *et al.* (1988), based on modeling the raw radial velocities. By obtaining a three-dimensional, model independent view of the structure, we can refine the details of the picture and make it more visually comprehensible, on the one hand, and more appropriate for theoretical analysis using *a priori* statistics on the other. But most important, the application of POTENT gives the picture a physical basis: apart from measurement and sampling uncertainties, these *are* the dynamical fields underlying the observed radial velocities, if the structure has evolved by *gravity*!

The comparison with the *IRAS* analysis and with the distribution of optical galaxies in redshift surveys is just beginning, and the results described above are very preliminary. The comparison with theoretical scenarios is even more premature; because the output of POTENT should allow, in principle, statistical quantitative comparison, this analysis deserves extreme care. We hope for rapid progress in the near future.

The results presented here are quite satisfactory given the present data, but they are just a humble beginning of a long-term project. The sparseness of the sampling, in view of the distance measurement errors, is the current major source of error, so with additional data we hope for a gradual improvement in the quality of our results. A comparison of the resolution in the different parts of the volume between the results obtained with the fixed window smoothing and the varying window (Figure 3), shows where we currently have the best data and where we need more data. The Perseus-Pisces region is an obvious case where data is urgently needed. The near future promises new data, for spirals, S0s and ellipticals, that will cover some of the poorly-sampled regions and/or increase the density of objects over large volumes. POTENT can deal with the new, heterogeneous, extended data sets as soon as they become available; we encourage observers to continue and extend the so fruitful hunt for more distances.

ACKNOWLEDGEMENTS

Most of the research described here has been done in collaboration with Sandy Faber and Alan Dressler. We thank Dave Burstein for his help in compiling the various data sets. We thank Amos Yahil, Marc Davis, and Michael Strauss for providing the *IRAS* density map and pursuing the comparison with us, and Ofer Lahav for ongoing collaboration on the comparison with the optical data. This research was supported by the NSF and NASA and by the US-Israel Binational Science Foundation.

REFERENCES

Aaronson, M., Bothun, G. D., Cornell, M. E., Dawe, J. A., Dickens, R. J., Hall, P. J., Sheng, H. M., Huchra, J. P., Lucey, J. R., Mould, J. R., Murray, J. D., Schommer, R. A., and Wright, A. E. 1989, *Ap. J.*, **338**, 654. (A89).

Aaronson, M., Bothun, G. D., Mould, J. R., Huchra, J. P., Schommer, R. A., and Cornell, M. E. 1986, *Ap. J.*, **302**, 536. (A86).

Aaronson, M., Huchra, J. P., and Mould, J. R. 1979, *Ap. J.*, **229**, 1.

Aaronson, M., Huchra, J. P., Mould, J. R., Tully, R. B., Fisher, J. R., van Woerden, H., Goss, W. M., Chamaraux, P., Mebold, U., Siegman, B., Berriman, G., and Persson, S. E. 1982a, *Ap. J. Suppl.*, **50**, 241. (A82).

Aaronson, M., Huchra, J. P., Mould, J. R., Schechter, P. L., and Tully, R. B. 1982b, *Ap. J.*, **258**, 64.

Bertschinger, E., and Dekel, A. 1990, *Ap. J. (Letters)*, **336**, L5. (BD).

Bertschinger, E., and Dekel, A. this volume, p. 67 (Paper I).

Bertschinger, E., Dekel, A., Faber, S. M., Dressler, A., and Burstein, D. 1990, *Ap. J.*, **364**, 370 (BDFDB).

Bertschinger, E., and Juszkiewicz, R. 1988, *Ap. J. (Letters)*, **334**, L59.

Blumenthal, G. R., Dekel, A., and Primack, J. R. 1988, *Ap. J.*, **326**, 539.

Burstein, D. this volume, p. 39.

Burstein, D., Faber, S. M., and Dressler, A. 1990, *Ap. J.*, **354**, 18

Davis, M., Efstathiou, G., Frenk, C. S., and White, S. D. M. 1985, *Ap. J.*, **292**, 371.

Dekel, A. 1988, in *Large Scale Motions in the Universe: A Vatican Study Week*, ed. V. C. Rubin and G. V. Coyne, S.J. (Princeton: Princeton University), p. 559.

Dekel, A., Bertschinger, E., and Faber, S. M. 1990, *Ap. J.*, **364**, 370 (DBF).

Dekel, A. and Rees, M. J. 1987, *Nature*, **326**, 455.

Djorgovski, S., and Davis, M. 1987, *Ap. J.*, **313**, 59.

Djorgovski, S., de Carvalho, R., and Han, M. S. 1989, in *The Extra-Galactic Distance Scale*, ed. S. Van den Bergh and C. J. Pritchet, *A.S.P. Conf. Ser.* **4**, (Provo: Astronomical Society of the Pacific), p. 329.

Dressler, A. 1988, *Ap. J.*, **329**, 519.

Dressler, A. this volume, p. 19.

Dressler, A., and Faber, S. M. 1990, *Ap. J.*, **354**, 13 (DF).

Dressler, A., Lynden-Bell, D., Burstein, D., Davies, R. L., Faber, S. M., Terlevich, R. J., and Wegner, G. 1987, *Ap. J.*, **313**, 42.

Faber, S. M., and Burstein, D. 1988, in *Large Scale Motions in the Universe: A Vatican Study Week*, ed. V. C. Rubin and G. V. Coyne, S. J., (Princeton: Princeton University), p. 115.

Faber, S. M., Wegner, G., Burstein, D., Davies, R. L., Dressler, A., Lynden-Bell, D., and Terlevich, R. J. 1989, *Ap. J. Suppl.*, **69**, 763.

Górski, K., Davis, M., Strauss, M. A., White, S. D. M., and Yahil, A. 1989, *Ap. J.*, **344**, 1.

Górski, K., Hoffman, Y., and Bertschinger, E. 1989, preprint.

Groth, E. J., Juszkiewicz, R., and Ostriker, J. P. 1989, *Ap. J.*, **346**, 558.

Kaiser, N. 1984, *Ap. J. (Letters)*, **284**, L9.

Lahav, O., Edge, A. C., Fabian, A. C., and Putney, A. 1989, *M.N.R.A.S.*, **238**, 881.

Lucey, J. R., and Carter, D. 1988, *M.N.R.A.S.*, **235**, 1177.

Lynden-Bell, D. this volume, p. 1.

Lynden-Bell, D., Faber, S. M., Burstein, D., Davies, R. L., Dressler, A., Terlevich, R. J., and Wegner, G. 1988, *Ap. J.*, **326**, 19 (7S).

Lynden-Bell, D., Lahav, O., and Burstein, D. 1989, *M.N.R.A.S.*, **241**, 325
Ostriker, J. P., and Suto, S. 1990, *Ap. J.*, **348**, 378
Peebles, P. J. E. 1987, *Nature*, **327**, 210.
Scaramella, R., Baiesi-Pillastrini, G., Chincarini, G., Vettolani, G., and Zamorani, G. 1989, *Nature*, **338**, 562.
Silk, J. 1989, *Ap. J. (Letters)*, **345**, L1.
Strauss, M., and Davis, M. this volume, p. 53.
Tully, R. B., and Fisher, J. R. 1977, *Astr. Ap.*, **54**, 661.
Tully, R. B., and Fisher, J. R. 1987, *Nearby Galaxies Atlas*, (Cambridge: Cambridge University).
Yahil, A. 1988, in *Large-Scale Motions in the Universe: A Vatican Study Week*, ed. V. C. Rubin and G. V. Coyne, S. J., (Princeton: Princeton University), p. 219.

DISCUSSION

M. Strauss: You see infall into Virgo as part of a cylindrical flow along the Virgo-Ursa Major 'Sausage'. Is this in agreement with Dave Burstein's picture of Virgo infall *not* being dominated by a single gravitating point?

A. Dekel: The cylindrical flow in the Local Supercluster region is a result of the Virgo-Ursa Major 'Sausage', plus the Great Attractor, plus the local void, plus everything else around us. We are testing now how good the simple spherical Virgo infall model is in determining Ω.

D. Burstein: The data Dekel and collaborators are using are the same peculiar-velocity data I use to make my graphs. At some point, what we conclude from looking at graphs must agree with how we model motions. I will be interested in what the POTENT method can conclude about infall within the Local Supercluster.

S. White: Could you clarify whether the 'Cosmic Mach Number', V/σ, depends on the amplitude of the fluctuation spectrum and so on the strength of biasing?

A. Dekel: Thank you for bringing this up. V/σ is independent of biasing! So, smoothed over 1200 km s^{-1}, the observed Cosmic Mach Number is consistent with CDM! They still might disagree when smaller smoothing is applied, as argued by Suto and Ostriker.

N. Kaiser: I would have thought that your smoothing would have tended to cause you to overestimate the Cosmic Mach Number. If so, would not a reduction in smoothing radius improve the agreement between theory and observation?

A. Dekel: I don't think so. The same smoothing should be taken into account in the theoretical calculation, so both, the experimental and theoretical Mach numbers, would be overestimated with more smoothing.

P. Shapiro: I would like to address the implications of your result that a CDM model with large bias factor ($b > 1$) is inconsistent with the large-scale velocity field data but CDM with small bias ($b \sim 1$) is okay. One of the original reasons for making the so-called 'standard' CDM model be one with large bias ($b = 2.5$) was the requirement that the *small-scale* peculiar-velocity data be well matched. If the amplitude of the initial density fluctuation spectrum is set high enough to correspond to a value of the bias factor as small as unity, I recall the CDM model *cannot* satisfy the *small-scale* velocity-field data. How is it possible to reconcile *all* of the velocity data, therefore, with a CDM model with b = 1? Can we rule the CDM model out on this basis?

S. White: The small-scale velocity-field statistics are dominated by contributions from nonlinear clumps. As a result the theoretical estimates of the CDM preferred values are much less certain than estimates for large-scale statistics. (Observational uncertainties are, of course, much greater in the latter). However, if you take standard interpretations of both large-scale (*i.e.* coherent flows of amplitude $\sim$500 km s^{-1}) and small-scale (*i.e.* typical *rms* galaxy velocities of perhaps 300 km s^{-1}) statistics, they appear quite incompatible with CDM for any normalization.

LARGE-SCALE PECULIAR VELOCITIES AND GRAVITATIONAL CLUSTERING

A. KASHLINSKY
Astronomy Unit, School of Mathematical Sciences, Queen Mary College
University of London, Mile End Road, London E1 4NS, England

ABSTRACT If the large-scale structure in the universe formed by gravitational clustering of $n < 0$ density fluctuations, the correlation function of massive objects/clusters of the clustering hierarchy would be increased over the underlying mass auto-correlation function. This means that if light traces mass, the peculiar gravitational field on the average would be stronger when measured by galaxies associated with such clusters. This in turn would lead to larger *rms* peculiar velocities deduced from the motions of such samples. Ellipticals are known to populate preferentially dense regions (rich clusters), and it is therefore suggested that this mechanism is responsible for the large peculiar motions observed recently by the 'Seven Samurai' in a sample of elliptical galaxies.

INTRODUCTION

Large-scale dynamics of systems of galaxies can provide us with direct information about the distribution of peculiar gravity on various scales and, hence, impose significant constraints on galaxy formation theories. Therefore, it is very important to map the distribution of galaxies in peculiar velocity space, particularly since it directly reflects the *true* mass distribution independently of whether 'light traces mass' or not. Unfortunately, at present only two peculiar velocities are thought to be known. One comes from the dipole anisotropy of the microwave background, which is interpreted as motion of the Local Group (on a scale of $\sim 5\ h^{-1}$ Mpc) at 630 km s^{-1} with respect to the comoving frame. The other is the Seven Samurai (7S) measurement of peculiar motions of ~ 600 km s^{-1} on a scale $\sim 50\ h^{-1}$ Mpc using a sample of ~ 400 elliptical galaxies and a modified Faber-Jackson distance indicator (Dressler *et al.* 1987, Lynden Bell *et al.* 1988).

These two peculiar velocity measurements can be used to constrain the spectrum of the primordial density fluctuations (PDF) and, consequently, to narrow the range of possible scenarios for galaxy formation (Vittorio *et al.* 1986). Indeed, the continuity equation for irrotational flow gives the Fourier transform of the peculiar-velocity field as

$$\mathbf{V}_k = i\, H_0\, f(\Omega)\, \mathbf{k}\, \delta_k / k^2, \quad \text{where } f(\Omega) = \partial(\ln \delta_k) / \partial(H_0 t) \simeq \Omega^{0.6}.$$

The *rms* peculiar velocity on scale *r* can now be calculated by choosing a suitable window (filtering) function to filter out the small-scale contribution, *i.e.*

$$\langle V^2 \rangle = \int d^3k\, |V_k|^2\, W_v(kr) / (2\pi)^3$$

Figure 1 shows theoretical predictions for $\langle V^2 \rangle^{1/2}$ for hierarchical models with $P(k) \propto |k|^n$

for n = -1, 0, 1, and for Ω = 0.1 (solid lines) and Ω = 1 (dashed lines). For simplicity the window function used here corresponds to a step function which is unity on scales $\leq 5\ h^{-1}$ Mpc and zero otherwise. Note that more sophisticated filtering would not change this curve noticeably (*cf.* Kaiser 1983, Vittorio *et al.* 1986). The (two) data points for 5 and 50 h^{-1} Mpc are shown with their error bars, and it is easy to see that none of the spectra can reproduce the 7S velocity on 50h^{-1} Mpc.

We will argue (see also Kashlinsky 1989) that the large peculiar velocity measured by 7S is a result of the 'bias' in the velocity field naturally produced by gravitational clustering of $n < 0$ PDF, but before we go into detailed theory in the next section, let us note two trends in Figure 1: i) as n increases, the *rms* peculiar velocity decreases; and ii) as Ω increases, $\langle V^2 \rangle$ increases for $n < 0$; does not change for $n = 0$; and decreases if $n > 0$, such as the spectrum of the Cold Dark Matter (CDM) model.

It is easy to understand these trends through the primeval mass auto-correlation function $\xi_N = \int d^3k\ P(k) \exp(-i\mathbf{k}\cdot\mathbf{r})\ /\ (2\pi)^3$, which measures the excess probability of finding another turned-around object at a distance r from the given one. Indeed, if $n = 0$, $\xi_N = 0$ and, hence, the *average* peculiar mass in the vicinity of a given turned-around structure is zero. In this case peculiar gravity in the vicinity of a given turned-around mass is dominated by the central mass, leading to a Keplerian falloff in $\langle V^2 \rangle$ at large r. If n is negative, ξ_N is positive, and the peculiar mass/gravity on a given scale is greater than in the white-noise ($n = 0$) case, leading to a slower decrease of $\langle V^2 \rangle$ with r. On the other hand, if $n > 0$, such as the CDM spectrum on large scales, the primeval mass/gravity on any scale is smaller than in the white-noise case and the *rms* peculiar velocity decreases faster with r. Note, however, that because $\xi_N \propto r^{-(n+3)}$ decreases very steeply with r for positive n PDF, the decrease in $\langle V^2 \rangle$ is not much more dramatic than for the white-noise spectrum. The trends with Ω are now also apparent: the larger the Ω, the greater is the contribution of the peculiar gravity and, hence, the above tendencies get amplified.

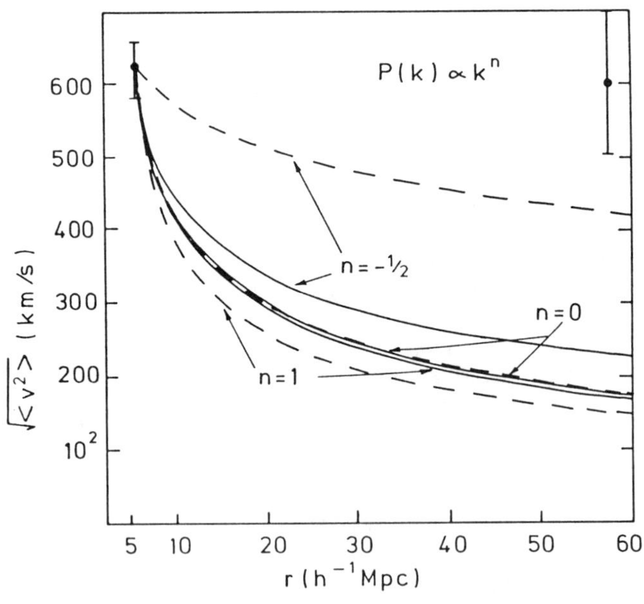

Fig. 1. *Rms* peculiar velocities for various n.

Suppose now that 'light traces mass' *everywhere* on large scales, but that we select the galaxies from structures/clusters that for some reason have an amplified correlation function. Then from the previous discussion it is clear that if 'light traces mass,' such a sample would *on the average* map regions in the universe with large(r) peculiar gravity and, hence, galaxies in such a sample should exhibit larger-than-average peculiar velocities. On the other hand, from the galaxy morphology-density relationship one knows that ellipticals populate mainly dense systems, *i.e.* (rich) clusters of galaxies (Dressler 1980). Hence it is reasonable to suppose that by sampling large-scale motions of elliptical galaxies one actually measures dynamics of clusters of galaxies, which these ellipticals are known to populate preferentially. Hence one deals here with relative motions of more massive and rare members of the clustering hierarchy. On the other hand, if structures in the universe formed by gravitational clustering of $n < 0$ PDF, so that the primeval mass auto-correlation function is positive on large scales, the more massive (rare) members would have an amplified correlation function (Kashlinsky 1987). In other words, in the gravitational clustering picture, massive members of the hierarchy cluster together, as is implied by the larger correlation length of clusters of galaxies (Bahcall and Soneira 1983, Shectman 1985, Postman *et al.* 1986, Sutherland 1988). Therefore, if the primeval mass auto-correlation function is positive, as inferred from the correlation function of clusters of galaxies, more massive members of the hierarchy would *on the average* be located in the regions of higher density, and their large-scale dynamics would exhibit larger *rms* peculiar velocities than those of the 'typical' most-abundant hierarchy members such as groups of galaxies.

GRAVITATIONAL CLUSTERING AND LARGE-SCALE VELOCITIES

Consider the evolution of PDF with a power spectrum $P(k) = |\delta_k|^2 \propto k^n$ at small k (large scales) where density fluctuations are still linear and where the (primeval) mass distribution is still unaffected by gravity. The underlying mass auto-correlation function for such PDF is given by the Fourier transform of $P(k)$:

$$\xi_N(r) = \int d^3k\, P(k) \exp(-i\mathbf{k}\cdot\mathbf{r}) \,/\, (2\pi)^3 = -\sin(\pi n/2)\,(r \,/\, \lambda_s)^{-(n+3)}, \qquad (1)$$

where $\lambda_s \simeq 5\ h^{-1}$ Mpc as normalized by galaxy clustering (Peebles 1980). For $n = 0$ the distribution is a purely Poissonian white noise and $\xi_N = 0$; for $n < 0$ the correlation function is positive, and a sphere of a given size would on the *average* contain more mass than in the white-noise case, whereas if n is positive $\xi_N < 0$, and a sphere of any size would contain less gravity than in the $n = 0$ case. It is further assumed that PDF are gaussian, or that the probability to measure $\delta \equiv \delta\rho \,/\, \langle\rho\rangle$ on a given mass scale M is

$$p(\delta)\, d\delta \propto \exp[-\delta^2 \,/\, 2\delta_*^2(M)]\, d\delta$$

with $\delta_*^2(M) = \langle\delta^2\rangle = m^{-2\alpha}$ for PDF whose power spectrum $\propto k^n$. Here $\alpha = 1/2 + n/6$ and $m = M/M_0$; M_0 being the mass scale for which $\langle\delta^2\rangle = 1$, or roughly the mass within λ_s, $\simeq 3\times10^{13}\, M_\odot$. Any object with δ greater than some critical δ_c, $[(3\pi/4)^2$ in the Einstein-de Sitter universe] would have turned around by now and formed a cluster. Hence the probability that an object of mass M has turned around at any time in the past is

$$P_M = \int p(\delta)\, d\delta \text{ (integrated from } \delta_c \text{ to } \infty)$$

and the probability that a cluster of mass in the range $(M, M+dM)$ has formed now, *i.e.* the fraction of independent objects in $(M, M+dM)$, is $n(M)\,dM = (\partial P_M/\partial M)\,dM$ (Press and Schechter 1974).

Most clusters in the hierarchy thus produced would have mass $\simeq M_0$, while the fraction of clusters that turned around on scales $>> M_0$ would be small, and their correlation function would be amplified if $n < 0$ (Kaiser 1984, Kashlinsky 1987). Indeed, the joint probability density for finding simultaneously δ_1 on scale M_1 and δ_2 on scale M_2 is

$$p_{12}\,[\delta_1, \delta_*(M_1);\ \delta_2, \delta_*(M_2)] = \tag{2}$$

$$\{(m_1 m_2)^\alpha / [2\pi(1-\xi_N^2)^{1/2}]\} \exp\{-[m_1^{2\alpha}\,\delta_1^2 + m_2^{2\alpha}\,\delta_2^2 - 2\,\xi_N\,(m_1\, m_2)^\alpha\, \delta_1\, \delta_2]/2(1-\xi_N^2)\}.$$

Consequently, the probability of finding two objects/clusters of different masses M_1 and M_2 separated by r that have turned around at any time between now and $t = 0$ is

$$P_{M_1 M_2} = \int p_{12}\, d\delta_1\, d\delta_2 \qquad \text{(integrated from } \delta_c \text{ to } \infty),$$

and the joint probability to find two clusters, one in the mass range (M_1, M_1+dM_1) and the other in (M_2, M_2+dM_2), that have turned around now is

$$n(M_1, M_2)\, dM_1\, dM_2 = (\partial^2 P_{M_1 M_2} / \partial M_1\, \partial M_2)\, dM_1\, dM_2.$$

Hence, the correlation function between clusters of different masses in the clustering hierarchy is (Kashlinsky 1987)

$$\xi_{M_1 M_2} = [n(M_1, M_2) / n(M_1)\, n(M_2)] - 1$$

$$= \{\exp\,[\xi_N\,(m_1 m_2)^\alpha / (1 - \xi_N^2)] / (1 - \xi_N^2)^{1/2}\} - 1. \tag{3}$$

As one can see, the correlation function between massive ($m > 1$) members of the clustering hierarchy would be amplified (reduced) if $\xi_N > 0$ ($\xi_N < 0$). The data on the cluster-cluster correlation function indicates that $\xi_{M_1 M_2}$ is positive on scales up to 60 - 100 h^{-1} Mpc, and thus requires the power spectrum of PDF to have $n < 0$. In this case, as equation (3) implies, massive members of the clustering hierarchy 'cluster' together, leading on the average to a larger gravitational field/mass when sampled by them. Therefore, one would expect to find larger average peculiar velocities when measured from a sample of (more) massive members of the hierarchy. For $n > 0$, the situation would be reverse, while for $n = 0$ there would be no difference between peculiar motions of clusters of clusters of different masses.

If one denotes the large-scale density field associated with mass M (*i.e.* it is equal to $\delta(\mathbf{r})$ where there is a mass concentration M and zero otherwise) as

$$\delta_M(\mathbf{x}) = \int d^3k\, \delta_{\mathbf{k},M} \exp[-i\mathbf{k}\cdot\mathbf{x}]\, /(2\pi)^3.$$

equation (3) can be rewritten as

$$\xi_{M_1 M_2}(r) = \langle \delta_{M_1}(\mathbf{x})\, \delta_{M_2}(\mathbf{x} + \mathbf{r}) \rangle = \int d^3k\, \delta_{\mathbf{k},M_1}\, \delta^*_{\mathbf{k},M_2} \exp[-i\mathbf{k}\cdot\mathbf{r}]\, /(2\pi)^3. \tag{4}$$

In the linear limit ($\delta << 1$) the continuity equation gives the Fourier transform of peculiar velocity of mass M as

$$\mathbf{V}_{k,M} = i\, H_0\, f(\Omega)\, \mathbf{k}\, \delta_{k,M} \,/\, k^2, \quad \text{where } f(\Omega) = H_0^{-1}\, \dot{\delta}_k \,/\, \delta_k \simeq \Omega^{0.6}.$$

To quantify the effect of the increased $\xi_{M_1M_2}$ on the velocity field, we consider the velocity correlation function (Peebles 1987) $v(r) = \langle \mathbf{V}(\mathbf{x}) \cdot \mathbf{V}(\mathbf{x}+\mathbf{r}) \rangle$. If we choose to study peculiar motions of masses M_1 and M_2 only, the velocity correlation function for this velocity field is

$$v_{M_1M_2} = \langle \mathbf{V}_{M_1}(\mathbf{x}) \cdot \mathbf{V}_{M_2}(\mathbf{x}+\mathbf{r}) \rangle = \int d^3k\, \mathbf{V}_{k,M_1} \cdot \mathbf{V}^*_{k,M_2} \exp[-i\mathbf{k}\cdot\mathbf{r}] \,/\, (2\pi)^3$$

$$= H_0^2 f^2(\Omega) \int d^3k\, \delta_{k,M_1}\, \delta^*_{k,M_2} \exp[-i\mathbf{k}\cdot\mathbf{r}] \,/\, k^2\, (2\pi)^3. \qquad (5)$$

Since $\nabla^2 \exp[-i\mathbf{k}\cdot\mathbf{r}] = -k^2 \exp[-i\mathbf{k}\cdot\mathrm{r}]$, equation (5) reduces to

$$\nabla^2 v_{M_1M_2} = -H_0^2 f^2(\Omega)\, \xi_{M_1M_2}(r). \qquad (6)$$

It is obvious from equations (6) and (3) that if for some reason we were to select subsamples of massive ($m > 1$) objects of the clustering hierarchy and measure their *rms* peculiar velocity field, we would measure a larger coherence length for this field if $n < 0$ ($\xi_N > 0$); no difference from that of $m = 1$ objects if $n = 0$ ($\xi_N = 0$); and a smaller coherence length if $n > 0$ ($\xi_N < 0$). The intuitive reasons for this have been explained above.

DISCUSSION

Once the (peculiar) velocity correlation function is known, one can determine the *rms* peculiar velocity field for a given spectrum of PDF. The *rms difference* between peculiar velocities of two adjacent points separated by r is $2^{1/2}$ [v(0) – v(r)], so that an increase in the coherence length of $v(r)$ leads to larger peculiar velocities. Indeed, the *rms* peculiar velocity on scale r is $2^{1/2}\, v(r)$, and $2v(0)$ is effectively the squared central velocity dispersion of the turned-around systems. We are interested in the large-scale ($>>\lambda_s$) peculiar velocity field where equation (1) is applicable; on scales $\leq \lambda_s$ PDF have been affected by gravity and (1) is no longer relevant for determining the *rms* peculiar velocities. In order to determine the asymptotic behavior of $v_{M_1M_2}$ on large scales, we have to solve (6) coupled with (3), and with peculiar velocity normalized to some (known) velocity v_0 at some distance $r_0 << r$: for example, one can choose v_0 to be 600 km s^{-1} at $r_0 \simeq 5\, h^{-1}$ Mpc to be consistent with the dipole anisotropy of the microwave background. (As we will see, the amplitude of $\langle V^2 \rangle$ on scales $>> r_0$ is rather insensitive to the precise value of v_0 for $n < 0$, the case of most relevance to the model considered here.)

It is easy to see now that for $n < 0$ the *rms* peculiar velocity at sufficiently large r, where $\xi_{M_1M_2} \simeq (m_1m_2)^\alpha\, \xi_N < 1$, tends asymptotically, almost independently of v_0, to

$$\langle V^2_{M_1M_2} \rangle^{1/2} \simeq 700\, (\delta_c A_n^{1/2}\, \Omega^{0.6})\, (m_1m_2)^{\alpha/2}\, (\lambda_s/5h^{-1}\,\mathrm{Mpc})(r/5h^{-1}\,\mathrm{Mpc})^{-1/2(1+n)}\ \mathrm{km\ s^{-1}} \qquad (7)$$

Here $A_n = \sin(\pi n/2) \,/\, (n^2+n)$ is positive for $-1 < n < 2$, and in evaluating (7) we have used the same 'step' window function as in the Introduction. A more sophisticated and realistic filtering may change the quantitative result, but it is clear that the qualitative trend would remain: namely, that for $n < 0$ ($\xi_N > 0$), even for a low Ω universe, one can measure high large-scale peculiar velocities by preferentially sampling motions of galaxies located in massive and, hence, relatively rare members of the clustering hierarchy. For example, for $n = -1/2$ and $\Omega = 0.3$, the value implied by observations (Peebles 1986), equation (8a) gives

$\langle V^2 \rangle^{1/2} = 600$ km s^{-1} on a scale of 50 h^{-1} Mpc if one were to sample peculiar motions of galaxies selected from clusters only $\simeq 8\delta_c^{-12/5}$ times more massive than the typical hierarchy members, such as groups of galaxies ($\simeq 10^{13} M_\odot$). Decreasing n further would decrease α, the exponent of m, but would also decrease the slope of $\langle V^2 \rangle$ with r. At any rate, what is clear is that in the gravitational clustering theory with a negative n power spectrum, one can always 'choose' large enough m to be consistent with the large-scale velocities reported by the 7S without violating other constraints on Ω and $P(k)$.

The interpretation of this result is fairly straightforward as simply a generalization of the multiplicity function theory of Press and Schechter (1974) to regions of a given overdensity $\delta << \delta_c$. Just as for turned-around objects ($\delta \geq \delta_c$), there would be a certain distribution of size/peculiar-mass/peculiar-gravity at a given linear density contrast δ, and what in effect has been shown here is that for $n < 0$ PDF, the regions of unusually high peculiar mass/velocity dispersion would on the average be composed of more massive hierarchy members, such as clusters of galaxies. Finally, let us note that the theory of the origin of peculiar velocities developed here is strongly supported by the recent findings of Scaramella *et al.* (1989, this volume), who find unusually rich concentration of (rich) clusters of galaxies in the direction of the Great Attractor and suggest that it may be related to the latter. As we have shown, rich/massive clusters should, indeed, 'cluster together' and hence mark regions of unusually large peculiar gravity if the structure in the universe formed by gravitational clustering of $n < 0$ PDF.

REFERENCES

Bahcall, N. and Soneira, R. M. 1983, *Ap. J.* **270**, 20.

Dressler, A. 1980, *Ap. J.* **236**, 351.

Dressler, A., Faber, S. M., Burstein, D., Davies, R. L., Lynden-Bell, D., Terlevich, R. J., and Wegner, G. 1987, *Ap. J. (Letters)*, **313**, L37.

Kaiser, N. 1983, *Ap. J. (Letters)*, **273**, L17.

Kaiser, N. 1984, *Ap. J. (Letters)*, **284**, L9.

Kashlinsky, A. 1987, *Ap. J.*, **317**, 19.

Kashlinsky, A. 1989, *Ap. J. (Letters)*, **343**, L5.

Lynden-Bell, D., Faber, S. M., Burstein, D., Davies, R. L., Dressler, A., Terlevich, R. J., and Wegner G. 1988, *Ap. J.*, **326**, 19.

Peebles, P.J.E. 1986, *Nature*, **321**, 27.

Peebles, P.J.E. 1987, *Nature*, **327**, 210.

Postman, M., Geller, M. J., and Huchra, J. P. 1986, *A.J.*, **91**, 1267.

Press, W. H. and Schechter, P. L. 1974, *Ap. J.*, **187**, 425.

Scaramella, R., Baiesi-Pillastrini, G., Chincarini, G., Vettolani, G., and Zamorani, G. 1989, *Nature*, **338**, 562.

Scaramella, R., Baiesi-Pillastrini, G., Chincarini, G., Vettolani, G., and Zamorani, G. this volume, p. 13.

Shectman, S. A. 1985, *Ap. J. Suppl.*, **57**, 77.

Sutherland, W. 1988, *M.N.R.A.S.*, **234**, 159.

Vittorio, N., Juszkiewicz, R., and Davis, M. 1986, *Nature*, **323**, 132.

DISCUSSION

A. Dressler: I'd like to clear up a common misconception about the morphology-density relation. While it is true that a typical cluster has a high fraction of ellipticals, 20 to 30%, the typical elliptical does *not* live in a rich cluster (because few galaxies are found in rich clusters). The E/S0 samples of the Seven Samurai and the Dressler/Faber sample are mostly not in Abell or Zwicky clusters. Most are found in poor or loose groups or are completely isolated. It is simply not true that these samples trace the distribution of rich clusters preferentially. This 'bias' is actually quite small. For the new E/S0 data in the Great Attractor region by Dressler and Faber, only less than about 15% belong to the Centaurus cluster, the only cluster in the region populous enough to be an Abell cluster.

A. Kashlinsky: You do not need particularly rich or massive clusters to produce large enough velocity on $\sim 50\ h^{-1}$ Mpc scale. All you need is that the structures you sample are more massive than the average ($\sim 10^{13}\ M_{\odot}$) hierarchy member. For example, as I showed, for $n = -1/2$ and $\Omega \sim 0.3$ you need to sample galaxies that live preferentially in clusters/groups just ~ 8 times more massive than the average. If $n < -1/2$ this number would decrease further.

A. Dekel: We get high enough bulk motions even for *rms* fluctuations if $\Omega \sim 0.3$ and $n = -1$ (*e.g.* see Blumenthal, Dekel, and Primack 1988 or Peebles 1988). So what is new here?

A. Kashlinsky: It has indeed been obvious for a long time that for $n = -1$ peculiar velocities go logarithmically with scale, and one can find this explicitly and implicitly in textbooks by Weinberg (*Gravitation and Cosmology*) and Peebles (*Large-Scale Structure...*). The results here are, however, *qualitatively* different. Let me repeat the summary again: If the primordial density-fluctuation spectrum is such that the primeval mass auto-correlation function is positive, then gravitational clustering naturally leads to amplified correlation between more-massive-than-average members of the clustering hierarchy (*Ap. J.* 1987, **317**, 19). This in turn means that more massive clusters/groups live on the *average* in regions of higher peculiar gravity. Hence, this would lead to large coherence length in the peculiar velocity field deduced from motions of galaxies associated preferentially with such clusters/groups.

A. Dekel: My intuition tells me that a coherent motion on a scale $\sim 50\ h^{-1}$ Mpc would require a large positive ξ_{cc} on even larger scales. Is the smoothing done properly in your calculation? By the way, I don't think we have evidence for $\xi_{cc} > 0$ beyond 50 h^{-1} Mpc, or even 30 h^{-1} Mpc.

A. Kashlinsky: As you can see from the numbers and equations I presented, the coherence length of the velocity correlation function is determined by that of the mass auto-correlation function. Thus, in order to explain the *amplitude* of the Seven-Samurai velocity on $\sim 50\ h^{-1}$ Mpc scale, it is enough to have $\xi_{cc} \sim 0.1$, consistent with the data. Of course, that probably indicates that ξ_{cc} may be finite and positive on a still larger scale, but I do not think that one needs to have ξ_{cc} positive on a scale as big as 100 h^{-1} Mpc (see also the reply to Matarrese).

A. Dekel: Superclustering, *i.e.* high ξ_{cc}, can be a statistical effect (*e.g.* biasing of Gaussian peaks à la Kaiser) *not* associated with real mass-density fluctuations of high amplitude. It is not necessarily associated with large-scale motions (the continuity equation is not valid for these clusters). Aren't you assuming that clusters trace mass?

A. Kashlinsky: Of course, the stronger cluster-cluster correlation function can be explained in terms of biased galaxy formation, *i.e.* galaxies form on density peaks and light does not trace the mass (Kaiser *Ap. J. (Letters)*. 1984, **284**, L9). This explanation would *not* produce the bias in the velocity field. On the other hand, I have shown (*Ap. J.* 1987, **317**, 19) that *both* the stronger cluster-cluster correlation function and its dependence on the cluster richness can be explained in terms of gravitational clustering of $n < 0$ fluctuations. In this case light traces mass everywhere on large scales. If and only if this is the correct interpretation of the (stronger) cluster-cluster correlation function would the velocity field have a larger coherence length when mapped preferentially by motions of more-massive-than-average members of the clustering hierarchy.

S. Matarrese: Since one needs filtering over the scale M, the $n = 0$ spectrum would imply $\xi > 0$ everywhere and $\langle \delta M \rangle^{1/2} \propto M^{-1/2}$.

A. Kashlinsky: I agree. The results I presented are equivalent to the window function equal to the step function, *i.e.* it excludes scales less than $\lambda_s \sim 5\ h^{-1}$ Mpc and is one on greater scales. With more sophisticated (and realistic) filtering, the results may change quantitatively, but the qualitative trend would remain the same - namely if the primeval mass auto-correlation function is positive, *rms* motions of more massive members of the clustering hierarchy should exhibit a larger coherence length.

RECONSTRUCTING THE DENSITY FIELD FROM PECULIAR VELOCITIES: A MAXIMUM PROBABILITY TECHNIQUE

NICK KAISER AND ALBERT STEBBINS
CIAR Cosmology Program, Canadian Institute for Theoretical Astrophysics
McLennan Laboratories, University of Toronto, Toronto, Canada M5S 1A1

INTRODUCTION

If large-scale structure has grown by gravitational instability, the peculiar velocity should be related to the mass density contrast by

$$\nabla \cdot \mathbf{v} = -\Delta\rho/\rho, \tag{1}$$

and the curl of the velocity should vanish. In linearly biased models (see Bardeen *et al.* 1986), we expect $\Delta\rho/\rho = (1/b)\,(\Delta n/n)_{\text{galaxies}}$. By comparing estimated peculiar velocities of galaxies inferred from Tully-Fisher or $D_n - \sigma$ techniques with the distribution of galaxies from redshift surveys one can hope to study the dynamics of large-scale structure growth and test these theoretical ideas. If there *is* a correlation between the inferred velocity and the inferred gravity, this firstly gives us confidence that the distance estimates are reliable, and secondly gives some support to the idea that gravity is indeed the dominant force driving structure growth. The slope of the correlation then tells us the constant bias factor b. If one could show independently that the distance estimates are reliable (possible in principle by studying much more distant galaxies, though difficult in practice), one could then simply read off b(**r**) as a function of position.

CHOICE OF METHOD

One is faced at the outset with two possibilities: either (1) use velocities to predict $\Delta\rho/\rho$ and compare with $\Delta n/n$, or (2) Use $\Delta n/n$ (and some assumed bias factor b) to infer $\Delta\rho/\rho$, predict the gravity, and compare with the observed velocities. Given perfect data, both approaches would yield identical results. Realistically, we can expect differences, and both are well worth pursuing. An advantage of method (1) is that the relation between $\nabla \cdot \mathrm{v}$ and $\Delta\rho/\rho$ is a local one. If one knows the velocity field within some region, then this fully determines the density field in that region, and one need not worry about the effect of density fluctuations outside the sample since the divergence of the velocity is blind to tides. The major disadvantage of this method is that one is starting from what are typically the noisier and patchier data, and so one must worry about the propagation of errors. An advantage of method (2) is that, particularly with the advent of the *IRAS* galaxy sample, one starts with a uniform sample covering nearly the whole sky. The main disadvantages with method (2) are the influence of mass concentrations beyond the sample, and the problem of correcting for the distortion of the clustering pattern in redshift space. In comparison, since method (1) gives both the real-space density field and the velocity field, this can readily be converted to a prediction for the density in redshift space as is observed. Here we will describe an approach to method (1).

Having chosen to proceed from velocity data to predict the density field, there are still various possible approaches. Given perfect data, one could simply integrate the line of sight velocity along each line of sight to obtain (minus) the Newtonian potential $\phi(\mathbf{r})$, and then derive the density from Poisson's equation. In practice however, one is presented with noisy and patchy data, and the method must somehow accommodate this. One possibility (pioneered by Bertschinger and Dekel 1989) is to smooth the velocity with some kind of moving filter to make a continuous velocity field. If the filter radius is chosen to be large enough to straddle the holes in the data, and grows with increasing distance to keep the statistical noise under control, this seems to give promising results. Historically, the preferred alternative has been to fit some kind of parameterized model to the velocity field. In the early days, a single spherical overdensity was used to describe the Virgo supercluster, and this modelling has become more sophisticated with *e.g.* the modelling of the Seven-Samurai data with the Great Attractor. A disadvantage of such modelling is the restriction of the assumed geometry – spherical blobs in most cases.

Our approach is to fit a model with a very large number of parameters, and thereby avoid imposing any restrictive assumptions about the spatial geometry. The model we fit is

$$\Delta\rho/\rho = \sum a_k \cos(\mathbf{k}\cdot\mathbf{r}) + b_k \sin(\mathbf{k}\cdot\mathbf{r}), \tag{2}$$

and we assume that the density field is periodic inside some cubical box of side L. The line of sight velocity field for such a model is readily found to be

$$u_{\rm model}(\mathbf{r}) = \sum \mathbf{k}\cdot\mathbf{r}\,[a_k \sin(\mathbf{k}\cdot\mathbf{r}) - b_k \cos(\mathbf{k}\cdot\mathbf{r})] \,/\, (k^2 r) \tag{3}$$

and the coefficients a_k and b_k are to be adjusted to best fit the observed velocities. One might be tempted to do this by maximizing the log-likelihood

$$\log\,[P(\mathrm{data}\,|\,\{a_k,b_k\})] \propto \sum\, (u - u_{\rm model})^2 \,/\, \sigma^2, \tag{4}$$

where the sum is over the galaxies with observed velocities and the σ's are the statistical uncertainties. For m Fourier modes, this would give a $2m$ by $2m$ matrix equation for the a_k, b_k. Maximizing likelihood has the great advantage that it tells us how to assign statistical weights to the galaxies. However, a problem is that as the number of parameters becomes large, the inversion becomes badly conditioned. Clearly, if the number of parameters exceeds the number of data, the matrix is singular, but even for a smaller number of parameters this direct inversion will give bad results. The problem arising here is a general one facing model fitters: if the model fitted has any propensity to 'blow up', then in the absence of constraining data it will almost certainly do just that. In the present application we know that the data become few and noisy at large distance, so we would expect a density map derived in this way to show large spurious fluctuations at large distances.

REGULARIZED INVERSION

Some kind of regularization is called for to render this problem well conditioned. The solution we have adopted is to maximize the probability of the model coefficients subject to the data, and some assumed prior model for the probability distribution for the coefficients. In particular, we assume that the coefficients are drawn from a Gaussian distribution with some power spectrum $\langle a_k^2\rangle = \langle b_k^2\rangle = P(k) = P_0\, k^n$. Technically, this has the effect of modifying the ML matrix equation by adding a diagonal stripe $\delta_{ij}/P(k)$, and this renders the matrix non-singular for an arbitrary number of parameters.

This approach is very similar in spirit to maximum entropy reconstruction. A convenient mental model for maximum entropy reconstruction (Gull and Daniell 1978) of an image given some constraining data, is to imagine a roomful of monkeys making pixel images at random, comparing each trial image with the data and, if compatible, saving the image. After a large number of trials, the image which has been chosen most frequently is adopted as the 'most probable' image. This process is most similar to ours if we were to adopt $n = 0$ *i.e.* a 'white noise' power spectrum. In the maximum entropy solution it seems not to be practical to allow the spectral index to vary, and, perhaps in consequence, it is often argued that a white-noise prior is in some way a maximally 'non-committal' prior assumption. In our application, it becomes clear that no unambiguously optimal spectral index can exist, since it is up to the beholder to decide whether we are reconstructing the velocity field or the density field, and what is 'white' noise in density corresponds to 'pink' noise in the velocity. If the data were sufficiently good, there would be no practical distinction, since the likelihood would then be very sharply peaked around some point in model space, and different choices of power spectrum would, within reason, give essentially identical results. Were that the case, however, it is unlikely that one would resort to Bayesian techniques anyway, and it is manifestly not the case here.

One qualitative distinction between maximum probability or maximum entropy maps and maximum likelihood maps is that while the latter tend to blow up in the absence of data, the former tend to relax towards uniformity – while a uniform density field is extremely improbable given our prior probability distribution, it is nonetheless more probable than any other distribution. We feel that this is an advantage, though it is a point which must be borne in mind when interpreting the maps – if one sees that the density contrast becomes small in some region of space, it probably does not mean that the data require this but more probably reflects the paucity of data in that region. Ideally we would place some error bar on the density at each point, and this we hope to do in the future. For the time being, we have concentrated on varying the parameters P_0 and n and seeing what influence this has on the resulting maps. If a feature in the map is largely independent of such changes in our prior belief then that feature is required by the data and should be taken seriously.

RESULTS

In our reconstructions we are computationally limited to a finite number of Fourier modes, so we truncate the series of equation (2) by limiting the maximum values of $|\mathbf{k}|$ we include. For a fixed number of modes the resolution of our reconstructions is then proportional to L. If we choose L much larger than the region in which there is good data then the reconstruction will relax to uniform density at the edges of the box and the effect of the imposed periodic boundary conditions will be small. So we would like to choose L large in order to avoid spurious boundary effects. On the other hand, we would like to choose L small in order to increase the resolution of our reconstruction. In practice one must balance these two considerations.

The peculiar velocities are taken from the Mark II release of galaxy peculiar velocities compiled by D. Burstein. This includes data from Aaronson *et al.* (1982), Aaronson *et al.* (1986), Aaronson *et al.* (1989), Bothun *et al.* (1984), Dressler and Faber (1990), de Vaucouleurs and Peters (1984), Lucey and Carter (1988), Lynden-Bell *et al.* (1988), and other unpublished data. We use the Malmquist-bias corrected peculiar velocities of the Mark II release as well as the distance-estimator errors given there, and include all galaxies contained within our box.

More details of the reconstruction technique will be presented elsewhere, here we just present the results. In the figures we present the density field constructed using 1418 Fourier modes and $L = 200\ h^{-1}$ Mpc centered on the Local Group. This volume includes 1176 measured peculiar velocities, and our effective resolution is $\sim 18\ h^{-1}$ Mpc. All of the velocity reconstructions we present here are good fits to the peculiar velocity data, in some sense better fits than one would expect for the actual velocity field. We have made no corrections for nonlinearity in going from the velocity field in redshift-space to the density field in real-space. However, the range of models which fit the data probably lead to uncertainties at least as large as the corrections for nonlinearity. The amplitude of the prior distribution, P_0, is chosen such that the expected 3-D velocity dispersion in linear theory is the fixed value given in the figure captions. The power spectrum index, n, is also given there. A prior distribution with spectral index $n = 0$ prejudices the fit to generate the peculiar velocities on small scales, while $n = -1$ is agnostic as to the scale at which the velocities are generated, and $n = -2$ prefers larger scale velocities.

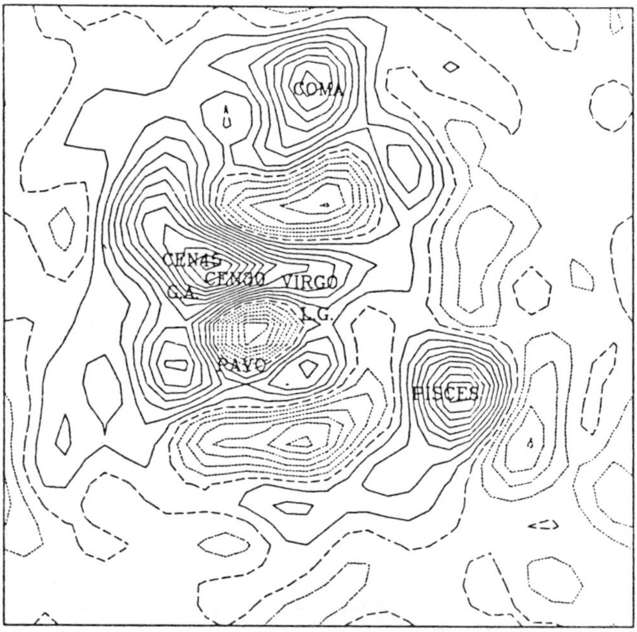

Fig. 1. A contour map of the reconstructed $\delta\rho/\rho$ in the Supergalactic plane. The local group is in the center and the box has dimensions of $200\ h^{-1}$ Mpc on a side. The dashed curve represents the 0 contour while the solid (dotted) curves show positive (negative) overdensity in increments of 0.2. Selected galaxy clusters within 15 h^{-1} Mpc of the Supergalactic plane are indicated, their names centered on their projected positions. Also shown is the projected position of the local group (L.G.) and the Great Attractor (G.A.). The map was constructed with a prior distribution with an expected velocity dispersion of 1000 km s^{-1} and a spectral index of 0. The small overdensities at the edge of the maps is indicative of poor data coverage in these regions.

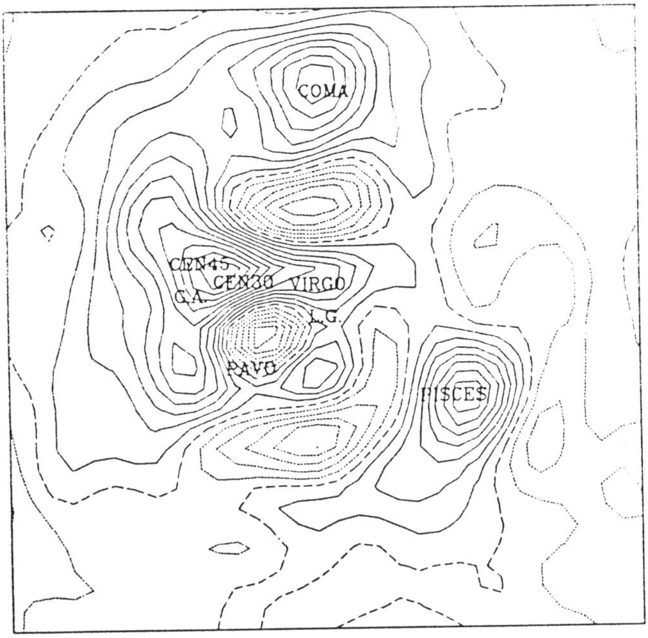

Fig. 2. Same as for Figure 1 except the spectral index is -1.

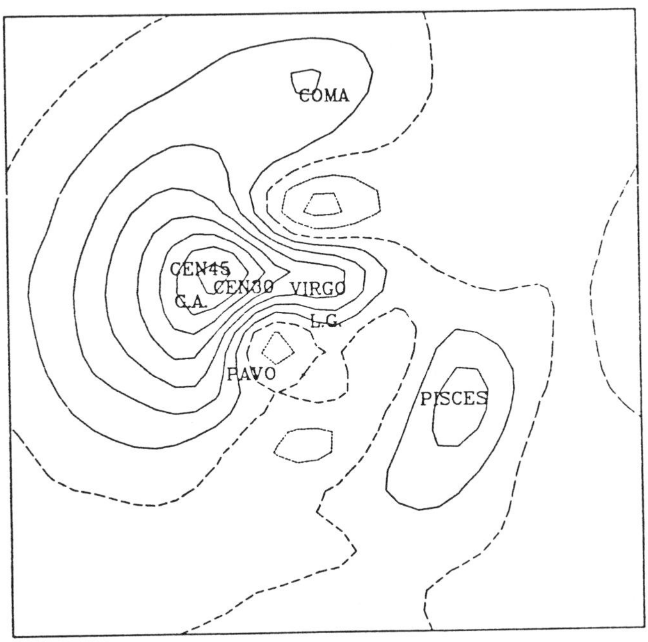

Fig. 3. Same as for figure 1 except the expected velocity dispersion is 500 km s^{-1} and the spectral index is -2.

The figures themselves are the reconstructed density field in the Supergalactic plane for various prior probability distributions. The difference between Figures 1 and 3 indicate the broad range of models which fit the data. Hence we have a ways to go before this method can truly determine the density field (except perhaps averaged over very large scales). Nevertheless there are certain robust features common to all the maps as the reader can see for herself. We have also indicated the position of some galaxy clusters which are near the Supergalactic plane. There does seem to be some correspondence between inferred overdensities and known features in the galaxy distribution. However we caution the reader that no correction has been made for inhomogeneous Malmquist bias which also might lead to this effect.

CONCLUSIONS

We have developed a new technique for reconstructing the cosmological density field from measured galactic peculiar velocities. This method of model fitting which maximizes the probability distribution is well behaved for models with a large number of parameters (i.e. 'non-parametric' fits) in contrast to the maximum likelihood method which is not. The method does not produce a unique density field from the data since one has the freedom to choose a prior probability distribution for the density field and different priors lead to different fits. This is because the present peculiar velocity data set is not of sufficient quality to determine the density field. We have calculated the maximum probability fit for a variety of prior distributions, obtaining a set of models which fit the data well but are either more or less smooth. Yet all of the fits show some correspondence between the inferred distribution of mass and the observed distribution of luminous matter.

At present we have ignored the effects of nonlinearity and of inhomogeneous Malmquist bias. We have also have not yet calculated the uncertainties in our fits. We hope to correct these short-comings in the near future.

REFERENCES

Aaronson, M., Bothun, G. D., Cornell, M. E., Dawe, J. A., Dickens, R. J., Hall, P. J., Sheng, H. M., Huchra, J. P., Lucey, J. R., Mould, J. R., Murray, J. D., Schommer, R. A., and Wright, A. E. 1989, *Ap. J.*, **338**, 654.

Aaronson, M., Bothun, G. D., Mould, J. R., Huchra, J. P., Schommer, R. A., and Cornell, M. E. 1986, *Ap. J.*, **302**, 536.

Aaronson, M., Huchra, J. P., Mould, J. R., Schechter, P. L., and Tully, R. B. 1982, *Ap. J.*, **258**, 64.

Bardeen, J. M., Bond, J. R., Kaiser, N. Szalay, A. S. 1986, *Ap. J.*, **304**, 15.

Bertschinger, E., and Dekel, A. 1989, *Ap. J. (Letters)*, **336**. L5.

Bothun, G. D., Aaronson, M., Schommer, R. A., Huchra, J. P., and Mould, J. R. 1984, *Ap. J.*, **278**, 475.

de Vaucouleurs, G., and Peters, W. L. 1984, *Ap. J.*, **287**, 1.

Dressler, A., and Faber, S. M. 1990, *Ap. J.*, **354**, 13.

Gull, S. F., and Daniell, G. J. 1978, *Nature*, **272**,686.

Lucey, J. R., and Carter, D. 1988, *M.N.R.A.S.*, **235**, 1177.

Lynden-Bell, D., Faber, S. M., Burstein, D., Davies, R. L., Dressler, A., Terlevich, R. J., and Wegner G. 1988, *Ap. J.*, **326**, 19.

DISCUSSION

A. Fairall: Your very last comment, 'CfA shells are caustics', is that just because the Coma cluster is present? What if we show you shells where there are no rich clusters?

N. Kaiser: One would expect the caustics to form in overdense regions where the Hubble expansion has been reversed. These need not necessarily contain rich clusters but should have a density contrast >5.

D. Burstein: As will hopefully be made clearer in my paper for this meeting, we *do* see infall within the local supercluster, but of the form $v(r) \propto r^{+1}$, not $v(r) \propto r^{-n}$. This infall is not directed towards either the Virgo or Ursa Major Clusters *per se*, but appears to be directed towards the middle of the supercluster, where these two clusters sit. If I look at your 'density' plot, it appears that the 'center' of what you called the Virgo Cluster is offset by approximately 700 km s^{-1} in x, relative to the real direction of Virgo.

N. Kaiser: I think the direction to the densest spot in the Local Supercluster agrees quite well with observations. At a detailed level there are certainly discrepancies, but I am more impressed by the agreement.

A. Dekel: Please help us get convinced that the b is indeed as small as 1.4. In particular, how do you reconcile this weak biasing with the arguments that led originally to high biasing, with b ~ 2.5, such as the fit to the galaxy correlation function of the *N*-body simulations by Davis *et al.*? Can we really make big empty voids with small b? How about the small-scale velocities?

N. Kaiser: Regarding the clustering of galaxies, it seems to me that one can play off a smaller bias against a larger amplitude for the primordial fluctuations. The dark cloud on the horizon is the small separation pair-wise velocities, but predicting these is fraught with problems, not least of these being our lack of understanding of how galaxies form.

A. Dekel: Your density maps are indeed very encouraging, but I recommend caution with the quantitative interpretation. Isn't it true that the contours you show actually convolve real local density with the quality of the local data? Isn't this the reason for the fact that Virgo looks like a more dramatic peak than the Great Attractor in your map? Can't this effect distort the real density contours within the local supercluster?

N. Kaiser: You are right that our method will tend to underestimate the amplitude of distant features. I think this may account for the low contrast of the Perseus-Pisces supercluster. The Great Attractor is relatively well sampled on this side at least - and seems quite robust.

A NEW METHOD FOR THREE-DIMENSIONAL HYDRODYNAMICAL *N*-BODY SIMULATIONS OF LARGE-SCALE COSMOLOGICAL FLOWS

PAUL R. SHAPIRO
Department of Astronomy, University of Texas
Austin, Texas 78712

HYSESUNG KANG
Department of Astronomy, University of Minnesota
Minneapolis, MN 55455

JENS VERNER VILLUMSEN
Department of Astronomy, Ohio State University
Columbus, OH 43210

ABSTRACT The development of a new, 3-D, cosmological, numerical hydrodynamics/*N*-body method for the simulation of large-scale flows and the formation of structure in the post-recombination Friedmann universe is described. The method is a merger of Smoothed Particle Hydrodynamics (SPH) with the Hierarchical Particle Mesh (HPM) method for solving Poisson's equation and the equation of motion. We argue that all such methods should pass the tough 1-D, pancake-collapse test proposed here since such structures are the simplest generic structures to result from the nonlinear growth of a 3-D spectrum of cosmological density fluctuations in a wide range of cases. We demonstrate that our method is the first to achieve reasonable qualitative and quantitative agreement with the known approximate analytical and 1-D numerical solutions of this problem.

INTRODUCTION

We shall describe a new three-dimensional (3-D), cosmological, hydrodynamical *N*-body method for the numerical simulation of large-scale flows and the formation of galaxies and large-scale structure in a post-recombination Friedmann-Robertson-Walker universe. There are several important reasons that such hydrodynamical simulations are necessary if we are to understand the observed large-scale spatial distribution and motions of galaxies. First, the process of galaxy formation, which couples gravitational collapse and gas dynamical effects, 'biases' the distribution and motions of the observed galaxies *relative* to the total mass distribution, including dark matter, and its motion. We must study the coupling of gravitational and hydrodynamical effects in order to learn what the 'bias' is. Second, such hydrodynamical effects result in *other* observables of large-scale structure and peculiar velocities besides the distribution and motion of galaxies. These include such things as X-ray emission from shock-heated or adiabatically compressed gas, quasar absorption lines, the Sunyaev-Zel'dovich effect on the temperature of the cosmic microwave background, and the Gunn-Peterson test for the presence of neutral H atoms in the intergalactic medium. Third, the galaxy formation process, itself, may release enough energy, as in the explosive galaxy formation model, that the peculiar velocities of galaxies may have a non-gravitational origin.

The method we have developed is a merger of the Smoothed Particle Hydrodynamics (SPH) method (*e.g.*, see Monaghan 1985) for solving the hydrodynamical conservation equations and the Hierarchical Particle-Mesh (HPM) *N*-body method (Villumsen 1989) for solving the Poisson equation and the equations of motion for the baryon-electron plasma (*i.e.* the intergalactic medium, or IGM) and a gas of collisionless dark-matter particles coupled to the IGM by gravity, all in an expanding, matter-dominated Friedmann universe. In the next section we briefly describe the numerical method. Then we discuss the importance of accurately representing the shocks and caustic surfaces associated with pancake formation as a generic nonlinear outcome of the growth of a spectrum of 3-D, cosmological density fluctuations. Finally, we describe a test of our SPH/HPM code against the pancake collapse problem.

THE SMOOTHED PARTICLE HYDRODYNAMICS/HIERARCHICAL PARTICLE-MESH *N*-BODY METHOD

The SPH method is a Lagrangian numerical hydrodynamics method which replaces the continuous baryon-electron fluid by a set of discrete gas 'particles' which carry mass and thermal energy and move with the local flow velocity. The particles are essentially moving centers of interpolation for representing the continuous flow variables. As such, they eliminate the need for a numerical grid or the interpolation back and forth between particles and grid. The principal virtue of this method is that, as a Lagrangian method, it has numerical resolution which dynamically adjusts so as to follow the compression, expansion, and distortion of the flow. In principle, this thereby provides the large dynamic range necessary to treat the density contrasts which arise, for example, during gravitational collapse, shock-heating, and radiative cooling.

The generalization of the SPH method to the case of cosmological hydrodynamics described here is one of several independent efforts along these lines (*cf.* Shapiro 1989, Evrard 1988, Hernquist and Katz 1989). One of the major distinctions amongst these different efforts is the choice of the *N*-body gravitational dynamics method which is coupled to the SPH method. We employ the HPM method, which utilizes particles and a cubic lattice grid to solve the Poisson equation by a Fast-Fourier-Transform (FFT) technique. The HPM method achieves very high resolution and dynamic range by adopting a hierarchy of cubic grids, one within the other, each with the same total number of cells and particles. The top grid is the coarsest, being of side L, grid spacing Δx, and mass per particle, m. This top grid has periodic boundary conditions. Within this top grid, a cubic sub-grid of side $L/4$ is located, with grid spacing $\Delta x/4$ and particle mass $m/64$. This sub-grid does not have periodic boundary conditions. In principle, the hierarchy can be extended to another sub-grid within the first sub-grid, this one of side $L/16$, grid spacing $\Delta x/16$, and particle mass $m/4096$, and then to yet another sub-grid, and so on.

In a discussion this brief, it will not be possible to give details concerning the SPH/HPM calculations. However, there are a few key details which we must relate in order to explain below our relative success in reproducing the solution to the pancake collapse test problem. The reader is referred to Monaghan (1985) for a more detailed description and for definitions of the SPH terminology. The resolving power of the SPH method is related to the so-called 'smoothing length', h, used for its interpolations, a kind of 'zone of influence' for each particle. Initially this smoothing length is just h_0, the mean interparticle spacing. In order for the resolution to adjust dynamically to accommodate a changing gas density, the smoothing length must vary in time and space as a function of that density. Ideally, the variations of h should be such as to maintain a constant number of nearest-neighbor particles which are only a smoothing length away or closer to a given particle. The interpolations which the SPH method performs in order to evaluate fluid quantities anywhere within the computational volume are based on the so-called kernel function W_{ij}, which is a function of this smoothing length. We have assumed a Gaussian kernel function W_{ij} which can be either *isotropic* or *anisotropic*. The *isotropic* kernel is given by

$$W_{ij}(h) = \pi^{-3/2} h^{-3} \exp(-r_{ij} / h^2), \tag{1}$$

where h is a scalar variable smoothing length given by one of two choices, according to

$$h = h_p = h_0 (\rho_p / \rho_0)^{-1/3}, \quad p = i \text{ or } j \tag{2}$$

where ρ_i and ρ_j are the density evaluated at the location $\mathbf{r}_i = (x_i, y_i, z_i)$ of particle i and the location $\mathbf{r}_j = (x_j, y_j, z_j)$ of particle j, respectively, $r_{ij} = |\mathbf{r}_i - \mathbf{r}_j|$ and ρ_0 is the initial mean density. The 'nearest-neighbor' particles for a particle i in this case are taken to be all those within a distance $r_{ij} \leq 3h$ of particle i. The *anisotropic* kernel is given by

$$W_{ij}(\mathbf{h}) = \pi^{-3/2} h_x^{-1} h_y^{-1} h_z^{-1} \exp\{-(x/h_x)^2 - (y/h_y)^2 - (z/h_z)^2\} \tag{3}$$

where $x^2 = (x_i - x_j)^2$, etc. and $\mathbf{h}$ is the anisotropic, *vector* smoothing length. For example, for a 1-D, planar flow directed along the z-direction, we can set $h_x = h_y = h_0$ constant, while allowing h_z to vary according to

$$h_z = h_{z,p} = h_0 (\rho_p / \rho_0)^{-1}, \quad p = i \text{ or } j. \tag{4}$$

Such an anisotropic kernel and smoothing length would, in this latter case, properly maintain a constant number of nearest-neighbor particles and optimize the accuracy and resolution of the SPH approach.

THE PANCAKE COLLAPSE TEST

Why the Pancake Test?

Before we can apply our new method to solve realistic large-scale cosmological flow and galaxy formation problems, we must demonstrate that it can successfully reproduce known solutions for a range of good test problems. Our SPH scheme can successfully match the analytical solution of the shock tube, or Riemann problem, for example. We focus in what follows on a much tougher test, that of the gravitational collapse of a 1-D plane-wave density fluctuation in a universe comprised of baryons and collisionless dark matter. This is the cosmological pancake problem, in which an initially linear amplitude density fluctuation grows to nonlinear amplitude, forms a caustic in the dark-matter distribution located in the plane of symmetry of the pancake and strong accretion shocks, one on each side of this central plane, followed by continued infall, phase mixing of the dark matter, and radiative cooling of the shocked baryon-electron plasma. Detailed, 1-D numerical solutions for this problem already exist, as do approximate analytical solutions, for comparison (Shapiro, Struck-Marcell, and Melott 1983; Shapiro 1984; Bond *et al.* 1984; Shapiro and Struck-Marcell 1985; Struck-Marcell and Shapiro 1989). In the absence of radiative cooling, a dynamic range in density of about 10^3 is required to resolve this flow. *With* radiative cooling, this dynamic range can increase to 10^5.

Why should we insist that our SPH/HPM code satisfy this pancake test? Such 1-D, pancake-like structures are a generic feature of the nonlinear growth of a 3-D spectrum of density fluctuations. This has been well known for some time for the case of a universe dominated by hot dark matter (HDM) (*e.g.* Arnold, Shandarin, and Zel'dovich 1982; Shapiro, Struck-Marcell, and Melott 1983). General arguments suggest that this is also expected for more general kinds of fluctuation spectra, a simple reflection of the natural tendency for initially small anisotropy in the mass distribution to amplify during gravitational collapse. Recently it has been demonstrated numerically by very high resolution, 2-D, collisionless particle, gravitational clustering simulations that, even without the imposition of a resolved small wavelength cutoff in the initial power spectrum *below* which power is absent, pancake formation is a general outcome of nonlinear density fluctuation growth (Melott and Shandarin 1989). These 2-D simulations demonstrate that pancake-like structures are expected even in situations involving hierarchical clustering, at least for the case of initially random, Gaussian fluctuations with power-law power spectra $<\delta_k^2> \propto k^n$ with spectra flatter than $n = 0$. A simple description is that, at each epoch, the matter forms pancake-like structures which fill space with a kind of cellular structure where the characteristic cell and pancake size at that epoch is of the order of the wavelength at which the amplitude of the power spectrum is just becoming nonlinear. As times goes on, in this hierarchical clustering situation, small pancakes merge into larger pancakes, and a cell structure of increasing comoving scale remains evident.

In order to illustrate in 3-D that 1-D, pancake-like structures are a generic feature of the nonlinear growth of a 3-D spectrum of density fluctuations, we have used the HPM method to perform a purely collisionless (*i.e.* no hydro included) calculation of the formation of structure in an HDM model universe. This calculation demonstrates that the large dynamic range afforded by the HPM scheme is *crucial* to solving cosmological hydrodynamics problems correctly. Figure 1 shows projected particle positions, viewed along one axis of the calculational grid, for a single time-slice of a two-level HPM calculation. In this run, each of the two grid levels has 64^3 cells and initially 32^3 particles, and each side of the top grid is 6 pancake wavelengths λ_p across. Inside the sub-grid, this simulation has the equivalent in resolution and dynamic range of a one grid calculation (*i.e.* top grid only) with 256^3 cells and 128^3 particles. The first box in Figure 1 shows the top grid and top grid particles. The second box is a 'blow-up' of a one-quarter depth projection of the sub-grid of size $(3/2)\lambda_p$ on a side, centered on a dense cluster, but which contains only the coarse resolution of the top grid particles. The third box represents the same volume as the second box, but displaying, instead, the higher resolution of the actual sub-grid calculation. The higher resolution sub-grid calculation dramatically reveals that the dense cluster found in the coarse resolution top grid is actually the intersection of multiple pancakes!

The SPH/HPM Pancake Test

We have tested our SPH/HPM code on the pancake problem by focusing on the growing mode of a 1-D, linear, sinusoidal, adiabatic, plane-wave density fluctuation, *i.e.* initial mass density $\rho = \langle\rho\rangle\{1 - \delta \cos [2\pi\, z/\lambda_p]\}^{-1}$ and velocity $v_z = -(\delta/2\pi) \sin [2\pi z/\lambda_p]$ at some initial epoch with scale factor $a_i = 1$ at which $\delta << 1$. We consider a two-level calculation, with a top grid of size $L = \lambda_p$ on a side and with a sub-grid of size $L/4$ on a side centered within the top grid, with 32^3 cells and 16^3 equal mass particles in each level initially. In addition, there are another 12288 particles of sub-grid mass which are initially outside the sub-grid but are possibly destined to infall into the sub-grid. We take 1/8 of all the top grid and sub-grid particles to be baryonic. All of the sub-grid-mass baryonic particles which either start inside the sub-grid or are destined to cross into the sub-grid eventually (*i.e.* 2048 particles) are treated as SPH particles. The mass per SPH particle is

$m_1/64$, where m_1 is the mass per particle of the top grid particles. The initial value of the SPH smoothing length in this case is $h_0 = 10^{-1.5}\,\lambda_p$. The higher resolution of the sub-grid calculation is equivalent to a one grid-level calculation with 128^3 cells and 64^3 particles total, with $32^3 = 32768$ SPH particles distributed throughout the entire top grid.

We chose the amplitude δ so that caustic and shock formation should occur at scale factor $a = a_c = \delta^{-1} = 4$ (corresponding to redshift $z = 6$), with $\lambda_p = 40$ Mpc, $\Omega_{total} = 1$, and $h = 0.75$. This run did not include radiative cooling. A selection of our results for a single time-slice after caustic and shock formation, at $a = 4.84$, are shown plotted in Figures

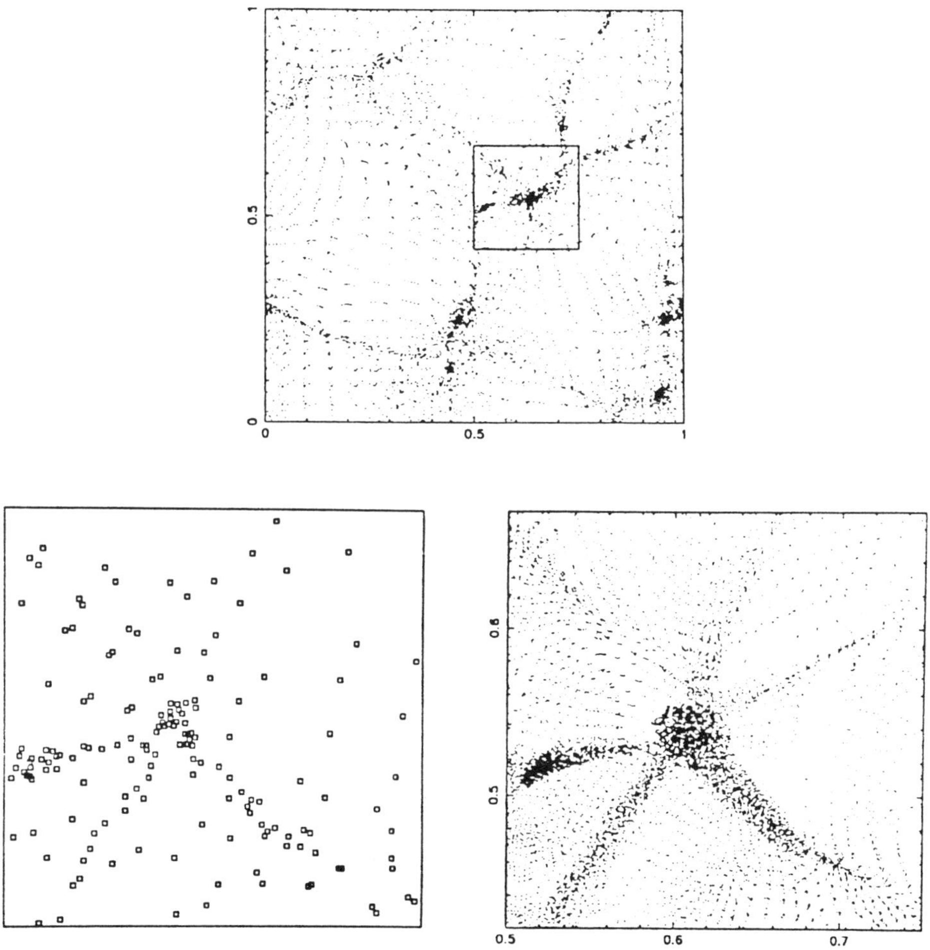

Fig. 1. A time-slice of HDM simulations. Top: particle positions are shown projected onto one face of the cubic top grid for top grid particles only. The box within the box indicates the second-level sub-grid. As for all three boxes shown, projection depth is one-quarter the width of the box. Lower left: projection of coarse top grid particles only, onto one face of the sub-grid box. Lower right: projection of sub-grid particles in same volume as shown in lower left box.

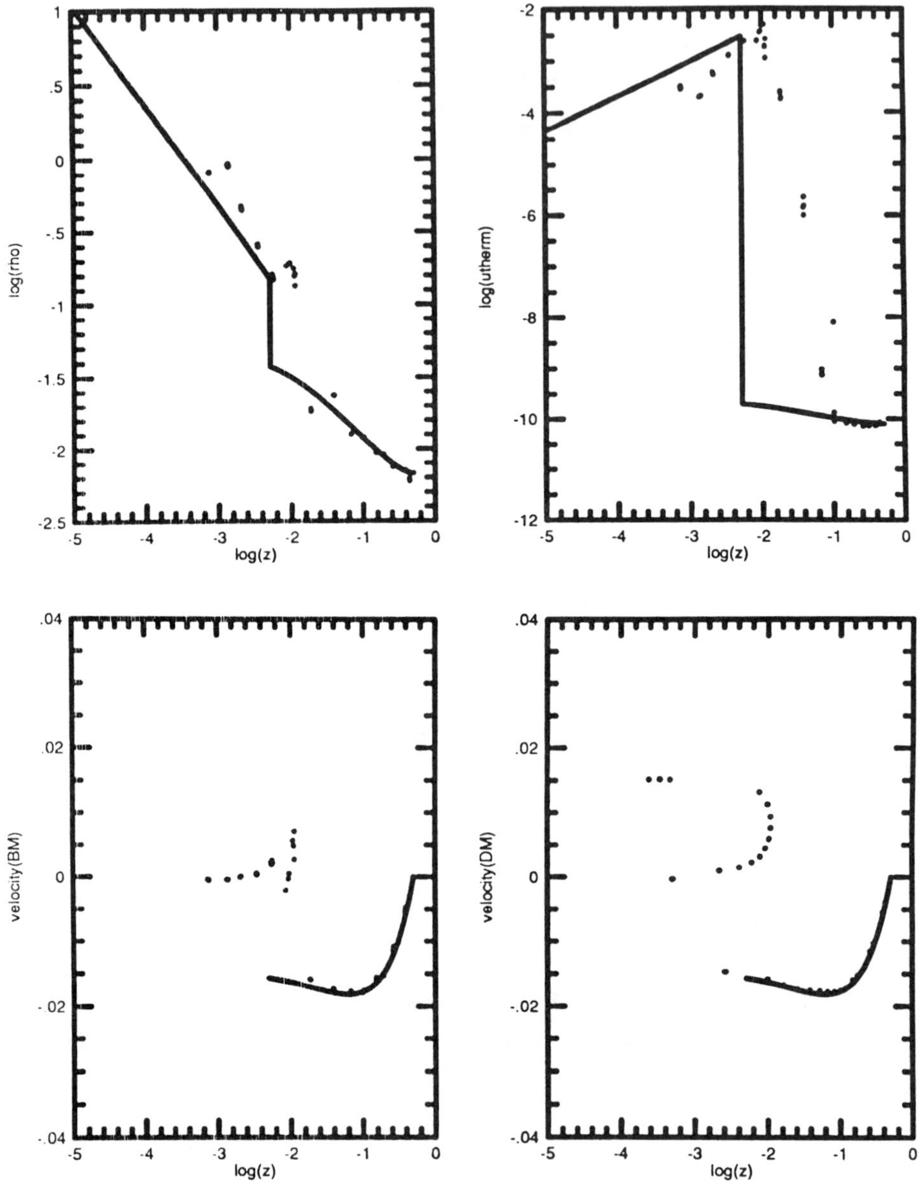

Fig. 2. SPH/HPM simulation results for 1-D, sinusoidal plane-wave pancake at $a = 4.84$. Normalized gas density (upper left), temperature (upper right), gas velocity (lower left), dark matter velocity (lower right) versus distance z (in units of $\lambda_p/2$) measured from the central plane of the pancake (x-y plane at $z = 0$). Dots are sub-grid simulation particles. Solid curves are the analytical solution.

2 and 3. The quantities plotted are those of only a subset of the particles, those close to the *z*-axis, which runs through the center of the cubic grid, perpendicular to the plane of symmetry of the pancake. The same plots show the approximate analytical solution for density, temperature, and pressure for the gas, as well as the analytical infall velocity profile for the dark matter and the gas outside of the predicted shock location (Struck-Marcell and Shapiro 1989). We have plotted these results against distance from the pancake central plane (the *x*-*y* plane which bisects the cubic grid) in logarithmic coordinates (in units of $\lambda_p/2$) to emphasize the high resolution and dynamic range of these calculations.

The numerical infall-velocity profiles calculated for the gas and dark matter outside of the shock are in good agreement with the analytical solution, as are the thermodynamic quantities well outside the shock. The dark matter, in fact, shows the expected caustic formation and subsequent 'winding up' of the spiral in phase space. The numerical location of the shock, identified with the position of the temperature peak, for example, is within 50% of the predicted shock location of the approximate analytical solution. This is quite remarkable, since this predicted location is at a distance $< 10^{-2}$ of the total grid size, several times smaller than the initial mean interparticle spacing of the SPH particles. Even more remarkable is the rough success in reproducing the amplitude and shape of the analytical post-shock profiles of density, temperature, and pressure, with immediate post-shock values which are within a factor of two of the analytical values. In order to reproduce the predicted shape of these post-shock profiles, including the expected rough isobaricity, a resolution of order 10^{-3} times the total grid size is required, which is more than 1.5 orders of magnitude smaller than the initial mean interparticle separation. We note that these numerical results

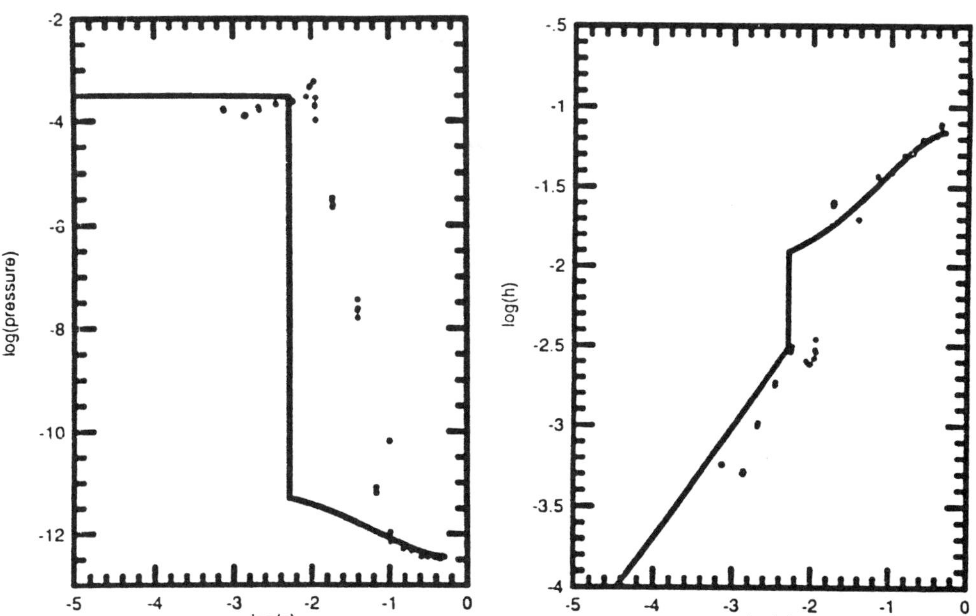

Fig. 3. Same as Figure 2, except normalized pressure (left), and smoothing length h_z (right). Dots are numerical values, solid curve for h_z is the 'ideal' h_z using analytical solution.

are much closer to the analytical solution than our earlier preliminary results reported in Shapiro (1989).

In order to achieve this level of agreement between the numerical and analytical solutions, we have discovered empirically that the following are necessary modifications of the traditional SPH approach. First, it is essential that the 1-D, planar-flow anisotropic kernel and variable smoothing length of equations (3) and (4) be used, rather than the isotropic quantities of equations (1) and (2). A measure of the resolving power of this approach is shown in Figure 3 in the plot of h_z, both for the numerical solution (the points) and for the *ideal* case (the solid curve) in which the gas density used in equation (4) is that of the analytical solution. Apparently, the numerical h_z is generally *less* than the ideal value for the post-shock flow, and the quantity h_z/z is less than one, so numerically resolving the post-shock flow is, in principle, possible. Second, it is necessary to turn artificial viscosity off in the energy equation during the pre-shock pancake collapse phase which ends the instant the shocks form at $a = a_c = 4$. Otherwise, the artificial *heating* which occurs in connection with artificial viscosity during this *pre-shock* collapse phase results in the appearance of the shock much further away from the central plane and spurious post-shock profiles of the thermodynamic variables.

We are very encouraged by these results to continue to develop the SPH/HPM method. The next step will be the automation of these 'special-handling' features, the anisotropic kernel and the removal of pre-shock artificial viscous heating. This will enable us to apply our code to a genuinely 3-D case in which the geometry of collapse and the shock formation epoch are not known *a priori*. Eventually, we will address the inclusion of radiative cooling, which drives the demand for large dynamic range significantly higher, and the extension to more than the two levels of the grid hierarchy.

ACKNOWLEDGEMENTS

We would like to thank J. J. Monaghan for graciously providing us with a copy of his SPH code at an early stage in this work. This work was supported in part by Robert A. Welch Foundation Grant F-1115, Texas Advanced Research Program Grant 4132, and an Alfred P. Sloan Foundation Fellowship in Physics for P.R.S. Some of our numerical work was performed on the Cray X/MP at the UT Center for High Performance Computation, the Cray 2 at the Minnesota Supercomputer Institute, and the Cray X/MP at Ohio State University. J.V.V. and P.R.S. are grateful for the hospitality of the Aspen Center for Physics in June 1988.

REFERENCES

Arnold, V. I., Shandarin, S. F., and Zel'dovich, Ya. B. 1982, *Geophys. Ap. Fluid Dyn.*, **20**, 111.

Bond, J. R., Centrella, J. Szalay, A., and Wilson, J. R. 1984, *M.N.R.A.S.*, **210**, 515.

Evrard, A. E. 1988, *M.N.R.A.S.*, **235**, 911.

Hernquist, L., and Katz, N. 1989, *Ap. J. Suppl.*, **70**, 419.

Melott, A. L., and Shandarin, S. F. 1989, *Ap. J.*, **343**, 26.

Monaghan, J. J. 1985, *Comp. Phys. Reports*, **3**, 71.

Shapiro, P. R. 1984, in *Clusters and Groups of Galaxies*, ed. F. Mardirossian, G. Giuricin, and M. Mezzetti (Dordrecht: Reidel), p. 447.

Shapiro, P. R. 1989, in *Fourteenth Texas Symposium on Relativistic Astrophysics*, ed. E. J.

Fenyves, *Ann. N. Y. Acad. Sci.*, in press.
Shapiro, P. R. and Struck-Marcell, C. 1985, *Ap. J. Suppl.*, **57**, 205.
Shapiro, P. R., Struck-Marcell, C., and A. L. Melott 1983, *Ap. J.*, **275**, 413.
Struck-Marcell, C. and Shapiro, P. R. 1989, in preparation.
Villumsen, J. J. 1989, *Ap. J. Suppl.*, **71**, 407.

DISCUSSION

E. Bertschinger: When I first saw your one-dimensional pancake collapse SPH simulation at a previous meeting in Toronto, I was discouraged by your failure to get the correct shock location with a nice smooth and rapid jump. However, I did not then fully appreciate how close the shock is to your boundary. It is asking too much of almost any numerical method to be able to resolve discontinuities and structures with a scale length of 10^{-3} of the total size. Placed in this context, the SPH results are more impressive to me now.

S. White: How do you envisage implementing your anisotropic smoothing kernel in situations where the orientation and nature of the local geometry of the flow are not known *a priori*?

P. Shapiro: Of course, it will be necessary to 'automate' the process of choosing the anisotropic smoothing length. We have several ideas about how to do this dynamically during the simulation, which we are currently exploring. One idea is to use the Lagrangian density changes calculated numerically after each time step to calculate the local deformation tensor for each particle. The eigenvalues and eigenvectors of this deformation tensor then indicate the amount of compression or expansion along the principal axes of the anisotropic volume change for each particle. This tells us how to orient and rescale the vector smoothing length $\mathbf{h} = (h_x, h_y, h_z)$ so as to keep the number of nearest-neighbor particles roughly constant in time for every particle.

ON DETERMINING THE D_n – logσ RELATION FOR ELLIPTICAL GALAXIES USING RED CCD IMAGES

G. WEGNER
Department of Physics and Astronomy
Dartmouth College, Hanover, NH 03755

ROGER L. DAVIES
Department of Astrophysics, University of Oxford
Keble Road, Oxford, OX1 3RH, England

MATTHEW COLLESS
Institute of Astronomy, University of Cambridge
The Observatories, Madingley Road, Cambridge CB3 0HA, England

DAVID BURSTEIN
Department of Physics and Astronomy
Arizona State University, Tempe, AZ 85287-1504

EDMUND BERTSCHINGER
Department of Physics, Massachusetts Institute of Technology
Cambridge, MA 02139

ROBERT K. MCMAHAN
Department of Physics and Astronomy, Phillips Hall CB 3255
University of North Carolina, Chapel Hill, NC 27599

ABSTRACT Details of the R-band filters being used to determine the D_n – logσ relation for elliptical galaxies and examples of the growth curves are given.

INTRODUCTION

This is a progress report on a project to employ the D_n – logσ relation in order to determine deviations from the Hubble flow in groups containing elliptical galaxies approximately to distances giving cz = 12,000 km s^{-1}. An earlier discussion giving some of the rationale for this project is given in Burstein (1989). Briefly, this is a natural extension of Lynden-Bell *et al.* (1988). However, objects located in a selection of groups and clusters have been chosen, as all-sky coverage to this distance would be prohibitive with currently available instruments. The final sample consists of 700 galaxies, and red CCD images have now been obtained for all of them.

Although the present observations are being conducted to twice the distance studied in Lynden-Bell *et al.* (1988), we expect that combining galaxies in groups and using the advantages of red-sensitive CCD images for determining the D_n – logσ relation will permit determining the deviations of these systems from the Hubble flow. Burstein *et al.* (1987)

have discussed how use of CCD images leads to an increased accuracy in the determined growth curves for the ellipticals relative to that achieved with B-band aperture photometry. This paper defines our red CCD photometric filter system that is being employed in the present study and discusses some preliminary results.

DEFINITION OF THE PHOTOMETRIC SYSTEM

The motivations for choosing a photometric system are complex, and many such systems have employed CCD's. Here it was desired to conform as well as possible with the Kron-Cousins (KC) *RI* system (Cousins 1976, Landolt 1983) for the 1.3-m McGraw-Hill telescope of the Michigan-Dartmouth-MIT Observatory employing a RCA CCD. It was further desired to have a set of glass filters that would not degrade with time during the course of a long-term observing project. After considerable trial and error, the *BVRI* filter set listed in Table I was derived. All filters have been cemented using ocalith cement of Rolyn Optics Co.

TABLE I Specification of the Broad Band Filters

filter	glass components
B	Hoya: 2.5mm L-39 + 1.5mm CM-500; Schott: 1mm BG-12
V	Hoya: 2.5mm Y-50; Schott: 2mm BG-18
R_{KC}	Hoya: 3.0mm HA-30 + 2.5mm O-56
I_{KC}	Hoya: 2.5mm RT-830

The combined responses of the MDM 1.3-m telescope, RCA CCD, and each filter are shown in Figure 1. The CCD plus telescope response was measured by Boroson (1988) and the filter transmissions were determined both at Dartmouth and NOAO using Perkin-Elmer laboratory spectrophotometers. A substantial amount of data has also been taken with the NOAO No. 1 0.9-m telescope using these same filters, but with a Tektronics CCD, producing similar spectral responses. Note that for the I_{KC} filter, the longward response is determined largely by the CCD sensitivity. The transmission for the R_{KC} filter defined above is further compared with a Mould *R* interference filter and the KC response given by Cousins (1976) in Figure 2, where it can be seen that the present *R* filter closely follows the KC filter.

RESULTS ON ELLIPTICAL GALAXIES

Figure 3 shows examples of the growth curve for circular apertures centered on the galaxy NGC 7647 and the residuals of this curve from the de Vaucouleurs $r^{1/4}$ law. The program GRASP, originally written by Cawson (1987), has been used to extract the data. The vertical lines in the right top side of Figure 3 indicate positions for the three times *FWHM* of the seeing and 10% of the sky background. In order to utilize a larger portion of the data, improved fitting of the profiles is achieved using seeing-convolved $r^{1/4}$ profiles and adjusting for errors in the sky level. Results for three independent frames of the galaxy are shown in the bottom panels of Figure 3, where the dotted lines refer to the extent of the derived seeing and the 10% sky level, while the lowest curves show residuals to the fits of each image. Preliminary results indicate that order 75% of the program galaxies yield

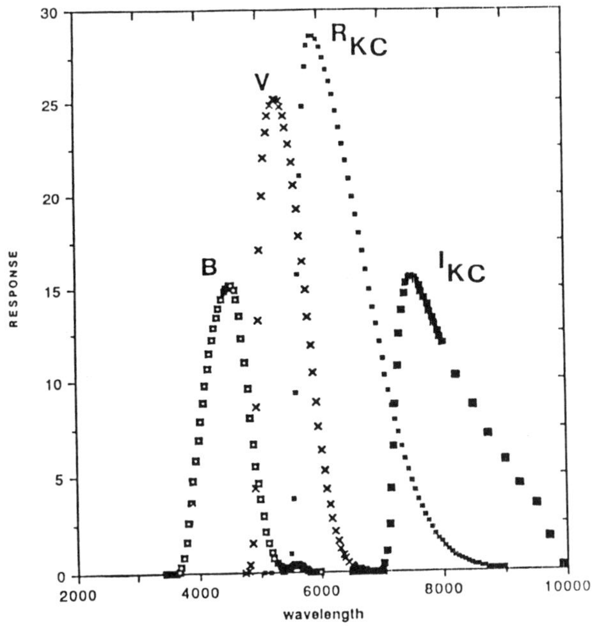

Fig. 1. Responses of the broadband *BVRI* filters described in Table I for use with galaxy photometry. Shown are the transmissions of the filters convolved with the response of the MDM 1.3 m plus RCA CCD detector system.

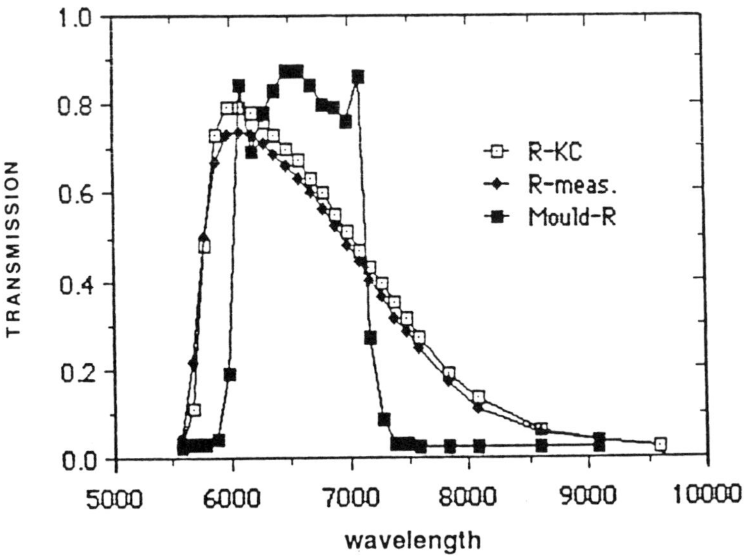

Fig. 2. Comparison of the transmissions of the *R* filter in Table I, Mould, and Kron-Cousins *R* filters.

excellent overall fits to the $r^{1/4}$ law, from which D_n and departures from this relation can be investigated. Further comparisons of this sort from MDM 1.3-m data are given in Burstein (1989).

CONCLUSIONS

The broadband filter and instrumental combinations reported here satisfactorily match the Kron-Cousins *RI* bands. The *R* filter makes use of the high sensitivity of the CCD and increased brightness of the ellipticals in this spectral region and has relatively high throughput. Thus it is capable of producing excellent profiles for elliptical galaxies that can yield log D_n and in turn be used for investigations of large-scale cosmological structure requiring distances.

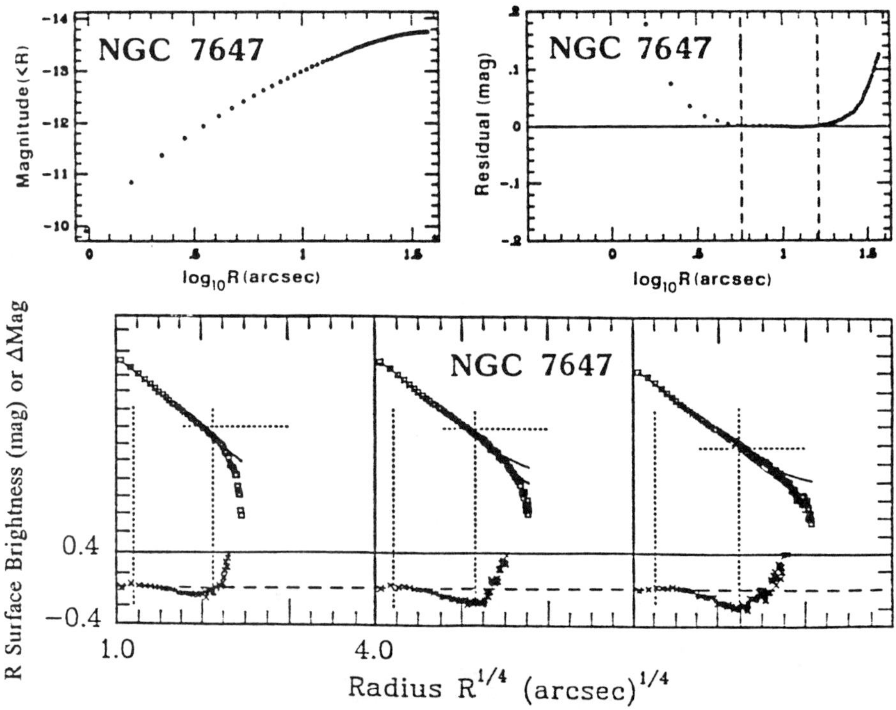

Fig. 3. (Top) A typical growth curve for the E galaxy NGC 7647 in *R* through circular apertures (left) and residuals from the $r^{1/4}$ law (right) using raw data. (Bottom) Fits of observed surface brightness to the seeing convolved $r^{1/4}$ laws and their residuals for three independent images of NGC 7647. The vertical lines are described in the text.

REFERENCES

Boroson, T. Y. 1988, private communication.

Burstein, D., Davies, R. L., Dressler, A., Faber, S. M., Stone, R. I. S., Lynden-Bell, D., Terlevich, R. J., and Wegner, G. 1987, *Ap. J. Suppl.*, **64**, 601.

Burstein, D. 1989, in *Les Mondes des Galaxies*, ed. H. C. Corwin and L. Bottinelli, in press.

Cawson, M. G. M. 1987, GRASP Documentation, (Tucson: University of Arizona).

Cousins, A. W. J. 1976, *Mem. R.A.S.*, **81**, 25.

Landolt, A. U. 1983, *A. J.*, **88**, 439.

Lynden-Bell, D., Faber, S. M., Burstein, D., Davies, R. L., Dressler, A., Terlevich, R. J., and Wegner, G. 1988, *Ap. J.*, **326**, 19.

NEARBY REDSHIFT SURVEYS

L. NICOLACI DA COSTA AND PAULO S. PELLEGRINI
Departamento de Astronomia, CNPq/Observatório Nacional
Rua General José Cristino, Rio de Janeiro (RJ), Brazil

ABSTRACT Here we present a status report on nearby redshift surveys showing that for moderate depths, namely samples including galaxies brighter than $m_{B(0)} = 14.5$, we are rapidly approaching the maximum sky coverage possible from optical samples. We indicate that future work on large-scale structures should concentrate in generating a homogeneous all-sky sample, taking advantage of the large database of redshifts presently available in both hemispheres, and in carrying out slice surveys, probing deeper in selected regions of the sky. A brief review of the slice surveys currently in progress in both hemispheres is presented. Finally, we review some of the properties of the spatial distribution of galaxies as revealed by the several wide-angle redshift surveys of the nearby universe and comment on some of the quantitative results obtained from the analyses of the CfA Redshift Survey (CfA1) and the Southern Sky Redshift Survey (SSRS) samples.

INTRODUCTION

The past decade has certainly been a very special period for cosmology, as progress has been made on several different fronts and has stimulated active development of theories of formation and evolution of structures in the universe. An important contribution to this development has come from new theoretical ideas like inflation, which may account for some fundamental cosmological facts (*e.g.* homogeneity, isotropy, and flatness of the observed universe) and cosmic strings, which may be an alternative way of producing density fluctuations. Particle physics has also made its contribution, suggesting new dark-matter candidates such as massive neutrinos, photinos and axions. Finally, several astrophysical processes have been investigated which may lead to a biased galaxy formation, with galaxies forming preferentially in high-density regions of the underlying matter distribution. All these ideas have been incorporated into the theories of the origin of large-scale structures, and their consequences are currently being examined. Equally exciting has been the progress achieved by the elaborate *N*-body simulations carried out over the past few years, which have allowed exploration of the non-linear evolution of structures starting from initial conditions representative of different cosmogonic scenarios.

From the observational side an important contribution has come from the completion of several wide-angle redshift surveys carried out during the past decade, that have helped improve our understanding of the distribution of galaxies in the nearby universe. The database from these surveys has been used extensively for statistical studies of the galaxy distribution, with the goal of unravelling the nature of the clustering properties of galaxies. The data have also been used for direct confrontation with model distributions obtained from *N*-body simulations of different cosmogonic scenarios. This program, if successful, may tell us something not only about the origin of structures in the universe but also about the nature of dark matter, which may in turn impact the development of particle-physics theories. Here we present a brief review of the present status of the ongoing surveys of the nearby universe, with particular attention to the southern hemisphere, where we have been working for the past several years.

THE NEARBY UNIVERSE

Wide-Angle Surveys

The modern era of galaxy redshift surveys began around 1981 with the completion of the original CfA Redshift Survey (Huchra *et al.* 1983), which includes 2400 galaxies distributed over an area of 2.66 steradians in the northern hemisphere. This was the first wide-angle survey to go deep enough to probe beyond the Local Supercluster. Although relatively shallow, this survey provided a first glimpse of the complexity of the galaxy distribution, showing that galaxies lie preferentially in 'filamentary' structures, which occupy a small fraction of the total volume and form interconnected structures with characteristic scales comparable to the survey depth. It also showed the existence of large underdense regions apparently devoid of bright galaxies.

Although this sample was used extensively in statistical analyses and in the comparison with the results of numerical models, it was clear that it did not represent a fair sample of the universe at large, motivating the undertaking of several other large area surveys, such as the Southern Sky Redshift Survey (SSRS, da Costa *et al.* 1988), comprising about 2000 objects, and the IRAS Redshift Survey, an all-sky survey of 2300 objects. We should also mention other extensive surveys of large but more specific areas of the sky, such as the Arecibo Survey (*e.g.* Haynes and Giovanelli 1986) conducted to study the long, filamentary Perseus-Pisces Supercluster and its environs, and the Supergalactic Plane Survey (SPS, Dressler 1988), a low galactic latitude survey carried out to search for the 'Great Attractor'. The above compilation is not intended to be complete but to illustrate some of the most important wide-angle, moderately deep surveys carried out in the past decade.

In Figure 1(a) we show the sky-projected distribution of galaxies, in Galactic coordinates, of the CfA1 (including the small patch in the southern hemisphere) and SSRS samples. Note that in the northern Galactic hemisphere the sample is limited at $b = 40°$ and that there is a large uncovered region from $\ell = 240°$ to $\ell = 360°$ due to the declination cutoff $\delta = 0°$ adopted for the CfA1. In the southern Galactic hemisphere there is an uncovered strip which corresponds to the region defined by $b < -30°$ and $-17.5° \leq \delta \leq -2.5°$.

In order to complete these regions observations have been carried out from the Smithsonian's F. L. Whipple Observatory in Arizona, the South African Astronomical Observatory near Sutherland, the European Southern Observatory at La Silla, and the Complejo Astronomico on el Leoncito in Argentina. In the northern Galactic cap, in the region $b \geq 30°$ and $\delta < 0°$, this extension comprises 834 galaxies brighter than the nominal magnitude limit of $m_{B(0)} = 14.5$, drawn from the ESO/Uppsala Catalog (Lauberts 1982) and Morphological Catalogue of Galaxies (MCG, Vorontsov-Velyaminov and Arhipova 1963-74), out of which we have measured redshifts for 668 galaxies. In the southern Galactic cap we have just completed the measurements of radial velocities for the 441 galaxies brighter than 14.5 (in the $B(0)$ system) drawn from the MCG catalog.

The samples plotted in Figure 1(a) and (b) illustrate the sky coverage currently available for magnitudes brighter than 14.5. As can be seen, the sky-coverage is approaching the limit of what can be done from optical samples (about 50% of the celestial sphere). We should emphasize, however, that the figure is somewhat misleading in the sense that there are severe problems of uniformity across the sky. Due to the lack of photometry, galaxy samples in the southern hemisphere were drawn primarily from the diameter-limited ESO/Uppsala catalog, except north of −17.5° where we have used the MCG. Attempts to define a common magnitude system have been made, but the uncertainties are still large. The next major step would be to define a common photometric system which would allow us to combine the data from the northern and southern hemispheres into a single homogeneous catalog. A preliminary attempt is currently being made to obtain CCD photometry of equatorial galaxies drawn from the MCG, with the selected sample overlapping with both the Zwicky *et al.* (1961-68) and the ESO/Uppsala catalogs. In this manner we hope to establish a common photometric system which can be used to derive statistical corrections to be applied to the magnitudes and diameters coming from the different catalogs. This is an essential step for analyses that require a uniform selection criterion in both hemispheres.

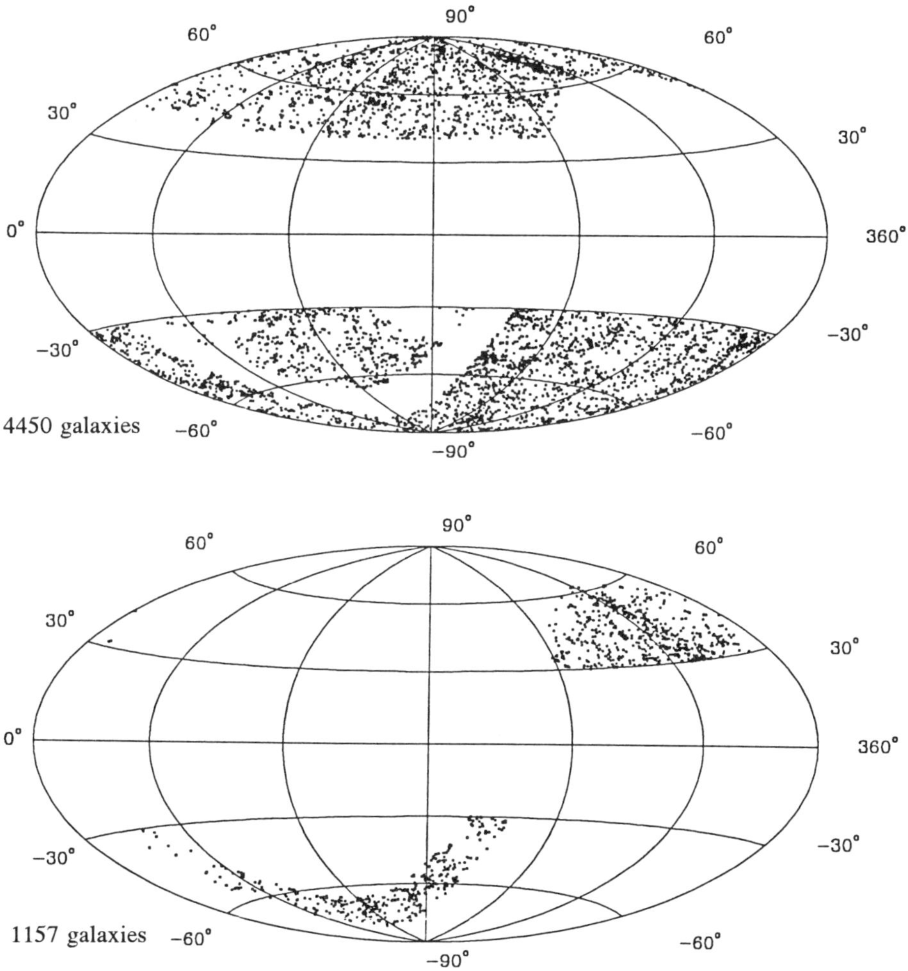

Fig. 1. Distribution of galaxies projected onto galactic coordinates: (a) CfA1 and SSRS samples; (b) MCG and northern Galactic cap extensions.

Finally, we should mention that we are currently planning to observe the dwarf galaxies in the ESO/Uppsala Catalog in the declination range covered by the 140-foot radio telescope in Green Bank to analyze the distribution of these objects relative to the brighter galaxies. This is particularly interesting in the south due the closeness of the large voids.

Deeper Slice Surveys

The large scale of the inhomogeneities observed in the nearby surveys, comparable to the survey depth, indicates that the currently available surveys do not probe volumes sufficiently large to be considered as representative of the universe at large. This in turn implies that one should consider going to fainter magnitude limits. However, extending the existing surveys to a fainter magnitude limit like 15.5 would require observing over 15000 objects in a point-and-shoot mode, which with the presently-available facilities would represent an extremely long term project. Therefore, despite the relative loss of information, the alternative is to carry out two-dimensional surveys, a strategy originally adopted by de Lapparent *et al.* (1986).

Such a program has been underway for several years in the northern hemisphere with the CfA Redshift extension (CfA2) which is extending the CfA1 survey in selected regions, essentially to the magnitude limit of the Zwicky catalog. At the present time there are three 6°-wide slices completed. Another important ongoing program at the CfA is the so-called Century Survey, which is expected to complete a redshift survey of a 1° x 100° strip down to a magnitude limit of 17.5. Further details of the work being carried out at the CfA can be found in Huchra and Geller (this volume).

We should point out that the observed large coherence length of the large-scale structures implies that the selected slices should be well separated on the sky in order to represent statistically independent samples. Towards this goal we are currently carrying out observations in two slices: the first, centered at $\delta = -35°$, 10° wide in declination and 135° in right ascension from 20^h to 5^h (*e.g.* da Costa *et al.* 1989). This slice contains some of the most prominent structures observed in the southern hemisphere, including the large void at about 5000 km s^{-1}. This region is also of interest, as Kirshner and collaborators are presently carrying out a very deep survey in the same declination range. The second slice is the 15° strip in the declination range $-17.5° < \delta < -2.5°$ in the equatorial region, where observations can be conducted from both hemispheres, which should expedite its completion.

Several technical problems must be solved before reliable statistical samples become available from these slices. In the equatorial region the most severe problems are the unreliable magnitudes of the MCG catalog as well as its known incompleteness at magnitudes fainter than about 15.0. In order to improve the catalog it would require a major effort of scanning plates in the region in order to generate a homogeneous sample to a well-defined magnitude limit, which will probably be done by the APM group soon. At the same time the CCD observations being carried out in this region could be used for calibration. This problem is somewhat alleviated in the southernmost strip because of the availability of the APM survey, which identifies galaxies to a faint magnitude limit ($m < 21.5$). However, the uncertainties in the individual magnitudes are relatively large for the brighter objects.

RESULTS

Qualitatively, both the SSRS and deeper southern slices have revealed several important features of the distribution of galaxies. For instance, the SSRS has demonstrated remarkably well the existence of extended two-dimensional flat structures crossing the observed volume and forming an intricate network with an apparent sponge-like topology, where voids form multiply-connected amorphous structures. The deeper slice has shown that the existence of conspicuous voids with diameters of the order of 50 h^{-1} Mpc are quite frequent and represent a fundamental characteristic of the galaxy distribution. The extent of the voids does not seem to change significantly as one goes to fainter magnitudes, suggesting that they are devoid of all types of normal galaxies.

The availability of the SSRS probing a volume of space comparable to that of the CfA1 has enabled us (Pellegrini *et al.* 1990) to examine how general are the properties derived from the application of different statistical indicators to the CfA1 sample. This question of reproducibility is important since the CfA1 sample has been extensively used to compare and to calibrate the model distributions obtained in *N*-body simulations. In this comparison, volume-limited samples at different radial distances were used to compute different statistics, such as the two-point correlation function, multiplicity function, percolation properties and nearest-neighbor distributions. These are the most commonly used statistics to describe the properties of high- and low-density regions and to compare real and model distributions. In Figure 2 some of these indicators are compared for volume-limited samples 4000 km s^{-1} deep, drawn from the SSRS and CfA1. Figures 2(a), (b), (c) show the multiplicity function

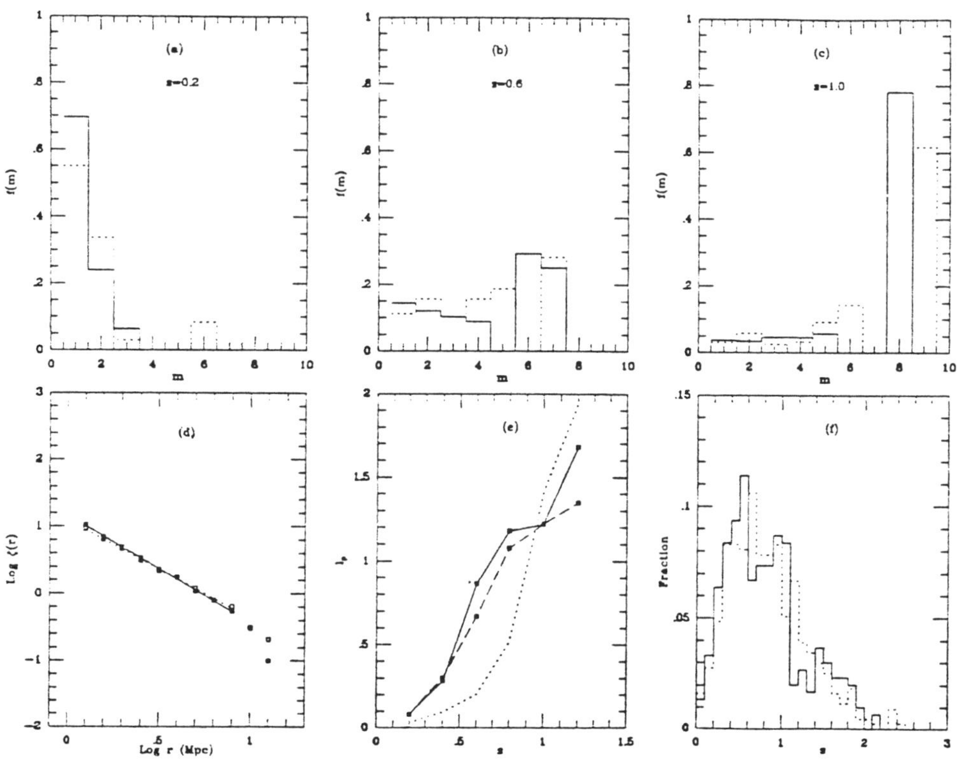

Fig. 2. Comparison between the SSRS (continuous lines, full symbols) and the CfA northern sample (dashed lines, empty symbols). (a), (b), and (c): Multiplicity functions for three values of the neighborhood radius s; (d): correlation functions and fits $\xi(r) = (r/r_0)^{-\gamma}$; (e): variation of the percolation parameter l_p for the largest system identified; and (f): histogram of the fraction of galaxies at a normalized distance s from a grid point (f).

f(m), which is the fraction of galaxies in systems of multiplicity m (see Einasto *et al.* 1984 for definitions) for three different values of the normalized percolation radius $s = r/r_0$, where r_0 is the mean interparticle distance of the sample; Figure 2(d) shows the growth of percolation parameter $l_p = l_{max}/L_{max}$, where l_{max} is the maximum dimension of the largest system detected at a given s (Einasto *et al.* 1984) and L_{max} is a characteristic dimension of the total volume surveyed; Figure 2(e) presents the two-point correlation function $\xi(r)$ for these samples calculated using the Monte-Carlo estimator and adopting a mean density internal to each sample and, finally, figure 2(*f*) shows the distribution of distances of the nearest neighbor galaxy to points in a grid superposed on each sample (Ryden and Turner 1984), which yields a measure of characteristic sizes of the voids. Based on the similarity of the results we may conclude that, although neither the SSRS nor the CfA1 are fair representations of the universe at large scales, the properties of the statistical indicators, which typically measure the clustering on scales less than about 10 h^{-1} Mpc are statistically the same, indicating that they must reflect a general characteristic of the galaxy distribution. Other characteristics, such as the void probability function and counts-in-cells, associated with higher-order moments of the correlation function, are also being examined, as well as the topological analysis developed by Gott and collaborators.

We have also attempted to investigate the possible spatial segregation of galaxies of different types by examining the dependence of the two-point spatial correlation function on morphology and on surface brightness (Santiago and da Costa 1990). Central values of the surface brightness of most galaxies in the SSRS were obtained from surface photometry carried out using a PDS microdensitometer at the Observatório Nacional. In Figure 3 we show the correlation functions estimated for different morphological classes and different surface brightnesses. We find from these Figures that not only are the early-type galaxies more clustered than spirals, confirming the earlier results of Davis and Geller (1976), but that spirals in different surface-brightness ranges also exhibit a segregation as strong as that observed for morphology. Unfortunately, the spatial distribution of the dwarf galaxies cannot be properly examined because of the paucity of redshifts in the southern hemisphere.

Finally, we have used the enlarged sky coverage to compute the peculiar velocity induced on the Local Group by the galaxy distribution in our surroundings. Although optical surveys will always neglect a substantial fraction of the sky, the results may be useful for

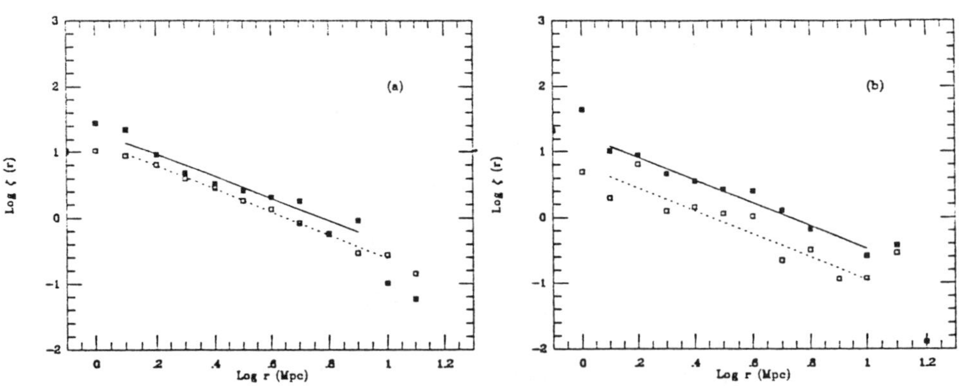

Fig. 3. Correlation functions for galaxies in the SSRS: (a) ellipticals and S0s (full symbols), and Spirals (empty symbols); (b) Spiral galaxies in the upper (full symbols) and lower (empty symbols) 33% brightness interval. Power-law fits are also shown.

calibrating the peculiar velocity computed using, for instance, the IRAS sample. The cumulative velocity induced by the galaxy distribution as a function of the radial distance is shown separately for the northern and southern samples in Figure 4 using the procedure and coordinate system described by Davis and Huchra (1982). The peculiar velocity calculated is somewhat higher than previously estimated, due to the repulsive contribution coming from the large voids in the southern hemisphere, reducing the value of Ω calculated by these authors (Pellegrini and da Costa 1990).

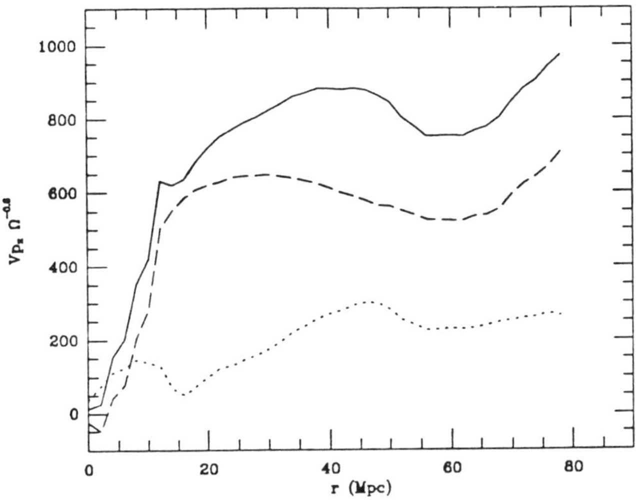

Fig. 4. Cumulative z-component of the peculiar velocity as a function of the radial distance. The dotted curve represents the contribution of the southern Galactic cap, the dashed curve the contribution of the northern Galactic cap, and the continuous curve the total z-component of the peculiar velocity induced on the Local Group.

FINAL REMARKS

Redshift surveys have been one of the main driving forces in the recent development of cosmology, in particular the wide-angle surveys like the CfA1, SSRS and IRAS. However, increasing the depth of these surveys is difficult due to the large number of galaxies involved. Therefore, it seems inevitable that in the near future, surveys are likely to be of the two-dimensional type. In any case, future progress in observations of large-scale structures will mean large amounts of telescope time, and progress will be slow unless a strong commitment is made to support this kind of program.

ACKNOWLEDGEMENTS

We would like to thank our several collaborators in the various surveys efforts that have been briefly described here which include H. J. Calderon, A. Fairall, M. Geller, J. Huchra, D. Latham, S. Lipari, B. Santiago, P. Vettolani, and C. Willmer.

REFERENCES

da Costa, L. N., Pellegrini, P. S., Sargent, W. L. W., Tonry, J., Davis, M., Meiksin, A., Latham, D. W., Menzies, J. W., and Coulson I. A. 1988, *Ap. J.*, **327**, 544.

da Costa, L. N., Pellegrini, P. S., Willmer, C., and Latham, D. W. 1989, *Ap. J.*, **344**, 20.

Davis, M., and Geller, M. J. 1976, *Ap. J.*, **208**, 13.

Davis, M., and Huchra, J. P. 1982, *Ap. J.*, **254**, 425.

de Lapparent, V., Geller, M. J., and Huchra, J. P. 1986, *Ap. J. (Letters)*, **302**, L1.

Dressler, A. 1988, *Ap. J.*, **329**, 519.

Einasto, J., Klypin, A. A., Saar, E., and Shandarin, S. F. 1984, *M.N.R.A.S.*, **206**, 529.

Haynes, M. P., and Giovanelli, R. 1986, *Ap. J. (Letters)*, **306**, L55.

Huchra, J. P., and Geller, M. J. this volume, p. 143.

Huchra, J. P., Davis, M., Latham, D. W., and Tonry, J. 1983, *Ap. J. Suppl.*, **52**, 89.

Lauberts, A. 1982, *The ESO/Uppsala Survey of the ESO(B) Atlas*, (Garching: European Southern Observatory).

Meiksin, A., and Davis, M. 1986, *A.J.*, **91**, 191.

Pellegrini, P. S., and da Costa, L. N. 1990, *Ap. J.*, **357**, 408.

Pellegrini, P. S., Willmer, C. N., da Costa, L. N., and Santiago, B. X. 1990, *Ap. J.*, **350**, 95.

Ryden, B. S., and Turner, E. L. 1984, *Ap. J. (Letters)*, **287**, L59.

Santiago, B., and da Costa, L. N. 1990, *Ap. J.*, **362**, 386.

Vorontsov-Velyaminov, B. A., and Arhipova, V. P. (1963-74), *Morphological Catalog of Galaxies*, (Moscow: Moscow University), Parts 2 to 5.

Zwicky, F., Herzog, E., Wild. P., Karpowicz, M., and Kowal, C., 1961-78, *Catalog of Galaxies and Clusters of Galaxies,* Vols. 1-6 (Pasadena: California Institute of Technology).

DISCUSSION

D. Lynden-Bell: While I am delighted to hear that we shall soon have redshifts of MCG galaxies in the $-2.5° \geq \delta \geq -17.5°$ strip, nevertheless the sky will still be very non-uniform. We already know that MCG picks many fewer galaxies there than ESO does in the neighboring circle. Complete redshift surveys whose finding charts are drawn from incomplete galaxy surveys gain little from their completeness. All this underlies the importance of a complete galaxy survey of the MCG strip - a task that we hope to start soon with the APM in Cambridge.

R. Giovanelli: I'd like to remind everybody that after the collapse of the 300-foot telescope, the only 21-cm machines available for the Southern Skies are the 64 m at Parkes and the 42 m at Green Bank (Bonn is really used at higher frequencies). This type of observation is important for the detection of low optical-surface-brightness, gas-rich objects, yet the now existing instruments fall far short of ideal: with their very large beams they yield observations with huge dilution factors (greater than 100). We need a more sensitive 21-cm machine in the South.

OBSERVATIONS OF LARGE SCALE STRUCTURE IN THE UNIVERSE: RECENT RESULTS FROM THE CFA REDSHIFT SURVEY

JOHN P. HUCHRA AND MARGARET J. GELLER
Harvard-Smithsonian Center for Astrophysics
60 Garden Street, Cambridge, MA 02138

ABSTRACT The second CfA Redshift Survey, comprising galaxy radial velocity data for a sample of ~ 15,000 galaxies to a well-defined magnitude limit in the northern sky, is now over 70% complete. We have finished four contiguous 6° slices in the southern Galactic cap, and three contiguous slices in the north, as well as the slice through the Virgo cluster. The intermediate scale geometry of the structures seen in the first slice persists; galaxies are distributed on thin surfaces of 'bubble-like' structures. 'Vertical' cuts through our declination slices clearly show surfaces closed in 3 dimensions. Voids are extremely common, come in a variety of sizes ranging up to 5,000 km s^{-1}, and are underdense by factors of ~ 5 with respect to the mean and factors of ~ 20 with respect to the density on the surfaces. Voids fill ~ 80% of the volume of the local universe. We present evidence for a very large coherent structure in the northern celestial hemisphere. There is a 'wall' of galaxies at about 10,000 km s^{-1} that extends through nearly 100° of right ascension and 30° of declination.

INTRODUCTION: THE CFA SURVEY

The goal of the Center for Astrophysics (CfA) Redshift Survey is to map the *general* galaxy distribution in the northern celestial hemisphere. When the survey is complete it will cover the sky at Galactic latitudes $b \geq 30°$ to a limiting magnitude $m_{B(0)} = 15.5$. The galaxy catalog is essentially the one made by Zwicky. The complete survey will include well over 15,000 redshifts and cover over 3 steradians. The typical external error in a redshift is 30 km s^{-1}, which is very small compared with the scale of any structure in the survey map.

We are also extending the survey into the Galactic plane, initially observing the brighter galaxies (*e.g.* Michel and Huchra 1988), but proceeding to fainter limits as observing time becomes available. This extension has been motivated in part by the *IRAS* galaxy surveys (Dey *et al.* 1990; Strauss *et al.* 1990), and by our desire to see how well the large structures seen in the northern hemisphere extend around the sky. We have also (in collaboration with da Costa's group in Brazil) been extending the survey into the southern hemisphere in order to eventually provide all-sky coverage to some moderate depth (Pellegrini *et al.* 1990). Finally, with the Dartmouth group, we have been extending sections of the survey to the magnitude limit of the Zwicky catalog (Thorstensen *et al.* 1989).

CURRENT STATUS

To date (March 1990), more than 11,000 galaxy redshifts in the survey have been measured, over 75% of these using the 1.5-m Tillinghast Reflector at the Whipple Observatory on Mt.

TABLE I Survey zone summary (to 5 March 1990)

Zone	α	δ	N_{gal}	N_{done}	
		North Galactic Cap			
+0	8^h to 17^h	+00.0° to +02.5°	533	172	
+6	8^h to 17^h	+02.5° to +08.5°	1716	508	
+12	8^h to 17^h	+08.5° to +14.5°	1674	1586	**
+18	8^h to 17^h	+14.5° to +20.5°	1379	749	
+24	8^h to 17^h	+20.5° to +26.5°	1015	689	**
+30	8^h to 17^h	+26.5° to +32.5°	1117	1107	comp
+36	8^h to 17^h	+32.5° to +38.5°	732	732	comp
+42	8^h to 17^h	+38.5° to +44.5°	726	717	comp
+48	8^h to 17^h	+44.5° to +50.5°	570	443	
+54	8^h to 17^h	+50.5° to +56.5°	594	267	
+60	8^h to 17^h	+56.5° to +62.5°	427	190	
Polar	8^h to 17^h	+62.5° to +90.0°	758	30	
North Subtotals			11241	7462	
		South Galactic Cap			
+0	20^h to 4^h	−2.5° to +6.0°	1217	876	
+9	20^h to 4^h	+6.0° to +12.0°	481	481	comp
+15	20^h to 4^h	+12.0° to +18.0°	496	494	comp
+21	20^h to 4^h	+18.0° to +24.0°	377	375	comp
+27	20^h to 4^h	+24.0° to +30.0°	466	466	comp
+33	20^h to 4^h	+30.0° to +36.0°	607	557	**
+39	20^h to 4^h	+36.0° to +42.0°	437	237	**
+45	20^h to 4^h	+42.0° to +48.0°	187	72	
Polar	20^h to 4^h	+48.0° to +90.0°	105	49	
South Subtotals			4373	3607	
Current Totals			15614	11069	
Galactic Plane Survey – to magnitude 15.0					
Summer	17^h to 20^h	+0.0 to +90.0	433	249	
Winter	4^h to 8^h	+0.0 to +90.0	560	415	

N_{gal} is the number of galaxies in the zone
N_{done} is the number of galaxies with redshifts
** are the strips currently being observed

Hopkins. We have four complete and one nearly complete 6° strips in the southern Galactic cap and three complete plus one nearly complete strip in the north. A summary of the survey zones and their status is given in Table I.

Analyses have been published for the luminosity function, correlation function and small scale galaxy motions (de Lapparent, Geller and Huchra 1988; 1989) and for the selection of and properties of galaxy groups (Ramella, Geller and Huchra 1989; 1990) from the first two northern strips. The luminosity function is similar to that found for the first CfA survey (to m_{Zw} = 14.5, Huchra *et al.* 1983), with M^* = −19.15 (H_0 = 100 km s^{-1}), α = −1.2, and $\phi^* \sim$ 0.025 galaxies Mpc^{-3} mag^{-1}. The correlation function has an amplitude described by s_0 = 7.5 Mpc and a slope of 1.6. More detailed descriptions of these results can be found in the above papers and in Geller and Huchra (1988). The data paper for the first slice will be published this spring (Huchra *et al.* 1990), complete with morphological

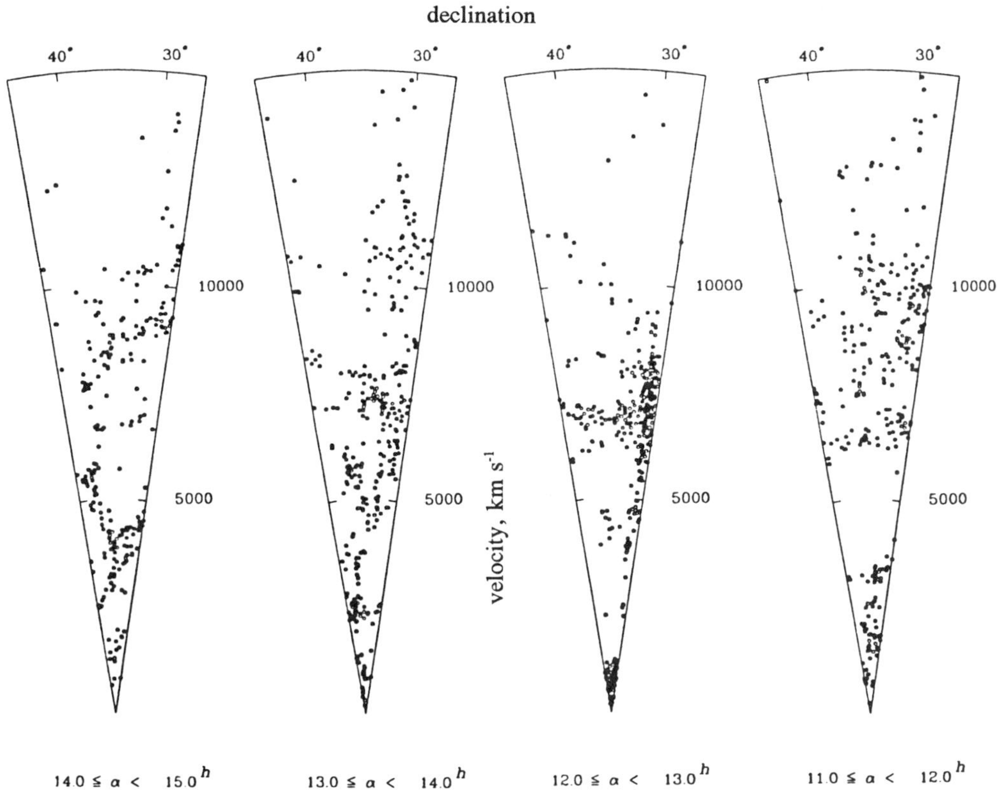

Fig. 1. Cuts made in declination, at constant right ascension, through the three contiguous slices. The four cuts are each 1^h thick. The 11-12^h wedge clearly shows the Coma-Virgo Void around 5000 km s^{-1}. The 12-13^h wedge is dominated by Coma with a structure rising almost 'vertically' (in declination) from the cluster. The 13-14^h and 14-15^h wedges both cut through significant voids at 3500 and 7000 km s^{-1}, respectively, as seen in Figure 2. Both show the voids closing off at higher declination – the 'top side' of the voids seen in Figure 2's declination pie slice.

types determined by H. Corwin. Aside from the very dense core of the Coma Cluster, galaxies of all types trace the structures seen similarly. The velocity field on small scales is moderately quiescent, with *rms* peculiar velocities ~ 300 km s^{-1} on 1 Mpc scales.

RECENT RESULTS

As the survey has become increasingly more complete, we have begun to assemble full sky maps of the galaxy distribution. We also have begun to test the hypothesis originally proposed (de Lapparent, Geller and Huchra 1986) that the structures seen are 'bubble-like' as opposed to the conjecture by Gott, Melott and Dickinson (1986) that the topology is more sponge-like. Figure 1 shows vertical slices (in declination) taken through our first three contiguous northern strips, shown in Figure 2 in the opposite projection. Where the data are available, at least two large voids show 'circular' profiles when 'sliced' in perpendicular directions. The large voids at 12.5^h and 3500 km s^{-1}, and at 14.5^h and 7000 km s^{-1} appear to close off at higher declination, supporting the view that at least some voids are completely enclosed.

One of our first full sky maps is shown in Figure 3. In this projection of a full 360°, by 20° thick wedge, one can see some of the evidence for the 'Great Wall' (a name attributed in part to Avishai Dekel). This structure, which goes across most of the northern sky, can also be seen in the wedge between 0° and 20°, and contains several thousand galaxies. Its apparent extent (limited in some directions only by the extent of our survey) is 60 × 170 × 10 Mpc. If Ω is taken to be 0.2 (the dynamically derived value for groups

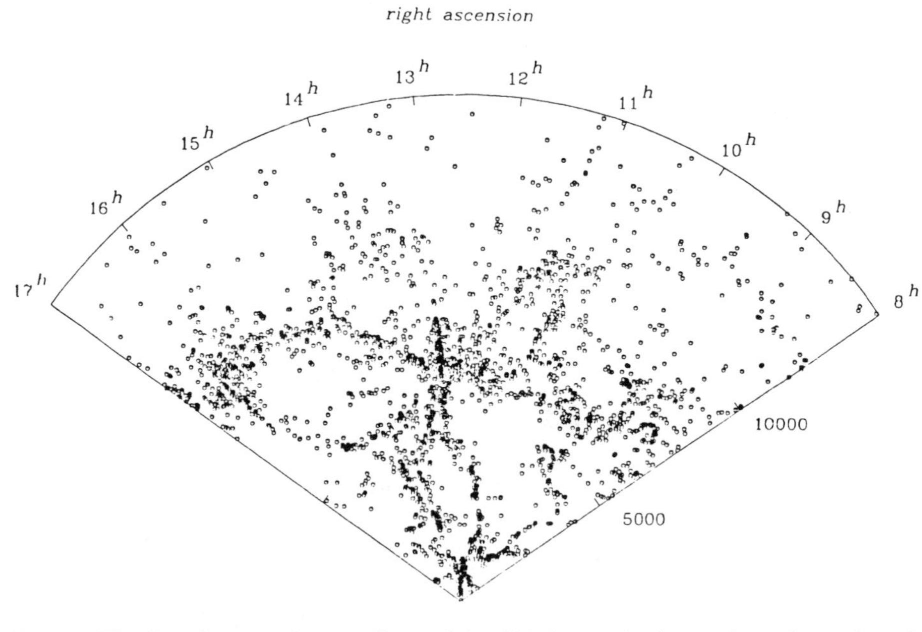

Fig. 2. The first three contiguous slices of the CfA Survey in the northern sky projected in right ascension. The wedges in Figure 1 cut vertically through this slice at the respective right ascensions. The three 6° slices combined contain 2535 galaxies in the velocity range 0 - 15,000 km s^{-1} and brighter than 15.5 magnitude.

and clusters of galaxies), then the 'Great Wall' weighs in at about $2 \times 10^{16}\ M_{\odot}$. This is comparable to the estimated mass of the 'Great Attractor' and about 10 times the mass of the Local Supercluster. A more complete description of the recent results of the CfA Redshift Survey can be found in Geller and Huchra (1989).

We would like to thank the CfA redshift survey team, S. Tokarz, J. Peters, E. Horine and A. Michel, for their great efforts. The program to produce large pie slices was developed by Marc Postman. This research is supported in part by NASA Grant NAGW-201 and by the Smithsonian Institution.

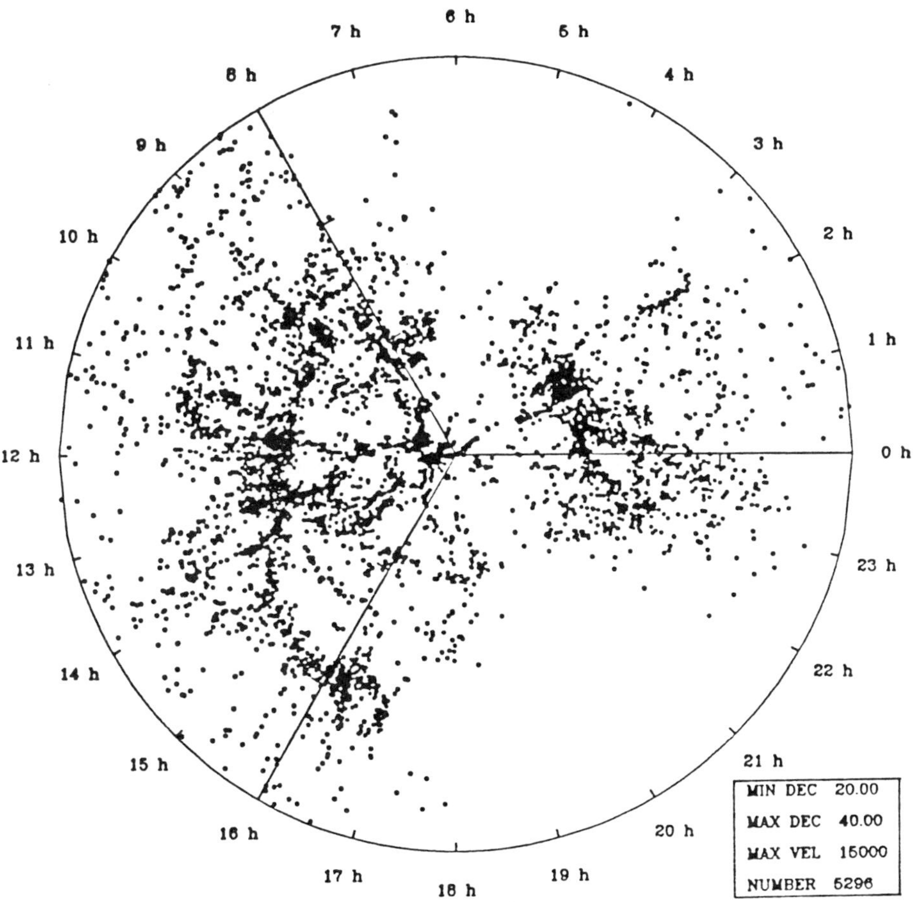

Fig. 3. Full circle 360° view, 20° thick, out to 15,000 km s^{-1}. The 'Great Wall' is seen extending between approximately 9^{h}, 7000 km s^{-1}, and 17^{h}, 10,000 km s^{-1}. On the opposite side of the sky, the Perseus-Pisces supercluster is readily seen. All galaxies with redshifts are plotted; the blank regions are obscured by the Galactic Plane.

REFERENCES

de Lapparent, V., Geller, M. J., and Huchra, J. P. 1988, *Ap. J.*, **332**, 44.
de Lapparent, V., Geller, M. J., and Huchra, J. P. 1989, *Ap. J.* **343**, 1.
Dey, A., Strauss, M. A., and Huchra, J. P. 1990, *A.J.*, **99**, 463.
Geller, M. J, and Huchra, J. P. 1988, in *Large Scale Motions in the Universe: A Vatican Study Week*, ed. V. C. Rubin and G. V. Coyne, S.J., (Princeton: Princeton University), p 3.
Geller, M. J, and Huchra, J. P. 1989, *Science*, **246**, 897.
Gott, J. R., III, Melott, A. L., and Dickinson, M. 1986, *Ap. J.*, **306**, 341.
Huchra, J. P., Geller, M. J., de Lapparent, V., and Corwin, H. E. 1990, *Ap. J. Suppl*, **72**, 433.
Michel, A., and Huchra, J. P. 1988, *P.A.S.P.*, **100**, 1423.
Pellegrini, P. S., da Costa, L. N., Huchra, J. P., Latham, D. W., and Willmer, C. 1990, *A.J.*, **99**, 751.
Ramella, M., Geller, M. J., and Huchra, J. P. 1989, *Ap. J.*, **344**, 57.
Ramella, M., Geller, M. J., and Huchra, J. P. 1990, *Ap. J.*, **353**, 51.
Strauss, M., A., Davis, M., Yahil, A., and Huchra, J. P. 1990, *Ap. J.*, **363**, 49.
Thorstensen, J., Wegner, G., Hamwey, R., Boley, F., Geller, M. J., Huchra, J. P., and Kurtz, M. J. 1989, *A.J.*, **98**, 1143.

DISCUSSION

A. Dekel: I am impressed by the large correlation length, $s_0 \sim 7.5$, the scatter about it (5 to 12 h^{-1} Mpc), and the line-of-sight velocity dispersion being large, 300 $\pm$100 km s^{-1}. How good are these determinations? Note the 'Great Wall' is huge!

S. White: I think if you wish to determine the 'true' value of the correlation length, the most economical way is to assign a few nights of large telescope time (with a fiber spectrograph) to Steve Maddox. His APM survey provides a correlation function for a volume three orders of magnitude larger than that studied by these slices. There is almost no statistical uncertainty in the amplitude on scales corresponding to a few Mpc, and surely this volume should be fair. A few hundred redshifts would determine the selection function empirically for this survey to an accuracy sufficient to determine the amplitude of the three-dimensional correlations to good 'astrophysical' accuracy.

R. Opher: Your survey shows various evidences for voids and empty spaces. Can we characterize the voids more quantitatively? Can we write down a correlation function for the voids as we do for the galaxies?

E. Bertschinger: I don't think that anyone has attempted to characterize the void distribution using correlation functions, and I don't think that correlation functions would be very useful. The voids are certainly anticorrelated in the sense that if two voids were to overlap they would probably be considered one larger void instead. Moreover, as Richard Gott and his collaborators note, the voids are likely to be interconnected with a sponge-like topology.

INTERCONNECTIONS BETWEEN VOIDS AND OTHER PROBLEMS ADDRESSED BY USING AVAILABLE SOUTHERN REDSHIFTS

A. P. FAIRALL, A. MAURELLIS, A. JONES, G. KAUFFMANN
D. MATRAVERS, AND G. F. R. ELLIS
Departments of Astronomy and Applied Mathematics
University of Cape Town, Rondebosch, 7700 South Africa

ABSTRACT This paper describes various investigations into voids, and other large-scale structures, being carried out by the Cosmology Group at the University of Cape Town. We argue that it is possible to map large-scale structure using all available redshifts (as well as 'controlled' redshift samples), since voids are void of all galaxies. The greater quantity of data makes a three dimensional examination possible. Suitable software has been developed for establishing void centers, sizes and shapes, as well as examining void interfaces. We have examined seven interfaces between southern voids. One interface shows a clear two dimensional sheet of galaxies separating the voids. Three other interfaces look similar, but the remaining three appear to have interconnecting holes (*i.e.* significantly larger than the mean galaxy spacing). Maps of the spatial distribution of emission-line galaxies and a possible scenario for the development of voids are also presented.

THE CASE FOR USING ALL AVAILABLE REDSHIFTS

The voids that exist in the large-scale distribution of galaxies in redshift space are three-dimensional regions that are void of all galaxies, regardless of galaxy luminosity, size or type. This is apparent in the separate plots for elliptical, spiral and low-luminosity galaxies presented at this conference by Huchra (see also Bothun *et al.* 1986, Thuan *et al.* 1987). The question then is whether we could not map large-scale structures and voids using all available redshifts as well as those from controlled surveys such as the Southern Sky Redshift Survey (SSRS - da Costa *et al.* 1988). Obviously, many quantitative parameters can only be extracted using statistically controlled data, but the revelation of foam-like structures only came as the quantity of data increased. Using all available redshifts greatly increases the quantity of data.

For the past several years, the first author and collaborators have built up a catalog of optical redshifts, available from the literature or preprints, for galaxies south of Declination $-17.5°$ (Fairall and Jones 1990). Although these data may be subject to direction selection effects, Figure 1 shows that nevertheless sky coverage is reasonable. Since 1983, we have used this database to map superclusters and voids. Figures 2 and 3 are representative 'slice' plots extracted from the most recent version of the catalog and plots (Fairall and Jones, 1988). Confidence in the reality of the structures in the plots is provided in two ways. First, as we have added data to these plots over the years, it has only served to make the structures more sharply defined - it has not removed or significantly changed the structures. In particular, the voids have not been filled in, nor have their sizes changed, only their boundaries have been made more definite. The two main voids in Figure 2 (Eridanus and

Sculptor) were identified in the first version of our plots (Winkler 1983). Second, Figure 2 can be directly compared against the equivalent plots of SSRS (da Costa *et al.* 1988). Although our catalog currently includes only about a quarter of the SSRS data, there is complete agreement in the structures revealed. This is obvious when plots are superposed.

The 1988 version of our catalog, containing entries for some 6000 galaxies, enables us to plot 'foam-like' structure out to $cz \sim 10{,}000$ km s^{-1} and to identify major superclusters to $cz \sim 25{,}000$ km s^{-1}.

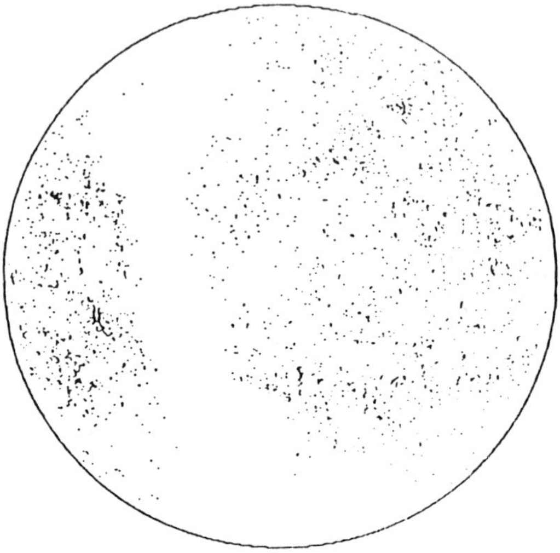

Fig. 1. The sky distribution (Declination −17.5° as periphery) of catalog galaxies, with cz < 10,000 km s^{-1}, is similar to that of ESO catalog galaxies (Lauberts 1982).

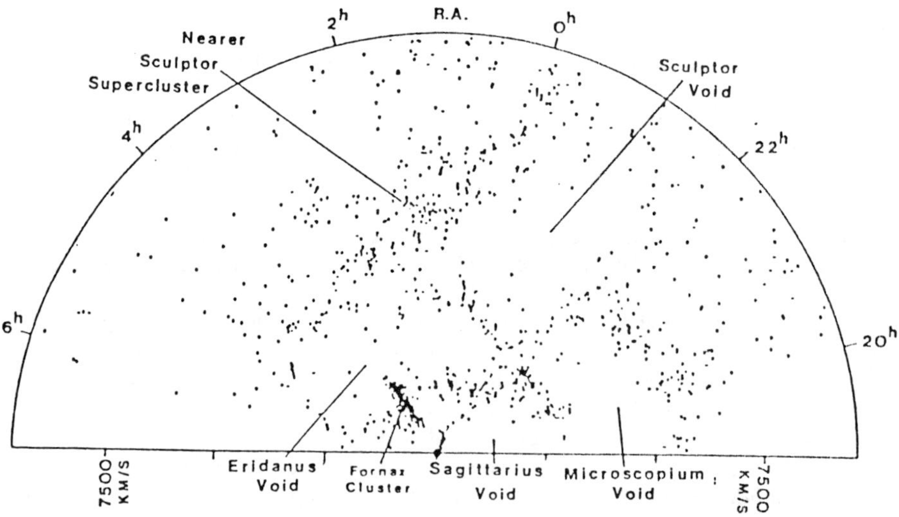

Fig. 2. Redshift plot of galaxies in the southern Galactic hemisphere between declinations −17.5° and −47.5°. Most of the structures happen to run perpendicular to the somewhat thick 'slice'.

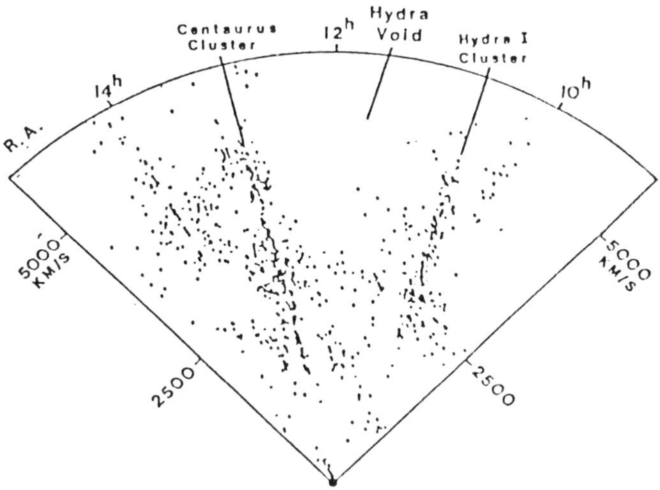

Fig. 3 Redshift plot of galaxies in the northern Galactic hemisphere between declinations −22.5° and −42.5°.

TOOLS FOR EXPLORING VOIDS

Unlike well-known thin slices through foam-like structures (*e.g.* de Lapparent *et al.* 1986), which are effectively two-dimensional, we use three-dimensional data. One of us (A. M.) has developed a computer program COELIS that converts the data into rectangular coordinates and then allows us to manipulate, examine, and find fits for voids in three dimensions.

The first aspect this program had to address (in early 1987) was whether a void that showed up in a two dimensional cross-section was truly a void in three dimensions. The software enabled us to cut through a prospective void in three perpendicular sets of slices and to establish visually an approximate center. From this trial center, galaxies were counted in spherical shells of increasing radius and a histogram constructed (see the example in Figure 4). The histogram also identified the direction of the galaxies seen from the center, in terms of eight sectors; this enabled us to improve on the trial center. Thus we found our southern voids to be empty at the centers (typically to a radius of ≥ 1000 km s^{-1} in redshift space) with a distinct concentration of galaxies in a surrounding, roughly-spherical, shell. There also tended to be a few galaxies lying inward from the surrounding shell. The histogram routine can also be run to show galaxy density (per unit volume of redshift space).

This approach, however, was constrained by the use of spherical shells. A more generalized approach was to fit tri-axial ellipsoids, by means of least squares and eigen-vectors, to the surrounding shell (Maurellis 1990). A related problem, the spectrum of void sizes, is also being addressed (by G. K. - following suggestions by L. Kofman). Again the data are transferred into rectangular coordinates but counted within a grid of cells (usually 300 km s^{-1} on a side). Computer algorithms then search for voids within this lattice. Since smaller voids may often be artifacts of data spacing, the same procedures are tried on a large number of random distributions having the same number of galaxies and the same radial selection effects as the observed data set, and the random component subtracted.

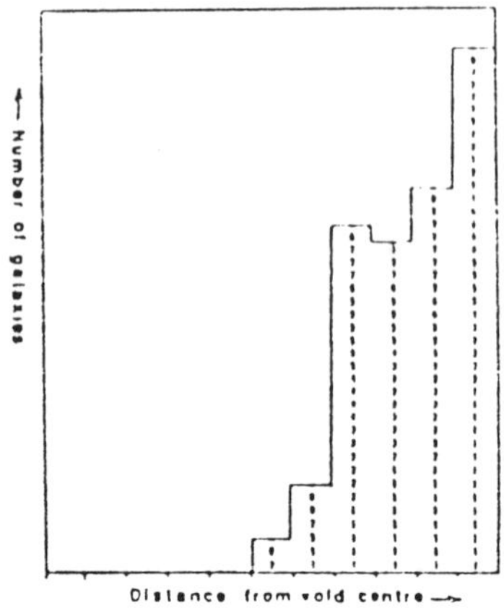

Fig. 4. Histogram for southern Eridanus void.

TOOLS FOR EXPLORING VOID INTERFACES

The nature of the interfaces between voids (*i.e.* soapsud-like walls or sponge-like interconnections) cannot be examined properly in 2-dimensional slices. However the COELIS software allows us to shift the origin and rotate the coordinates and to step from one void to another - see the example in Figure 5. In particular, we find a thin planar sheet of galaxies to separate the Eridanus and Sculptor voids. There is no significant connecting hole. A detailed presentation of this work, including statistical analysis of the galaxy distribution in the interface, and a listing of the galaxies involved, is presented elsewhere (Maurellis *et al.* 1989).

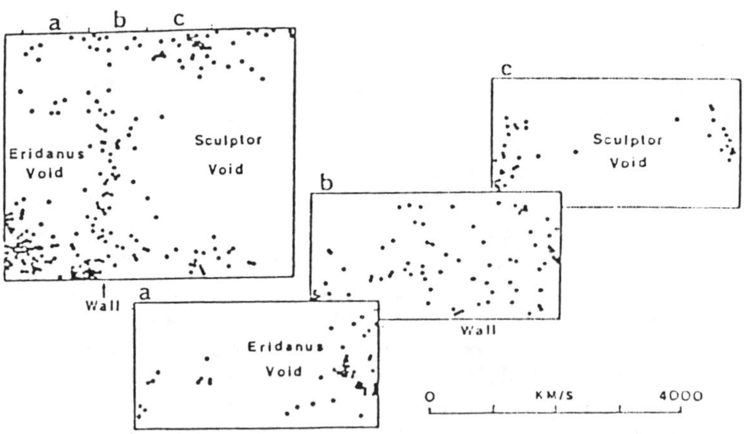

Fig. 5. The interface between two voids shown in rectangular coordinates. The main diagram shows a view 'from above'; the smaller diagrams show sequential slices, as viewed from its left.

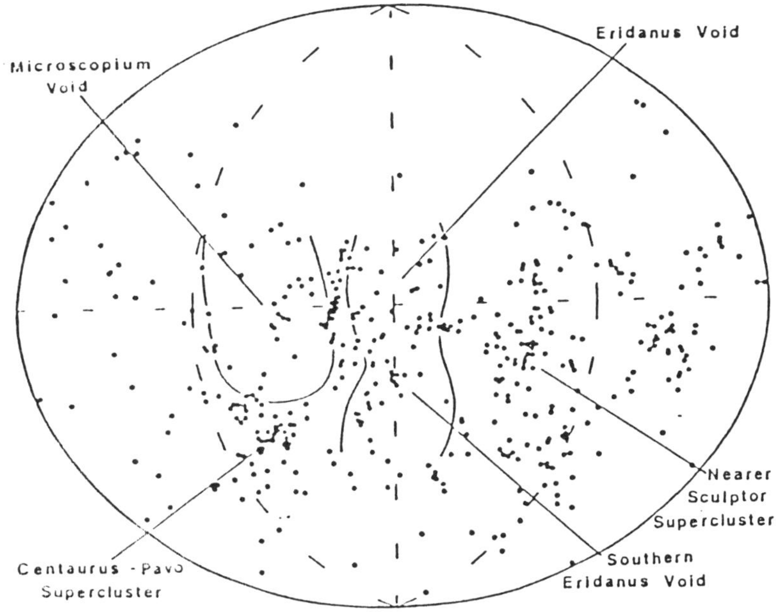

Fig. 6 The view from the center of the Sculptor void.

A complementary feature of the COELIS software is demonstrated in Figure 6, which shows us an all-sky equi-area view from the center of the Sculptor void. The data points include only those galaxies within 3500 km s^{-1} of the center of the void (the software also allows for stepping through spherical shells) so to show only the galaxies that immediately surround the void. Also included are outlines of voids that lie beyond these surrounding galaxies. Note that the data is sparser both towards the top of the figure, since one is looking out of the data set, and generally towards the periphery since one is looking at greater redshifts. Such a plot is useful for exploring interconnections between voids. Again the sheet of galaxies lying in front of the Eridanus void can be seen, but similar sheets do not appear to exist in front of the neighboring Microscopium and Southern Eridanus voids. Instead, fairly large ($\sim$1000 km s^{-1}), but offset, interconnecting holes are apparent. Their sizes are significantly larger than the average galaxy spacing.

VOID INTERCONNECTIONS - SWISS CHEESE OR BATH SPONGE

Aside from the Eridanus Void - Sculptor Void interface, we have used COELIS rectangular coordinate slices and all-sky plots to examine all interfaces between the northern and southern Eridanus, Sculptor, Microscopium and Sagittarius voids. For these seven interfaces, the indications are that three (northern-southern Eridanus and those above) show evidence of interconnecting holes. While sheets of galaxies do separate some voids, there is not conclusive evidence of any void being totally isolated from all its neighbors. This result is consistent with all low-density regions being interconnected (Gott this volume).

DISTRIBUTIONS OF EMISSION-LINE GALAXIES

The software (written by A. J.) that produces the plots from our galaxy catalog is also capable of identifying Seyfert and other emission-line galaxies - a sample plot is given in Figure 7. While such galaxies generally follow the same distribution as ordinary galaxies, they tend to avoid dense clusters, while certain regions seen to have a higher proportion of emission objects, in particular that at $\alpha \sim 22^h$, $\delta = -30°$, $cz = 2000$ km s^{-1}.

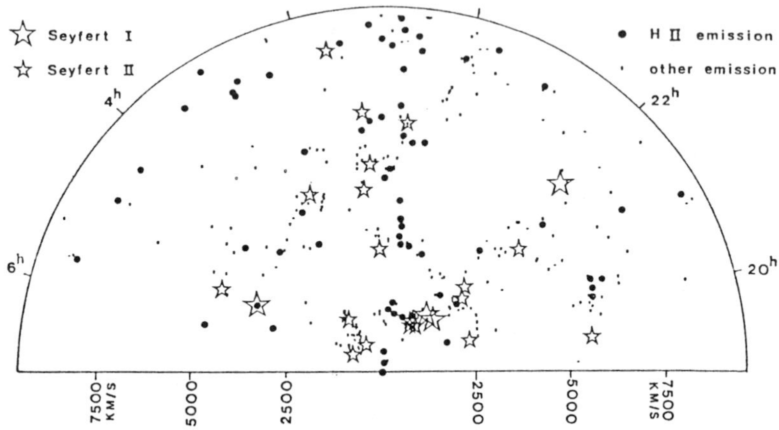

Fig. 7. The distribution of emission-line galaxies in the declination −17.5° to −47.5° slice (*cf.* Figure 2).

DEVELOPMENT OF VOIDS

We have also considered various scenarios for the development of voids (see Fairall 1987, Ellis and Fairall 1988, Fairall 1988). Figure 8 suggests that voids were initiated at various points in space, either by explosions or any other mechanism that has a 'cosmological Alka-Seltzer' effect. Where our scenario differs from that of other investigators, is that we consider the expansion of the voids to have been arrested by the expanding bubbles of material impinging upon one another. Further, such compression of material between the bubbles would initiate galaxy formation. We hope to explore such scenarios in future investigations.

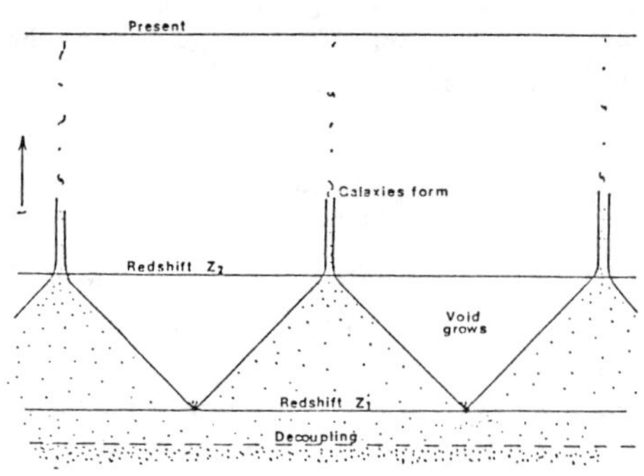

Fig. 8. The development of voids in comoving coordinates.

REFERENCES

Bothun, G. D., Beers, T. C., Mould. J. R., and Huchra, J. P. 1986, *Ap. J.*, **308**, 510.

da Costa, L. N., Pellegrini, P. S., Sargent, W. L. W., Tonry, J., Davis, M., Meiksin, A., Latham, D. W., Menzies, J., and Coulson, I. 1988, *Ap. J.*, **327**, 544.

de Lapparent, V., Geller, M. J., and Huchra, J. P. 1986, *Ap. J.*, **302**, L1.

Ellis, G. F. R., and Fairall, A. P. 1988, *Ap. J.*, **327**, 34.

Fairall, A. P. 1987, *M.N.A.S.S.A.*, **46**, 47.

Fairall, A. P. 1988, in *Morphological Cosmology*, ed. Duerbeck and Flin (Dordrecht: Kluwer Academic Publishing) in press.

Gott, J. R., III this volume, p. 173.

Lauberts, A. 1982, *The ESO/Uppsala Survey of the ESO (B) Atlas*, (Garching: European Southern Observatory).

Maurellis, A., Fairall, A. P., Matravers, D. and Ellis, G. F. R. 1989, *Astr. Ap.*, submitted.

Thuan, T. X., Gott, J. R., and Schneider, S. E. 1987, *Ap. J. (Letters).*, **315**, L93.

DISCUSSION

M. Strauss: I have a question about the distribution of Seyfert galaxies in your catalog. As Alex Filippenko, among others, has shown, many seemingly normal galaxies hide mini-Seyferts in their nuclei. Do the galaxies you've labelled as Seyferts include some mini-Seyferts? It would be interesting to look at the space distribution of LINERS and mini-Seyferts in space.

A. Fairall: I am familiar with Alex Filippenko's work, but none of the Seyfert galaxies in these plots are such mini-Seyferts.

M. Strauss: Has anyone looked at the distribution of active galaxies in the CfA Redshift Survey?

R. Giovanelli: J. Salzer (Michigan) is doing a thesis on exactly this question (*e.g.*, see Salzer *et al. A.J.* Nov. 1988).

A. Dekel: I don't fully agree with your 'alkaseltzer' model: thin walls can form and persist even without dissipation as a result of the expansion of the universe, especially if $\Omega \sim 1$. Galaxies don't have to form in the walls - they can be pushed there, and stick. Did you try to quantify 'voids' and 'walls' in the distribution of galaxies, to calculate statistics that would be useful to compare with theory?

A. Fairall: It is very dangerous to extract statistical information from uncontrolled data. But what the data can give is the size and shape of voids, thicknesses and extents of walls, *etc.*

R. Opher: You described the voids as Swiss cheese or as a bath sponge. Can we describe the voids similarly as we describe density fluctuations? Can we describe the voids, for example, as a void noise spectrum? This would mean that there are lots of little voids which grow and can join up with one another. It would also appear to me that the distribution of the voids would be a function of the limiting magnitude used in the study.

A. Fairall: A spectrum of void sizes is possible. There is no problem identifying the larger voids, but recognition of smaller voids depends on the quantity of data (*i.e.* limiting magnitude). However, it is possible to count apparent small voids in the sample, but then to subtract an average count made in random distribution (with the same number of points per distance interval built in) so the true number emerges.

E. Bertschinger: You have studied voids in the real galaxy distributions as much as anyone. Can you suggest quantitative measures of some properties of observed voids that might be applied to *N*-body simulations? I'm asking you to apply your specific knowledge of real voids, as opposed to a theorist's conception of a void.

A. Dekel: What could we do about the non-uniform selection of such uncontrolled redshift surveys?

E. Bertschinger: If the voids are as underdense as they seem, maybe little control is needed! Naturally, a method applied to an *N*-body simulation must account for the observational selections.

A. Fairall: The voids in my maps are not just underdense, but the centers are empty. Their detection does not seem to depend on the non-uniform selection of data. For instance, many of the voids here were identified in the first version of these plots - by my student Hartmut Winkler in 1983. Since then, much more data has been added, but the same voids are there. Perhaps the most useful quantitative tool to compare against *N*-body simulations would be a form of void-size spectrum.

R. Gott: Drs. Fairall and White have summarized the situation quite well. In a sponge-like topology the voids are connected in the sense that if you live in a void you can go and visit other voids by low-density tunnels (holes). There may be a wall between you and a neighboring void. (In a random phase sponge-like topology one would expect this 50% of the time.) The question is can you get to the neighboring void by going around? If you can get to the other voids by going along *some* low-density path (perhaps round about) then the voids form one connected structure as occurs in a sponge-like topology. In the Swiss cheese topology the voids are completely surrounded by walls, so if you live in a void you are completely trapped in that void and cannot visit *any* of the other voids without crossing a high-density wall.

R. Giovanelli: Fry and others have applied void probability functions to the comparison of data and numerical simulations. They seem to succeed in distinguishing between models. Would you like to comment whether that is a viable technique along the lines your earlier question suggested?

E. Bertschinger: Lachieze-Rey and collaborators have also applied this method. It certainly looks promising, but I am concerned that treatment of the density field as discrete points might make the void probability function sensitive to either catalog selection effects and incompleteness or to the astrophysical processes ('biased' galaxy formation) that force us to sample the underlying mass distribution with discrete objects. Can someone design a topology-like statistic that is sensitive to properties of voids other than their genus?

A. Fairall: I need only add that my student, Guenevere Kauffmann, has also tried applying the void probability function to our southern data.

GALAXY FORMATION IN VOIDS

JENS VERNER VILLUMSEN
Department of Astronomy, Ohio State University, Columbus, OH 43210

ABSTRACT I have studied the formation of galaxies in the Cold Dark Matter (CDM) scenario in voids using a Hierarchical Particle Mesh (HPM) *N*-body code. In voids such as the Boötes void, CDM predicts that the formation of dwarf galaxies up to $\sim 5 \times 10^9\ M_\odot$ is unbiased, but that the formation of brighter galaxies is strongly suppressed.

INTRODUCTION

A universe dominated by Cold Dark Matter (CDM) has become the standard scenario against which observational data and theoretical studies are measured. Large efforts have been invested in disproving CDM, so far with limited results. I will refer to a CDM model with an initial Zel'dovich spectrum, $\Omega_0 = 1$, $H_0 = 100\ h$ km s^{-1} Mpc^{-1}; $h = 0.5$ and a bias factor, b, of $\delta\rho/\rho$ (8 h^{-1} Mpc) = 1/b = 1/2.5 as the standard model. This model has been very successful in describing the universe on the mass scales of galaxies and on the length scales of galaxy clustering (White *et al.* 1987). Problems have arisen on large scales where there is too little power to explain easily large-scale clustering and velocity fields. These problems include the clustering of Abell clusters (Bahcall and Soneira 1983) and the coherent velocity field associated with The Great Attractor (Dressler 1989). The standard model is normalized on galaxy clustering scales, and with this normalization there is too little power on large scales to drive the Abell clustering and the large-scale velocity fields. CDM can be saved by assuming that the bias factor, b, is not scale invariant but decreases with increasing scale. This should not be seen as a weakness for CDM which tells us only how density fluctuations evolve. We don't know how galaxies form, they are essentially put in by hand in *N*-body simulations, while in analytical work galaxies are made by some prescription out of the density fluctuations. The variation of the bias factor with scale tells us that the formation of luminous objects is not as simple as we would have liked.

The large-amplitude clustering of Abell clusters seems to have been put on hold. The contaminations discussed by various authors (*e.g.* Blumenthal 1989) strongly question the statistical significance of the clustering amplitude, and we must await an X-ray-selected sample of clusters.

Redshift surveys such as Kirshner *et al.* 1987 and de Lapparent *et al.* 1986 have shown large regions of space nearly devoid of galaxies. The Boötes void (Kirshner *et al.* 1987) has a diameter of 3000 km s^{-1}, while the CFA survey (de Lapparent *et al.* 1986) shows bubble-like structures on scales of 2500 km s^{-1}.

These voids or bubbles are deficient in galaxies but not necessarily in mass; we can measure the light but not the mass. These voids have been held up as disproving Gaussian fluctuations and the CDM scenario. However, it has been shown (Gott 1989) that the topology of the large-scale structure is consistent with Gaussian fluctuations. CDM does not tell us about the light distribution but rather about the mass distribution, so unless we can infer the mass distribution and show that it is inconsistent with CDM, we have not disproved anything.

This is where *N*-body simulations come in handy. You run a simulation starting from initial conditions prescribed by CDM until the present. Galaxy halos can be found and you can then study the properties and spatial distribution of these halos.

SIMULATIONS

I have run a number of *N*-body simulations to test the biasing of galaxy formation in voids. The purpose is to find the density of galaxy halos of a certain mass in voids and in 'typical' regions to ascertain how the background density affects the formation of halos. *N*-body simulations cannot tell us about galaxy luminosities; we have to use some prescription of halo mass-to-luminosity ratios to infer galaxy luminosities. An *N*-body simulation has to be able to resolve the relevant length and mass scales. The simulation has to be large enough to encompass a Boötes void, but it also has to be able to resolve halo scales. It is not enough to be able to resolve halo length scales, it is also necessary to resolve halo mass scales. In an *N*-body code where the near field interaction is done by straight particle-particle interaction you can in principle reach scales as small as you want (*e.g.* Bouchet and Hernquist 1988). However, if your particle mass is larger than a halo mass, all this length resolution does not do you any good.

I have used the *N*-body code HPM (Villumsen 1989) to attack this problem. HPM is based on a standard Particle-Mesh code (*e.g.*, see Efstathiou *et al.* 1985) where Poisson's equation is solved on a uniform cubical grid with periodic boundary conditions using Fast Fourier Transforms. The equations of motion are solved in comoving coordinates using a time-centered leapfrog method. HPM uses a hierarchy of grids, boxes within boxes, each with the same number of grid cells and a similar number of particles, to achieve very high resolution and large dynamic range. Sub-grids do not have periodic boundary conditions and particles can enter and exit. This code is described in excruciating detail in Villumsen (1989), where it is also compared to other *N*-body codes.

A realization of a CDM power spectrum (Davis *et al.* 1985) is generated on a 64^3 grid with 32^3 particles where the length of the unit cube is 600 Mpc for $h = 0.5$. This system is evolved to the present, which is determined by calculating the velocity correlation function on linear scales and comparing it to the theoretical CDM function normalized to b = 2.5. On these scales the displacements from the uniform grid are small and the mass correlation function is not meaningful, so the velocity normalization is used. The density distribution is smoothed with a Gaussian with a σ of 60 Mpc, and the highest underdensity is found. I generate a sub-grid of width 1/6 centered on the void and generate initial conditions taking into account that particles can enter and exit. This system is then evolved to the present and a sub-grid of width 1/36 is entered. From the simulation it is known from which volume of space particles enter the sub-grid, and for this volume initial conditions are generated which are then evolved to the present. The bootstrap is continued to one more level where three grids are started. This four-level calculation is then carried to the present. The dynamic range in length is thus 1.4×10^4 from 600 Mpc down to 43 kpc and in mass 3.3×10^{11} from $1.6\times10^{19}\ M_\odot$ down to $5\times10^7\ M_\odot$. The number of waves that can be specified in each dimension is 3456.

In CDM the amplitude of the density fluctuations increase with decreasing scale, so it is necessary to start the simulation at high redshift so that the displacements on the smallest scales are less than the mean interparticle spacing. The initial power spectrum is shown in Figure 1, together with the theoretical power spectrum (Davis *et al.* 1985). In each grid the

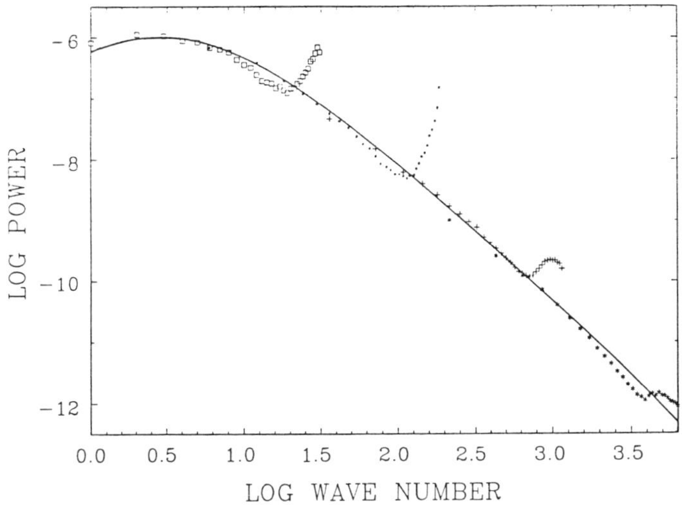

Fig. 1. Theoretical and generated CDM power spectrum.

generated spectrum traces the theoretical spectrum well up to the Nyquist frequency, so the spectrum is well fitted over 3.5 orders of magnitude. In the top grid the white noise level is at −4.5, so the displacements in the top grid are very small. In the bottom grid the white noise level is at −11.5, so the displacements at the Nyquist frequency are still significantly less than the mean interparticle spacing. The hooks in the spectrum are induced by the mass assignment to the grid and do not grow.

In Figures 2-6 are shown slices of the particle distribution after 17 expansion factors at b = 2.5. All slices have the width of the grid that they are in, and their depth is a quarter of the width. The squares in Figures 2-4 indicate in projection the position of a sub-grid. In all these slices there should be 8192 particles if the local density equals the critical density. Looking at the top grid in Figure 2, it is not at all obvious where the Boötes Void is located, the particle displacements are quite small and the particle mass is 5×10^{14} $M_\odot$, approximately that of a large cluster. Grid 2 in Figure 3 has a length of 100 Mpc and a particle mass of 2×10^{12} $M_\odot$, approximately that of a large galaxy halo. There is plenty of virialized structures on this scale that are not resolvable by the top grid. The mean density in this slice is 0.86. Grid 3 in Figure 4 has a length of 17 Mpc and a particle mass of 10^{10} $M_\odot$, and the mean density in this slice is 0.13. There are very few virialized structures on this scale, and the initial particle grid is visible. The level 4 grids have a length of 3 Mpc and particle masses of 5×10^7 $M_\odot$. Slices from two of the level 4 grids are shown in Figures 5 and 6. With mean densities 0.15 and 0.17, these slices show that there are dwarf galaxies forming even in these low density region.

It is important to do 'calibration' runs to see how the halo formation in these low-density regions compares with regions where the background density is the critical density. For this purpose I run simulations with only one grid but with the exact same waves in the initial conditions as before. There is no external field and the self-gravity has periodic boundary conditions, so the mean density is forced to be the critical density. The level 4 grids are simulated this way, and the results can be compared to the real simulations. This comparison is not completely honest; the mass correlation length for b = 2.5 is 6 Mpc, which is twice the size of the lowest sub-grids. In the calibration the long waves will be

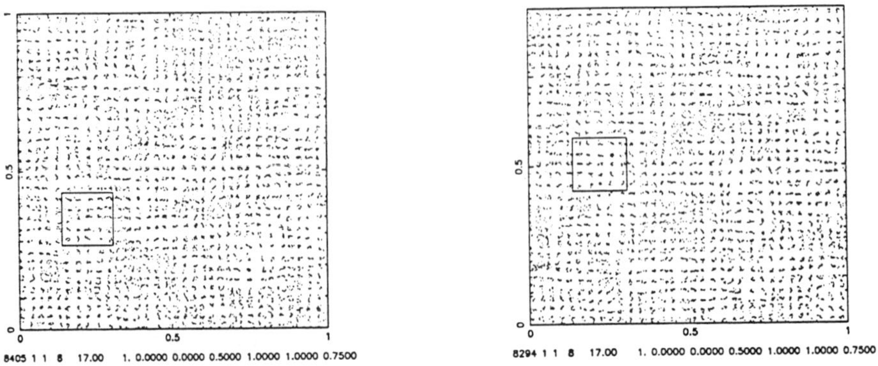

Fig. 2. Slice of top grid, 600 Mpc across.

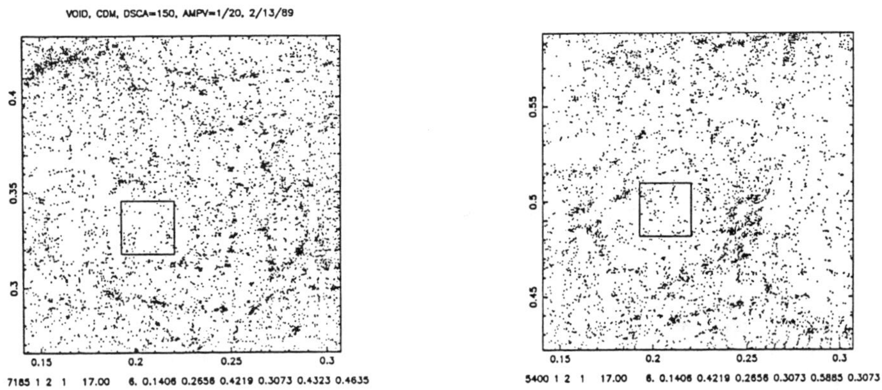

Fig. 3. Slice of second grid, 100 Mpc across.

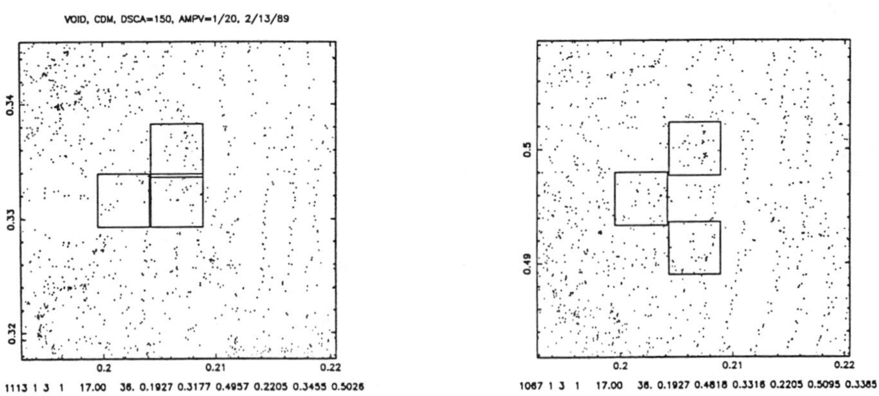

Fig. 4. Slice of third grid, 17 Mpc across.

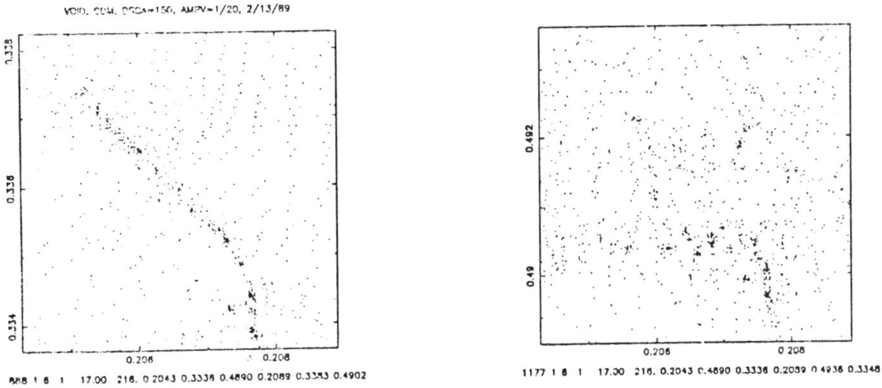

Fig. 5. Slice of fourth grid, 3 Mpc across.

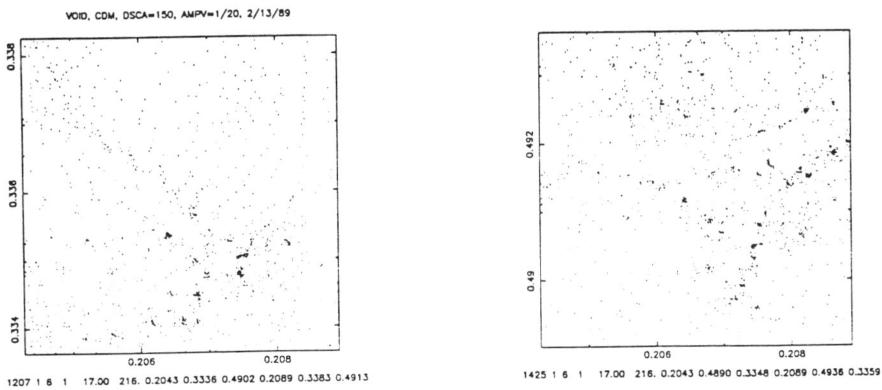

Fig. 6. Slice of fourth grid, 3 Mpc across.

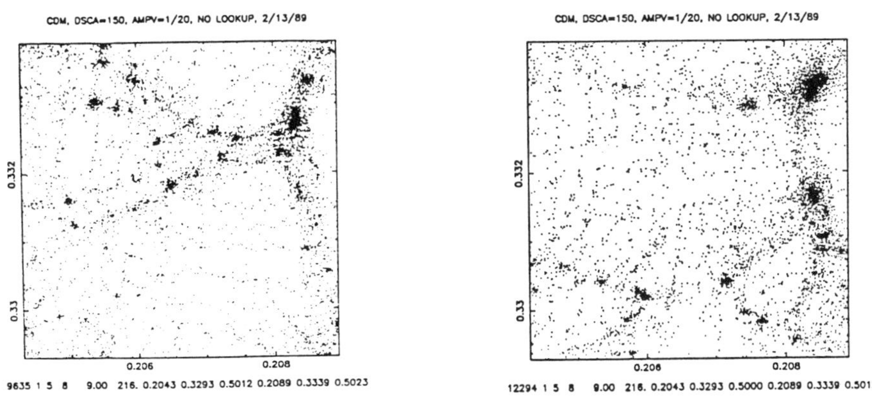

Fig. 7. Slice of calibration run, 3 Mpc across.

non-linear, which will influence the large-scale structure, so it is not to be trusted. The small-scale structure, which is the main interest in these simulations, is not expected to be influenced greatly by this. Figure 7 shows a slice of the calibration at the same epoch as before, and we see large galaxies that do not show in the real simulations.

RESULTS

I find galaxy halos by the friend-of-friends algorithm generously donated by Davis *et al.* (1985), that finds neighbors within a fraction B of the mean interparticle spacing in the initial conditions. A group of particles found this way will typically have an overdensity $B^{1/3}$ if $B < 1$. If B becomes comparable to unity, the groups will percolate through the simulations. This algorithm is a well-defined but not necessarily very physical way of finding halos.

Figure 8 shows a log-log plot of the multiplicity function in both the real and the calibration runs for $B = 0.1, 0.2, 0.3$. The data in three level 4 grids has been combined. The multiplicity function is defined as the fraction of the total mass in groups in a logarithmic interval of number of particles or mass. As expected (Davis *et al.* 1985), the multiplicity function is essentially flat for the calibration runs up to $\sim 10^4$. The curves for the three values of B differ little except at the high end, where the lower values of B cut off earlier. This algorithm with a large B can lump two very large nearby galaxies together which would be separate with a lower B. The real simulations show a multiplicity function that for $B = 0.2, 0.3$ agree with the calibration up to a multiplicity of about 100, which corresponds to a mass of $\sim 5\times10^9\ M_\odot$, but unlike the calibration runs, the curve for $B = 0.1$ is significantly below the other curves. The curves cut off at a multiplicity of ~ 100, so there are no halos above $\sim 5\times10^9\ M_\odot$. The underdensity in mass in the void thus equals the underdensity in the number of dwarf galaxies, so dwarf galaxies are an unbiased tracer of

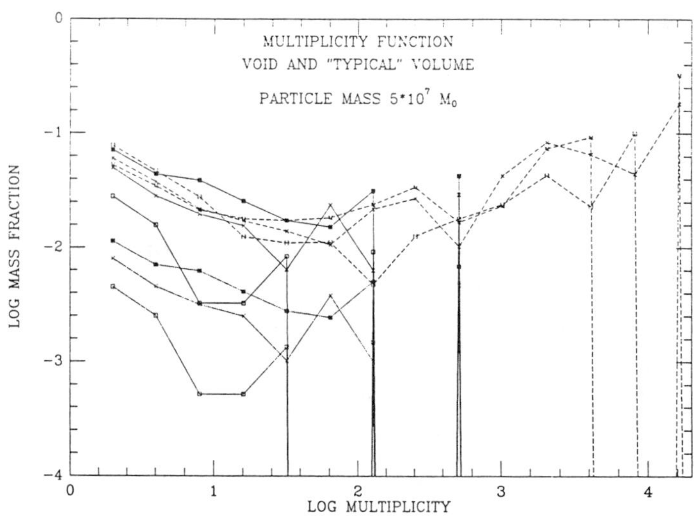

Fig. 8. Multiplicity function for real simulation (thick curves) and calibration run (thin curves) for $B = 0.1, 0.2, 0.3$ (full, dotted, dashed).

the large-scale mass distribution, but bright galaxies are strongly biased. The mean density in the real simulations is 1/6th critical density, so the number density of dwarf galaxies is 1/6th the universal mean. The curve for $B = 0.1$ is significantly below the other curves, which indicates that the dwarf galaxies in the void tend to be low surface brightness. The cutoff in the multiplicity function is not due to lack of particles; at the end of the real simulation there are half as many particles left in the small grids as there are in the calibration run. In the calibration run the multiplicity function cuts off 2 orders of magnitude higher than the real run.

These results can be understood in a qualitative sense if it is assumed that galaxies are formed at high peaks of the initial Gaussian random density field smoothed on the appropriate mass scales (Bardeen *et al.* 1986). In a void, the density fluctuations on galaxy scales are modulated by larger-scale waves that form the void. In CDM the amplitude of density fluctuation is a decreasing function of scale, so the modulation by the large-scale waves becomes more important with increasing scale. It is thus expected that dwarf galaxies will be less biased than large galaxies that form from lower amplitude peaks. This kind of behavior is generic to scenarios where the density fluctuations decrease with increasing scale, and is, for example, not expected for hot dark matter.

These kinds of simulations are useful for testing the scenarios where galaxies form at the peaks of the density distribution. The large-scale waves modulate the small-scale density distribution, and the effects of this modulation on galaxy formation can be studied. Work is continuing on these issues.

REFERENCES

Bahcall, N. A., and Soneira, R. M. 1983, *Ap. J.*, **270**, 20.

Bardeen, J. M., Bond, J. R., Kaiser, N., and Szalay, A. S. 1986, *Ap. J.*, **304**, 15.

Bouchet, F. R., and Hernquist, L. 1988, *Ap. J. Suppl.*, **68**, 521.

Davis, M., Efstathiou, G., Frenk, C. S., and White, S. D. M. 1985, *Ap. J.*, **292**, 371.

Dressler, A., this volume, p. 19.

Efstathiou, G., Davis, M., Frenk, C. S., White, S. D. M. 1985, *Ap. J. Suppl.*, **57**, 241.

Gott, J. R., III, Miller, J., Thuan, T. X., Schneider, S. E., Weinberg, D. H., Gammie, C., Polk, K., Vogeley, M., Scott, J. S., Bhavsar, S. P., Melott, A. L., Giovanelli, R., Haynes M. P., Tully. R. B., and Hamilton, A. J. S. 1989, *Ap. J.*, **340**, 625.

Kirshner, R. P., Oemler, A., Schechter, P. L., and Shectman, S. A. 1987, *Ap. J.*, **314**, 493.

de Lapparent, V., Geller, M. J., Huchra, J. P. 1986, *Ap. J. (Letters)*, **302**, L1.

Primack, J. R., Olivier, S., Blumenthal, G. R., Dekel, A. this volume.

White, S. D. M., Frenk, C. S., Davis, M., and Efstathiou, G. 1987, *Ap. J.*, **313**, 505.

Villumsen, J. V. 1989, *Ap. J. Suppl.*, **71**, 407.

DISCUSSION

M. Davis: It would be easier to interpret your results if you plotted the mass function versus circular velocity, not multiplicity.

J. Villumsen: I agree, but I haven't had the courage to do it. The problem is that my resolution is 40 Kpc, so it may be misleading.

A. Dekel: How do you 'find' galaxies?

J. Villumsen: I use a 'friends-of-friends' algorithm that was generously donated by the 'Gang of Four'. It searches for neighbors within a certain fraction, B, of the mean interparticle spacing and generates groups.

A. Dekel: Could you please summarize your results and how they differ from the results of White *et al.*?

J. Villumsen: Very briefly, the underdensity of galaxies up to $5\text{x}10^9$ to $10^{10}\ M_\odot$ halo mass is the same as the mass underdensity in the voids. There should be very few galaxies over $10^{10}\ M_\odot$ in halo mass. The Gang of Four did not have the resolution to do the job.

A. Dekel: Can you tell whether your 'dwarfs' indeed have low 'surface brightness'?

J. Villumsen: I expect that my dwarfs have low surface brightness. The multiplicity function is significantly higher for $B = 0.3$ than for $B = 0.2$ than for $B = 0.1$. This is in contrast to the 'typical' calibration volume, where there was little difference, so I have a lot of low surface-brightness dwarfs in the void.

S. White: The smallest halos you are testing would have to correspond to galaxies with at most $10^7\ L_\odot$. This is ten magnitudes fainter than L_* and thus much fainter than the kinds of dwarfs which we normally discussed.

J. Villumsen: The faintest dwarfs are not observable at the distance of the Boötes void, but the most massive 'unbiased' dwarfs at $10^{10}\ M_\odot$ should be observable. There are other voids closer to us than Boötes, so the unbiased part of the luminosity function should be observable.

M. Strauss: What is the shape of your void? Is it spherical? Is it connected to other voids?

J. Villumsen: The void is nearly spherical if you use the zero-overdensity contour after smoothing the data with a Gaussian with a σ of one mean-interparticle spacing. The void is connected to several other low-density volumes.

THE DISTRIBUTION OF OVER 2000 GALAXIES IN THE ZONE OF AVOIDANCE NEAR HYDRA

RENEÉ C. KRAAN-KORTEWEG
Astronomisches Institut der Universität Basel
Venusstrasse 7, CH-4102 Binningen, Switzerland

ABSTRACT A galaxy search in the southern galactic plane ($|b| \leq 10°$) revealed over 2200 mainly unknown galaxies with $D \geq 0.2'$ in 15½ fields of the ESO/SERC survey. The innermost 6° of the dust plane remains nearly opaque. But the distribution of the detected galaxies in the more transparent regions discloses a relative overdensity around $\ell \sim 275°$ which can not be explained by the patchy extinction. It can be traced across the galactic plane and seems to be part of a structure which possibly envelops the Antlia and Hydra clusters as well. A feature of this size at the distance range of Hydra will add significantly to the peculiar velocity of the Local Group.

INTRODUCTION

The anisotropy in the cosmic microwave background (CMB) radiation is interpreted as a motion of the Local Group (LG) with respect to a CMB frame, induced by the total gravitational acceleration of the irregular distribution of mass. In the absence of biasing, either the known distribution of galaxies, galaxy clusters, and voids will enable an estimate of the total gravitational vector, or the total extragalactic light flux can expose the gravitational field (Lynden-Bell this volume). Both methods require whole sky coverage. But due to the foreground absorption of the Galaxy, a large part of the extragalactic sky remains hidden (~15%). Therefore, determinations of the gravitational field have had to resort to various filling mechanisms of the plane.

Part of the peculiar velocity of the LG is attributed to the gravitational force of the Virgo cluster (but, see Burstein this volume). Subtracting this component leaves about 500 km s^{-1} towards $\ell = 274°$, $b = 12°$ to be explained (Sandage and Tammann 1984, Shaya 1984). The direction points extremely close to the dust equator and interesting galaxy structures might indeed be obscured in this little-known area. Partly-concealed suspects in this area are the Hydra and/or Centaurus cluster (*cf.* Lynden-Bell 1988) and the Great Attractor (*cf.* Dressler 1988, this volume).

Yet, the optical waveband can give further insight about galaxy structures behind the galactic plane. This was demonstrated by Weinberger (1980) who found an excess number of galaxies in the galactic plane exactly at the location of the Perseus-Pisces supercluster on one side and A569 on the other side of the plane (Focardi *et al.* 1984, Hauschildt 1987).

A similar search with a broader latitude range was begun in the south using the ESO/SERC survey. The apex of the CMB corrected for the Virgocentric component was chosen as starting point. It is close to the Antlia and Hydra clusters (Hopp and Materne 1985). Moreover, the distribution of all galaxies in Lauberts (1982), as well as the distribution of extragalactic optical light (Lynden-Bell this volume) and *IRAS* galaxies (Meurs and Harmon 1988) indicates an extension of these clusters into the galactic plane.

THE DATA

The search is being done by a systematic visual inspection of the IIIa-J film copies of the ESO/SERC survey, using a viewer with a magnification of $f = 50$. The area searched so far encompasses 15½ fields (F91-F93, F125-F129, F165-F170, F211-F213). These roughly 400 square degrees span a range in galactic coordinates of $\ell \sim 266°$ to $296°$ and $b \sim -10°$ to $+8°$ (respectively $\alpha \sim 8^h\ 30^m$ to 12^h, $\delta \sim -47.5°$ to $-67°5$).

A lower diameter limit of 0.2′ was imposed to avoid confusion with faint stars, stellar blends, and to warrant consistency in the identification of the galaxies. Depending on the surface brightness, the smallest galaxies have magnitudes in the order of $B_J = 18 - 19.5$ mag. 1939 galaxies were discovered as well as 334 candidates. Only 85 galaxies were previously recorded in the ESO/Uppsala Survey of the ESO(B) Atlas (Lauberts 1982), hence less than 4% have diameters over 1′ (but they are heavily absorbed!).

Diameters were measured, average surface brightness estimated, and the morphological types described. The diameters in combination with the average surface brightness give a magnitude estimate.

The positions were accurately determined with the OPTRONICs at ESO in Garching. A preliminary comparison of the positions of galaxies from overlapping borders of the fields suggests that the precision lies well below 1 arcsec.

An assessment of the reliability of the diameter and magnitude estimates was made possible by MacGillivray, who extracted the images determined by the automated measuring machine COSMOS for the visually detected galaxies of field 213. The correlation between the diameters is satisfactory with decreasing scatter towards smaller diameters. On average the present diameters are 10% smaller; they correspond to an isophote of about 24.5 mag arcsec^{-2}. This is confirmed independently by the comparison of the 85 galaxies in common with Lauberts. The relation between the estimated magnitudes and the COSMOS values is surprisingly good, with no deviation from linearity up to the faintest galaxies and a scatter of only $\sigma = 0.47$ mag.

Due to the heavy front absorption, the outer parts of the galaxy images - foremost the spiral structure - are partly obscured and often only the bulges remain distinct. Because of the local variability of obscuration, a homogenous type classification could not be achieved. Still, for about 2/3 of the galaxies, at least a rough morphological classification is made. The majority of the galaxies are spirals (~ 75%).

A catalog for the first 10 fields (*cf.* Kraan-Korteweg 1988) listing the above mentioned properties of the galaxies is in preparation (Kraan-Korteweg 1989).

THE DISTRIBUTION OF THE GALAXIES

The distribution of the detected galaxies in galactic coordinates is illustrated in Figure 1. The outlined area encloses the search fields. The apex of the CMB dipole, corrected for Virgocentric infall, is marked with an arrow ($b = 12°$) and coincides with the direction to the Antlia ($b = 18°$) and Hydra ($b = 25°$) clusters. The dots signify the certain galaxies ($N = 1939$); the crosses mark the possible galaxies ($N = 334$).

It is quite obvious from the distribution that the actual dust equator at these longitudes is offset to the south by about 1.5 - 2°. This was already established for longitudes between $\ell = 200° - 330°$ from the hydrogen column densities (Kerr and Westerhout 1965). The detection of galaxies very close to the true galactic equator ($\leq \pm 3°$) is still not very successful.

The more transparent regions demonstrate that the number density of galaxies in the outlined area not only depends on galactic latitude but also on galactic longitude: below the plane the number density distinctly increases from $\ell = 290°$ to around $\ell = 272°$, whereas on the other side a highest density at constant galactic latitude is reached around $\ell = 280°$. The few extremely faint galaxies across the plane seem to connect the relative overdensities.

The number counts of galaxies are, of course, no objective measure of densities since they are strongly influenced by the patchiness of the galactic dust clouds. However, an overlay of the galaxy distribution with the dark clouds and their different opacities (as analyzed by Feitzinger and Stüwe 1984) clearly exhibits that the denser areas are not due to extremely transparent regions. This is independently confirmed by the numerous extinction values in the galactic plane (Neckel and Klare 1984) and at $b = 10°$ (Burstein and Heiles 1982). On the contrary, the extinction values and dark clouds - taken on average - show an increasing gradient towards $\ell = 270°$. This is supported by the appearance of the galaxies themselves: below the plane the galaxies around $\ell = 285°$ are well resolved and the disks of the spirals distinct, whereas in the densest region ($\ell \sim 272°$) merely the bulges of the galaxies remain visible. Thus the evident overdense regions would actually be more pronounced when corrected for extinction.

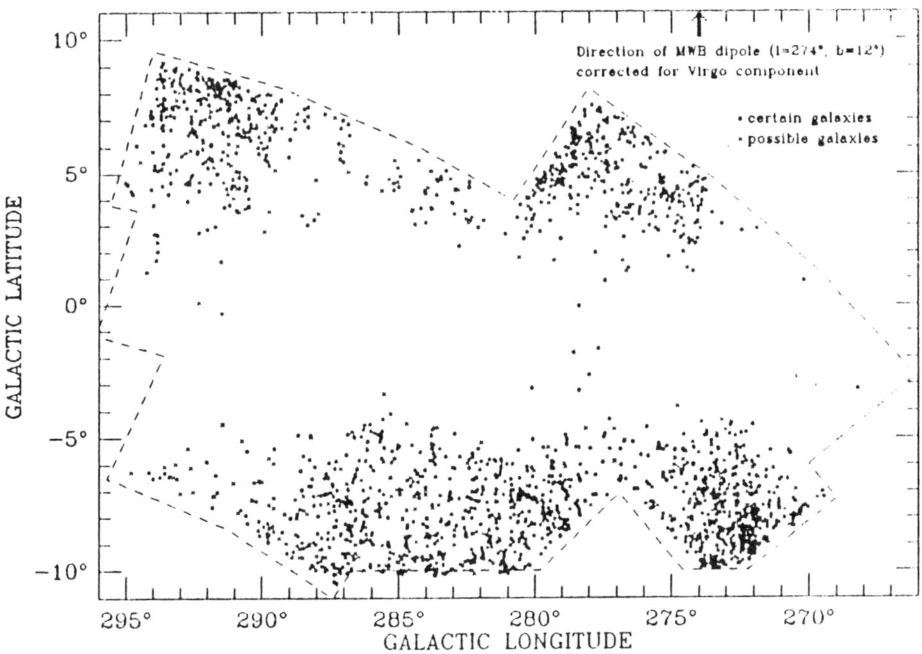

Fig. 1 Distribution of the optically detected 2273 galaxies in the galactic plane.

PROPERTIES OF THE DETECTED GALAXIES

Diameters, Magnitudes and the Effect of Absorption

The galaxies appear on average quite small ($\langle D \rangle \sim 0.4'$) and faint ($\langle B_J \rangle \sim 18$ mag). However, already the galaxies around $b = \pm 10°$ suffer an absorption of about $A_B = 1.0$ to 1.5 mag (A_B is assumed to be equal to A_J), which increases for decreasing latitudes. This is apparent in plots of magnitude or diameter versus cosec(b): while at the highest latitudes a broad range in diameter and magnitude is found, a systematic cut-off discriminates more and more against apparently large and bright galaxies for decreasing latitudes (*e.g.* around $b \sim 3°$ the brightest galaxies have $B_J \geq 18$ mag and $D \leq 20''$).

The effect of extinction at these latitudes is important. When magnitudes are measured at a certain isophote, the absorption A_B will reduce the average surface brightness $\langle SB \rangle$ marginally (if A_B is not too large), but the observed area of the galaxy will shrink considerably. As the magnitude is proportional to $[\langle SB \rangle \cdot \text{area}]$, it will be subjected to an additional dimming Δ (*i.e.* $B_J^0 = B_J - A_B - \Delta$), due to the loss in size, and to a lesser extent to the decrease in $\langle SB \rangle$.

These effects have been studied in detail by Cameron (1989, *cf.* Figure 2), who artificially absorbed the intensity profiles of various galaxies in Virgo. He deduced analytical functions which satisfactorily describe these absorption effects. The functions are distinct for spirals as compared to E and S0 galaxies. For typical absorption values in the plane the effects are significant [*e.g.* for $A_B = 2.5$ mag, $D^0 = 2.9D_{obs}$ ($2.6D_{obs}$) and the additional dimming $\Delta = 0.7$ (0.4) for spirals (E and SO); these values increase dramatically for increasing A_B].

Applying these laws and adopting an idealized absorption law (neglecting the patchiness in extinction) a preliminary correction of the absorption effects on the average parameters of the present galaxy sample yield a mean diameter of $\langle D_{25}^0 \rangle = 0.8'$ and a mean magnitude of $\langle B_J^0 \rangle \sim 16.3$. These values correspond roughly to the expectation values of Hydra and Antlia cluster members.

Radial Velocities

The redshift information compiled from the literature is sparse and restricted mainly to higher latitudes ($|b| \geq \pm 8°$). No reliable distance estimate for the overdensity across the galactic plane is therefore feasible. But the available velocities seem to support the notion that it is at a similar velocity as Antlia and Hydra: around longitudes of $\ell \sim 280° \pm 5°$ the measured velocities in *and* adjacent to the search area ($N \sim 60$) show a strong preponderance for values around $v_0 \sim 2800$ km s^{-1} with very small scatter ($\Delta v \leq \pm 200$ km s^{-1}) - that holds on both sides of the obscured plane.

IRAS Data

Part of the galaxies were rediscovered in the *IRAS* Point Source Catalogue. A cross identification is in principle viable even for the most absorbed galaxies, because the infrared fluxes hardly suffer from extinction, although the increasing level in the background radiation and the high source counts also impedes the qualification as reliable point source at these low latitudes.

Still, 111 of the 2273 galaxies, having reasonable colors ($f_{60} > f_{25}$ or $f_{12}, f_{100} > f_{60}$) were identified in the *IRAS* PSC. In addition, the Reject Catalogue (RejC) was also searched and another 61 identifications are plausible. (The data from the RejC generally has only 1-hour-confirmed sightings and therefore low flux quality. The existence of an optical image at the exact location however enhances the reality of the source. At a later stage the co-added fluxes will enable a confirmation of the tentative identifications.)

At first glance the percentage of detections in *IRAS* seems relatively low (4.1% in the PSC plus 2.7% in the RejC). Considering that identifications were only possible for optically detected galaxies, the actual area in which identifications are possible reduces to about 250 square degrees. The *IRAS* galaxies then have an average density of $\rho = 0.7$ galaxies per square degree. Quoted densities for *IRAS* galaxies with $m_B^0 < 18$ mag and $f_{60} > 0.5$ Jy range between $\rho = 0.25$ galaxies per square degree (Soifer *et al.* 1984), and $\rho = 0.37$ (Wolstencroft *et al.* 1986, hereafter [W86]). These values are not directly comparable. But if the here present *IRAS* galaxies are equally restricted to $f_{60} > 0.5$ Jy one finds $\rho = 0.4$ galaxies per square degree; the corresponding true magnitude limit is certainly brighter than $m_B^0 = 18$ mag, therefore, the density of *IRAS* galaxies with optical identifications in the here regarded highly obscured area is at least equivalent, if not higher than the counts at the South Galactic Pole (W86). The distribution of the *IRAS* galaxies is illustrated in Figure 2. As for all distributions of *IRAS* galaxies the clustering is less pronounced than for the optical survey, because early-type galaxies are hardly sampled in the *IRAS* bands. Adding, though, the color-selected galaxies from the *IRAS* PSC (as defined by Meurs and Harmon 1988 (crosses)) a connection across the plane is exhibited, apparently uniting the optically detected higher density regions.

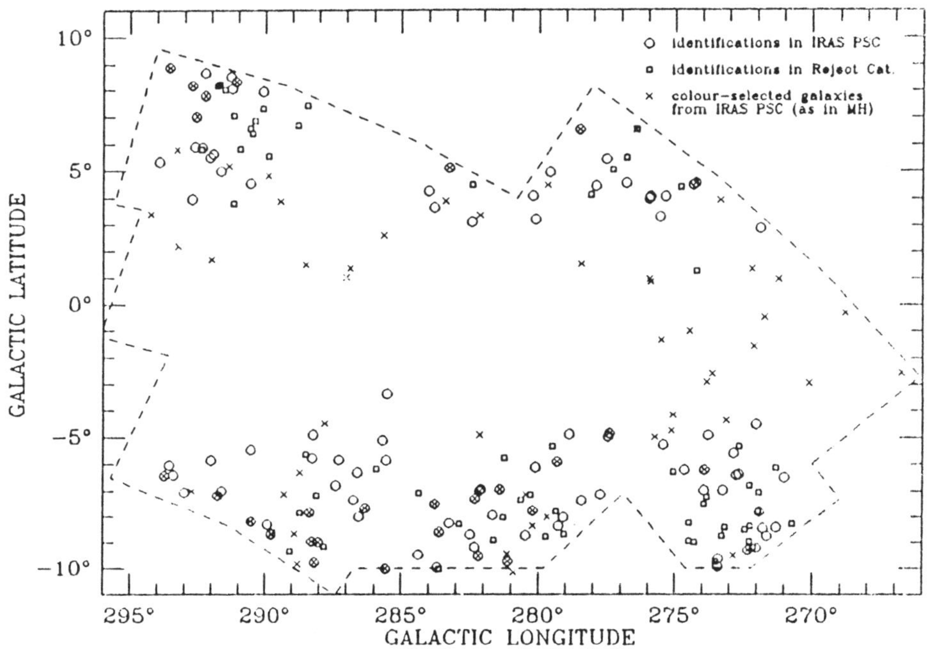

Fig. 2 Galaxies identified in the *IRAS* PSC and color-selected *IRAS* galaxies (not detected in the search).

CONCLUSIONS AND FUTURE PLANS

The distribution of the optically detected galaxies in the galactic plane reveals a relative overdensity around $\ell \sim 275°$, reaching all the way across the plane. This feature does not seem to be an artifact of local variations in the opacity of the galactic plane. The search will be extended at least to $-10° \leq b \leq +8°$ for the longitude range of Figure 1.

Corrections of the diameters and magnitudes for absorption make the majority of the galaxies comparable to Hydra and Antlia cluster members. It is planned to do CCD photometry in the *R* band for the majority of the galaxies, which will allow a better determination of the intrinsic properties of these galaxies. The addition of some *B* photometry will lead to estimates of extinction at specific locations in the plane.

The few available velocity measurements in and adjacent to the detected overdensity support the idea that it stems from a similar distance as the Hydra and Antlia clusters. Optical redshift measurements for some of the denser regions, using the multi-fiber system OPTOPUS of ESO, are in preparation, as well as observations in HI.

The complementation of the galaxies with color-selected galaxies from the *IRAS* PSC discloses a bridge connecting the higher density regions. An indication of this bridge across the plane is also visible in the optical (*cf.* Figure 1), though at slightly larger longitudes. The available data so far suggests that the detected overdensity forms one large structure with Hydra and Antlia, hence extending over at least 35°. It is the only 'nearby' large structure which does not lie in the supergalactic plane. It is expected to disturb the Hubble flow at the location of the LG with a significant component away from the supergalactic plane, towards the observed CMB dipole. The distribution of the total light from galaxies, which uses the cloning method to simulate the galaxy distribution in the galactic plane (Lynden-Bell this volume) does actually reproduce a structure like this; its total gravitational acceleration does approximate the observed apex of the CMB better than any other previous determinations. Future observations will allow a confirmation of this structure, if real.

ACKNOWLEDGEMENTS

I would like to thank H. MacGillivray, who derived the diameters and magnitudes from COSMOS scans for the visually detected galaxies of field F213, E. Meurs for a list of the color-selected *IRAS* galaxies in the relevant area, and B. Madore for the *IRAS* cross identifications at IPAC. Without the financial support of the Swiss National Science Foundation this project could not have been pursued.

REFERENCES

Burstein, D. this volume, p. 39.
Burstein, D., and Heiles, C. 1982, *A. J.*, **87**, 1165.
Cameron, L. M. 1989, preprint.
Dressler, A. 1988, *Ap. J.*, **329**, 519.
Dressler, A. this volume, p. 19.
Feitzinger, J. V., and Stüwe, J. A. 1984, *Astr. Ap. Suppl.*, **58**, 365.
Focardi, P., Marano, B., and Vettolani, G. 1984, *Astr. Ap.*, **136**, 178.
Hauschildt, M. 1987, *Astr. Ap.*, **184**, 45.
Hopp, U., and Materne, J. 1985, *Astr. Ap. Suppl.*, **61**, 93.
Kerr, F. J., and Westerhout, G. 1965, in *Galactic Structure*, (Chicago: University of

Chicago), p. 186.
Kraan-Korteweg, R. C. 1988, in *Large Scale Structure: Observations and Instrumentation*, ed. C. Balkowski and S. Gordon (Meudon: Observatoire de Paris), p. 155.
Kraan-Korteweg, R. C. 1989, in preparation.
Lauberts, A. 1982, *The ESO/Uppsala Survey of the ESO(B) Atlas*, (Garching: European Southern Observatory).
Lynden-Bell, D. this volume, p. 1.
Lynden-Bell, D., Faber, S. M., Burstein, D., Davies, R. L., Dressler, A., Terlevich, R. J. and Wegner, G. 1988 *Ap. J.*, **326**, 19.
Meurs, E. J. A., and Harmon, R. T., *Astr. Ap.*, **206**, 53, (MH).
Neckel, Th., and Klare, G. 1980, *Astr. Ap. Suppl.*, **42**, 251.
Sandage, A. and Tammann, G. A. 1984, in *Large Scale Structure of the Universe, Cosmology and Fundamental Physics, First ESO-CERN Symp.*, ed. G. Setti and L. van Hove, (Garching: European Southern Observatory), p. 127.
Shaya, E. J. 1984, *Ap. J.*, **280**, 470.
Soifer, B. T. *et al.* 1984, *Ap. J. (Letters)*, **278**, L71.
Weinberger, R. 1980, *Astr. Ap. Suppl.* **40**, 123.
Wolstencroft, R. D., Clowes, R. G., Kalafi, M., Leggett, S. K., MacGillivray, H. T. and Savage, A. 1986, in *Light on Dark Matter*, ed. F. P. Israel (Dordrecht: Reidel), p. 425, (W86).

DISCUSSION

D. Burstein: In the Seven-Samurai survey, we found that extinctions measured by HI column density alone (as in the case south of $\delta = -23°$) are about two times too big in the region $230° < \ell < 310°$, $-15° < b < 18°$. We have corrected our reddenings via the Mg_2 *versus* (*B-V*) relation for ellipticals in this region (Burstein *et al.* 1987).

You have done an outstanding job on this survey. As we have discussed before, I believe this is one of the most important galaxy surveys currently being done. Is it possible that you can extend this survey around the Galactic equator? After all, your eyes are better calibrated than anyone else's.

R. Kraan-Korteweg: As I mentioned in my talk, I will definitely fill in the area between $+8°$ and $-10°$ in Galactic latitude within the area given in Figure 1, to get rid of the nasty corners. After this I plan to continue also further along the Galactic Plane. However, since doing this search is an extremely time-consuming project, I really do not want to state at this point of time, to how far I will extend it.

D. Lynden-Bell: First, congratulations on heroic, or should I say heroinic work. It should be possible to get surface densities of 21-cm emission to compare with your estimates of absorption. All this region has been measured, but the integrated emission will only be available from the observers.

D. Burstein: Unfortunately HI measures at $|b| < 10°$ are significantly affected by HII regions and molecular hydrogen. This leads both to large fluctuations on small size scales and in HI gas-to-dust ratio. I would prefer to use the Mg_2 *versus* color relation for ellipticals to measure reddening, and assume $A_B = 4 \cdot E(B\text{-}V)$. This should be accurate to ~ 0.02 or 0.03 mag in *E(B-V)*. Thus I would greatly encourage Reneé, in her redshift survey, to get good signal-to-noise spectra of ellipticals, from which Mg_2 can be accurately measured (as well as velocity dispersion, of course).

M. Strauss: You said that there were a number of *IRAS* galaxies from the Meurs and Harmon survey that don't appear in your sample. Have you gone back to the ESO plates to try to identify these objects? We have also done identifications of *IRAS* galaxies close to the plane in the Hydra region. We can send you our identifications to cross-correlate with your lists, if you'd like.

R. Kraan-Korteweg: I received the *IRAS* galaxies in the plane near Hydra just recently from Meurs. I did not yet check the fields again, but I certainly intend to do that. I am naturally interested in receiving your identifications. It would allow some check on the completeness of the present search.

R. Giovanelli: For how many of these objects can you reliably tell the morphological type?

R. Kraan-Korteweg: I cannot quote exact numbers, but there are about 25% without morphological types. Of the remaining galaxies the majority are spirals (~75%). Whenever the resolution was sufficient - respectively the extinction was low - a more detailed classification besides E or S was attempted. This was possible for about 25% of the detected galaxies. The catalog in preparation will include a column referring to the uncertainty of the galaxian type and/or subtype.

TOPOLOGY OF LARGE-SCALE STRUCTURE: RECENT RESULTS

J. RICHARD GOTT, III
Department of Astrophysical Sciences, Peyton Hall
Princeton University, Princeton, NJ 08544

ABSTRACT This conference on large-scale structure and motions in the universe is in part to celebrate the completion of the Southern Sky Redshift Survey (da Costa *et al.* 1988). In this paper I shall show some of our first results analyzing the topology in this sample (Park, Gott, da Costa, and Melott 1989). Also, I will discuss some preliminary results from our 2-D analysis of the CfA slice (Park, Gott, and Melott 1989) and relate these to our previous studies.

In a recent series of papers (Gott, Melott, and Dickinson 1986; Gott, Weinberg,and Melott 1987; Hamilton, Gott, and Weinberg 1986; Melott, Weinberg, and Gott 1988; Weinberg, Gott, and Melott 1987), we have outlined a method for quantitatively measuring the topology of structures in 3-D. We first smooth the data with a gaussian smoothing function $W(r) \propto \exp(-r^2/\lambda^2)$, where λ is the smoothing length. Then we define density contour surfaces by the fraction f of the volume on the high density side where ν is defined by

$$f = 2(\pi)^{-1/2} \int_{\nu}^{\infty} \exp(-t^2/2)\, dt \tag{1}$$

Thus contours are labeled by ν. We have shown that for random phase initial conditions (as occur in the standard model where the structure we see today grows from small fluctuations due to random quantum noise in the early universe) as long as λ is larger than the mean particle separation and the correlation length, the topology we see today reflects that of the initial conditions. The topology of a density contour surface is measured by its genus defined as

$$G_{3\text{-D}} = \text{Number of Holes} - \text{Number of Isolated Regions} \tag{2}$$

For random phase initial conditions the genus per unit volume is

$$g_{3\text{-D}} = N_3\, (1 - \nu^2) \exp(-\nu^2/2), \tag{3}$$

where

$$N_3 = (2\pi)^{-2} \left(\langle k^2 \rangle /3\right)^{3/2} \tag{4}$$

is the appropriate moment over the smoothed initial power spectrum (Hamilton, Gott, Weinberg 1986; Bardeen *et al.* 1986; Doroshkevich 1970; Adler 1981). Note that for a random phase distribution, which is what we expect from the standard model, the form of the genus curve $g_{3\text{-D}}(\nu) \propto (1 - \nu^2) \exp(-\nu^2/2)$ is independent of the power spectrum, while its amplitude depends on the power spectrum.

For a 2-D slice, the genus may be defined as

$$G_{\text{2-D}} = \text{number of isolated high density regions} - \text{number of isolated low density regions} \quad (5)$$

For a random phase distribution the genus per unit area is given by

$$g_{\text{2-D}} = N_2 \, \nu \, \exp(-\nu^2/2) \quad (6)$$

where

$$N_2 = (2\pi)^{-3/2} \langle k^2 \rangle /2 \quad (7)$$

is an appropriate moment over the 2-D power spectrum (Melott *et al.* 1989).

For a 1-D pencil beam survey the genus may be defined as

$$G_{\text{1-D}} = \text{Number of High Density Regions} \quad (8)$$

and for a random phase distribution the genus per unit length is

$$G_{\text{1-D}} = N_1 \exp(-\nu^2/2) \quad (9)$$

where

$$N_1 = (2\pi)^{-1} (\langle k \rangle)^{1/2} \quad (10)$$

(Ryden 1988; Ryden *et al.* 1989).

We have recently completed an extensive study of the 3-D topology of a large number of observational samples (Gott *et al.* 1989). We find that for $\lambda \geq 1200$ km s^{-1} the distributions in redshift space follow the random phase law quite accurately. This is consistent with the standard model in which the structures we see today grew from random quantum noise in the early universe. For $\lambda \sim 600$ km s^{-1} we find a genus curve that is approximately random phase but which shows a slight shift to the left in the direction of a 'meatball topology.' CDM models show similar shifts but not quite as large as those shown in the observations.

We have now studied the topology of the Southern Sky Redshift Survey (SSRS), (Park, Gott, da Costa, and Melott 1989), at a smoothing length of $\lambda = 650$ km s^{-1}. Its genus curve is shown in Figure 1. The open squares show the raw data with uniform average density assumed outside the survey region. The filled circles with error bars are the statistical bootstrap estimates of the genus from the data (*cf.* Barrow, Bhavsar, and Sonoda 1984; Gott *et al.* 1989) which have been found from *N*-body simulations to give the most accurate measure of the topology and its measurement errors. It is quite consistent with the best fitting random phase topology curve, which is shown for comparison. Of our five observational samples at 600 km s^{-1}, two are now perfectly random phase, and three show slight meatball shifts. Biased CDM model simulations of the size of our observational samples for comparison appear random phase about 75% of the time and show slight meatball shifts about 25% of the time. The addition of the SSRS sample thus acted to improve the agreement with the biased CDM model. Both the data and the biased CDM model produce samples which are either random phase or show slight 'meatball' shifts.

Park, Gott, and Melott (1989) are conducting a 2-D topology study of the original

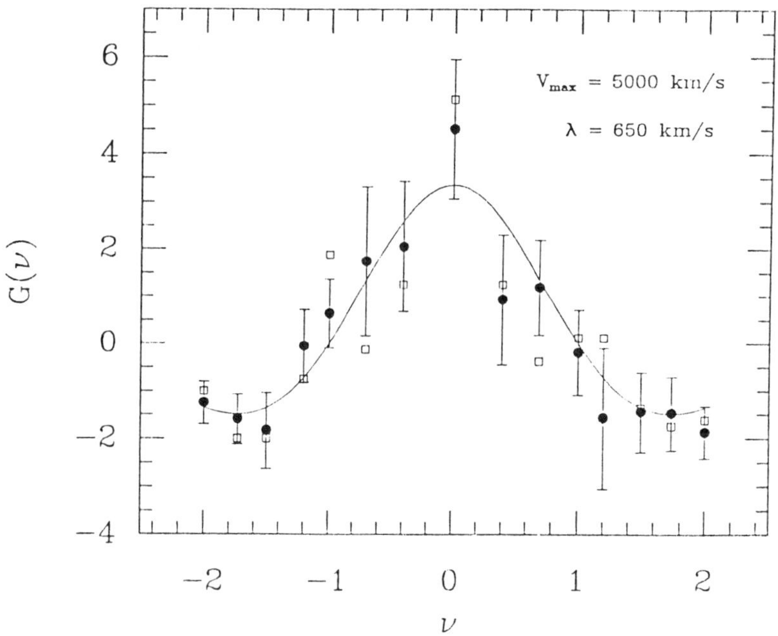

Fig. 1. The 3-D genus curve for the southern sky redshift survey out to a redshift of 5000 km s^{-1}, smoothed at $\lambda \sim 650$ km s^{-1}, and corrected for the selection function in the usual way. The open squares show the genus assuming average density in the external region while the filled circles show with error bars the result of 12 statistical bootstrap runs. The statistical bootstrap results assume a Poisson distribution of galaxies in the external region. Simulations show that in general the bootstrap means give a measure of the topology which is less subject to boundary conditions and shot noise, and the error bars give a reasonable estimate of the measurement errors. Also shown as a solid line is the best fit theoretical 3-D genus curve (equation 3) expected for a random phase distribution. The data appear to be consistent with that expected from a random phase distribution (from Park, Gott, da Costa, and Melott 1989).

de Lapparent, Geller, and Huchra (1986) CfA slice. The genus curve for this sample is shown in Figure 2 with the same symbols as in Figure 1. The random phase curve is shown for comparison. In the region $-1 < \nu < 1$ the data are consistent with a random phase topology. In the regions $\nu < -1$ and $\nu > 1$ the curve shows significant shifts in the meatball direction. The meatball shift in the $\nu > 1$ part of the curve is clearly due to the presence and prominence of the Coma cluster in the sample. We should caution that this is still a rather small sample (containing about 1100 galaxies) as compared with the combined 3-D samples (9000 galaxies) studied in Gott *et al.* 1989, and further note that this first slice by being centered on the Coma cluster may not be typical. However, the results certainly appear promising and encourage us to try more and larger 2-D samples.

Ryden *et al.* (1989) have studied the 1-D topology of the same samples that Gott *et al.* (1989) studied in 3-D. This was done by dividing the entire survey region into a set of pencil beam surveys. When the 3-D study of a sample indicated agreement with a random phase topology, so did the 1-D topology study, and when a slight meatball shift was seen in a sample in 3-D, it showed up also in the 1-D study. Thus we have two independent

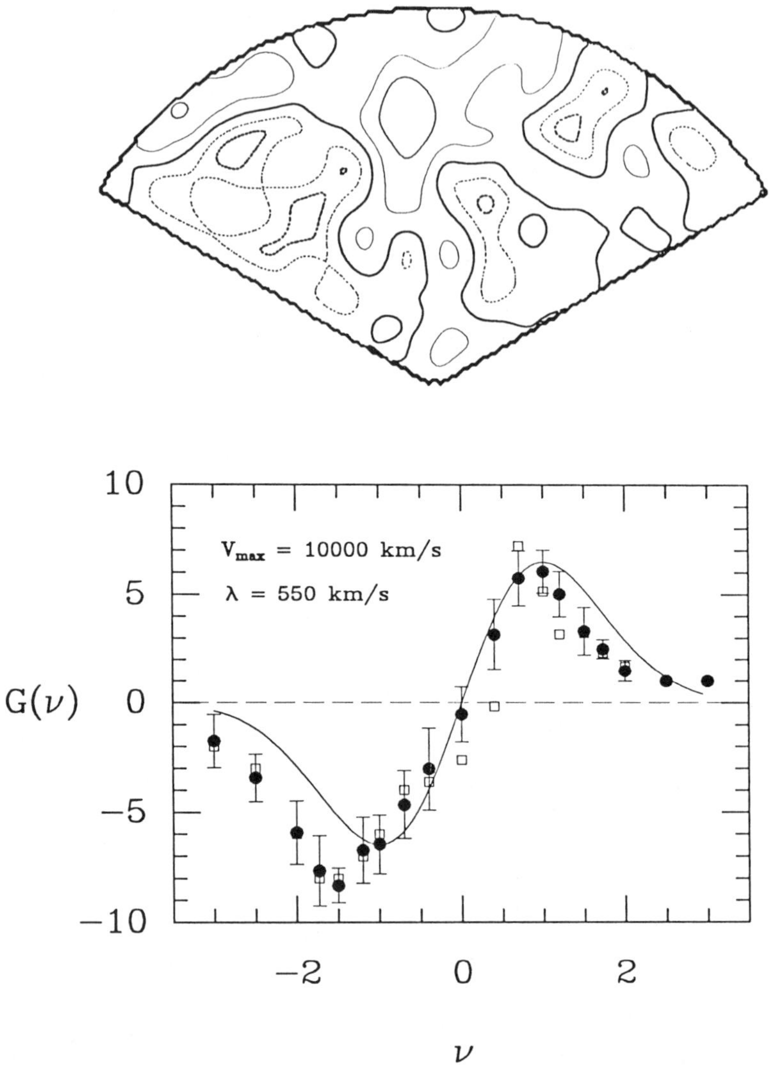

Fig. 2. (top) The first CfA slice out to a redshift 10,000 km s^{-1}, smoothed at $\lambda = 550$ km s^{-1} assuming average density in the external region, and corrected for the selection function in the usual way. The median density contour is shown as the heaviest solid line. The $+1\sigma$ contour (by area fraction) is shown as the lightest solid line, while the $+2\sigma$ contour (primarily enclosing the Coma cluster) is shown as a solid line of intermediate thickness. The -1σ and -2σ contours are shown as light and heavy dotted lines respectively. The contours follow rather nicely what the eye sees in the data. (bottom) The 2-D genus curve for the CfA slice, with raw and statistical bootstrap data points as described in Figure 1. Also shown for comparison is the best fit theoretical 2-D genus curve (equation 6) expected for a random phase distribution. Over the range $-1 < \nu < 1$ the data are consistent with that expected for a random phase distribution. But for $\nu < -1$ and $\nu > 1$ the data show a significant shift to the left toward a 'meatball' topology (from Park, Gott, and Melott 1989).

methods which give the same results.

To summarize, we have studied the topology of various observational samples with 3-D, 2-D, and 1-D techniques. In all cases the results are either consistent with random phase or show slight shifts in the meatball direction, a pattern also shown in biased CDM models.

We acknowledge support from NSF Grant AST87-21484 and NASA Grant NAGW-765. We thank Marc Postman for giving us the CfA slice data. It is also a particular pleasure to thank L. N. da Costa for hospitality in Brazil.

REFERENCES

Adler, R. J. 1981, *The Geometry of Random Fields*, (Chichester: Wiley).

Bardeen, J. M., Bond, J. R., Kaiser, N., and Szalay, A. S. 1986, *Ap. J.*, **304**, 15.

Barrow, J. D., Bhavsar, S. P., and Sonoda, D. M. 1984, *M.N.R.A.S.* **210**, 19P.

da Costa, L. N., Pellegrini, P. S., Sargent, W. L. W., Tonry, J., Davis, M., Meiksin, A., Latham, D. W., Menzies, J. W., and Coulson I. A. 1988, *Ap. J.*, **327**, 544.

de Lapparent, V., Geller, M. J., and Huchra, J. P. 1986, *Ap.J. (Letters)*, **302**, L1.

Doroshkevich, A. G. 1970, *Sov. Ap.*, **6,** 320 (translated from *Astrofizika*, **6**, 581).

Gott, J. R., III, Miller, J., Thuan, T. X., Schneider, S. E., Weinberg, D. H., Gammie, C., Polk, K., Vogeley, M., Scott, J. S., Bhavsar, S. P., Melott, A. L., Giovanelli, R., Haynes M. P., Tully. R. B., and Hamilton, A. J. S. 1989, *Ap. J.*, **340**, 625.

Gott, J. R., III, Melott, A. L., and Dickinson, M. 1986, *Ap. J.*, **306**, 341, (GMD).

Gott, J. R. III, Weinberg, D. H., Melott, A. L. 1987, *Ap. J.*, **319**, 1, (GWM).

Hamilton, A. J. S., and Gott, J. R., III 1988, *Ap. J.*, **331**, 641.

Hamilton, A. J. S., Gott, J. R., III,, and Weinberg, D. H. 1986, *Ap. J.*, **309**, 1, (HGW).

Melott, A. L., Cohen, A. P., Hamilton, A. J. S., Gott, J. R., III, and Weinberg, D. H. 1989, *Ap. J.*, **345**, 618.

Melott, A. L., Weinberg, D. H., and Gott, J. R., III 1988, *Ap. J.*, **328**, 560.

Park, C., Gott, J. R., and Melott, A. L. 1989, in preparation.

Park, C., Gott, J. R., da Costa, L. N., and Melott, A. L. 1989, in preparation.

Ryden, B. S. 1988, *Ap. J. (Letters)*, **333**, L41.

Ryden, B. S., Scherrer, R. J., Melott, A. L., Craig, D. A., Gott, J. R., III, Weinberg, D. H., Bhavsar, S. P., and Miller, J. M. 1989, *Ap. J.*, **340**, 647.

Weinberg, D. H., Gott, J. R., III, and Melott, A. L. 1987, *Ap. J.*, **321**, 1, (WGM).

DISCUSSION

S. Shandarin: Did you calculate the density distribution function after smoothing? Was it always Gaussian?

R. Gott: The density distributions after smoothing at $\lambda \sim 600$ km s^{-1} were somewhat non-Gaussian (the median density was somewhat below the mean). We are smoothing at basically a correlation length where $\delta\rho/\rho$ is of order unity.

S. White: In your talk you emphasized that the data differ from the CDM model in the sense of having more pronounced 'meatball' topology. This seems to suggest that high-density structures seen in the data are less connected than in the model. How does this square with the usual criticism that CDM produces less coherent large-scale structure than is observed?

R. Gott: A 'meatball shift' in the data means that the clusters are more prominent than the voids. In the Giovanelli and Haynes sample, the single connected structure of the Perseus-Pisces Supercluster is so prominent that it pushes the curve in the meatball direction. It should be noted that this is only a shift from the random-phase model. Even in the Giovanelli and Haynes sample the median density contour is sponge-like, so the high density regions are still all connected.

S. White: Can you get useful topological information from the projected galaxy distribution? This would allow you to make use of data sets containing millions of galaxies.

R. Gott: Yes, we can study the two-dimensional topology to determine whether a projected distribution on the sky has random phase at larger scales. We are working on this now.

S. Matarrese: When you compare the random-phase theoretical curves with the data, what fixes the height of the theoretical curve? Do you use the same fitting radius that you used for the data?

R. Gott: For each figure we produce a theoretical curve whose amplitude gives the best fit to the data (produces the best χ^2 fit). You can see that the best-fit amplitude we get for our combined data (including the Southern Sky Redshift Survey) is $G_s \sim 20$, with a smoothing length of $\lambda = 600$ km s^{-1}, while the predicted amplitude at that smoothing length from the CDM model is $G_s \sim 23$, in excellent agreement with the data.

N. Kaiser: I have a comment on the 'bootstrap' error bars. It seems to me that whenever one is trying to infer some statistical property of the universe from a survey of finite volume there are two types of sampling error. One source of error is that the density field may not be representative, and the other is that we sample the density with a set of points. The bootstrap method only takes account of the latter, yet experience from other studies of large-scale structure shows that this is usually not dominant.

R. Gott: Yes, the bootstrap error tells just the statistical measurement error, the errors we make in measuring the structure we see because we are sampling a finite series of points. The other source of error, whether the sample we have is a fair sample of the universe, may be estimated from counting statistics, since we are basically counting structures (holes). If the amplitude of the genus curve is N, then this fair sampling error is of order $(N)^{1/2}$. We have also estimated the extent to which the peak in the genus curve might be shifted due to sampling errors. Our key observed samples (Giovanelli and Haynes, CfA) are large enough that the meatball shifts we have observed in these samples are two or three times larger than expected from sampling fluctuations.

A. Dressler: Suppose the cell walls in the CfA Slice are in fact thin. Are you giving up anything in terms of diagnostics by smoothing with the mean interparticle spacing (which includes the void volume as well)?

R. Gott: Thin walls are naturally produced by non-linear effects (caustics) in the standard model, as Shandarin will discuss in his talk. We are interested in the linear-regime effects. We are interested in where the walls are, not how thin they are. Using smoothing we are trying to recover the topology of the fluctuations in the linear regime which could have grown into the structure we see today. We can then test to see if this topology has random phase as expected for the standard model.

D. Burstein: Let me ask the 'naive' observer's question: to what extent do you and others take into account 'Fingers of God' caused by the velocity dispersion in clusters? In particular, the dispersion in dense clusters like Coma and Perseus?

R. Gott: As you can see in the CfA study, the only effect of the Finger of God in the Coma Cluster is to make its contour somewhat elliptical, pointed towards earth. It changes the genus not at all. In general we have shown, using *N*-body simulations, that since we are using a smoothing length of $\lambda \sim 600$ km s^{-1} or more, and since peculiar velocities are generally ≤ 600 km s^{-1}, their effect on the measured topology is negligible.

SOME INTERESTING PROPERTIES OF THE GALAXY DISTRIBUTION

DIEGO G. LAMBAS AND HERNÁN MURIEL[1]
Observatorio Astronómico Córdoba, Laprida 854, 5000 - Córdoba, Argentina.

ABSTRACT We report on the results of new statistics designed to study anisotropies in the distribution of galaxies around brightest cluster galaxies and ellipticals. The results indicate that the alignment effects related to the position angles of both brightest cluster galaxies and ellipticals are consistent with the presence of filaments or sheets of galaxies rather than with a single overdensity on one side of the objects as would be expected in a tidal picture for the origin of alignments.

INTRODUCTION

Previous studies regarding the alignment of galaxies and cluster major axes with the large-scale galaxy distribution have been extensively analyzed (Carter and Metcalfe 1980, Binggeli 1982, Djorgovski 1987, Argyres *et al.* 1986 [AGPS]; Lambas *et al.* 1988a,b [LGPa,b]). In these last works there is convincing evidence of alignment of position angles of ellipticals (Es) and brightest cluster galaxies (BCGs) with the counts in the Lick maps. These effects are expressed using an anisotropy parameter, b, in the two-point correlation function

$$\omega(\theta,\phi) = \omega(\theta)\,[1 + b\cos(2\phi)\,]$$

where ϕ is the azimuthal angle with respect to the object's position angle (PA). In LGPb this parameter is found to be remarkably similar for clusters and ellipticals, $b \sim 0.10 \pm 0.05$.

Several questions arise from these results. One of them is the geometry associated with these alignment effects. If the alignments come from enhanced counts on both sides of the objects the two models studied by AGPS may apply, namely filaments or sheets of galaxies. However, if the counts are high only on one side of the objects it is possible to speculate for a tidal origin of the alignment effects, where no filaments or sheets would be required.

Many authors have analyzed the possible existence of a filamentary structure in deep galaxy catalogs (Groth and Peebles 1977, Kuhn and Uson 1982, Moody *et al.* 1983, Dobrzycki 1989). The detection of a filamentary structure in a clustered pattern is highly sensitive to the method applied. There is a great variety of results in the literature, and one cannot extract firm conclusions regarding the isolation of filaments in the clumpy galaxy distribution.

In this work we apply statistics using the information contained in the position angles, unbiased by the galaxy distribution, in the search for the presence of filaments.

The statistics and the results are described in the next section. We conclude that the alignment effects are real, consistent with a filamentary pattern.

---

[1]Consejo Nacional de Investigaciones Científicas y Técnicas

THE DATA

The data consist of magnetic versions of the Uppsala General Catalog of bright galaxies (Nilson 1973), Catalogs of rich clusters of galaxies (Abell 1958, Shectman 1985) and the Lick maps (Shane and Wirtanen 1967, Seldner *et al.* 1977).

The PAs of galaxies were taken from the Uppsala catalog, where we have added those given in table I of LGPb for ellipticals. The PAs of Shectman clusters were derived directly from the counts in the Lick maps in a circle of 0.5° radius centered on the objects. The PAs of Abell clusters were taken as the PAs of the brightest cluster member from Struble and Peebles (1985).

ANALYSIS

We have used a cross-correlation analysis between objects taken from the catalogs (centers) and the counts in the Lick maps. A list of the samples of center objects (galactic latitude greater than 40°) is given in table I.

TABLE I Number of Centers in the Samples

E	S(b/a<0.3) (S)	S0(b/a<0.3) (S0)	Abell (A)	Shectman (Sh)	E+Abell (E+A)
162	472	261	134	409	296

Alignment analysis

We have applied a similar procedure to that of AGPS and LGP. We fit the counts around a typical object of a given sample with:

$$N(\theta,\phi) = \langle N\rangle[1 + \omega(\theta,\phi)] = \langle N\rangle\{1 + \omega(\theta)\,[1 + b\cos(2\phi)\,]\,\} \qquad (1)$$

where $\langle N\rangle$ is the average count per Lick cell (10′ × 10′).

We have binned the counts in circular sectors of 15° and, as in AGPS and LGP, in 6 rings (see Table II). This table shows the values of the alignment parameter b obtained for all the samples. It is seen in the table that Es and BCGs are aligned with the neighboring galaxy distribution, consistent with AGPS and LGP ($b \sim 0.1$). Also, one notices that the samples of edge-on spirals and S0s show no alignment effects and that the strong anisotropy imposed by the adopted PAs in Shectman clusters drops to zero at $\theta > 1°$.

TABLE II Rings Used to Calculate the Alignment Parameter

Sample	0.25-0.5°	0.5-1.0°	1.0-2.0°	2.0-4.0°	4.0-8.0°
E	.13 ±.04	.13 ±.04	.09 ±.03	.09 ±.03	.03 ±.04
S	.04 ±.05	.07 ±.03	.02 ±.02	−.03 ±.02	−.02 ±.02
S0	.08 ±.05	.06 ±.04	.03 ±.03	−.03 ±.03	−.03 ±.03
A	.16 ±.05	.09 ±.05	.11 ±.05	.08 ±.03	−.03 ±.03
Sh	.25 ±.02	.06 ±.02	.03 ±.02	.01 ±.02	.01 ±.02

The lack of an alignment signal ($\theta > 1°$) in Shectman clusters is encouraging in the sense that possible bias in the Lick maps would be affecting the selection criteria of these clusters as well as their PAs. On the contrary, Abell clusters which have a significant alignment signal are independent of the Lick maps. This gives additional support to the physical link between PAs of early-type galaxies and the distribution of galaxies on large scales.

Single Overdensity and Filament Models

Using the counts around each object we proceeded as follows: we defined $N_1(\theta, \mathrm{j})$, $N_2(\theta, \mathrm{j})$ the mean density in the ring θ, centered in the jth object, within 45° in the direction of the object's PA (parallel direction) and N_3, N_4 the corresponding densities in the perpendicular direction. We normalize using

$$n_i(\theta, \mathrm{j}) = N_i(\theta, \mathrm{j}) \,/\, n(\theta, \mathrm{j}) \quad i = 1,4$$
$$n(\theta, \mathrm{j}) = 1/4 \sum N_i(\theta, \mathrm{j}) \quad i = 1,4 \tag{2}$$

where the normalization $n(\theta,\mathrm{j})$ gives equal weight to all centers j and rings θ. This procedure excludes the possibility that a few centers dominate the statistics.

Due to the presence of the alignment effects we will argue that the perpendicular alignment (3,4) has typical counts and that the parallel alignment (1,2) may be generated from (3,4) by adding either a single overdensity on one side, in which case the alignments could be associated with tides, or (1,2) might be derived from (3,4) by the addition of counts on the two sides, which would be directly related to the presence of filaments or sheets of galaxies. With

$$n_1(\theta, \mathrm{j}) = \max\{n_1(\theta, \mathrm{j}), n_2(\theta, \mathrm{j})\}$$
$$n_2(\theta, \mathrm{j}) = \min\{n_1(\theta, \mathrm{j}), n_2(\theta, \mathrm{j})\}$$
$$n_3(\theta, \mathrm{j}) = \max\{n_3(\theta, \mathrm{j}), n_4(\theta, \mathrm{j})\} \tag{3}$$
$$n_4(\theta, \mathrm{j}) = \min\{n_3(\theta, \mathrm{j}), n_4(\theta, \mathrm{j})\}$$

we have in these models

$$n_1(\theta) = n_3(\theta) + \Delta_1$$
$$n_2(\theta) = n_4(\theta) + \Delta_2 \tag{4}$$

where $n_i(\theta) = \langle n_i(\theta, \mathrm{j})\rangle$ is the average across clusters. Thus, if $\Delta_1 \sim \Delta_2$ we have a pure filament, and if $\Delta_2 \sim 0$ a pure single overdensity.

Results

It can be seen from Table III that in the samples E, A, and E+A where the alignment effects are strongly detected ($\theta < 4°$), $n_2 > n_4$ by a similar amount as $n_1 > n_3$, although we found that in general $n_2 < (n_3 + n_4)/2$.

Two facts might affect our simple filament model of equation (1), decreasing the differences in $\Delta_2 = n_2 - n_4$. First, filaments in the sky have a finite extension (detected alignment at $\theta < 4°$) and second, filaments might be fragmented. We can see in Table III, sample E+A, that Δ_2 is slightly smaller than Δ_1. This could be produced by the mentioned effects.

TABLE III Mean Differences Δ_1 and Δ_2

Sample	0.25 - 0.5°		0.5 - 1.0°		1.0 - 2.0°		2.0 - 4.0°		4.0 - 8.0°	
	Δ_1	Δ_2	Δ_1	Δ_2	Δ_1	Δ_2	Δ_1	Δ_2	Δ_1	Δ_2
E	.10	.09	.07	.01	.05	.06	.04	.05	−.02	.02
S	.04	−.01	.01	.00	.01	.00	.00	−.02	−.01	−.01
SO	.05	.02	.07	−.01	.06	−.01	−.03	−.01	−.02	−.03
A	.16	.10	.09	.09	.08	.04	.10	.02	.01	.34
Sh	...	...	.04	.03	.03	.01	.02	.00	.03	−.01
E+A	.13	.10	.08	.05	.07	.05	.07	.04	−.01	.17
E+A 45°	−.06	.02	−.07	−.02	−.02	.00	.00	−.02	.03	.01
E+A Rnd	−.05	−.05	.01	−.02	.00	.00	.01	.00	.02	.01

It is remarkable that Shectman clusters, which are directly obtained from the Lick counts and their PAs derived from the counts in the inner 0.5°, show no strong alignment nor filament signals even in the closer rings ($\theta > 1°$). We conclude from these results that PAs are powerful predictors of the surrounding structure at large scales and also indicate that possible bias in the Lick maps could hardly produce the alignment and filament signals detected.

Confidence Tests

We have added 45° to all PAs (equivalent to rotating the distribution 45°). We expect a null result either for the alignment or filament effects ($\Delta_1 \sim 0$, $\Delta_2 \sim 0$). As shown in Table III the results for sample E+A (45°) are consistent with this picture. Typical fluctuations in Δ_1, Δ_2 around the mean are 0.02 on scales $1° < \theta < 4°$, and 0.03 at scales $< 1°$.

Assigning random PAs to the center galaxies we expect null results for the alignment and filament tests. Table III shows the results of this test applied to sample E+A. It can be seen in this table that this test also gives support to the reality of the results ($\Delta_1 \sim 0$, $\Delta_2 \sim 0$).

Monte-Carlo Error Estimates

We generated 1000 fiducial samples corresponding to each real sample with random PAs. Next, we counted the number of times M_1, M_2, M_3 that a random fiducial sample gives mean values Δ_1, Δ_2 and $\Delta_1+\Delta_2$ greater than those obtained for the real sample, respectively. The confidence in the detection of alignments (P_3), and filaments (simultaneously P_1, P_2) are therefore

$$P_i = 100(1 - M_i/1000)\ \%.$$

The confidence values derived for different samples are shown in Table IV. We can see in this table the high level of confidence of both alignment and filament effects in samples A, E and E+A on scales $\theta < 4°$. In contrast with this situation E(45°), A(45°), A+E(45°), and S exhibit a lack of systematics reflected in the low confidence values.

We conclude from these tests that the alignments effects are real and they are related to a large-scale filamentary pattern.

TABLE IV Confidence Estimates (*via* Monte-Carlo Simulation) %

Sample	0.25 - 0.5°			0.5 - 1.0°			1.0 - 2.0°			2.0 - 4.0°			4.0 - 8.0°		
	P_1	P_2	P_3	P_1	P_2	P_3	P_1	P_2	P_3	P_1	P_2	P_3	P_1	P_2	P_3
E	95.7	99.4	99.1	94.0	70.6	92.4	92.6	99.4	99.0	92.4	99.6	99.4	11.5	92.3	46.9
E 45°	12.7	66.8	26.1	43.4	21.9	32.8	39.5	45.3	40.2	70.5	4.6	35.5	91.5	87.3	93.9
A	99.0	99.6	99.9	97.4	99.9	99.6	98.4	96.0	99.3	99.8	85.4	99.8	61.1	90.4	89.8
A 45°	22.6	70.8	36.4	0.3	17.2	0.6	21.4	42.8	25.7	21.8	22.1	17.7	82.9	47.7	63.5
E+A	99.9	99.9	99.9	99.0	99.4	99.9	98.8	99.9	99.9	99.9	99.6	99.9	28.3	91.6	88.1
E+A45°	9.6	75.7	26.4	1.6	12.5	2.1	21.9	41.5	26.3	45.6	4.3	21.8	93.5	65.1	78.7
S	84.8	34.0	71.9	66.0	45.2	60.2	67.5	45.1	60.3	41.3	5.5	15.7	27.0	6.6	12.4
S0	85.6	80.4	88.0	99.0	27.9	91.8	98.7	31.7	90.7	9.7	37.5	12.7	8.6	2.1	2.6

DISCUSSION

In this work we have shown that alignment effects are statistically significant. Several tests indicate that they cannot be produced by bias in the Lick maps.

Using a new statistical procedure we have succeeded in relating the alignment effects of Es and BCGs with two-armed structures of linear dimension of at least 30 h^{-1} Mpc. These 2-D structures can be interpreted as a projection of filaments or sheets of galaxies in space.

The results of our statistical analyses can be summarized as follows:

1) We confirm the alignment effects reported by AGPS and LGPa,b that BCG and E galaxies tend to be aligned with the large-scale structure in the Lick maps. Both the scales, 4° for BCGs and 2° for Es, and the alignment parameter $b \sim 0.11 \pm 0.04$ are consistent with their results. We found no evidence for alignments in the samples of Spirals and S0s. The Δ_1, Δ_2 test reveals no peculiarities in the galaxy distribution around these objects. Unexpectedly, we found no evidence for an alignment signal in the Shectman cluster sample beyond 1°.

2) From the statistics of Δ_1, Δ_2 described above we find that the alignment effects of BCGs and Es are consistent with a model in which there are overdensities on both sides in the direction defined by the object's PA. This filamentary 2-D structure is weak, and the clustering pattern dominates. This fact is related to the small (10%) anisotropy in the angular cross-correlation function ($b \sim 0.1$).

Several models might account for the origin of the alignment and filament phenomena. We can think of primordial or evolutionary pictures. For instance, merging of galaxies in clusters could produce an alignment of Es' major axes within the cluster potential. However, as West, Dekel and Oemler (1989) point out, the alignment effects might put strong constraints to the models for the origin of large-scale structure. The fact that the mass distribution is in the linear regime on scales greater than $\sim 10\ h^{-1}$ Mpc and that BCG alignments extend beyond this scale is an indication that some of the alignments are primordial.

ACKNOWLEDGEMENTS

We are indebted to Dr. E. J. Groth for providing the magnetic version of the Lick maps as well as useful software. DGL has greatly benefitted from discussions with Dr. P. J. E. Peebles on the issues of this work. This work was partially supported by a grant from the Consejo Nacional de Investigaciones Científicas y Técnicas de la República Argentina

(CONICET) and the Consejo de Investigaciones Científicas y Tecnológicas de la Provincia de Córdoba (CONICOR).

REFERENCES

Abell, G. O. 1958, *Ap. J. Suppl.*, **3**, 211.

Argyres, P. C., Groth, E. J., Peebles, P. J. E., and Struble, M. F. 1986, *A. J.*, **91**, 471 (AGPS).

Binggeli, B. 1982, *Astr. Ap.*, **107**, 338.

Carter, D., and Metcalfe, N. 1988, *M.N.R.A.S*, **191**, 325.

Djorgovski, S. 1987, in *Nearly Normal Galaxies*, ed. S. Faber, (New York: Springer), p. 227.

Dobrzycki, A. 1989, *M.N.R.A.S.*, **236**, 403.

Groth, E. J. and Peebles, P. J. E. 1977, *Ap. J.*, **217**, 385.

Kuhn, J. R., and Uson, J. M. 1982, *Ap. J. (Letters)*, **263**, L47.

Lambas, D. G., Groth, E. J., and Peebles, P. J. E. 1988, *A. J.*, **95**, 975 (LGPa).

Lambas, D. G., Groth, E. J., and Peebles, P. J. E. 1988, *A. J.*, **95**, 996 (LGPb).

Moody, J. E., Turner, E. L., and Gott III, J. R. 1983, *Ap. J.*, **273**, 16.

Nilson, P. 1973, *Uppsala General Catalog of Galaxies*, (Uppsala, University of Uppsala).

Seldner, M., Siebers, B., Groth, E. J., and Peebles P. J. E. 1977, *A. J.*, **82**, 249.

Shane, C. D., and Wirtanen, C. A. 1967, *Publ. Lick Obs.* **22**, part 1.

Shectman, S. A. 1985, *Ap. J. Suppl.* **57**, 77.

Struble, M. F., and Peebles, P. J. E. 1985, *A. J.* **90**, 582.

West, M. J., Dekel, A., and Oemler, Jr., A. 1989, *Ap. J.* **336**, 46.

DISCUSSION

S. White: Would you like to speculate what this alignment means?

D. Lambas: Our results indicate that the angular-momentum vector is much less correlated to the neighboring galaxy distribution than the direction of largest anisotropy of an elliptical (assuming ellipticals are not supported by rotation). Perhaps this is connected directly to initial conditions.

E. Bertschinger: It is interesting that ellipticals, not late-type galaxies, show significant alignment with the clusters. Could it be that ellipticals form by mergers, and that these occur primarily in the plane of a flattened cluster because of the cluster velocity anisotropy?

D. Lambas: I think this is a possibility which should be considered. It would be very interesting to have predictions from theoretical models in order to compare them with the observations.

A. Szalay: I would like to advertise a project in its final phases with Dr. J. Bond, where we calculate alignment effects of clusters, and asymmetries of the densities around clusters. The idea is that the long-wavelength plane waves (even if they have a small amplitude) modulate the density of galaxies, and as the density is squashed normal to the wave front, such alignment arises quite naturally. The nice thing about it is that it is extremely sensitive on the large-wavelength part of the fluctuation spectrum, even if it is in the linear regime, so we can test that part separately.

A. Dekel: I'd separate the question into two: 1) why are the clusters aligned with their environments, and 2) why are the ellipticals (which are preferentially in clusters) aligned with their parent clusters? For the first, West, Oemler, and I, in our N-body simulations, found strong alignments which grow as a function of the power on large scales in the initial power spectrum. The comparison has interesting theoretical implications. The latter has also to do with elliptical formation in clusters, mergers, dynamical friction, and maybe some dissipation.

D. Lambas: The fact that ellipticals and brightest-cluster-galaxies have similar values of the parameter b (~ 0.1) on very different scales might not be easy to explain with completely different mechanisms, but of course this is a possibility.

S. White: I think a relatively general implication of this effect is that at least some ellipticals must have formed quite recently. If all ellipticals collapsed at redshifts above 10, it is hard to see why they would pick orientations which correlate with present large-scale filamentary structure. This could be used to argue for a merger origin within an elongated clustering pattern (the usual explanation advanced for strong alignment of cD galaxies with their parent clusters).

D. Lambas: I agree with you that it is not easy to imagine how early-type objects can be aligned with the distribution of galaxies at large scales in the present unless the alignments were originated when density contrasts on those scales were significantly large.

M. Davis: Have you considered studying ellipticals separately in and out of cluster centers? It is possible you will find different b values, since the cluster ellipticals could be explained by mergers along the major axis, whereas ellipticals in filamentary structures might be aligned by the anisotropic collapse of matter.

D. Lambas: Certainly if we had a larger sample of ellipticals, we could consider different subsamples, and in particular this issue could be addressed.

D. Burstein: The number of ellipticals you used, 183, seems about two times too small, relative to the number in the UGC with $b > 40°$. Why?

D. Lambas: Many ellipticals are too round.

D. Burstein: Ellipticities of elliptical galaxies in the UGC are very sensitive to two systematic effects: 1) smaller ellipticals have more errors in b/a than larger ellipticals - as much due to 'quantization' of diameters as anything else, and 2) round or nearly-round ellipticals can have position angles determined by eye more a function of the surrounding density than of the galaxy. Have you tested for differences in b as a function of either major-axis diameter, or axial ratio?

D. Lambas: There is not a large enough number, but the signal of alignment could hardly come from nearly round objects, simply because their position angles were not measured.

THEORY OF ADHESION FOR LARGE-SCALE STRUCTURE

SERGEI SHANDARIN
Institute for Physical Problems, Kosygina 2, 117334 Moscow, USSR

ABSTRACT I discuss a model for the formation of Large-Scale Structures in the universe, in the framework of the model of adhesion. The model, approximately describing the nonlinear stage of gravitational instability, is based on Burgers' equation, which has an analytic solution. Contrary to the widely-accepted practice of using linear density perturbations for the analysis of structure formation, the model deals with the initial conditions in terms of gravitational potential perturbations at the linear stage $\varphi(\mathbf{r})$. Basically, the model approximately predicts the velocity field at any chosen time directly from the initial potential, eliminating all intermediate stages. It contains one free parameter, representing a coefficient of viscosity ν. At present there are two approaches to structure formation in the framework of the model of adhesion. One of them assumes that ν is small but finite. In this case, in order to construct the structure one needs to integrate the motion of the matter over time. The other assumes ν to be infinitesimal. In this case only the skeleton of the structure can be constructed. However, it can be constructed at arbitrary time t directly from $\varphi(\mathbf{r})$ without intermediate steps, by means of a special geometrical technique based on constructing paraboloids tangential to the hypersurface $\varphi(\mathbf{r})$. Here I briefly discuss both approaches and compare the model with direct N-body simulations computed with identical initial conditions. The proposed model is shown to be in satisfactory agreement with the results of N-body simulations. This confirms the potential merits of the adhesion model for a wide class of cosmological scenarios including the Cold Dark Matter (CDM) model.

INTRODUCTION

The problem of the origin and evolution of Large-Scale Structure (LSS) in the universe is one of the fundamental issues of current cosmology. It is closely related to the problem of dark matter dominating the mean density of the universe, as well as the problem of generating primordial perturbations in the very early universe. It is also extremely interesting as an example of a complicated nonlinear dynamical process.

Hereafter, the LSS is assumed to form as a result of the growth of initially small density perturbations due to gravitational instability. Primordial perturbations are believed to be generated from random vacuum fluctuations during an inflationary stage. Thus the cosmological density perturbations at the linear stage are assumed to be a random field. They can be of a Gaussian or non-Gaussian type, depending on the assumptions about the inflationary stage. The inflationary model also predicts that the universe is flat, $\Omega = 1$. For comparison with numerical simulations, it is assumed that $\Omega = 1$, but the discussed technique is assumed to operate in both open and closed universe models equally well (Gurbatov, Saichev, and Shandarin 1989; Shandarin and Zel'dovich 1989).

At present the most straightforward way to study the evolution of the LSS is to carry out N-body simulations. During the last decade this has produced the most valuable results concerning the formation of the LSS. The possibilities of this approach are mainly restricted by the capabilities of modern computers, because the dynamical range in terms of length increases very slowly with the growth of the number of particles: $\ell \propto N^{1/3}$.

The evolution of density perturbations has been well understood theoretically only at the linear stage. However, linear theory has not proven to be sufficient to explain many features of the LSS. A very important step towards a theoretical understanding of the gravitational instability in the nonlinear regime was contributed by Zel'dovich (1970), who suggested an approximate solution for the growing mode of potential perturbations. His solution describes the evolution of perturbations in terms of particle motion from the unperturbed position with Lagrangian coordinate $\mathbf{q}$

$$\mathbf{r}(t,\mathbf{q}) = a(t)\,[\mathbf{q} - b(t)\,\nabla\Phi_0(\mathbf{q})], \tag{1}$$

where $\mathbf{r}$ is the Eulerian coordinate of the particle at the time t; $a(t)$ is a scale factor; $b(t)$ is the growing solution of the linear theory of gravitational instability. In a matter-dominated flat universe ($\Omega = 1$), $b(t) \propto a(t) \propto t^{2/3} \propto (1 + z)^{-1}$, where z is the redshift. The spatial structure of the growing mode of the density perturbations is defined by the potential vector field $\nabla\Phi_0(\mathbf{q})$. Using equation (1), one can easily find the density and velocity fields by means of a well-known analytic technique (Zel'dovich 1970, Shandarin and Zel'dovich 1989).

Pancakes are typical structures in Hot Dark Matter (HDM) models and have sizes of superclusters of galaxies. In CDM models the situation is much more complicated. But, if the initial spectrum possesses a cutoff on small scales, then the first objects to form at the nonlinear stage are pancakes of the cutoff sizes. Later on the gravitational instability continues as a process which is basically hierarchical clustering. However, there are some theoretical arguments (Gurbatov, Saichev, and Shandarin 1985; 1989) as well as evidence from numerical simulations (Melott *et al.* 1983, White *et al.* 1987, Melott and Shandarin 1989) that the distribution of clumps of matter resembles filamentary and pancake structures when the initial spectrum is steeper than $\delta_k^2 \propto k^{-1}$.

Unfortunately, approximation (1) grows worse quite soon after the formation of the first pancakes and becomes invalid soon after. Therefore, it cannot be directly used for the study of CDM models.

However, a rather simple modification of the Zel'dovich approach which allows an analytic solution was suggested (Gurbatov, Saichev, and Shandarin 1985; 1989; Shandarin 1988; Kofman and Shandarin 1988; Shandarin and Zel'dovich 1989). This new model invokes gravitational adhesion or sticking of the matter in pancakes, filaments, and clumps, which approximately describes the results of numerical simulations with gravitating collisionless particles (Doroshkevich *et al.* 1980). The adhesion qualitatively describes the fact that the pancakes remain rather thin in comoving coordinates. The one-dimensional numerical simulations (Doroshkevich *et al.* 1980, Kotok and Shandarin 1987), as well as analytical considerations (Dekel 1983), show that the thickness of pancakes, after a short period of quite rapid growth, stabilizes and increases very slowly with time. 2-D and 3-D simulations (Doroshkevich *et al.* 1980, Klypin and Shandarin 1983, White *et al.* 1987, Melott and Shandarin 1989, Bertschinger and Gelb 1989) showed that the thickness of pancakes and filaments is considerably smaller than their sizes and the distances between them (Figure 1).

Figure 1 also shows very complicated structures arising inside pancakes at late nonlinear stages. The adhesion model can only indicate the positions of filaments and clumps and cannot describe the internal distribution of matter.

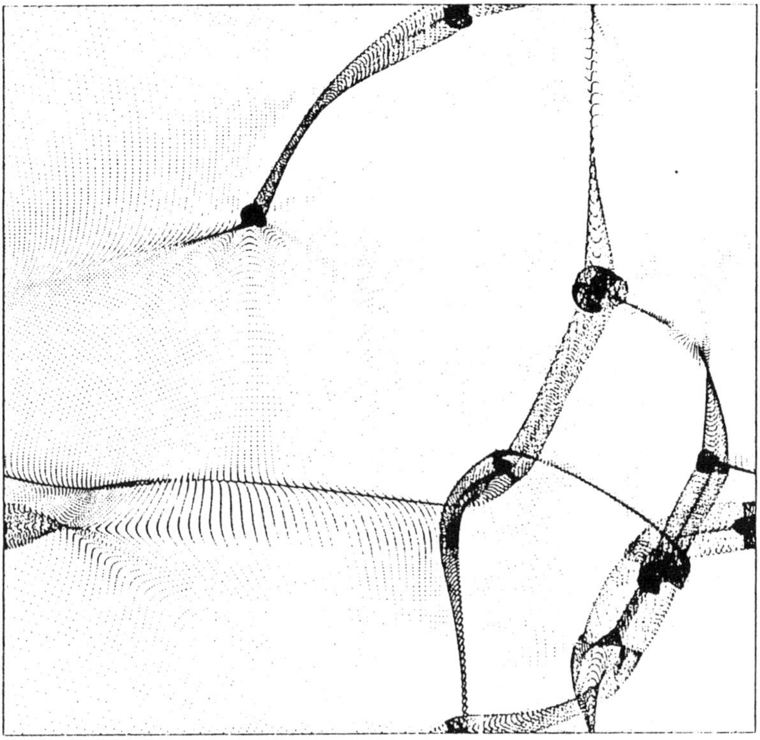

Fig. 1. The particle distribution in a 2-D high resolution *N*-body simulation.

Mathematically, the adhesion model is based on Burgers' equation. The principal feature in the adhesion model is the initial gravitational potential, φ. From the point of view of dynamics the model can deal with either Gaussian or non-Gaussian random fields. The gravitational potential does not change its spatial structure during the linear stage of gravitational instability and contains all information about all structures at every stage. Using Burgers' equation, one can approximately describe the formation of pancakes of arbitrary sizes as well as their transformation into filaments and clumps, and later the hierarchial clustering of clumps.

This talk aims to outline the adhesion model, discuss different numerical techniques for the realization of the model, compare it with the available results of numerical simulations made with the identical initial conditions, and discuss its principal advantages and drawbacks.

ADHESION MODEL

Here I briefly discuss the mathematical basis of the adhesion model. One can find the full derivation of the final equation in the papers by Shandarin (1988) and by Gurbatov, Saichev, and Shandarin (1989). The collisionless matter motion is described in comoving coordinates $\mathbf{x} = \mathbf{r}/a(t)$, using the peculiar velocity $\mathbf{u} = \mathbf{U} - H\mathbf{r} = a\, dx/dt$. The basic system of equations is (Peebles 1980)

$$d\mathbf{u}/dt + H\,\mathbf{u} = -(1/a)\,\nabla\varphi, \tag{2a}$$

$$\nabla^2\varphi = 4\pi G\, a^2\, (\rho - \langle\rho\rangle), \tag{2b}$$

$$\partial\rho/\partial t + 3H\,\rho + (1/a)\,\nabla(\rho\mathbf{u}) = 0. \tag{2c}$$

Here $a(t)$ is, as usual, a scale factor; $H = \dot{a}/a$ is the Hubble parameter; $\rho = \rho(\mathbf{x},t)$ is the density; $\langle\rho\rangle(t)$ is the mean density; φ is the perturbation of the Newtonian gravitational potential. One can calculate the gravitational acceleration from the Zel'dovich solution (1) and put it into equation (2a). This assumes that the motion of the matter outside pancakes, filaments, or clumps can be approximated by equation (1). Introducing new variables

$$n = a^3\rho, \tag{3a}$$

$$\mathbf{v} = \mathbf{u}/a\dot{b}, \tag{3b}$$

and using $b(t)$ instead of t, and introducing an additional viscosity term which qualitatively describes the gravitational adhesion, the resulting equation for velocity becomes

$$\partial\mathbf{v}/\partial b + (\mathbf{v}\nabla)\,\mathbf{v} = \nu\nabla^2\mathbf{v}. \tag{4}$$

(During the workshop D.Lynden-Bell made a comment that despite the fact that the viscosity term on the right-hand side of equation (4) looks artificial, in the case of potential motion it has the general form of a viscosity term with a constant coefficient of viscosity).

One additional very important simplifying assumption is usually made, that ν is small. The value of ν controls the thickness of pancakes, filaments, and clumps. The smaller ν, the thinner the structure elements mentioned above. On the other hand, the smaller ν, the smaller the influence of the viscosity term $\nu\nabla^2\mathbf{v}$ outside the density concentrations, permitting the use of the Zel'dovich approximation in these sparse regions.

Equation (4) is basic in the adhesion model. It is a 3-D generalization of Burgers' equation, well-known in the theory of turbulence (Burgers 1974). However, contrary to the vortex motion of the usual turbulent liquid, in our case the motion is purely potential,

$$\mathbf{v}(\mathbf{x},b) = -\nabla\Phi(\mathbf{x},b). \tag{5}$$

In this case the nonlinear equation (4) has an exact analytic solution (Burgers 1974),

$$\mathbf{v}(\mathbf{x},b) = \frac{\int d^3q\, [(\mathbf{x} - \mathbf{q})/b]\, \exp\{1/(2\nu)\, S(\mathbf{x},\mathbf{q},b)\}}{\int d^3q\, \exp\{1/(2\nu)\, S(\mathbf{x},\mathbf{q},b)\}} \tag{6}$$

where

$$S(\mathbf{x},\mathbf{q},b) = \Phi_0(\mathbf{q}) - (\mathbf{x} - \mathbf{q})^2/2b, \tag{7}$$

and $\Phi_0(\mathbf{q})$ is the initial velocity potential, $\Phi_0(\mathbf{q}) = \Phi(\mathbf{x},0)$. One can find a discussion of the techniques for obtaining exact solutions in some detail in several papers: Gurbatov, Saichev, and Shandarin (1985, 1989); Shandarin and Zel'dovich (1989); and Nusser and Dekel (1990). One can readily see that at small ν the main inputs into the above integrals come from the regions where the function $S(\mathbf{x},\mathbf{q},b)$ (considered as a function of $\mathbf{q}$ at constant $\mathbf{x}$ and

b) has maxima.

In some simple cases solution (6) allows us to study analytically the structures arising at late nonlinear stages (Gurbatov, Saichev, and Shandarin 1985, 1989; Kofman and Shandarin 1988; Doroshkevich and Kotok 1988). However, I begin by outlining some numerical techniques recently developed in the framework of the adhesion model.

NUMERICAL TECHNIQUES FOR THE ADHESION MODEL

Two kinds of numerical methods have been developed for the adhesion model. One of them assumes that ν is small but finite (Weinberg 1989, Weinberg and Gunn 1990, Nusser and Dekel 1990). In this case the velocity field is continuous, although at small ν it varies rapidly across pancakes and filaments. The pancakes, filaments, and clumps have finite thickness depending on ν.

The other model assumes that ν is infinitesimal (Kofman, Pogosyan, and Shandarin 1990; Pogosyan 1989). In this case the pancakes, filaments, and clumps become surfaces, lines of infinitesimal thickness, and points respectively; the velocity field is discontinuous on them. These surfaces, lines, and points are parts of the structure which we call 'the skeleton' of the structure.

The different assumptions about the value of ν result also in different numerical techniques used in the model.

$\nu \neq 0$

In this case the Eulerian position of a particle is given by the numerical solution of the integral equation,

$$\mathbf{x}(b,\mathbf{q}) = \mathbf{q} + \int_0^b db' \ \mathbf{v}[\mathbf{x}(\mathbf{q},b'), b'], \tag{8}$$

(Nusser and Dekel 1990). The Zel'dovich approximation and the adhesion model with a non-zero coefficient of viscosity are not identical even outside pancakes and filaments, although they become similar at some distance from them if ν is small. The value of ν determines the size of the region in Lagrangian space that provides the main contribution to the integrals (6). In practice Nusser and Dekel computed the integrals only over regions satisfying the condition $\exp\,[S_{max}(\mathbf{x},b) - S(\mathbf{x},\mathbf{q},b)]\,/\,(2\nu) > 1/\epsilon$, where ϵ is a small number which sets the desired accuracy (I would like to stress that here the function S has the opposite sign to that in the paper by Nusser and Dekel, resulting in a change of the form of the relation). Thus, as long as the Zel'dovich approximation is applicable, one can use it to find the Eulerian position of the particle. Whenever a particle approaches a filament it becomes necessary to switch to a time integration using the integral solution (6).

To compute the velocity field (6), Weinberg and Gunn (1990) used a technique based on a FFT routine. In this case the whole region under consideration contributes formally to the integrals. The method has the advantage of being very fast computationally. However, it is restricted when dealing with very small ν, because the integrand depends on ν exponentially.

$\nu \to 0$

Assuming $\nu \to 0$, one can use the method of steepest descent to show that the main contribution into integrals (6) in most cases comes just from one point of the global maximum of $S(\mathbf{x},\mathbf{q},b)$, considered as a function of $\mathbf{q}$ at fixed $\mathbf{x}$ and b. This means that the particle with current Eulerian coordinate **x** comes from Lagrangian points **q** satisfying the above condition of being maxima of S. This particle brings its velocity to **x** because in the variables (3) the velocity **v** does not change with time until the particle runs into a pancake, filament, or clump. A $\nu \to 0$, solution can be interpreted in terms of a rather simple geometrical construction as follows: i) let us define a paraboloid

$$P(\mathbf{x},\mathbf{q},b) = (\mathbf{x} - \mathbf{q})^2/2b + h, \tag{9}$$

and ii) adjusting h, construct the paraboloid to be tangent to the $\Phi_0(\mathbf{q})$ hypersurface at a chosen point **q**, provided that there is no crossing of the hypersurface $\Phi_0(\mathbf{q})$ at any other point, then iii) the projection of the apex of the paraboloid **x** corresponds to the Eulerian position of this particle. Figure 2 illustrates the procedure.

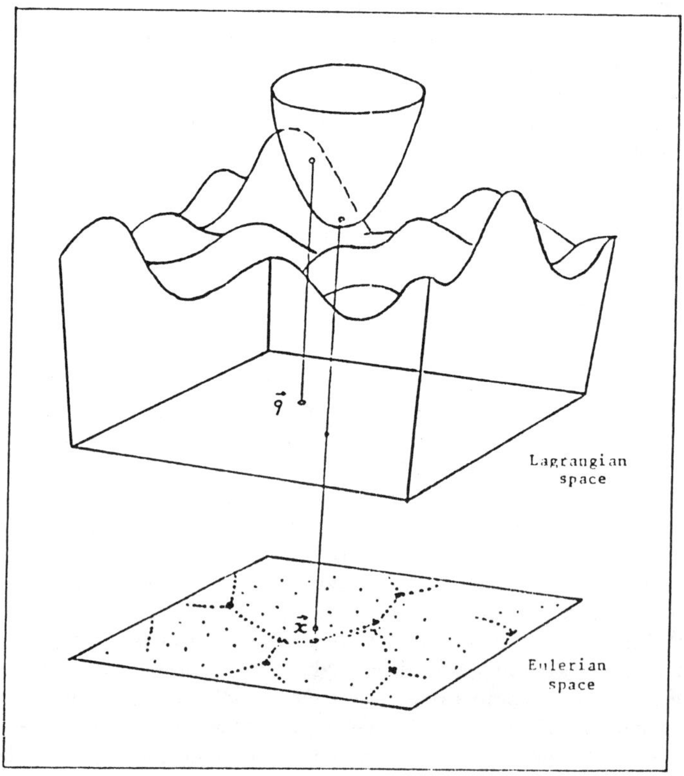

Fig. 2. Illustration of the geometrical procedure to construct solution (6) in a two-dimensional model. The paraboloid $P = (\mathbf{x} - \mathbf{q})^2/2b + h$ is constructed touching the random Gaussian surface $\Phi_0(\mathbf{q})$ at the point **q** by adjusting the parameters **x** and h. The condition that the paraboloid does not cross the surface Φ_0 at any point is assumed. The point **q** is interpreted as the place where the particle comes from, and the point **x** is the place where the particle is located at this time.

Solution (1) and this geometrical construction are equivalent until the particle runs into a pancake. However, when this happens, different approximations give different predictions. According to the Zel'dovich solution (1) the particle keeps moving as if nothing has happened. This results in unrealistically rapid growth of the thickness of the pancakes and contradicts the results of *N*-body simulations. Solution (6) cannot be easily interpreted in terms of the particle positions at this stage. So, let us restrict our discussion by considering the positions of pancakes and other elements of the structure.

At early stages the curvature of the paraboloid is large (more precisely: it is greater than that of Φ_0 at any point), and the geometrical technique discussed above results in a one-to-one correspondence between points of Lagrangian and Eulerian spaces. Geometrically it means that the paraboloid can be tangential to Φ_0 at every point $\mathbf{q}$ without crossing it at another point. At this stage there are neither pancakes nor filaments nor clumps. Later the pancakes and other structures form (I recall that as $\nu \to 0$ the pancakes are infinitesimally thin). In terms of paraboloids it means that there are particles (points in $\mathbf{q}$-space) for which one cannot construct the corresponding paraboloid (tangential to Φ_0 at this point) without intersecting the Φ_0 surface somewhere else. This happens because the curvature of the paraboloid decreases with time and the paraboloid becomes wide. All particles of this kind have already stuck in the pancakes or other structures by this time. At this stage three additional types of paraboloids can appear which are tangential to Φ_0 in a few points simultaneously: (i) in two points ($\mathbf{q}_1$, $\mathbf{q}_2$), (ii) in three points ($\mathbf{q}_1$, $\mathbf{q}_2$, $\mathbf{q}_3$), and (iii) in four points ($\mathbf{q}_1$, $\mathbf{q}_2$, $\mathbf{q}_3$, $\mathbf{q}_4$). In all cases the paraboloid is assumed not to cross Φ_0 at any point. (In 2-D space only the first two types of paraboloids are possible at a generic moment of time.) One can find a discussion of technical questions concerning the implementation of this method in some detail in papers by Kofman, Pogosyan, and Shandarin (1990); Pogosyan (1989).

An evident disadvantage of the model with $\nu \to 0$ is that it cannot describe the internal structure of the pancakes, filaments, and clumps. Generally speaking it is also far from evident that it reproduces the skeleton of the structure with reasonable accuracy. A way to examine the model is to compare the structure which it predicts against the structure developed from the identical initial perturbations according to *N*-body simulations. In the following section I present the result of such a comparison.

THE ADHESION MODEL VS *N*-BODY SIMULATIONS

Here I briefly discuss the results of a comparison of the adhesion model with direct 2-D and 3-D *N*-body simulations carried out with the identical initial conditions.

$\nu \neq 0$

Nusser and Dekel (1990) compared the adhesion model with 2-D *N*-body simulations. They acknowledged that at late times the adhesion model is a better approximation of the *N*-body results than the Zel'dovich solution. The cost of one time step of their code in terms of computer time is proportional to N_p^2, where N_p^2 is the number of particles in filaments and clumps at that time. They mentioned also that their code was not 'fully optimized', but that "it allows a fully parallel treatment of particles based on the full initial conditions but independent of the evolution of the other particles".

A full 3-D code which uses Gaussian convolution to solve the numerical integrals in (6) has been developed by Weinberg and Gunn (1990). They used it for computing the structure in a Cold Dark Matter model. They found the adhesion model to be "quite accurate on scales of a few Mpc and up, and the adhesion code runs $\sim$20-100 times faster than N-body codes for the same numerical parameters". The adhesion calculation used 120^3 particles on an equal grid to simulate a volume 360 h^{-1} Mpc on a side.

$\nu \to 0$

Here I discuss in some detail the result of a comparison of the adhesion model, assuming ν to be infinitesimal, with N-body simulations done by Kofman, Pogosyan, and Shandarin (1990). The N-body simulations used a particle-mesh code and were performed with 64^2 particles on an equivalent mesh with periodic boundary conditions. In order to calculate the skeleton of the structure at an arbitrary stage the present adhesion code takes roughly the same amount of computer time as a standard particle-mesh code with equivalent space and mass resolution to calculate one step in time.

The initial perturbation was a random Gaussian field with a spectrum of density perturbations $\delta_k^2 \propto k^2$, or equivalent gravitational (velocity) potential spectrum $\Delta_k^2 \propto \delta_k^2 k^{-4} \propto k^{-2}$ with sharp cutoffs both at the longest possible wavelength corresponding to $k_{min} = 2\pi/64$ and at the shortest wavelength corresponding to $k_{max} = 2\pi/8$. In a sense, the chosen density perturbation spectrum in 2D is equivalent to $\delta_k^2 \propto k$ in 3-D. The initial amplitude at $b = 1$ is $(\delta\rho/\rho)_{in} \sim 0.3$ on the mesh scale. We assumed an Einstein-DeSitter matter-dominated universe, which means that $b(t) = a(t)$. The results of the calculations are illustrated in Figure 3, corresponding to the epoch $a \sim 20$, when $(\delta\rho/\rho)_{lin.theor.} \sim 6$. To facilitate the comparison between the particle distribution and the skeleton of the structure, the two are combined in Figure 3c.

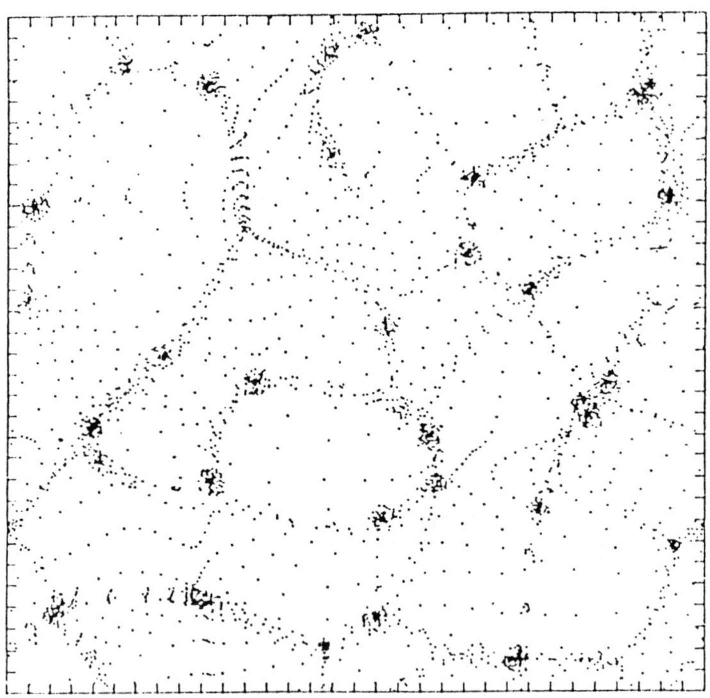

Fig. 3a The distribution of particles calculated in a 2-D N-body simulation at epoch $a \sim 20$, when $(\delta\rho/\rho)_{lin.theor.} \sim 6$.

Obviously, Figure 3 demonstrates very good agreement between the skeleton and the particle distributions obtained by means of an *N*-body simulation. Slight differences between a few filaments and clumps are definitely within the mesh size, which is basically the precision of both codes. It is worth stressing that this Figure shows a quite advanced nonlinear stage when the extrapolation of the Zel'dovich approximation predicts a very different distribution.

However, the agreement between the two codes gradually grows worse. Our preliminary results show that at the epoch when $(\delta\rho/\rho)_{\text{lin.theor.}} \sim 15$ the discrepancy between our adhesion code and *N*-body simulations becomes noticeable. So far the cause of this disagreement has not been completely understood. Figure 3 shows that the code based on Burgers' equation works quite well at rather late nonlinear stages when $(\delta\rho/\rho)_{\text{lin.theor.}} \sim 6$. The agreement between the two codes is also very good when $(\delta\rho/\rho)_{\text{lin.theor.}} \sim 10$. Remember that the calculation of the skeleton at every moment of time is completely independent of the previous stages.

One possible source of the disagreement might be connected with the adhesion code itself, in particular with the crudeness of the calculation of the tangential points which lie on the boundaries separating 'free' (*i.e.* points outside filaments and clumps) and 'stuck' points (*i.e.* points stuck into filaments or clumps at previous stages) in Lagrangian space. The

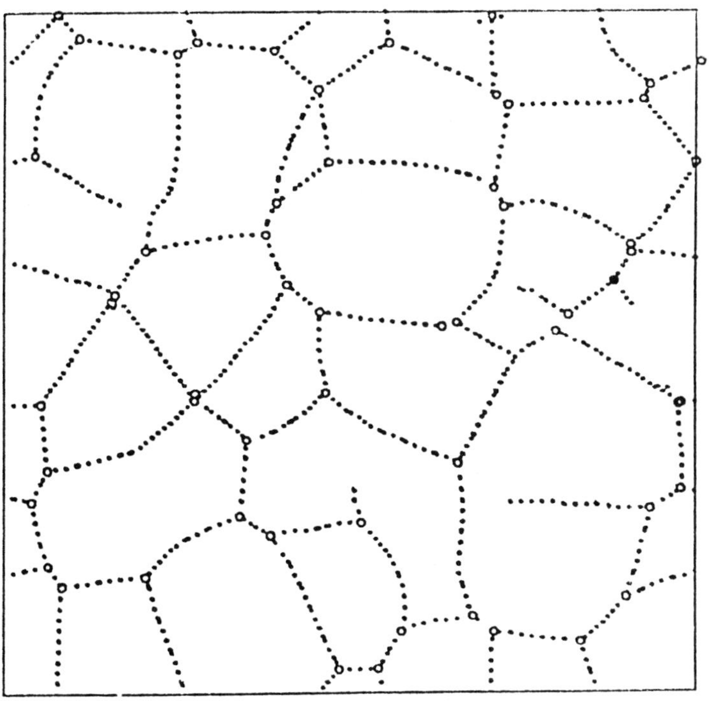

Fig. 3b. The skeleton of the structure constructed in the framework of the adhesion model. The dots show the positions of the paraboloid apexes when they are tangential to the initial velocity potential at two points simultaneously (filaments). Small circles indicate the positions of triple touching (clusters). The initial conditions and epoch are identical with Figure 3a.

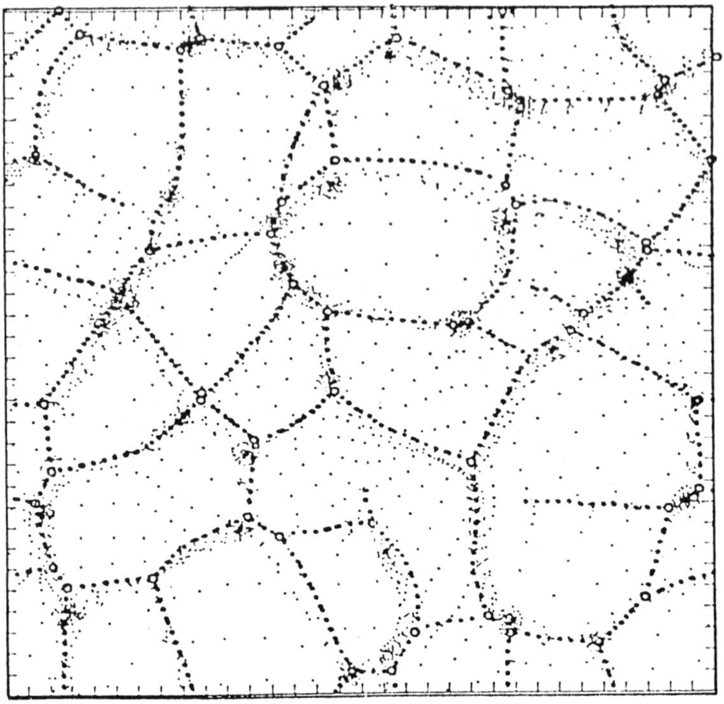

Fig. 3c. The combined representation of the adhesion model together with the N-body simulation.

influence of discreetness is definitely worse at later stages when the number of free and boundary particles become smaller at a given grid.

Another cause could be related to the N-body simulation itself. The code was not designed for such long runs. Actually, it is far from simple to test its accuracy at very late stages without doing simulations with a much larger number of particles and mesh cells. The comparison of the adhesion model with considerably larger 2-D N-body simulations (512^2 particles on an equivalent mesh, Melott and Shandarin 1989) is in progress.

STATISTICAL CHARACTERISTICS OF THE GRAVITATIONAL POTENTIAL PERTURBATIONS

It can be shown that the initial velocity potential Φ_0 in equation (1) is connected with the Newtonian gravitational potential φ in the Poisson equation (2b) (Kofman and Shandarin 1988) (Ω is assumed to be equal to 1);

$$\varphi = -\dot{\Psi} - 1/(2a^2)\,(\nabla\Psi)^2, \tag{10a}$$

where

$$\Psi = (\dot{a}a^2)\,\Phi. \tag{10b}$$

At the linear stage Φ and Ψ are both small, and the second term on the right-hand side of equation (10a) can be dropped. One can also see from equation (2b) that the gravitational potential does not change with time at the linear stage; therefore, $\varphi \propto \Phi_0$. In turn, the fluctuations of the gravitational potential can be simply related to the adiabatic perturbations of the space-time metric (Peebles 1980, Chibisov and Mukhanov 1983),

$$ds^2 = (1 - 2\varphi)\, dt^2 - a^2(t)\,(1 + 2\varphi)\,\gamma_{\alpha\beta}\, dx^\alpha\, dx^\beta. \tag{11}$$

Thus, the problem of the formation of LSS in the framework of the adhesion model can be formulated as a construction of the structure as described above directly from the potential perturbations generated presumably during the inflationary stage. An interesting question immediately arises: is it possible to reverse the problem and reconstruct the field φ_0 from observations of the LSS using an adhesion model? Let us return to the statistics.

A random Gaussian field can be characterized by the moments (Adler 1981, Bardeen *et al.* 1986)

$$\sigma_j^2 = (1/2\pi) \int d^3k\, k^{2j}\, |\varphi_k|^2. \tag{12}$$

Of course, in the 2-D model the change from d^3k to d^2k is necessary ($|\varphi_k|^2 \propto \Delta_k^2$ was used above). One of the most important parameters characterizing the random field is the dimensionless parameter $\gamma_\varphi = \sigma_1^2 / (\sigma_0\, \sigma_2)$ (not to be confused with $\gamma_{\alpha\beta}$.) In the case of the gravitational potential it depends very weakly on the scale of the short wavelength cutoff, differing considerably from the case of the mass-density field. For the standard HDM model, potential $\gamma_\varphi \sim 0.56$, and for the standard CDM model, $\gamma_\varphi \sim 0.04$.

The parameter γ_φ lying in the range [0,1] is in a sense a measure of the 'temperature' of the model: it is rather small in the CDM model, and significantly larger in the HDM model. Obviously, γ_φ and other parameters can be strongly changed in the case of nonflat spectra (Kofman and Linde 1987, Kofman and Pogosyan 1988, Kofman *et al.* this volume).

The dimensional ratio $R_* = \sqrt{3}\, \sigma_1/\sigma_2$ characterizes the average distance between peaks of the gravitational potential, which is approximately given by $\ell \sim 4R_*$. In the HDM model $\ell \sim 48$ Mpc, and in the CDM model $\ell \sim 12$ Mpc (all numbers were calculated applying the cutoffs to the initial spectrum on scales of about 1 Mpc and the present horizon size).

The distance where the two-point correlation function $\xi(R) = \langle \varphi(\mathbf{r})\, \varphi(\mathbf{r} + \mathbf{R}) \rangle$ equals zero is approximately $R_0 \sim \sigma_0/\sigma_1$. This parameter characterizes the typical size of the regions where the potential is of the same sign. Since the ratio $R_*/R_0 \sim \gamma$, then the typical number of peaks per one region of constant sign scales as $N_p \propto \gamma^{-d}$ where d is the dimensionality of the space in question.

The potential in the HDM model has a rather simple structure: a typical region where the potential φ is positive contains only a few peaks. On the contrary, in the CDM model a typical region where φ is positive consists of very many peaks. This difference is very important from the point of view of the construction of the paraboloids, manifesting itself in apparently different evolution of the models.

In our calculations the parameter $\gamma \sim 0.7$, therefore, our model can be considered as a 2-D analogy of the HDM model.

RANDOM-PACKED CELLULAR STRUCTURE

The adhesion model dealing with the skeleton of the structure provides a natural connection between the problem of the LSS and studies of random-packed structures, like soap foam, which was first realized by Icke and Van der Weygaert (1987), but in the context of a different, purely statistical, model. They have introduced hypothetical random centers of repulsion and constructed the cellular structure dividing the space into cells with the aid of a procedure known as the Voronoi tessellation. Accepting completely the general idea of tessellation, I have to stress the principal differences between their model and the adhesion model. First of all, the model based on Burgers' equation is dynamical. It predicts the evolution of the structure. The model of Icke and Van der Weygaert is static and in present form is only descriptive. As far as we know there has not been an attempt to relate the properties of the structure to some physical process. Our model operates with a very well specified process: the primordial perturbation of the gravitational potential, which is the principal physical feature in any model of an inflationary universe. Needless to say, the geometrical procedures themselves used to construct the structure are very much different in these two models. However, some general ideas like the tessellation (mentioned above) or the construction of the so-called Delaunay triangles seems to be very productive in both models. Concerning the centers of repulsion in the model of adhesion, they are maxima of the gravitational potential perturbations. At late nonlinear stages only high positive peaks govern the evolution of the LSS. The skeleton of the structure plays the role of a convenient auxiliary construction to help study the LSS. As mentioned above, it relates the problem of the LSS formation to other branches of physics where a variety of structures are studied, including soap foam, the packing of atoms in disordered condensed systems, *etc.* (Ziman 1979, Weaire 1983).

In the present universe the knots of the skeleton can be related to clusters of galaxies, and the internal regions of the cells to voids of galaxies. The geometry of the cellular structure in the model of adhesion is determined by the paraboloids touching the random-packed structure of the peaks of the Gaussian field of gravitational potential φ.

Peaks, surrounded by circles of radius R_*, resemble the random-packed 'atoms' and can be distributed with a sort of a short-range order but without long-range order. In cosmology this short-range order can be actually quite large: for instance, in the CDM model it can be of the order of R_0 as mentioned above. It is worth stressing that the peak distribution strongly depends on the model (*i.e.* on the spectrum of linear perturbations). Generally speaking, at every chosen time t not all the peaks are involved in the procedure for constructing the structure skeleton, but only those peaks which are accessible (in the sense specified above) to the paraboloids with curvature $b(t)$.

It is interesting to mention some general features of 2-D structures, assuming that the structure can be roughly considered as a random-packing of polygonal cells (see Figure 3b). Let $f(n)$ be the distribution function of the number of sides of individual cells. It is well known that the mean value

$$\langle n \rangle = \sum_{n=2}^{\infty} n\, f(n) = 6,$$

which is a direct consequence of the Euler theorem: $n_{\rm knots} - n_{\rm filaments} + n_{\rm cells} = 1$, provided that the knots of the structure are of the generic type with only three filaments ending at them. There is positive correlation between the number of sides n of a cell and its area A. The mean area $\langle A \rangle$ of a cell with n sides depends linearly on n: $\langle A \rangle \propto n$. One can naturally speculate that the larger the void the more spherical its shape.

DISCUSSION

The first results of a comparison of the adhesion (sticking) model against 2-D and 3-D *N*-body simulations show that the adhesion model satisfactorily approximates the density distribution of collisionless matter (Nusser and Dekel 1990; Weinberg and Gunn 1990; Kofman, Pogosyan, and Shandarin 1990). It is worth stressing that the agreement has the character of a one-to-one correspondence in terms of the positions of clumps and filaments. The model has a definite advantage from the point of view of the computer time needed. These numerical results support the use of the adhesion model for analytical estimates made in other papers (Gurbatov, Saichev, and Shandarin 1989; Kofman and Shandarin 1988).

At very late stages, when the number of particles left outside the skeleton become small, the agreement between the adhesion model with $\nu \to 0$ and *N*-body simulations gradually worsens (Kofman, Pogosyan, and Shandarin 1990). I believe that this is caused partly by the influence of discreetness, which increases with time. The natural way around this problem is to populate the vicinities of the maxima of φ with more particles. Comparisons of the model with considerably better *N*-body simulations are now in progress.

Finally, I would like to discuss one more question. According to the adhesion model with $\nu \to 0$, the maximum points of the velocity potential Φ remain at rest all the time. They indicate the bottoms of voids. (Of course, as time passes low maxima will gradually go out of the game since the paraboloid cannot touch them. This corresponds to a merging of small cells; I do not discuss this effect now.) Therefore the adhesion model predicts that the maxima of the velocity potential do not move, or, if one wishes to be more realistic, move slowly enough. Actually, I would speculate that the higher the maximum the slower its motion. This speculation can be easily tested in *N*-body simulations.

ACKNOWLEDGMENTS

This talk was prepared during my visit to the Canadian Institute for Theoretical Astrophysics, Toronto, in April - May of 1989. I thank D. Bond, N. Kaiser, L. Kofman, and V. Sahni for very stimulating discussions and criticism. The written version of the talk was finished during my visit to the University of Kansas. I wish to thank my colleagues from the Department of Physics and Astronomy for help. This work was partly supported by NASA grant NAGW-1793.

REFERENCES

Adler, R. J. 1981, *The Geometry of Random Fields*, (New York: Wiley).
Bardeen, J. M., Bond, J. R., Kaiser, N., and Szalay, A. S. 1986, *Ap. J.*, **304**, 15.
Bertschinger, E., and Gelb, J. M. 1989, in preparation.
Blumenthal, G. B., Dekel, A., and Primack, J. 1988, *Ap. J.*, **326**, 539.
Burgers, J. M. 1974, *The Non-linear Diffusion Equation*, (Dordrecht: Reidel).
Chibisov, G. V., and Mukhanov, V. F. 1983, preprint of Lebedev Phys. Inst., Moscow.
Dekel, A. 1983, *Ap. J.*, **264**, 373.
Doroshkevich, A. G., and Kotok, E. V. 1988, preprint Inst. Appl. Math., Moscow.
Doroshkevich, A. G., Kotok, E. V., Novikov, I. D., Polyudov, A. N., Shandarin, S. F., and Sigov, Yu. S. 1980, *M.N.R.A.S.*, **192**, 321.
Gurbatov, S. N., Saichev, A. I., and Shandarin, S. F. 1985, *Sov. Phys. Doklady*, **30**, 921.
Gurbatov, S. N., Saichev A. I., and Shandarin, S. F. 1989, *M.N.R.A.S.*, **236**, 385.

Icke, V., and Van der Weygaert, R. 1987, *Astr. Ap.*, **184**, 16.
Klypin, A. A., and Shandarin, S. F. 1983, *M.N.R.A.S.*, **204**, 891.
Kofman, L. A., Blumenthal, G. R., Hodges, H., and Primack, J. R. this volume, p. 339.
Kofman, L. A., and Linde, A. D. 1987, *Nucl. Phys.*, **282B**, 508.
Kofman, L. A., and Shandarin, S. F. 1988, *Nature*, **334**, 129.
Kofman, L. A., and Pogosyan, D. Yu. 1988, *Phys. Lett.*, **214B**, 508.
Kofman, L., Pogosyan, D., and Shandarin, S. 1990, *M.N.R.A.S.*, **242**, 200.
Kotok, E. V., and Shandarin, S. F. 1987, *Sov. Astr.*, **64**, 1144.
Melott, A. L., Einasto, J., Saar, E., Suisalu, I., Klypin, A. A., and Shandarin, S. F. 1983, *Phys. Rev. Lett.*, **51**, 935.
Melott, A. L., and Shandarin, S. F. 1989, *Ap. J.*, **343**, 26.
Nusser, A., and Dekel A. 1990, *Ap. J.*, **362**, 14.
Peebles, P. J. E. 1980, *The Large-scale Structure of the Universe*, (Princeton: Princeton University).
Pogosyan, D. Yu. 1989, preprint.
Shandarin, S. F. 1988, in *Large Scale Structures of the Universe*, ed. J. Audouze, M. Ch. Peleton, and A. Szalay, (Dordrecht: Kluwer Academic Publishing), p.273.
Shandarin S. F., and Zel'dovich, Ya. B. 1989, *Rev. Mod. Phys.*, **61**, 185.
Weinberg, D. H. 1989, Ph. D. thesis, Princeton University.
Weinberg, D. H., and Gunn, J. E. 1990, *Ap. J. (Letters)*, **352**, L25.
White, S. D. M., Frenk, C. S., Davis, M., and Efstathiou G. 1987, *Ap. J.*, **313**, 505.
Zel'dovich, Ya. B. 1970, *Astr. Ap.*, **5**, 84.
Zel'dovich, Ya. B., and Novikov, I. D. 1983, *The Structure and Evolution of the Universe*, (Chicago: University of Chicago).

DISCUSSION

E. Bertschinger: In your high-resolution two-dimensional numerical simulations, are the substructures perpendicular to the pancake because the eigenvectors of the deformation tensor are orthogonal?

S. Shandarin: At present I cannot give you a complete theory explaining the complicated structures forming inside pancakes. However, I guess that the series of elongated substructures orthogonal to each other observed in some places can be related to the oscillation motion of the matter along the eigenvectors.

E. Bertschinger: Although the adhesion method breaks down when $(\delta\rho/\rho)_{\rm lin.theor.} > 10$, can you repair it by smoothing the initial field so that $(\delta\rho/\rho)_{\rm lin.theor.}$ is smaller at the final epoch?

S. Shandarin: I hope so. The work on this question is in progress.

E. Bertschinger: My unbiased Cold Dark Matter N-body simulation, filtered with a Gaussian of radius 1000 km s^{-1}, shows that the initial potential maximum coincides with the final *velocity* potential maximum. Thus, as you hypothesize, the bottom of the void does not seem to move. However, the final *gravitational* potential maximum does not coincide with the *velocity* potential maximum.

S. Shandarin: It is well known that at the linear stage the velocity and gravitational potentials are similar. However, generally speaking it is not true at the nonlinear stage. The problem is that the adhesion model predicts not only that the positions of velocity potential maxima do not change but that the positions of the corresponding particles also do not change. The latter is possible only if their positions coincide with the positions of the gravitational potential maxima and they do not move. Thus the hypothesis is only approximate.

R. Gott: You mentioned David Weinberg's work. He uses the topology technique to deduce from a set of observations (like the Giovanelli and Haynes sample) the most likely set of Gaussian initial conditions that could have produced it. Then he integrates forward using Burgers' equation to recover the non-linear effects. He can then iterate. The results are quite remarkable. He can reproduce a thin filament quite like the Perseus-Pisces superclusters in the Giovanelli and Haynes sample.

N-BODY SIMULATIONS WITH NON-GAUSSIAN INITIAL CONDITIONS

F. LUCCHIN, S. MATARRESE
Dipartimento di Fisica *Galileo Galilei*, Università di Padova
Via Marzolo 8, I-35131 Padova, Italy

A. MESSINA AND L. MOSCARDINI
Dipartimento di Astronomia, Università di Bologna
Via Zamboni 33, I-40126 Bologna, Italy

ABSTRACT Preliminary results are reported on *N*-body simulations of the large-scale structure of the universe starting from non-Gaussian initial conditions.

INTRODUCTION

Numerical simulations have been extensively applied to calculate the non-linear growth of structure in the universe assuming random-phase initial fluctuations (see, *e.g.*, Davis *et al.* 1985). Non-Gaussian initial conditions, on the other hand, have been studied in the framework of scenarios based on the percolation of cosmic explosions (Weinberg, Ostriker, and Dekel 1989; Yoshioka and Ikeuchi 1989) and on the accretion of matter around cosmic string loops (see, *e.g.*, Bennett and Bouchet 1988). We present here preliminary results obtained by *N*-body simulations on the evolution of large-scale structure in a flat universe, when some initial space-uncorrelated non-Gaussian distributions are assumed. Since our models are scale-free, they do not single out characteristic scales and times. An analysis of scale-free Gaussian models has been recently performed by Efstathiou *et al.* (1988).

NON-GAUSSIAN INITIAL DISTRIBUTIONS

During the last few years there has been increasing tendency in considering non-Gaussian primordial fluctuations (Peebles 1983) to explain peculiar properties in the large-scale structure of the universe. Moreover, many theoretical results support the non-Gaussian ansatz for primordial perturbations: recent analyses have shown that phase correlations may arise already during the linear evolution of perturbations, both of adiabatic and isocurvature type, generated by inflation in the early universe (see, *e.g.*, Matarrese, Lucchin, and Ortolan this volume; Kofman this volume; and references therein). It is then important to investigate the sensitivity of *N*-body simulations of large-scale structure to non-Gaussian features in the initial conditions. To this aim we consider three toy-model statistics. The first two belong to the so-called hierarchical distributions (see, *e.g.*, Fry 1985) these are:
i) the *thermodynamical model* by Saslaw and Hamilton (1984), for which the probability distribution of the mass density fluctuation δ_M is given by

$$p(\delta_M) = [(1 + \delta_M)^{-3/2} (2\pi)^{-1/2} / \sigma_M] \exp \{-\delta_M^2 / [2\omega (1 + \delta_M)] \} \qquad (1)$$

where w is the variance σ_M^2;
ii) the *negative binomial model* (Carruthers and Minh 1983):

$$p(\delta_M) = \sigma_M^{-2(1+1/w)} (1 + \delta_M)^{-1+1/w} \exp[-(1 + \delta_M) / w] / \Gamma(1 + 1/w) \quad (2)$$

iii) The third model is the *log-normal distribution*:

$$p(\delta_M) = (1 + \delta_M)^{-1} \exp\{-\ln^2[(1 + \delta_M)(1 + w)^{1/2}] / 2\ln(1 + w)\} / [2\pi \ln(1 + w)]^{1/2}, \quad (3)$$

where the density field is related to a Gaussian process F by the non-linear transformation $1 + \delta_M = (1 + w)^{-1/2} \exp(F)$, with $\langle F^2 \rangle = \ln(1 + w)$ (see, *e.g.*, Coles and Barrow 1987; Lucchin and Matarrese 1988).

In the following we shall refer to these models as *TH*, *NB* and *LN* respectively. As a reference Gaussian distribution we also evolve an initially white-noise *(WN)* realization. The three non-Gaussian distribution above share two common properties: they allow for the positive-mass constraint $\delta_M \geq -1$, and they are characterized by a positive skewness $\langle \delta_M^3 \rangle$. In this preliminary analysis we assumed that our random field takes independent values in different positions. *N*-body simulations starting from space-correlated non-Gaussian distributions, as those generated by inflation, are in progress (Messina *et al.* 1989).

NUMERICAL METHODS

We used a particle-mesh (PM) code which can satisfactorily simulate collisionless matter (Centrella *et al.* 1988). We use in these preliminary simulations a 32^3 grid, with one particle per cell on the average and a spline of second order. All particles have the same mass m. Moreover, a time step less than one twentieth of the stability limit is used to minimize the truncation errors in the 'leap-frog' integration scheme used to move the particles. With this treatment, the numerical results are reliable for large-scale structures.

We generate initial positions and velocities for non-Gaussian distributed particles directly in configuration space. By a standard rejection method, we generate a density-contrast field satisfying the assumed non-Gaussian distribution. The Zel'dovich algorithm can be then applied in Fourier space to perturb an initial uniform point distribution. The resulting positions are accepted as initial conditions (at t_i) for the numerical simulation when best-fitted by a Kolmogorov-Smirnov test to the assumed non-Gaussian δ_M distribution. The method also provides the correct distribution of peculiar velocities. The initial amplitude is assumed to be $\sigma = 0.3$, in order to avoid early shell-crossing due to the large tails of our non-Gaussian distributions. We then evolve our point distributions up to ten expansions $a(t) = 10$ $(a(t_i) = 1)$.

STATISTICAL TESTS

The large scale structure of the universe shows a remarkably complex structure with filaments, bubbles, large voids, ... It is then important to consider different and complementary statistical tests to analyze numerical results and observations. The tests used here are the following:

Cluster analysis

We look for clusters in our simulations by a percolation technique: one links all particles separated by less than a fraction b of the mean inter-particle distance ℓ; the linked structures are 'clusters'. The percolation curve, Figures 1a and 1b, shows the dependence of the maximum extension L_M of clusters on b, respectively at $a = 1$ and $a = 9$. The 'critical percolation parameter' b_c is defined so that $L_M(b_c)$ equals the box size L. All the non-Gaussian initial distributions percolate less ($b_c = 0.95$) than *WN* ($b_c = 0.86$): they have larger 'clumpiness' and lower 'filamentariness' than a Poisson distribution. The time evolution rapidly changes the percolation properties: at $a = 9$ the *WN* model percolates at $b_c \sim 0.62$, while the non-Gaussian models percolate more rapidly ($b_c = 0.60$ for *TH*, $b_c = 0.51$ for *NB* and $b_c = 0.52$ for *LN*), which seems to indicate that these distributions rapidly evolve toward a more filamentary geometry.

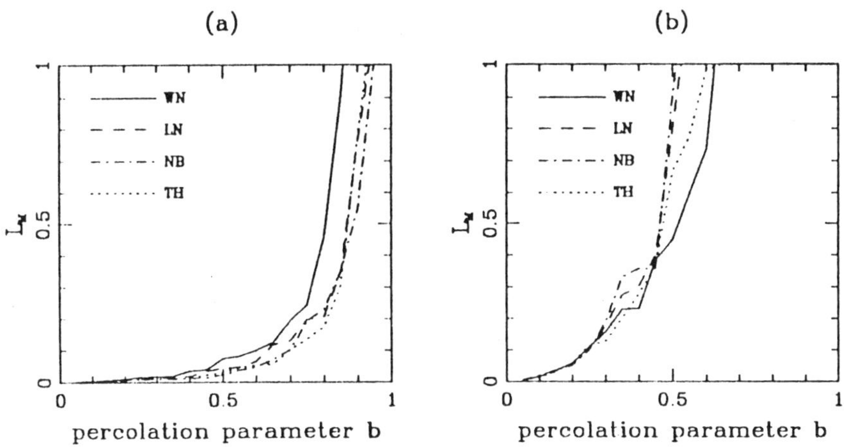

Fig. 1. Maximum length of percolation versus b for the four models at the initial time (a) and after 9 expansions (b).

Mass multiplicity

The percolation analysis provides a 'cluster' catalog for any value of b. The cluster mass multiplicity n_i is the number of clusters per unit volume containing $i = 1,2,\ldots$ particles. In order to avoid the large discreteness fluctuations which characterize the quantity n_i we prefer to consider the integrated mass function

$$F(>M) \equiv \sum_{j\geq i} j\, n_j \,/\, \sum_{j\geq 1} j\, n_j,$$

representing the fraction of mass in clusters with mass $M \geq im$. In the continuum limit the Press-Schechter (1974) theory predicts $F(>M) = \mathrm{erfc}\,[(M/M_{PS})^{\alpha}]$ (see, *e.g.*, Gott and Turner 1977), where erfc is the complementary error function, M_{PS} is a cutoff mass, and α is related to the spectral index n by $\alpha = (n + 3)/6$. In Figure 2 we plot the integrated mass function for the *WN* and the *LN* models at $a = 3, 6, 9$, obtained with $b = 0.25$. We also study the evolution with redshift z of the characteristic mass

$$M_c(z) \equiv \sum_{i\geq 1} i^2\, n_i(z) \,/\, \sum_{i\geq 1} i\, n_i(z)$$

(see also Colafrancesco, Lucchin, and Matarrese 1989). In the Press-Schechter model

$$M_c = \pi^{-1/2} \Gamma \, [\, (1 + \alpha) \, / \, 2\alpha \,] \, M_{PS}$$

evolves as $M_c(z) \propto (1 + z)^{-1/\alpha}$. A plot of M_c as a function of the expansion factor a is given in Figure 3 for $b = 0.25$ and $b = 0.5$. Both $F(>M)$ and M_c do not show appreciable sensitivity to the change of the initial statistics.

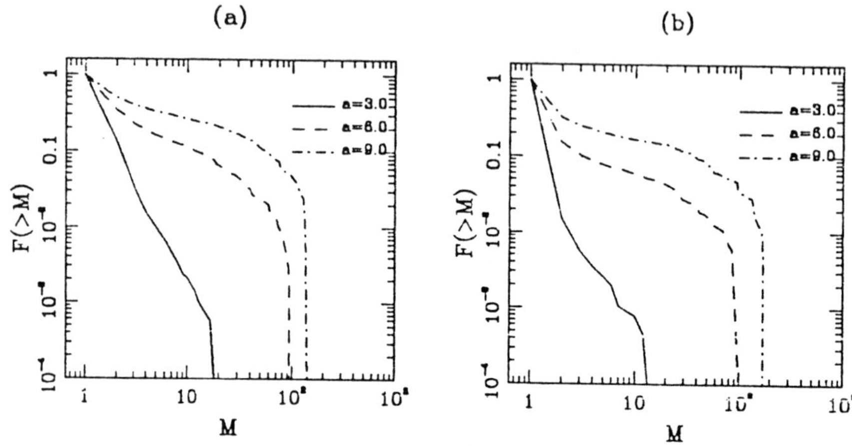

Fig. 2. Integrated mass function for the *WN* (a) and *LN* (b) at different expansion factors.

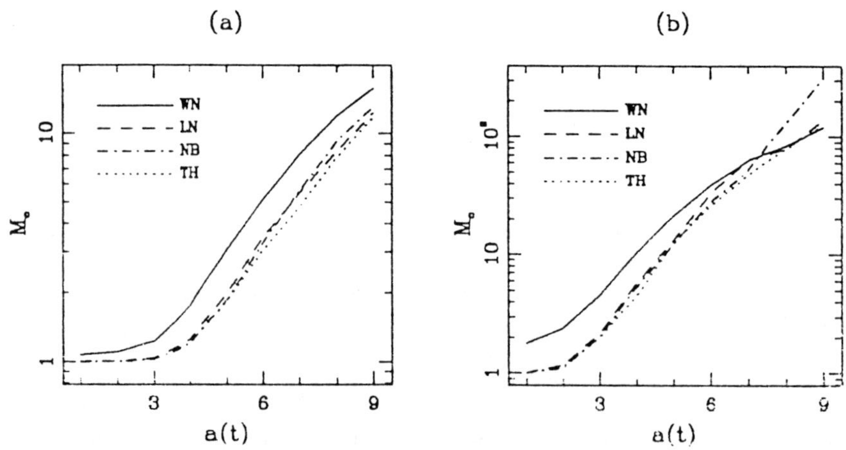

Fig. 3. Time evolution of the characteristic mass M_c with $b = 0.25$ (a) and $b = 0.5$ (b) for the four models.

Topology

We tested the 'genus statistics' describing the topology of the iso-density contours $\delta_M(x) = \nu \, \sigma_M$ of the matter distribution. We refer to Gott (1989) for definitions and references. In the Gaussian case the genus curve shows symmetry between high- and low-density regions. Departures from this symmetry are expected in the presence of phase correlations in the matter distributions both of primordial and of dynamical origin. The isodensity contours are calculated after suitable filtering over a scale λ, and the mean genus per unit volume is obtained using the program by D. Weinberg (1988). For the sake of comparison with the results by Gott *et al.* (1989) we do not modify the routine mapping the fractional volume above a certain threshold into a value of ν, which makes use of the complementary error function (this is, of course, only consistent with a Gaussian distribution; in a better treatment this routine should be consistently upgraded). In Figure 4 we show the genus evolution for *WN* and *TH* (*NB* and *LN* are similar to *TH*) with $\lambda \sim 1.1\ell$: the curves for the non-Gaussian models rapidly shift toward a 'meatball' topology [as emphasized by Gott *et al.* (1989) discreteness effects could be relevant at small filtering scales; we note however that comparing different models at the same filtering scale is still meaningful]. As Figure 5 shows, if one doubles λ the non-Gaussian genus curve recovers the initial nearly symmetric shape.

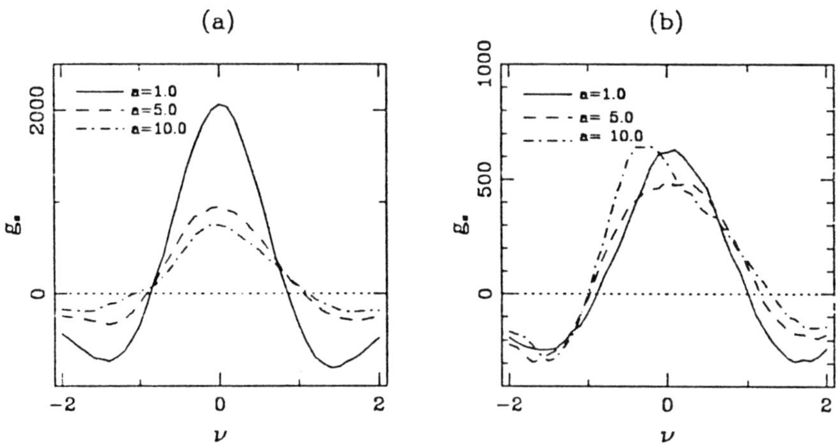

Fig. 4 Genus evolution for *WN* (a) and *TH* (b) with $\lambda \sim 1.1\ell$.

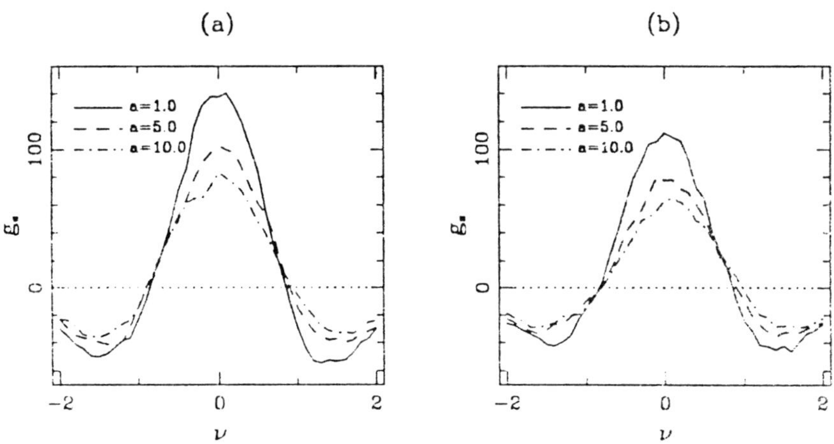

Fig. 5. Genus evolution for *WN* (a) and *TH* (b) with $\lambda \sim 2.2\ell$.

DISCUSSION

In this work we have only considered space-uncorrelated non-Gaussian toy models in order to test the sensitivity of non-linear evolution of large-scale structures to non-random phases in the initial conditions. We analyzed the results of our simulations with the three tests mentioned above and by means of further statistical methods: the traditional two- and three-point correlation functions, the nearest neighbor statistics (Ryden and Turner 1984), the filament-detecting statistics (Bhavsar and Ling 1988), and the quadrupole statistics (Vishniac 1986), which we do not present here. The aim of this detailed statistical 'screening' was to find out the most suitable tests for discriminating the initial particle distribution. From this preliminary results we conclude that the percolation test is the most promising one. We also plan to analyze the peculiar velocity distribution.

The most relevant issue, however, is that of building up space-correlated initial particle distributions based on a theory for the origin of primordial non-Gaussian fluctuations. More specifically we plan to apply our machinery (implemented on a 64^3 grid) starting from non-Gaussian initial conditions originated by inflation (Matarrese, Ortolan and Lucchin 1989). Such a problem can be successfully afforded because these non-Gaussian fluctuation fields are related, *via* a non-linear transformation, to a space-correlated Gaussian field, which makes it possible to generate them by standard algorithms in Fourier space (Messina *et al.* 1989).

REFERENCES

Bhavsar, S. P., and Ling, E. N. 1988, *Ap. J. (Letters)*, **331**, L63.

Bennett, D. P., and Bouchet, F. E. 1988, *Phys. Rev. Lett.*, **60**, 257.

Carruthers, P., and Minh, D.-V. 1983, *Phys. Lett.*, **131B**, 116.

Centrella, J. M., Gallagher III, J. S., Melott, A. L., and Bushouse, H. A. 1988, *Ap. J.*, **333**, 24.

Colafrancesco, S., Lucchin F., and Matarrese S. 1989, *Ap. J.*, **345**, 3.

Coles, P., and Barrow, J. D. 1987, *M.N.R.A.S.*, **228**, 407.

Davis, M., Efstathiou, G., Frenk, C. S., and White, S. D. M. 1985, *Ap. J.*, **292**, 371.

Efstathiou, G., Frenk, C. S., White, S. D. M., and Davis, M. 1988, *M.N.R.A.S.*, **235**, 715.

Fry, J. N. 1985, *Ap. J.*, **289**, 10.

Gott, J. R., III this volume, p. 175.

Gott, J. R., III, Miller, J., Thuan, T. X., Schneider, S. E., Weinberg, D. H., Gammie, C., Polk, K., Vogeley, M., Jeffrey, S., Bhavsar, S. P., Melott, A. L., Giovanelli, R., Haynes M. P., Tully. R. B., and Hamilton, A. J. S. 1989, *Ap. J.*, **340**, 625.

Gott, J. R., III, and Turner, E. L. 1977, *Ap. J.*, **216**, 357.

Kofman, L. A. this volume, p. 339.

Lucchin, F., and Matarrese, S. 1988, *Ap. J.*, **330**, 535.

Matarrese, S., Lucchin, F., and Ortolan, A. this volume, p. 353.

Matarrese, S., Ortolan, A., and Lucchin, F. 1989, *Phys. Rev.*, **D40**, 290.

Messina, A., Moscardini, L., Lucchin, F., and Matarrese, S. 1989, in preparation.

Peebles, P. J. E. 1983, *Ap. J.*, **274**, 1.

Press, W. H., and Schechter, P. 1974, *Ap. J.*, **187**, 425.

Ryden, B. S., and Turner, E. L. 1984, *Ap. J. (Letters)*, **287**, L59.

Saslaw, W. C., and Hamilton, A. J. S. 1984, *Ap. J.*, **276**, 13.

Vishniac, E. 1986, in *Inner Space/Outer Space*, ed. E. W. Kolb, M. S. Turner, Olive, K., Seckel, D., and Lindley, D. (Chicago: University of Chicago Press), 190.

Weinberg, D. H. 1988, *Pub. A.S.P.*, **100**, 1373.
Weinberg, D. H., Ostriker, J. P., and Dekel, A., 1989, *Ap. J.*, **336**, 9.
Yoshioka, S., and Ikeuchi, S. 1989, *Ap. J.*, **341**, 16.

DISCUSSION

S. White: As I understand your initial-condition algorithm, it seems that it must give effectively random-phase white-noise fluctuations on all scales significantly larger than the high-mass tail of the initial δ distribution. This is because the fluctuation in such large mass cells is the sum of a large number of independent, identically distributed (but non-Gaussian) distributions and so must itself be a Gaussian (provided the second moment of the individual distribution exists). Thus all your distributions should look approximately Gaussian, once evolution has proceeded well beyond the initial conditions. A 'more natural' non-Gaussian initial condition might attempt to be equally non-Gaussian on all scales.

F. Lucchin: We are aware of this problem: in fact, we tested for the moment non-correlated initial distributions. In any case, the initial non-Gaussian distribution, still after ten expansion factors, affects the time evolution and the overall properties (like the percolation parameter) of our models. We plan to test more physical non-Gaussian distributions allowing for initial correlations, in particular 'scale-invariant' distributions provided by inflation: such statistics show precisely the property that you mention.

J. Primack: You mentioned at the end that you plan to simulate CDM non-Gaussian fluctuations obtained from inflation. What sort of inflationary model generates the non-Gaussian Zel'dovitch fluctuations, and in what sense are they non-Gaussian? And how would you lay down the initial conditions for the simulations, with the necessary CDM spectrum?

F. Lucchin: In a recent paper by Matarrese, Ortolan, and myself we found that a stochastic approach to inflation reveals that non-Gaussian fluctuations (non-zero $n>2$ moments) are a quite natural output of inflation. In particular the inflationary models we analyzed generate a density distribution related to a Gaussian one by a non-linear transformation. We plan to test this kind of model by laying down the initial CDM spectrum *via* the underlying Gaussian statistics.

CORRELATIONS OF TWO MILLION GALAXIES

S. J. MADDOX, W. J. SUTHERLAND, G. EFSTATHIOU AND J. LOVEDAY
Department of Astrophysics, University of Oxford
Keble Road, Oxford, OX1 3RH, England.

ABSTRACT We have constructed a machine-measured survey of 20 million images brighter than $b_J = 22$ mag covering an area of 4300 square degrees. The surface brightness profile of each image is used to select a sample of 2 million galaxies brighter than $b_J = 20.5$ mag. The galaxy photometry is corrected to remove field effects on each plate, matched from plate-to-plate and calibrated by CCD photometry. We have measured the two-point angular-correlation function of the galaxies in six 0.5-mag slices. After scaling by Limber's formula there is good agreement between correlation functions of the different slices. At small angles the correlation functions are well approximated by a power law, but at larger angles they fall below the power law. The position of the break is similar to that in the Lick survey, but the slope beyond the break is shallower. The shallower break leads to more large-scale power than is consistent with the 'standard' Cold Dark Matter models.

INTRODUCTION

A deep, homogeneous sample of galaxies covering a large area of sky is necessary to make reliable measurements of large-scale galaxy clustering. We have used the Automatic Plate Measuring (APM) system (Kibblewhite *et al.* 1984) to digitize glass copies of 185 UKSTU J survey plates in the southern Galactic cap. We have currently analyzed an area of 4300 square degrees covering $b > -40°$ and $\delta < -20°$. In this area we have detected about 20 million objects from which we have selected a uniform sample of 2 million galaxies to a magnitude limit of $b_J = 20.5$ mag. At this limit the survey has an effective depth of $\sim 600\ h^{-1}$ Mpc. We are now extending the survey to include plates in the equatorial region with $0° > \delta > -20°$ and $b < -40°$. Some aspects of the APM Galaxy Survey have been described by Maddox *et al.* (1988), and more details are given in Maddox *et al.* (1990a,b).

THE APM MEASUREMENTS

The APM system in Cambridge is a high-speed microdensitometer linked to a special-purpose computer which performs on-line image detection and processing. During each scan the APM system locates all images with more than a certain number of pixels brighter than a predetermined threshold above sky. We set the minimum image area at 2 arcsec^2 and the threshold at $b_J = 24.5$ to 25 mag arcsec^{-2}, depending on the emulsion noise of the plate. A scan of a typical plate records 300,000 images to a limiting magnitude of $b_J \sim 22$ mag. The limit for uniform image detection over the whole survey is $b_J = 21.5$ mag, set by the shallowest plates.

For each image 15 parameters are calculated from the sky-subtracted densities. These parameters are the mean x and y coordinates of the image, the isophotal summed density, the density-weighted second-order moments, the peak density, and areas above 8 preset density levels. A more detailed discussion of the parameters is presented in the Appendix of Maddox *et al.* (1990a). The (x,y) positions are converted into right ascension and declination using standard transforms determined from astrometric stars on each plate. We define an APM magnitude for each image from the isophotal sum of pixel values,

$$m_{iso} = +2.5 \log_{10}(\Sigma D).$$

We use the area, A_1, together with m_{iso} to calculate an approximate total magnitude, m, for each image. This is approximately linear compared to standard b_J magnitudes, so $b_J = z - m$, where the zero-point $z \sim 29.4$ mag.

The APM treats all connected pixels above the local threshold as an image, and so spurious images are caused by satellite trails and the halos and diffraction spikes of bright stars. Measurements of images near step wedges, globular clusters and local dwarf galaxies are also unreliable. Therefore, we examined each of the fields by eye and excluded all such areas from our survey. The total area excluded amounts to 3.7% of the whole survey and leaves 4300 degrees^2 of useful sky.

We repeated about 5% of the successful scans in order to estimate the size of measurement errors in the data and to check the long-term consistency of the measurements. For each matched image we compute the difference in the x, y and m from the two scans. The *rms* errors are typically 0.1" in position and 0.04 mag.

We also compared copy plates with original plates of the same field to check that the ESO copying process does not introduce significant errors. Comparisons between survey copies and non-survey original plates of the same fields are essentially the same as comparisons between different original plates, with *rms* errors of 0.2" in positions and 0.1 mag. These measurements confirm other investigations of copy plates which show that surface photometry measured by the APM and COSMOS machines is not significantly degraded by using copies instead of original plates.

Overlapping areas of neighboring fields provide an additional check of the errors in the data. The random errors in relative image positions are $\sim$ 0.5", but small distortions over each field cause systematic position errors of $\sim$ 1 - 2". The combined effects of vignetting and differential desensitization (which we call 'field effects') cause large systematic magnitude errors in plate overlaps ($\sim$ 0.4 mag). We correct for these effects as described briefly below and in more detail by Maddox *et al.* (1990b).

STAR-GALAXY SEPARATION

Profile Residuals

The APM measures all images on each field, and we use the image parameters to decide which are galaxies. The surface brightness profile of a stellar image is determined only by the instrumental response and seeing conditions during the observation. Therefore the image parameters of stars lie in a well-defined region of parameter space which we exclude to leave a sample of non-stellar objects. Pairs of objects with very small separations on the sky are analyzed and recorded as a single image by the APM real-time software. The parameters for such merged images are non-stellar and so they contaminate the non-stellar sample. We use two further parameters to reject merged images and produce a clean sample of galaxies (see below).

The surface-brightness profile of each image is given by the peak density, P, the radius of gyration, K, and the image area, A_i, above each of eight density levels. For each parameter ($p_i = A_i$, $1 \leq i \leq 8$, $p_9 = P$ and $p_{10} = K$), we produce a scatter plot against m and locate the stellar locus as a function of magnitude, giving $p_i^*(m)$. For every image we calculate the difference between each parameter p_i and the stellar value $p_i^*(m)$ at the appropriate magnitude. We define the profile residual ψ for each image as the logarithm of the weighted sum of the residuals $\psi = \log_{10} \sum w_i (p_i - p_i^*)^2$. The weights are set so that stars lie near $\psi = 0$.

Images with $m < 8.5$ mag are blurred by seeing and emulsion noise, so no reliable classification can be made. For images with $8.5 < m < 9.0$ mag some distinction between stellar and non-stellar images can be made, but the confidence level for any image being non-stellar is small. The galaxy images are well separated from the stars for $9.5 < m < 12.5$ mag, and so we can reliably reject stellar images.

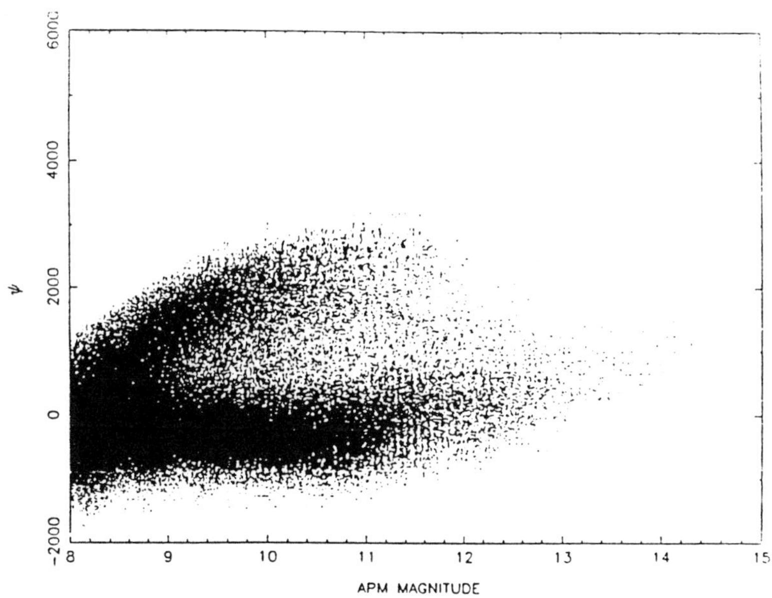

Fig. 1. The distribution of ψ for images on a typical field.

Comparisons With Visual Checks

In order to test the completeness and contamination of the galaxy sample produced by the ψ parameter, we made visual classifications of images. We classified all images brighter than $b_J \sim 21$ mag in six $8' \times 8'$ areas on each of 16 copies. AAT plates give more reliable visual classifications for fainter images, so we also used classifications from AAT plates (Drinkwater 1987; Morton, Krug and Tritton 1985).

For images with $b_J > 20.5$ mag we found that the distribution of stars as a function of ψ extends into the distribution of galaxies, so a complete and uncontaminated sample of galaxies cannot be selected from the UKSTU plates. For galaxies with $20.5 > b_J > 20.0$ mag our optimal sample is 90% complete and contains $\sim$12% contaminating images.

For images brighter than $b_J = 20.0$ mag the distributions of stars and galaxies are clearly separated, but the $8'$ areas and AAT plates do not contain enough galaxies to estimate reliably the completeness of the galaxy sample for $b_J < 19.5$ mag. Therefore we inspected six $16'$ square regions on each of 8 plates to a limit of $b_J = 21$ mag, and six $40'$ square regions on each of 6 plates to a limit of $b_J = 19$ mag. From the earlier checks, we estimate that $<1\%$ of galaxies with $b_J < 20.0$ mag have $\psi < 700$, and so we examined only images with $\psi > 700$ in these larger areas. The 1643 galaxies in these areas show that selecting images with $\psi > 1300$ gives a galaxy sample $>98\%$ complete for $b_J < 20.0$ mag and negligible contamination from stellar images. However, the sample is significantly contaminated by merged stars for $b_J < 19.5$.

Merged Images

The two parameters we use to identify merged images are based on the radius of gyration and the degree of saturation of each image. We compare the measured radius of gyration of an image, $K_{\rm true}$, to the radius of gyration, $K_{\rm ell}$, that would be expected for a single elliptical image with the same ellipticity and surface brightness profile. Most images are single; $K^2_{\rm true}/K^2_{\rm ell} \sim 1$. Images consisting of two components of roughly equal magnitude have a large value of $K^2_{\rm true}$ compared to the value of $K^2_{\rm ell}$ for the equivalent single elliptical image. We identify the stellar locus of $K^2_{\rm true}/K^2_{\rm ell}$ as a function of magnitude, $k^*(m)$, and define k for each image as $k = (K^2_{\rm true}/K^2_{\rm ell})/k^*(m)$.

A bright star with a faint companion object is more saturated than a galaxy with otherwise similar parameters. We estimate the fraction of saturated area for each image as

$$r = (0.2A_7 + A_8)/A_1.$$

This parameter was chosen in an *ad hoc* way to give a good separation between the visually checked galaxies and merged images. Then we locate the stellar locus of r as a function of magnitude, $r^*(m)$, and define $\mu = r/r^*(m)$.

The scatter plot of k against μ shows a clear separation of visually checked galaxies and merged objects. We define a boundary in the (k,μ) plane which separates the two classes of image, and then in the automated classification we classify images as galaxies or mergers according to whether they lie above or below the boundary. The residual contamination of merged objects in the final galaxy sample is $\sim 5\%$ for $17.0 < b_J < 19.5$ mag.

For images $b_J > 16.5$ mag the ratio of the number of galaxies to stars is less than 1/20, so an extremely reliable algorithm is required to reject stellar and merged images. Classifications from the APM image parameters alone do not produce an acceptable bright-galaxy sample so we have limited our present analysis to galaxies fainter than $b_J = 17.0$ mag. One of us (JL) has examined and individually classified all of the significantly non-stellar objects brighter than $m = 12.5$ mag, and assigned an approximate morphological type to each galaxy. This sample contains $\sim 17{,}770$ galaxies to a limit of $b_J = 16.5$ mag and will be published separately as the APM Bright Galaxy Survey (Loveday *et al.* 1990 and this volume).

PHOTOMETRIC CORRECTIONS

Corrections Within Each Field

The zero-point which relates standard b_J to APM magnitudes varies across each plate depending on vignetting and emulsion sensitivity. The vignetting of the UKST has been calculated theoretically and also measured from the telescope (Dawe and Metcalfe 1982). Positional variations in the sensitivity of the emulsion produce extra field effects which are different for each plate. To remove these field effects, we first measured the mean response function from the mean galaxy surface density averaged over the plates, and then used the variations in background sky density on each plate to calculate a magnitude correction.

Real structures in the galaxy distribution are not correlated with field boundaries, so they are smoothed out by averaging over many plates. Hence the mean galaxy density, N, as a function of position in the field follows the mean field effects. Our mean galaxy counts near the survey limit are approximated by $\log(N/N_0) = 0.45\Delta m$, so we can calculate the magnitude of the mean field effects.

Variations in desensitization between different plates make the field effects different for each plate, but if we assume the sky flux is constant across a plate, the background sky density, D_s, follows the field response across each plate. We calculate the central sky density, D_0, for each plate and plot $\Delta D = D_o - D_s$ against the Δm inferred from the mean galaxy density. Then a single quadratic fit to all of the fields gives the mean calibration of ΔD to Δm and allows us to calculate a field correction from the sky density on each plate. After applying this correction the averaged galaxy density varies by $<4\%$ over the field.

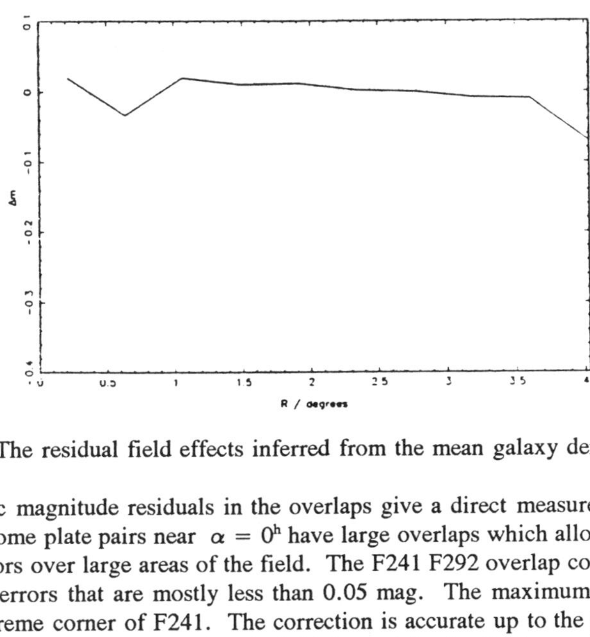

Fig. 2. The residual field effects inferred from the mean galaxy density.

Systematic magnitude residuals in the overlaps give a direct measure of the residual field effects. Some plate pairs near $\alpha = 0^h$ have large overlaps which allow us to measure the residual errors over large areas of the field. The F241 F292 overlap covers about $4° \times 6°$, and shows errors that are mostly less than 0.05 mag. The maximum error is ~ 0.15 mag, in the extreme corner of F241. The correction is accurate up to the plate edges and, since only the central 5° square of each field is used in the final survey, errors in the extreme corners of a field are not important.

Plate-to-plate Matching Of Photometry

The field corrections described above give magnitudes with a uniform zero-point across each plate, but each plate has a different zero-point. To adjust the magnitudes of each plate to a common zero-point we first estimate the difference between the two zero-points of every overlapping plate pair. For each image in an overlap we compare the magnitude measurements from the two plates, m_i and m_j, and then fit a straight line to a plot of $(m_i+m_j)/2$ against (m_i-m_j) for the overlap. The mean *rms* scatter of (m_i-m_j) about each fit is 0.22 mag, and the *rms* of the zero-point offsets, T_{ij}, is 0.26 mag. Since we have not accurately linearized the raw APM magnitudes, we allow a different slope in each fit. The *rms* residual between the slope for any single plate and unity is 0.05. For simplicity in the following discussion we treat the matching as if only zero-point offsets are used, but we have included slopes in our actual matching procedure.

Next we use the T_{ij} offsets to calculate a correction, C_i, for each plate which converts the measured magnitudes to a uniform matched magnitude system. We use an iterative routine to calculate the set of plate corrections C_i^e which minimizes the weighted sum of $(T_{ij}^e + C_j^e - C_i^e)^2$ over the overlaps for each individual plate. Our algorithm is equivalent to minimizing the quantity F of Seldner *et al.* (1977; hereafter SSGP) in their matching of the Lick survey.

The errors in the estimated plate zero points depend on the errors in the overlap coefficients. SSGP considered a model in which errors do not propagate across more than one overlap. Groth and Peebles (1986; hereafter GP86) extended the model to include the propagation of errors over many overlaps. If each plate has 4 overlaps, and the error in zero point for each plate is δC_i, and the error in each measured transform is δT_{ij}, then the SSGP model predicts that the variance of δC_i is given by $\beta_{ii}^2 = (4/25)\,\sigma^2$, where we have written var $(\delta T_{ij}) = \sigma^2$. The SSGP model also predicts that the covariance of the errors of neighboring plates is given by $\beta_{ij}^2 = -(1/25)\,\sigma^2$. The GP86 model including large-scale correlations of errors predicts $\beta_{ii}^2 = (\sigma^2/4)\,S_1$ and $\beta_{ij}^2 = (\sigma^2/8)\,S_2$, where S_1 and S_2 are constants which depend on the number of plates in the survey. Approximating our survey by a 13×13 array, we find $S_1 = 1.827$ and $S_2 = 1.666$.

We estimate σ^2 from the overlap residuals $\sigma^2 = (5/3)$ var $(T_{ij}^e + C_j^e - C_i^e)$. The scatter about the line of equality in Figure 3 implies $\sigma^2 = 3.5\times10^{-3}$. The SSGP model therefore gives $\beta_{ii}^2 = 5.6\times10^{-4}$, and $\beta_{ij}^2 = -1.4\times10^{-4}$. The GP86 model gives $\beta_{ii}^2 = 1.6\times10^{-3}$ and $\beta_{ij}^2 = 7\times10^{-4}$. The increase of 10^{-3} in β_{ii}^2 is from the variance of large-scale gradients in the zero-point errors.

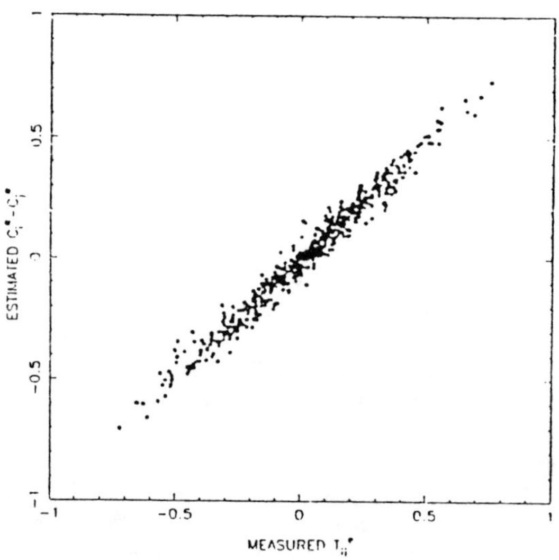

Fig. 3. T_{ij} plotted against $C_i^e - C_j^e$ for each overlap.

To improve the plate-to-plate uniformity of the ψ selection, we adjusted the ψ limit for each plate using overlaps in the same way as for the magnitude matching. The residual errors in this ψ matching increase the effective σ^2 to 5×10^{-3}, and so our overall estimate of errors in galaxy density are $\beta_{ii}^2 = 2.2\times10^{-3}$ and $\beta_{ij}^2 = 10^{-3}$.

CCD Calibrations And Surface Density Maps

We determined the overall zero-point of the matched magnitudes from 41 sequences of galaxies observed with the UCL CCD camera on the 1.0-m telescope at the South African Astronomical Observatory (Walker 1984). Each sequence consists of a small group of galaxies with a range of magnitudes typically $17 < b_J < 21.4$ mag. We measure the zero-point, $z = \langle b_J + m \rangle$ for each sequence. The mean value is $z = 29.4$ mag and the mean *rms* error in each estimate is 0.09 mag.

The overlap matching errors are ~0.04 mag, so they are not directly detectable by each individual sequence. However, systematic gradients can be quantified from the covariance of the sequence residuals as a function of angular separation. The amplitude of $\langle (z_i - \langle z \rangle)(z_j - \langle z \rangle) \rangle$ is $< 2 \times 10^{-3}$ for separations $\theta > 5°$, which is consistent with $\beta^2_{ii} = 2 \times 10^{-3}$ as measured above.

A map of the surface density of all the images with $20.5 > b_J > 19.5$ mag after correction and calibration is shown in Plate 1. The galactic gradient in stellar density can clearly be seen, with the galactic bulge towards the right. Also, the Fornax and Sculptor dwarf galaxies, and some globular clusters and bright stars are visible as small patches of very high image surface density. These areas were removed from the galaxy catalogue. Structure in the galaxy distribution can be seen, but is significantly masked by the smooth distribution of the stars. Plate 2 shows a similar surface density map for our galaxy sample with $20.0 > b_J > 19.5$ mag. In the next section we quantify the clustering in the galaxy distribution by measuring $\omega(\theta)$.

THE ANGULAR CORRELATION FUNCTION, $\omega(\theta)$

We use two estimators to measure $\omega(\theta)$. On scales $\theta < 0.5°$, we use the coordinates of each galaxy to count pairs and estimate $\omega(\theta) = F(DD/DR) - 1$, where DD is the number of data pairs with separations in the range $\theta \pm \delta\theta$. This is normalized by the term DR which is the pair count determined by cross correlating the data points with a set of points distributed at random within the boundary with F times the mean density of galaxies. On scales $0.1° < \theta < 30°$ we compute $\omega(\theta)$ from maps of the galaxy distribution in an equal-area projection using the estimator $\omega(\theta) = \langle N_i N_j \rangle / (\langle N_i \rangle \langle N_j \rangle) - 1$, where N_i and N_j are the counts in cells i, j, and the angle brackets denote an average over pairs of cells separated by an angle $\theta \pm \delta\theta$. We correct both of these estimates for the dilution of clustering from residual contamination by merged images, which we estimate from the visual checks.

We can approximate $\omega(\theta)$ by a power law at small angles; a least-squares fit to $\omega(\theta)$ in Figure 4 over the range $0.01° < \theta < 1°$ gives a slope of -0.668 and an amplitude of 2.84×10^{-2} at $\theta = 1°$. On large angles the slope of $\omega(\theta)$ gradually steepens. The break at ~3° is extremely significant given the scatter between zones. The inset in Figure 4 shows that $\omega(\theta)$ is very close to zero (to within $\sim 5 \times 10^{-4}$) for $\theta > 6°$.

The offset between the estimate of $\omega(\theta)$ counting only pairs of cells on the same plate (intra-plate) and the estimate counting only pairs of cells on neighboring plates (inter-plate) provides a measure of the variance of the plate matching errors (Groth and Peebles 1977, hereafter GP77). We find a mean offset of 1.7×10^{-3}, which is in good agreement with the offset of 1.2×10^{-3} expected from our error estimates.

The variation of $\omega(\theta)$ as a function of magnitude is a further test for systematic errors in our catalogue. We calculated $\omega(\theta)$ for each of six disjoint 0.5-mag slices in the range $17 < b_J < 20.5$ mag, as shown in Figure 5(a). Then we used the relativistic version of Limber's equation (GP77, Phillipps *et al.* 1978) to compute the shifts in $\log(\omega)$ and $\log(\theta)$ which scale our results to the Lick depth (see GP77 Figure 15). The filled points in Figure

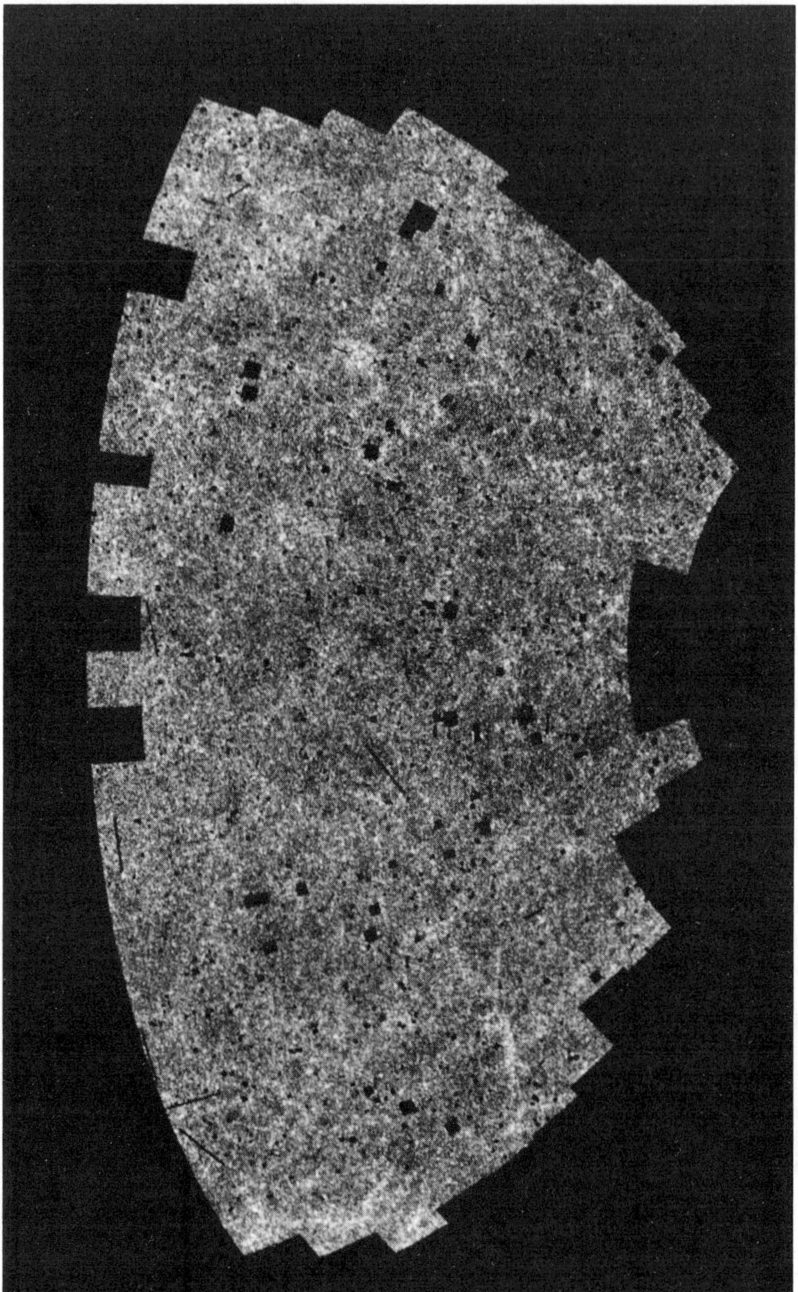

Plate 1. The distribution of all objects with $17 < b_J < 20.5$ mag in an equal area projection centered on the south Galactic pole.

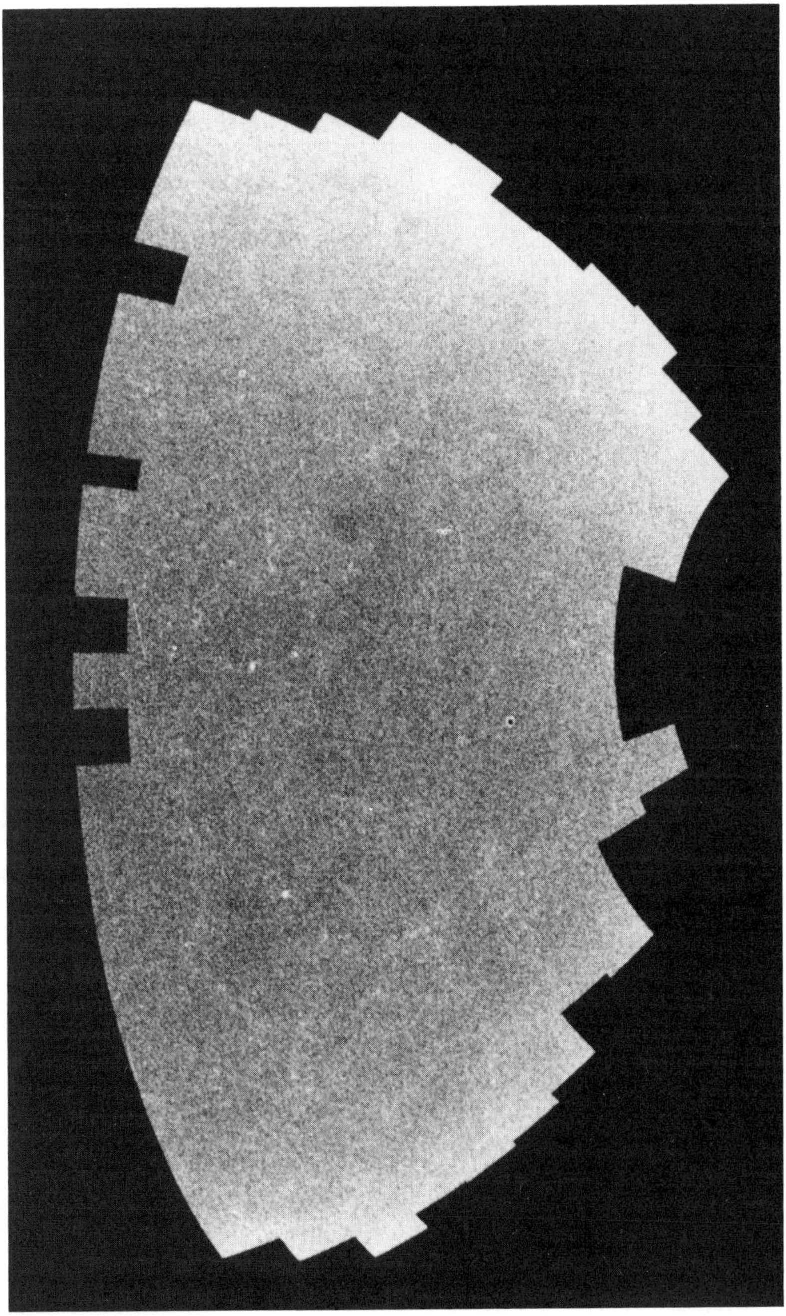

Plate 2. The corresponding distribution of galaxies with $17 < b_J < 20.5$ mag. The small empty patches in the map are regions that we have excluded around bright stars, nearby dwarf galaxies, globular clusters and step wedges.

5(b) show the $\omega(\theta)$ curves after scaling. The agreement between these curves is excellent, and the scatter is consistent with the errors expected from our plate-matching procedure. This test provides important evidence that our angular correlation functions are measuring real clustering in the galaxy distribution rather than clustering induced by systematic errors.

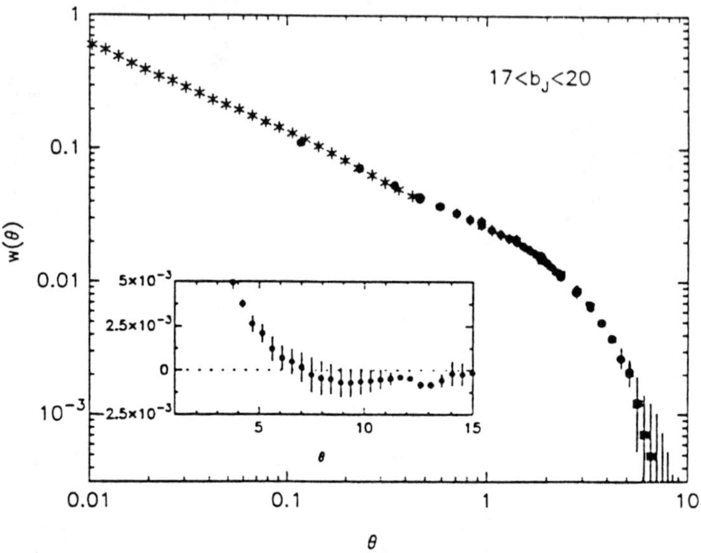

Fig. 4. The mean of $\omega(\theta)$ for galaxies with $17 < b_J < 20$ mag in each of four separate areas of sky covering roughly equal strips in right ascension. The error bars are computed from the scatter between the zones.

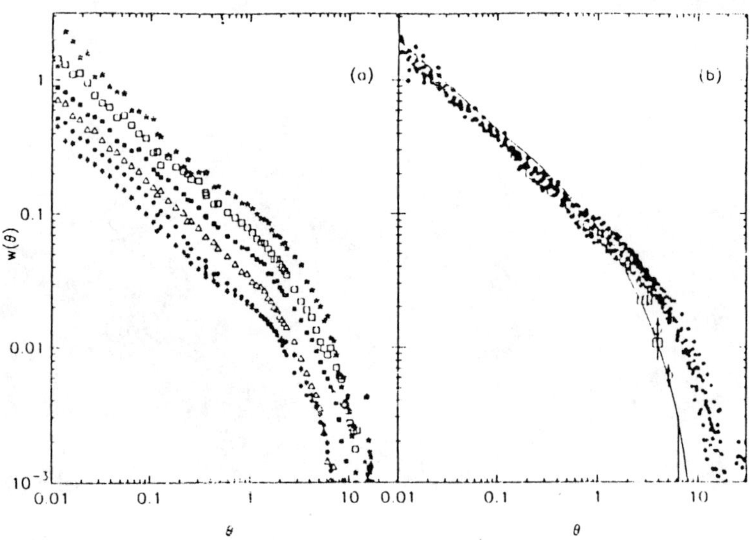

Fig. 5. Scaling of our $\omega(\theta)$ measurements, with the GP77 results and two CDM models.

The open points in Figure 5(b) show $\omega(\theta)$ measurements from the Lick survey (GP77). Our results are in good agreement on scales up to $\theta \sim 3°$ but disagree on larger scales. GP77 find that $\omega(\theta)$ breaks sharply from a power law on scales of $\sim 3°$, whereas we find a more gentle decline. However, GP77's estimate of $\omega(\theta)$ is based on filtered versions of the Lick map, and it possible that filtering has removed a component of intrinsic clustering. For the APM survey, we estimate gradients in the matching errors should be of marginal importance and so we have not filtered our maps. A full discussion of the effects of plate-matching errors on our results will be presented elsewhere (Maddox *et al.* 1990b).

In a previous article we estimated $\omega(\theta)$ from a small region of about 25 plates from the APM survey (Maddox *et al.* 1988) and found good agreement with GP77's results. This does not conflict with the results shown in Figures 4 and 5, but demonstrates that there are large sampling fluctuations in a 25-plate patch at these magnitude limits. Collins *et al.* (1989) have estimated $\omega(\theta)$ from a contiguous area of 24 Schmidt plates to $b_J = 20.5$ mag. However, our analysis implies that their region is too small for an accurate determination of $\omega(\theta)$ at $\theta > 1°$.

The two curves in Figure 5(b) show predictions for $\omega(\theta)$ based on the 'standard' biased Cold Dark Matter (CDM) model of Davis *et al.* (1985). We computed $\xi(r)$ from the numerical simulations described by White *et al.* (1987) using their prescription for biasing. The dotted line shows the result for $h = 0.5$; it provides a good fit to $\omega(\theta)$ on small scales but fails to match the Lick $\omega(\theta)$ on scales $> 1°$. The solid line shows the result of lowering the Hubble constant to $h = 0.4$, while retaining all other parameters fixed. This result matches the Lick data on scales $> 1°$ but predicts too high an amplitude on small scales. Neither model matches the results from the APM Galaxy Survey. Our survey provides strong evidence for large-scale power in the galaxy distribution which is difficult to reconcile with the standard CDM model of White *et al.* (1987). We emphasize, however, that the standard CDM model represents just one variant of the cold dark matter cosmogony and that it may be possible to construct variants which are compatible with our measurements.

The detection of small-amplitude clustering on large scales is extremely sensitive to systematic errors. The tests that we have applied to the APM Survey indicate that we are measuring real galaxy clustering, but the results clearly need to be confirmed by other surveys. Our redshift survey of bright APM galaxies (Loveday *et al.* 1990) and the galaxy catalog of Heydon-Dumbleton *et al.* (1989) are two ongoing projects that should provide important checks of our results.

ACKNOWLEDGEMENTS

We are extremely grateful for all the assistance provided by the APM group (M. J. Irwin, P. Bunclark and M. Bridgeland) over the last few years, and for numerous discussions with members of the Institute of Astronomy, Cambridge. We are also grateful to the University of Oxford for their support at a critical stage in the project. W. J. Sutherland and J. Loveday acknowledge the receipt of SERC studentships. S. J. Maddox has held an SERC studentship and research assistantship during the course of this work.

REFERENCES

Collins, C. A., Heydon-Dumbleton, N. H., and MacGillivray, H. T. 1989, *M.N.R.A.S.*, **236**, 7P.

Davis, M., Efstathiou, G., Frenk, C. S., and White, S. D. M. 1985, *Ap. J.*, **292**, 371.

Dawe, J. A., and Metcalfe, N. 1982, *Proc. Astr. Soc. Aust.*, **4**, 466.
Drinkwater, M. J. 1987, Ph.D. Thesis, University of Cambridge.
Groth, E. J., and Peebles, P. J. E. 1977, *Ap. J.*, **217**, 385 (GP77).
Groth, E. J., and Peebles, P. J. E. 1986, *Ap. J.*, **310**, 499 (GP86).
Heydon-Dumbleton, N. H., Collins, C. A., and MacGillivray, H. T. 1989, *M.N.R.A.S.*, **238**, 379.
Kibblewhite, E. J., Bridgeland, M. T., Bunclark, P., and Irwin, M. J. 1984, *Astronomical Microdensitometry Conference*, NASA Conf. Pub., **2317**, p. 277.
Loveday, J., Efstathiou, G., and Peterson, B. A. 1990, in preparation.
Maddox, S. J., Efstathiou, G., and Loveday J. 1988, in *IAU Symposium 130, Large Scale Structures of the Universe*, ed. J. Audouze, M.-C. Pelletan and A. Szalay, (Dordrecht: Kluwer), p. 151.
Maddox, S. J., Sutherland, W. J., Efstathiou, G., and Loveday J. 1990a, *M.N.R.A.S*, **243**, 692.
Maddox, S. J., Sutherland, W. J., Efstathiou, G. and Loveday J. 1990b, *M.N.R.A.S.*, **246**, 433.
Morton, D. C., Krug, P. A., and Tritton, K. P. 1985, *M.N.R.A.S.*, **212**, 325.
Phillipps, S., Fong, R., Ellis, R. S., Fall, S. M., and MacGillivray, H. T. 1978, *M.N.R.A.S.*, **182**, 673.
Seldner, M., Seibers, B., Groth, E. J., and Peebles, P. J. E. 1977, *A. J.*, **82**, 249 (SSGP).
Walker, A. R. 1984, *M.N.R.A.S.*, **209**, 83.
White, S. D. M., Frenk, C. S., Davis, M., and Efstathiou, G. 1987, *Ap. J.*, **313**, 505.

THE REALITY OF LARGE-SCALE FILAMENTARY STRUCTURE IN THE LICK MAP

R. M. SMITH, C. S. FRENK
Department of Physics, Science Laboratories, South Road
University of Durham, DH1 3LE, England

S. J. MADDOX,
Institute of Astronomy, University of Cambridge
The Observatories, Madingley Road, Cambridge, CB3 0HA, England.

M. PLIONIS
Astronomy Centre, University of Sussex, Brighton, BN1 9QH, England

ABSTRACT We present the results of automated measuring-machine scans of POSS plates covering two regions of sky that show prominent large-scale filamentary structure in the Lick map. Similar structures are seen in the new maps indicating that the morphology of the galaxy distribution in the Lick map is reproducible. Redshifts of randomly selected objects within these features suggest that coherent structures exist on scales of order 100 Mpc.

INTRODUCTION

Much of our present understanding of the distribution of galaxies on large scales stems from analysis of the Shane-Wirtanen (hereafter Lick) galaxy-count catalog (Shane and Wirtanen 1967) reduced by Seldner *et al.* (1977). This catalog of galaxy counts in $10' \times 10'$ cells covers all the sky visible from the Lick Observatory ($\delta \geq -23°$) and is complete down to $m_{pg} \sim 18.5$. Both the northern and southern Galactic caps show a prominent filamentary pattern, although that in the north is more pronounced (Barrow and Bhavsar 1987).

The Lick galaxy counts were made by eye, on plates not originally intended for studies of large-scale structure, and hence suffer from biases which led Seldner *et al.* to introduce several 'plate-correction factors' in an attempt to standardize all the plates. However, the accuracy of the Lick map has always been open to question. Geller *et al.* (1984) and de Lapparent *et al.* (1986) have claimed that small variations in the limiting magnitude and counting efficiency both across and between plates may lead not only to the 3° break in the two-point correlation function but also to the appearance of spurious filamentary structures.

The impact on our present knowledge, of the Lick map with its inherent uncertainties, compounded by the recent claims, has led us to reanalyze the large-scale galaxy distribution using an objective galaxy-counting technique and alternative plate material. The results of this study have encouraged us to undertake a redshift survey of galaxies randomly selected from prominent structures in the Lick map in an attempt to study their spatial distribution.

TECHNIQUES

The work of Geller *et al.* (1984) and de Lapparent *et al.* (1986) suggests that some of the large-scale structure in the Lick map could arise from plate inhomogeneities and observer-dependent biases. A test of the first possibility is to determine whether different plate material, with different plate variations and plate edges, shows the same structure as the Lick map. The Palomar Observatory Sky Survey (POSS) provides an easily-available source of such material. To overcome many of the difficulties and biases inherent in counting galaxies by eye requires the use of a rapid plate-scanning measuring machine such as the Automatic Plate Measuring machine (APM, Kibblewhite *et al.* 1983) at the Institute of Astronomy, Cambridge.

We selected two regions for detailed investigation from the plots of the Lick map displayed by Seldner *et al.* (1977). These correspond to two of the most prominent filaments in the map, one in the North Galactic Cap (NGC) and one in the South Galactic Cap (SGC). The northern filament is fairly continuous and mainly delineated by galaxies, while that in the south is part of the Aquarius-Capricornus Supercluster region (Batuski and Burns 1985, Tully 1987).

A total of 25 POSS plates covering the whole of the apparent extent of the filaments was scanned on the APM. Of these, 14 are in the NGC and 11 in the SGC. Since copies of the POSS plates were used, the image quality is not as good as on the originals. With this material the star/galaxy classification from the APM is reliable at about the 90% level. Matching the intensity scales between plates was difficult due to their small overlap. The method usually applied (Maddox 1988), fitting a linear function to a graph of $m_1 - m_2$ *versus* m_1, where m_1 and m_2 are the magnitudes of images in the overlap region on each plate, could not be used due to the poor image quality and large variations so close to the plate edges. Matching was therefore done by setting the APM instrumental magnitude limit such that the mean number density was the same as that of the corresponding region of the Lick map. This procedure propagates any intraplate gradients that may be present in the Lick map into the APM scans, but should not affect the morphology of high-contrast structures significantly. As the Lick counts were carried out visually, the catalog has not only a magnitude limit but also a minimum size criterion. The APM reduction provides an isophotal area for each galaxy in addition to its instrumental magnitude. To test the effect of a galaxy-size criterion, we also generated maps of galaxies larger than a threshold value, which was adjusted to give the required number density.

COMPARISON WITH THE SHANE AND WIRTANEN MAP

Qualitative Comparison

In Plate 1 we show grey-scale maps of the two regions scanned and compare these with the corresponding regions of the Lick map. Although the APM and Lick maps are not identical, they both show a similar structure. In particular, the thin filament in the NGC is reproduced in both maps. The filament in the SGC is more diffuse, but again the gross morphology of both maps is similar. Using a limiting area rather than a limiting magnitude criterion for the APM maps does not affect significantly the shape of the structure present. Subtraction of these two APM maps shows that the maximum deviation between them is 2 galaxies/cell, with no evidence of residual structure.

The APM scans allow the stellar distribution to be investigated in addition to the galaxy distribution. The stellar distribution is found to be quite uniform, with no evidence for filaments or clusters, implying that the observed galaxy distribution does not arise from

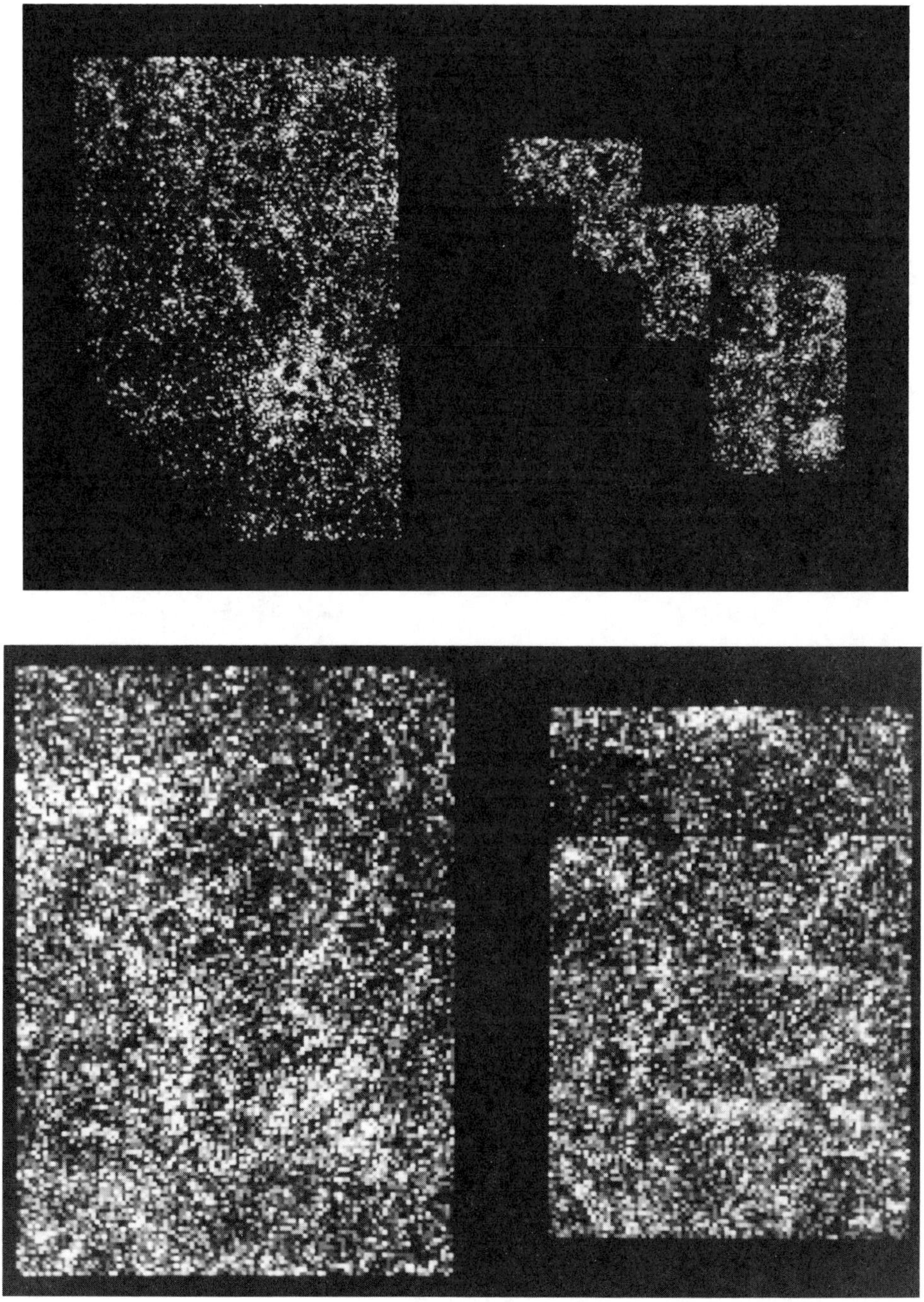

Plate 1. Grey-scale images of the gridded APM scans (left) and the corresponding regions of the Lick map (right).

obscuration in our own galaxy. Uniformity of the stellar distribution also indicates that the conclusions drawn above concerning the morphology of the galaxy distribution are not significantly affected by contamination of the APM galaxy maps by stars.

Quantitative Comparison

At the simplest level, the Lick and APM maps can be compared by plotting histograms of counts in cells of various sizes. Since the density in each 6° x 6° plate in the APM maps has been matched to that of the corresponding region in the Lick map, this comparison is only meaningful for cells much smaller than 6°. In Figure 1, we show histograms of counts in 10′ × 10′ cells and in 30′ × 30′ cells. There is a significant difference between the Lick and APM counts in the smaller cell size. At both low and high densities, the Lick maps show an excess number of cells compared to the APM maps. This effect is present in both our selected regions.

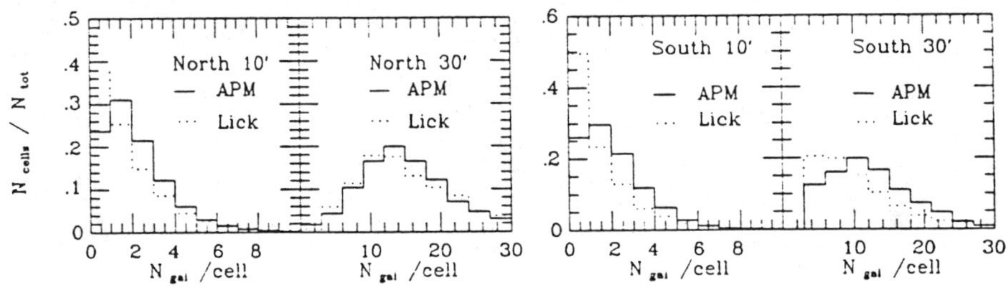

Fig. 1. Histograms of the density distributions in the APM and Lick maps both in 10′ and 30′ cells.

There are several possible reasons for the discrepancies in the 'high-resolution' counts. Firstly, although contamination by stars in the APM maps will not affect the morphological properties of the galaxy distribution, it can lead to a bias in the histogram of cell densities. A uniform stellar distribution could lead to a decrease in the number of zero- and low-density cells. A possible reason for the excess of high cell densities in the Lick map may be a tendency for the eye to reach fainter limiting magnitudes in regions of high density. Although in the lower-resolution (30′ × 30′) cells there are still some differences between the Lick and APM maps, these are much smaller than in the higher-resolution cells. There is a discrepancy of 4% in the number counts between the northern and southern scanned regions, which reflect a similar anisotropy present in the Lick map as a whole. A large discrepancy in the mean galaxy density between the northern and southern galactic hemispheres is well known but is usually attributed to the Local Supercluster. At the characteristic depth of the Lick map this feature should not be dominant and is therefore unlikely to be the cause of the excess in the North. This might be related to the dipole anisotropy seen in the microwave background and optical redshift surveys. Plionis (1988) and Lahav (1987) have also noted this discrepancy in two-dimensional maps of the galaxy distribution.

DO THE STRUCTURES EXIST AS SPATIAL FEATURES?

We have shown above that the filamentary structure seen in the Lick map is likely to be real in two-dimensions. We now investigate whether it is likely to reflect coherent spatial filaments or merely chance alignments in the sky of the clustered galaxy distribution.

We first identified filaments over the entire Lick map using the Minimal Spanning Tree technique (MST) described by Bhavsar and Barrow (1983). This technique delineates filamentary patterns by connecting neighboring points of a distribution in a unique way. To extend this method to the binned distribution of the Lick map, we first smoothed the counts on $40' \times 40'$ cells and then identified those cells which were local density maxima, using the algorithm of Moody, Turner, and Gott (1983). The MST connecting these cells was then created, 'pruned' and 'separated' to select the most prominent filaments. A detailed description of this procedure and the properties of filaments over the entire map will be presented elsewhere. Here we concentrate exclusively on the filaments identified by the MST in the regions of our APM scans. We have searched the literature for redshifts of clusters in the neighborhood of the two selected filaments. Figure 2 shows the filaments as defined by the MST together with the positions and redshifts of clusters in the region. The NGC filament contains 4 Abell clusters within it, and 3 of these have $\langle v \rangle = 23{,}840 \pm 100$ km s^{-1}. At this redshift, the filament found by the MST would have an extent of $\sim 200\ h^{-1}$ Mpc ($h = H_0/100$). The SGC shows a possible extended feature at $\sim 18{,}000$ km s^{-1} with weak evidence of superclustering at $\sim 24{,}000$ km s^{-1}.

The coincidences shown in Figure 2 encouraged us to undertake a redshift survey of randomly selected objects within the two structures. The core of the NGC structure was selected by eye and an area about that core chosen to include approximately 2000 galaxies. This area covered only the most overdense regions of the structure, where the densities were over 1.5 times the median density. A random, 1 in 50, sample was selected and the redshifts of most of these galaxies obtained with the IDS on the Isaac Newton Telescope on La Palma. We have also obtained redshifts of galaxies within clusters in both our regions; the clusters were identified with a friends-of-friends algorithm on the scanned POSS plates. These clusters are generally poorer than Abell clusters but have a minimum of 15 members. The

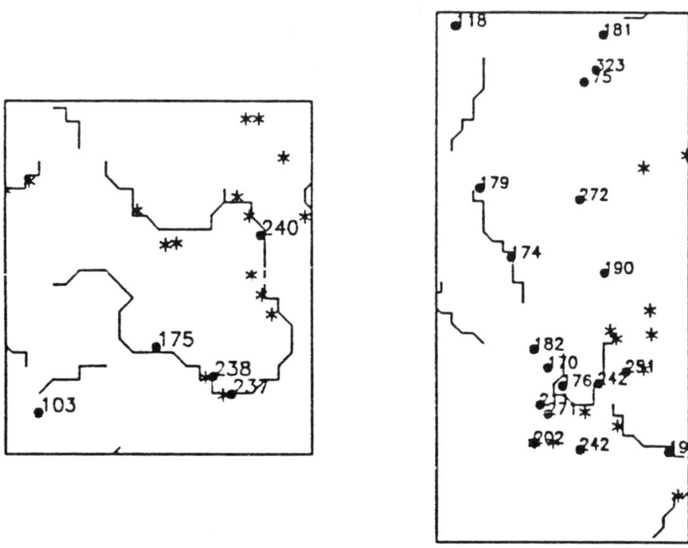

Fig. 2. Plot of the MST filaments found in the Lick map in the region of our APM scans. Superimposed are the positions and velocities of clusters with known velocities from the literature (•), together with the positions of all clusters in the region whose velocities are unknown (*).

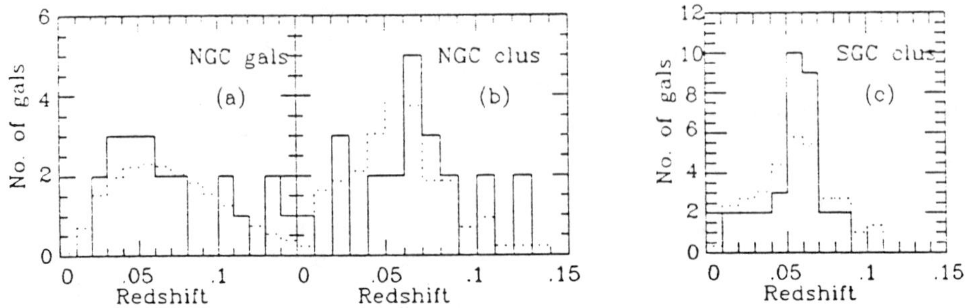

Fig. 3. Velocity histograms of the objects with measured redshifts. *Left*: randomly-selected galaxies on the NGC filament, together with the theoretical redshift distribution. *Middle*: clusters in the region of the NGC filament, with field cluster velocity distribution from Shectman. *Right*: same as *middle*, but for the SGC structure.

redshift survey is not yet complete, but in Figure 3 we show preliminary results for our sample of individual galaxies and clusters.

To determine the significance of any particular feature in Figure 3 requires an estimate of the field redshift distribution. For the sample of galaxies in the NGC we derived this by assuming a Schechter luminosity function for an unclustered sample of galaxies limited at an apparent magnitude of 18.5. For the sample of clusters we used the redshift distribution of clusters in the Lick map obtained by Shectman (1985). Both distributions have been normalized by the number of galaxies in our redshift survey. As our clusters are generally poorer than those selected by Shectman, we would expect their redshift distribution to be intermediate between that of the field galaxies and Shectman clusters. Since the difference between them is small, the significance of any features found is only weakly affected by the assumed background distribution.

The northern filament shows no strong excess over the field population at any redshift, either in the galaxy or cluster sample, suggesting that the filament may not exist as a spatial feature. The southern structure shows a much more prominent peak at a redshift of $\sim 18{,}000$ km s^{-1} which appears to be significant. Inspection of the redshift-position diagram shows that the clusters with similar redshifts are distributed over all the structure and are not confined to the supercluster region to the North. Although a more extensive redshift survey is required, our preliminary results suggest an elongated structure with a spatial extent of $\sim 130\ h^{-1}$ Mpc.

CONCLUSIONS

We have shown from APM scans of POSS plates that at least some of the structure seen in the Lick map is real in two-dimensions, and that long filaments do exist in the projected galaxy distribution. A preliminary redshift survey of selected galaxies and clusters within these structures shows that one of two selected filaments prominent in projection is likely to arise from chance alignments, whereas the other may be a true three-dimensional elongated structure with length of order 100 h^{-1} Mpc.

REFERENCES

Bhavsar, S. P., and Barrow, J. D. 1983, *M.N.R.A.S.*, **205**, 61P.

Barrow, J. D., and Bhavsar, S. P. 1987, *Q.J.R.A.S.*, **28**, 109.

Batuski, D. J., and Burns, J. O. 1985, *A.J.*, **90**, 1413.

de Lapparent, V., Geller, M. J., and Kurtz, M. J. 1986, *Ap. J.*, **304**, 585.

Geller, M. J., de Lapparent, V., and Kurtz, M. J. 1984, *Ap. J. (Letters)*, **287**, L55.

Kibblewhite, E., Bridgeland, M., Bunclark, P., Cawson, M., and Irwin, M. 1983, in *Astronomy with Schmidt-type Telescopes*, ed. M. Capaccioli, (Dordrecht: Reidel), p. 89.

Lahav, O. 1987, *M.N.R.A.S.*, **225**, 213.

Maddox, S. J. 1988, Ph. D. thesis, Cambridge University.

Moody, J. E., Turner, E. L., and Gott, J. R., III 1983, *Ap. J.*, **273**, 16.

Plionis, M. 1988, *M.N.R.A.S.*, **234**, 401.

Seldner, M., Siebers, B., Groth, E. J., and Peebles, P. J. E. 1977, *A.J.*, **82**, 249.

Shane, C. D., and Wirtanen, C. A. 1967, *Publ. Lick. Obs.*, **22**, Part 1.

Shectman, S. A. 1985, *Ap. J. Suppl.*, **57**, 77.

Tully, R. B. 1987, *Ap. J.*, **323**, 1.

DISCUSSION

M. Strauss: Your strong filament in the North Galactic Cap extends about one radian in length, thus it is likely to have a *variation* in distance from us comparable to its distance. So I'm not sure that you can rule out the three-dimensional nature of the filament based on a small number of galaxies with a wide distribution of redshifts.

R. Smith: The nearby filamentary structure seen in the galaxy distribution shows that these features are highly overdense. If the northern filament studied here is similar to those nearby structures, and it is aligned almost perpendicular to the line of sight, then it should be detectable from our redshift sample. However, the detectability of a filament depends on several things, such as the thickness, mean redshift, overdensity, and angle with respect to the line of sight. These may conspire in such a way that we could not detect a filament even from a large redshift survey. It is easier to detect a large feature than prove it doesn't exist. We therefore agree that the northern filament may yet be found to exist.

D. Lambas: Could we have an estimate of the fraction of real structure in the Lick maps? Certainly there is a lot of structure associated with plate borders.

R. Smith: It is very difficult to obtain an accurate estimate of the number of real structures seen in the Lick map. To answer the question we need to scan plates of the whole Lick region. In two dimensions the reality of the structure in the regions we have compared strongly suggests that the Lick map is morphologically correct and that most of the structure is real.

E. Bertschinger: Your map of filaments is reminiscent of Sergei Shandarin's 'skeleton of structure' arising from the evacuation of underdense regions and the piling up of matter into sheets, filaments, and clumps. In particular, the sharp changes of direction of your filaments look like the intersections of filaments at clumps in the theory of adhesion. Are clusters found preferentially where the filaments change direction?

R. Smith: There is no evidence that rich clusters of galaxies occur preferentially at sharp bends in the filaments. Our sample of poorer clusters also shows no excess at these positions in the northern filament. This is not surprising if, as our redshift data suggests, the filament is not a free spatial feature.

M. Davis: I find it quite remarkable that so many of these filaments (or sheets) are in the plane of the sky. Perhaps infall to the caustic surfaces is explaining the contrast in redshift space of those filaments aligned in the plane of the sky. However in the Southern Sky Redshift Survey we did find one 'Great Wall' aligned roughly 45° to the plane of the sky. You didn't see a strong peak in the redshift distribution in your northern filament, but perhaps this is because that filament is tilted.

R. Smith: If the filament is tilted, then there should be a uniform redshift variation along the structure. We find no strong evidence of this using our data.

D. Burstein: When I first started looking at large-scale structures in the UGC, some ten years ago, I was struck by both the apparent filamentary structure *and* hints of seeing 'sheets' oriented across the line of sight. I find it significant that these structures have persisted as real features as roughly ten times more redshifts have been obtained. It is worthwhile to bring up an old question, especially given the striking very large-scale structure we have seen here - at what point do we become convinced of the reality of coherent structures hundreds of Mpc long? (This question should really be credited to Brent Tully - he has been asking it for the past few years.)

R. Smith: The 'Great Wall' seen in the new galaxy redshift catalogs shows that a very large-scale structure exists in the nearby universe. However, this is only one example, and it is still necessary to investigate their frequency, morphology, and coherence lengths. Large-scale structures are seen in the Lick map but, as our data on the northern filament suggests, it is not certain that these features are present in three dimensions. The difference between the galaxy and cluster clustering properties shows that we need redshifts of both to properly map the large-scale structure. At present there is insufficiency of these data to make any definite conclusions about the reality of these two-dimensional structures. We shall require a large redshift survey of a number of the apparent two-dimensional filaments before we can be truly convinced either way.

R. Opher: From the redshifts, can you say whether the filamentary structure observed could be primordial, or must it necessarily be of recent origin?

R. Smith: Redshifts can only give distance information on the structures present. Whether the filaments are primordial or of recent origin depends on the interpretation of that information.

COMPARING THE ABELL AND ACO CATALOGS OF RICH CLUSTERS OF GALAXIES

R. P. OLOWIN
Department of Mathematical Sciences, Saint Mary's College of California
Moraga, California 94575

ABSTRACT The Abell all-sky catalogue of 4073 rich clusters of galaxies is discussed in order to explore the relationship between the northern section prepared by Abell in 1958 and its southern counterpart, the ACO Catalog prepared in 1989. In both catalogs a 'rich' cluster is operationally defined as one having at least 30 members within the magnitude range m_3 to $m_3 + 2$ (m_3 is the magnitude of the third brightest cluster member), and each with a nominal redshift less than 0.2. The southern data have been collected from a survey of United Kingdom 1.2-m Schmidt Telescope IIIa-J plates and films, and have been reduced to the systems defined by the northern data previously published by Abell. Particular attention is given to the 'overlap-zone', −17° to −27°, where the two catalogs share a common region of observation.

INTRODUCTION

The Abell (1958) catalogue of rich galaxy clusters has been a source list for studies of the large-scale structure of the universe for over 30 years. Much of modern cosmological thought has been stimulated by the studies of rich clusters of galaxies: their distribution in space, distances, and global properties; and the Abell catalog has provided an important observational component for these studies. After its publication the catalog remained virtually the only complete survey of rich clusters over the northern sky. The ensuing years brought both praise and criticism as well as an urgency to complete a southern extension of the survey, which was championed by both Abell and H. Corwin and G. Chincarini. The historical development of that process is described in Table I and has culminated with the publication of the all-sky version of the catalog by Abell, Corwin, and Olowin (1989).

Despite its usefulness, the original catalog has several limiting drawbacks: it was not an all-sky catalog; it suffered from redshift incompleteness and brightness-scale errors; and was not dynamically constituted, that is, clusters were discovered and classified 'operationally.' In the latter regard, a 'rich-cluster' is defined as one having at least 30 members within the magnitude range m_3 to $m_3 + 2$ (m_3 is the magnitude of the third brightest cluster member), within a radius of 1.5 Mpc (the so-called 'Abell-radius') and each with a nominal redshift less than 0.2. The practical nature of this definition bears some explanation.

The number count of a rich cluster having at least 30 members within a given magnitude range gives rise to Abell's original concept of 'richness.' Essentially, this is an attempt to count the cores of clusters assuming that they possess a common size or linear extent (1.5 Mpc). Note that the magnitude counts are done from the third brightest member; this is done to prevent possible contamination from foreground objects. Several authors ignore this problem and often use the first-ranked brightest galaxy in the cluster as a distance candle, but a warning is in order. Even experienced observers (Corwin and Olowin, in this

TABLE I History of Abell Catalog of Rich Clusters of Galaxies

Years	Participants	Activity
1958	Abell	Original Abell Catalog of Rich Clusters published
1975	Abell	Southern survey proposed
1975-1977	Abell & Corwin	Southern survey begins at Royal Obs. Edinburgh
1977-1981	Corwin	Southern survey continues
1983	Abell	Abell's death, October
1984	Chincarini	Completion of all-sky catalog urged
1985	Chincarini & Corwin	Joint NSF proposal
1985-1987	Olowin	Survey completed
1987-1988	Corwin & Olowin	All-Sky catalog prepared
1989	Abell, Corwin & Olowin	ACO Catalog published

case) have selected different objects as the first-ranked brightest galaxy in a cluster, and although the effect is stronger at smaller distance classes, mis-identification of cluster members is a real artifact of observation (Abell *et al.* 1989). It can only be resolved by redshift measurements, and those doing these are urged to sample as many cluster member candidates as possible.

Likewise, the distance class determinations have been made operationally by using the tenth brightest galaxy in the cluster as a distance indicator. Again, the tenth brightest member was chosen to insure membership in the cluster, barring, of course, the problems of superposition. The m_{10} magnitudes were calibrated with known redshifts to yield 'distance classes' but, in fact, because of the bandpass of the northern red POSS plates the northern magnitude estimates have almost no correlation with distance for $m_{10} > 17.0$. The southern data have been scanned using the SERC/ESO IIIa-J plates and films, which have a higher sensitivity to redshift than the POSS plates used in the north. This is discussed in more detail by Abell *et al.* (1989).

There is no doubt that chance superpositions occur in scanning plate fields. Several types of superpositions have been encountered in the southern survey that are worth mentioning. These can be designated using a core-halo description of clusters that can be useful in determining the quality of superposition. For example, halo-halo superpositions of nearby clusters will contaminate the counts in the overlap regions of the counting radius, especially if the clusters in question are at similar distances. In this case it has been the author's practice to count galaxies in each 'Abell radius' as if it were the only cluster in the field, including the members in the overlap region that may have belonged to the adjacent cluster. A note was then made, alerting users of the catalog, of count contamination for those clusters. The superposition problem is rendered more tractable, however, in situations where there is a greater distance differential, where the core of, say, a more distant cluster is seen imbedded in the core-halo structure of a nearby cluster. In this case the counts refer to the contribution of the more distant object and the nearby object separately, as far as is possible within the confines of the observing technique. There is no doubt, however, that superposition occurs in many other cases as well, in particular with those clusters showing 'substructure,' but the case can only be resolved by substantial redshift observations. As much as 25% of the northern catalog is estimated to be contaminated by such superpositions, and it is likely that the southern catalog suffers similarly.

THE OVERLAP ZONE: COMPARISON OF THE NORTHERN AND SOUTHERN SURVEYS

In order to preserve homogeneity between the original Abell northern catalog and the ACO southern catalog, special attention was given to the overlap zone −17° to −27°. Cluster identification and selection was done in the same manner as the northern sample, estimating galaxy counts and magnitudes so as to derive the reduction formulas necessary to correlate the southern data with the earlier survey material. For that purpose, any northern Abell cluster that was found in the overlap region was considered a newly discovered cluster and the new measures were thus able to be compared with the old.

275 northern clusters out of a total of 489 were found in the overlap zone. During the plate scanning, Corwin missed only two (A503 and A512, both weak clusters in dense star fields) of the 135 clusters he had in regions common with Abell's earlier northern survey. Corwin, however, would not have listed seven other 'Abell clusters' he scanned had they not been listed in Abell's northern catalog. Indeed, Corwin also discovered 45 new clusters in the overlap zone, and Abell himself a single one (A3205). The small number of clusters scanned or discovered by Abell is a result of his limited work on the plates and his death early on in the project.

Similarly, Olowin found 90 new clusters (richness class 1 or greater) in the overlap zone, eleven of which are in common with Corwin's lists. In addition, Olowin discovered 58 others, either less rich or beyond the $z = 0.2$ cutoff for the clusters.

The fact that so many new clusters were discovered comes as somewhat of a surprise. For instance, 6 new clusters were discovered with richness classes of 3 and split between distance classes of 5 and 6. Huchra (1987, private communication) has suggested that the northern survey is deficient of clusters in the overlap region. Perhaps this is a result of the high altitude of this region as seen from the SERC Schmidt telescope unit in Australia, and similar effects might exist in the southern data at lower altitudes as well. To be sure, Olowin, Chincarini and Corwin (1987) have found significant selection effects in the catalogs, yet when care is taken to define statistically complete samples (Batuski *et al.*, 1988) the two-point correlation function of the two catalogs are virtually identical.

STATISTICS OF COUNT AND MAGNITUDE COMPARISONS

Cluster membership was determined by including galaxies in the magnitude range described above, counted within a standard radius as determined by the tenth-brightest galaxy in the sample. This technique conforms to the criteria set for the northern catalog and allows comparison of the northern and southern data sets.

Counts from each of the observers tend to set apart zones of differing zero points for the bright and faint clusters, and strong slopes for clusters of intermediate brightness. These effects are most likely due to the different magnitude systems used for the surveys, because the southern survey estimated total V magnitudes, whereas for the northern survey Abell estimated red magnitudes.

Triangular comparisons were made between the observers and the data they had in common in the overlap zone, with the result that by taking a mean of the southern and northern comparisons and assuming that the standard deviations in Abell's northern and southern counts are the same, the errors in the counts of the observers are about the same, *i.e.* a standard deviation of ± 18. Although this corresponds to an uncertainty of nearly 30% in the counts for the clusters in the southern survey, it is not unrealistic considering the number of observers and differing magnitude and counting systems and plate material as compared to the north. It is thought by the observers that much of the uncertainty in the counts comes from the treatment of 'background' contamination. The southern counts were relieved of background counts by the application of a subtraction procedure derived from a 'universal' luminosity function, whereas the northern counts were corrected for background by the local subtractions which may have been more appropriate. Indeed, the former technique has produced clusters with negative counts when corrected in this fashion. Since even these were discovered as excesses above local areal density, the use of this procedure is questionable.

The magnitudes in the southern survey are total V magnitudes, estimated on the IIIa-J plate material, rather than red POSS magnitudes as estimated by Abell for the northern clusters. As a result, color corrections need to be accounted for when mapping the two magnitude systems to compare the surveys. There is a definite difference in the sensitivities of the plate emulsions to redshift, with the IIIa-J material showing a higher sensitivity than the POSS plate material. This may account for the disparate redshifts and the excess of clusters beyond $z = 0.2$ found in the north.

DISTRIBUTIONS OF CLUSTERS IN THE NORTH AND SOUTH

The distribution of clusters has been compared between the two catalogs (see Batuski *et al.* 1989, Couchman *et al.* 1989). Both show a decrease in cluster density with latitude towards the galactic plane. A decrease in cluster density towards high declinations is observed in the south and is attributed (without proof) to high airmass observations at these declinations or to regions of galactic obscuration; similar effects are noted for the north.

An underdense region of southern clusters is noted near galactic latitude −50°, while a relative and slight overdensity appears to occur for nearby clusters at distances less than 300 h^{-1} Mpc. Part of the latter may be due to systematic errors incurred by the method of estimating redshifts from the magnitudes of cluster members. Similarly, uncertainty in richness designations in the south arise from the method of background subtraction. Investigations of plate-to-plate variations show no significant effects within the expected uncertainties.

Correlation-function studies (Bahcall *et al.* 1988) suggest that the angular correlation function for the southern clusters is consistent with the northern-cluster correlation function. In addition, both catalogs satisfy the scaling relation between the correlations of distant and nearby samples, as would be expected for real spatial clustering, and thus provides further support for the intrinsic nature of large-scale structure in the universe.

REFERENCES

Abell, G. O. 1958, *Ap. J. Suppl.* **3**, 211.
Abell, G. O., Corwin, H. G., and Olowin, R. P. 1989, *Ap. J. Suppl.* **70**, 1.
Bahcall, N. A., Batuski, D., and Olowin, R. P. 1988, *Ap. J. (Letters)*, **333**, L13.
Batuski, D., Bahcall, N. A., Olowin, R. P., and Burns, J.O. 1989, *Ap. J.*, **349**, 599.
Couchman, H. M. P., McGill, C., and Olowin, R. P. 1989, *M.N.R.A.S.*, **239**, 513.
Olowin, R. P., Chincarini, G., and Corwin, H. G. 1987, *B.A.A.S.*, vol. **19**, No. 4.

DISCUSSION

A. Dekel: Should we be worried about the discrepancies in finding clusters in the overlap region between Abell and ACO? Abell defined his richness classes and distance classes assuming certain errors. Can we believe his error estimates? Can you specify where the disagreement lies? In $D = 6$ clusters? In $R = 0$? What would be the safe subsample of the Abell catalog to use?

R. Olowin: Should you be worried about the discrepancies in the catalogs? That depends on what you want to get out of them. Were Abell to do it all over again, I don't think he would do things in exactly the same way. The 'operational' definitions of richness and distance classes are an obvious weakness in both catalogs. Also, he probably never expected in 1958 that his catalog would be subjected to such robust statistical treatment as we have seen in the correlation-function analyses. This is a warning we make with the ACO Catalog – don't ask it to do more than what it can legitimately provide within the context of its limitations. For instance, we know that the emulsion sensitivities of the plates scanned for the northern and southern catalogs give different selection effects for cluster identification.

Were I to suggest a statistical sample, I would recommend selecting those clusters with $R > 0$ and $D > 6$ lying away from the Galactic Plane – as a start! The $R > 0$ class will hopefully avoid background subtraction errors, and $D > 6$ will avoid emulsion sensitivity problems.

You must remember, of course, that this is probably the last time a survey like this will be done 'by eye.' Steve Maddox *et al.*, and groups like his, will show the way to get consistent, well-understood and perhaps minimum- (or articulated-) biased samples.

M. Strauss: The APM survey of Maddox *et al.* offers an independent check of your cluster selection procedures. Have you compared your cluster list with those found in the APM survey, or used the APM survey to select clusters automatically using the same criteria you used?

R. Olowin: As of this date we have not done this comparison, though it is not clear how to easily quantify the 'human' cluster-selection algorithm. That, of course, should be done. I might point out that several different 'cuts' of the APM survey do correspond with our ACO cluster positions, but the APM clusters are defined differently from Abell clusters, so it is hard to say quantitatively (*i.e.* statistically) what that means.

M. Davis: I understand that Abell double-counted those galaxies in the overlap region of two Abell clusters. What did you do in this situation?

R. Olowin: I have done likewise when clusters were close enough for their 'Abell-radii' to overlap; but the case is somewhat complicated. For a close and distant 'pair' the segregation in magnitude is enough to separate each cluster, and the counts are made for each individually. In cases for density enhancements that are large and have condensations, operationally the definition of an 'Abell' cluster forces the observer to overlap counting circles and contamination occurs. In this case I *have* counted some of the same galaxies twice in the $m_3 + 2$ cores. When this occurs, we have flagged the clusters as having contaminated counts in the ACO Catalog notes.

AUTOMATIC GALAXY CLUSTER CATALOGS

W. J. SUTHERLAND, S. J. MADDOX AND G. EFSTATHIOU
Department of Astrophysics, University of Oxford
Keble Road, Oxford OX1 3RH, England

ABSTRACT We have applied a variety of peak-finding and percolation algorithms to define galaxy clusters in the APM Galaxy Survey. We have estimated the m_{10} and richness for each cluster using objective algorithms which mimic Abell's procedure with various different corrections for the background of field galaxies. We have evaluated the angular correlation functions $\omega_{cc}(\theta)$ for the various catalogs and find that the amplitude changes by a factor of ~ 3 depending on the selection parameters and method of background correction. We conclude that ω_{cc} does not provide useful information on the spatial correlation function of rich clusters ξ_{cc}, and hence that new redshift surveys are needed.

INTRODUCTION

The cluster correlation function ξ_{cc} is a powerful test of cosmological scenarios, but present measurements of ξ_{cc} and ω_{cc} are based on visually-selected cluster catalogs (Abell 1958, Abell, Corwin, and Olowin 1989). These catalogs are probably subject to systematic selection effects which are very difficult to quantify (Sutherland 1988, Dekel *et al.* 1988, Sutherland and Efstathiou 1989). Also, it is impossible to change the selection procedure without repeating the visual inspection of the plates, so it is difficult to quantify how the cluster sample depends on the selection procedure. The APM galaxy survey (Maddox *et al.* 1990a,b) contains accurate positions and magnitudes for two million galaxies over an area of 4300 square degrees. We use this survey to define objective cluster catalogs and study the changes in cluster properties caused by different selection procedures.

SELECTION OF CLUSTERS

As for existing visually-selected cluster catalogs, we have used two-stage procedures to select clusters. First we select possible cluster sites, and then we calculate a richness and m_{10} for each site and impose limits on m_{10} and richness for the final catalog. Here m_{10} is defined as the magnitude of the 10th brightest cluster galaxy, and the richness is defined as the count of galaxies above the background in a suitable magnitude interval. We used a low threshold to select the initial sites, so that all clusters satisfying the final m_{10} and richness limits are located in the first selection.

The initial sets of cluster sites were chosen using two methods. The first selection algorithm is based on that applied to the Lick catalog by Shectman (1985). We constructed counts of galaxies in cells in an equal-area projection, smoothed the counts with a suitable filter, and located peaks above a chosen threshold. The smoothing reduces the sensitivity of peak finding to the positioning of a cluster relative to the cell boundaries. The smoothing scale is ~0.1°, about the size of the most distant visible clusters. We repeated this process for several magnitude ranges to ensure sensitivity to the desired magnitude range, $16 \leq m_{10} \leq 20$.

Our second method of selecting cluster sites uses percolation. The distance between each galaxy and its near neighbors is calculated, and galaxies which are closer than a fixed distance are linked together. Galaxies linked to other linked galaxies form groups, and the mean position of each group is used as a cluster site. This method has the advantage that it does not preferentially select groups that are circular, which is the case when smoothed counts in cells are used. Since no smoothing over pixels is necessary, smaller clusters and closer cluster pairs are located by using this method rather than peak finding.

In the second stage of our cluster catalog construction we take the positions from the first stage and estimate the m_{10} and richness for each position by iteration. Starting at the initial cluster position, we select galaxies within a first approximation to the Abell radius, and we count down in magnitude to find the 10th galaxy *above the background*. The cluster radius is then redefined from the new m_{10}, and the process repeated until convergence or instability. Convergence generally occurs in 1 or 2 iterations for reasonably rich clusters (richness ≥ 30). For the clusters which converge to give an m_{10} and corresponding Abell radius, we calculated a richness. First we used the definition of Abell (1958), which is the count of galaxies within the Abell radius in a magnitude slice (m_3, m_3 + 2) above the background count in the same slice. We also used a measure of richness which is less sensitive to foreground contamination, defined as the count above the background within the Abell radius and in a magnitude slice ($m_{10} - 0.5$, $m_{10} + 1.5$).

We estimated the errors in the cluster parameters by comparing the pairs of measurements for clusters in plate overlaps. We found the scatter of the m_{10} estimates is ≈0.2 mag *rms*, and in the richness estimates is ~10 galaxies. The steep number-richness relation combined with this scatter means that ~50% of the clusters in any sample are within 1σ of the richness limit. Therefore the overlap between cluster catalogs which use different richness estimates is only ~50%.

We have investigated different methods of background correction used in the second stage of the cluster catalog construction. First we used a 'global' background, using the mean number density in the appropriate magnitude range averaged over the whole survey, and second a 'local' background, measured from an annulus 2° in radius outside each cluster. Figures 1a-d show sky maps of our four typical catalogs.

THE TWO-POINT CORRELATION FUNCTION

We have evaluated the angular correlation functions for each of the 4 final cluster catalogs, using an m_{10} cutoff of $b_J = 19$ and a richness cutoff of 30. The results are shown in Figures 2a and b. It is clear from the figures that the results are very sensitive to the cluster selection criteria; changing from a global to a local background correction typically reduces the amplitude by a factor ~2.

The reason for this difference becomes apparent on examining maps of the cluster distribution: in the cluster catalogs with a global background correction, several patches of very high cluster density are seen, corresponding to the positions of prominent high density 'clouds' in the galaxy maps. These patches are present in the locally-corrected maps but their density contrast is lower.

This effect is especially pronounced in the percolation-selected clusters: this appears to be due to the sensitivity of the percolation method to the local background. The large clouds of galaxies are split into many small groups by the percolation algorithm, and due to

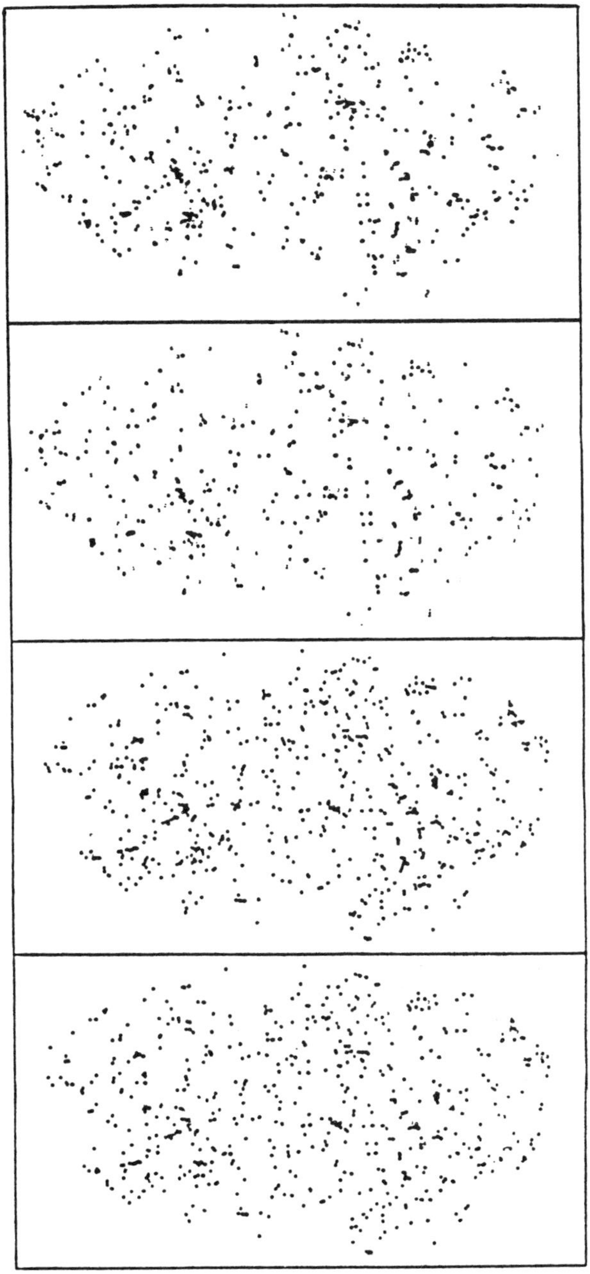

Fig. 1. The sky distribution of APM clusters with $m_{10} < 19$ and richness > 30: (a) percolation clusters, global background; (b) percolation clusters, local background; (c) peak-selected clusters, global background; (d) peak-selected clusters, local background.

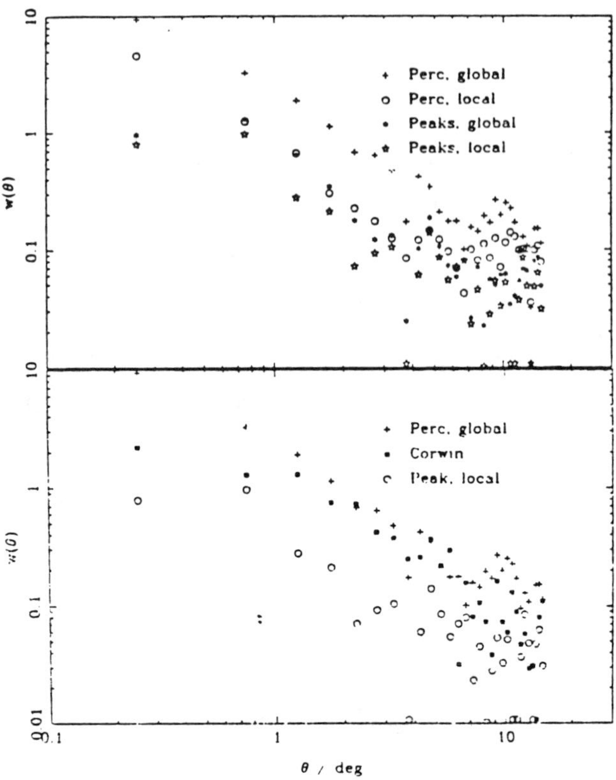

Fig. 2. (a) The angular correlation function, ω_{cc}, for several cluster samples. (b) The correlation functions for two of the above cluster samples, compared with that for Corwin clusters with similar distance and richness.

the high local galaxy density, many of these systems are 'rich' enough to be included in the final catalog. By rejecting close pairs of clusters, the percolation-selected cluster sample can be modified to give a correlation function that is very similar to the peak location. If two clusters are closer than 0.7 times the sum of their Abell radii, (*i.e.* there is significant overlap between them) we retained the richer cluster, but removed the poorer cluster from the sample. The removal of these pairs greatly decreases ω_{cc} at small angles, and makes it quite similar to the peak-located cluster sample.

DISCUSSION

The measurements of ω_{cc} described above show that the angular distribution of clusters is very sensitive to relatively small changes in the selection procedure. This is because the surface density contrast of even rich galaxy clusters is small in comparison to the clustered background of field galaxies, and because the number-richness relation is very steep, so the cluster catalogs are very sensitive to small changes in richness estimates.

It has been argued that the scaling of ω_{cc} between cluster samples with different estimated distances implies that the observed angular clustering of clusters faithfully represents the spatial clustering of clusters. We tested the scaling of ω_{cc} as a function of m_{10} for each of our cluster catalogs by calculating the correlation function for the samples with $m_{10} < 18$ and with $m_{10} < 19$, shown in Figures 3a and b.

The scaling relation predicts a change in amplitude of ~ 2.5 between the two magnitude limits. The results for $m_{10} < 18$ are rather noisy due to the small samples, but the change in amplitude between the two magnitude limits within each cluster sample is roughly consistent with the expected scaling relations. Scaling of the amplitude of ω_{cc} is consistent within *each* catalog, even though the amplitudes of ω_{cc} for the different catalogs differ by a factor of ~ 3.

Thus at least some (if not all) of these cluster catalogs must have systematic biases in the angular correlation functions, despite the approximate consistency with the scaling relation. We therefore conclude that it is not possible to derive a reliable ξ_{cc} result from angular data alone, and therefore further redshift surveys are needed, preferably combined with X-ray cluster selection.

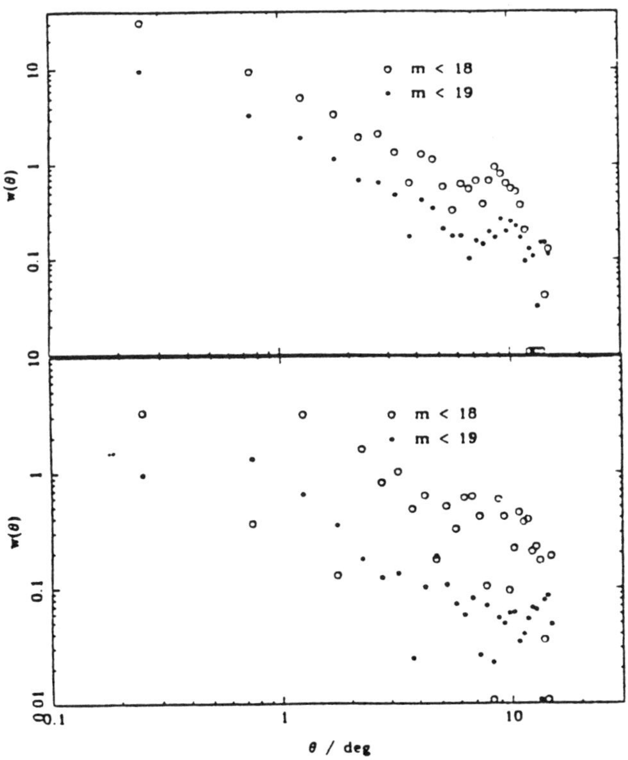

Fig. 3. (a) The correlation function for percolation clusters using global background, for two different magnitude limits. (b) The correlation function for peak-selected clusters using global background, for two different magnitude limits.

REFERENCES

Abell, G. O., 1958, *Ap. J. Suppl.*, **3**, 211.

Abell, G. O., Corwin, H. C., and Olowin, R. P., 1989, *Ap. J. Suppl.*, **70**, 1.

Dekel, A., Blumenthal, G., Primack, J., and Olivier, S., *Ap. J. (Letters)*, **338**, L5.

Maddox, S. J., Sutherland, W. J., Efstathiou, G., and Loveday J., 1990a, *M.N.R.A.S.*, **243**, 692.

Maddox, S. J., Sutherland, W. J., Efstathiou, G., and Loveday J., 1990b, *M.N.R.A.S.*, **246**, 433

Shectman, S. A., 1985, *Ap. J. Suppl.*, **57**, 77.

Sutherland, W. J., 1988, *M.N.R.A.S.*, **234**, 159.

Sutherland, W. J., and Efstathiou, G., 1989, in preparation.

DISCUSSION

P. Shapiro: Based on your conclusion that the cluster-cluster angular correlation function cannot reliably be determined by the existing cluster catalogs, would you say that the significant correlation reported by Primack in the intermediate angular range is not believable?

W. Sutherland: It is believable in the sense that all our samples are significantly above zero in this range (and a lot larger than ω_{gg}), but I don't think you can pin down a precise amplitude.

D. Burstein: If the intrinsic error in the number of galaxies per cluster is 10 to 20, does it make statistical sense to apply a guillotine to the data: *e.g.* $N_{gal} = 51 \equiv$ Rich, $N_{gal} = 49 \equiv$ Poor?

W. Sutherland: You have to draw the line somewhere! It won't affect the angular correlations IF the errors are random, but obviously it can cause a Malmquist bias in the number *versus* richness relation, *etc.*

R. Olowin: How many of the ACO clusters were found by your methods?

W. Sutherland: It obviously depends on your richness cutoff; we've tried using your centers and estimating m_{10} and richness, and it nearly always finds something there, though quite often it doesn't have richness > 30. If we take one of our catalogs with the same sort of selection criteria as ACO, there's only about 50% overlap just because of the noisy richness.

LARGE SHAPES OF VOIDS AT LATE TIMES

D. R. MATRAVERS AND A. MAURELLIS
Department of Applied Mathematics, University of Cape Town
Rondebosch, 7700 South Africa

ABSTRACT The peculiar expansion and the shear associated with the large-scale flow on the boundary of voids in the southern sky are estimated. The results show that the peculiar velocities are low and that the shear is of the same order as that in independent cluster-redshift data.

INTRODUCTION

While the topology of the large-scale structure of the universe has been the subject of considerable attention (see for instance Gott *et al.* 1989 and references they cite), the geometry of actual voids has not been studied in detail at all. A possible reason for this is the evidence that voids are not 'bubbles' (Gott *et al.* 1989), and therefore attempts to define boundaries and hence the shape is open to controversy. We have taken the view that, despite obvious difficulties, the evolution of the broad void shape, as defined by a tri-axial ellipsoid which best fits the galaxies and defines a galaxy-free region, will give information about the matter distribution in the neighborhood of the void. It must be emphasized that we do not claim that voids are 'closed up bubbles,' but that some of them (if not all) can be approximated usefully in that way.

Both explosion and perturbation scenarios for void formation in the early universe allow that in an otherwise homogeneous universe the voids would be spherical once the transients have decayed away. Deviations from sphericity occur because of interaction with a non-homogeneous background. We examine this in two ways.

In the first place, an ideal spherical void, if empty, should appear ellipsoidal in redshift space with major axis along the line of sight with an eccentricity which depends on Ω. The data for three southern-sky voids (Fairall and Jones 1988) supports either a small value of Ω, *i.e.*, a low-density universe, or the presence of unobserved matter in the void to reduce the inhomogeneity.

In the second place, we examine the ratios of the axes of the tri-axial ellipsoids which fit the data best and use these ratios to estimate the eigenvalues of the shear tensor in the neighborhood of the void. The values obtained are of the same order as those obtained by Peebles (1971) for a nearby cluster and so support a proposition that there is nothing special about the inhomogeneity in the matter distribution near a void. The shear is of the same order, although the eigen-directions change (see Figure 1).

The calculations we have done are based on a naive approach to the problem of the evolution of void geometry. However, they provide consistent results which are of some interest.

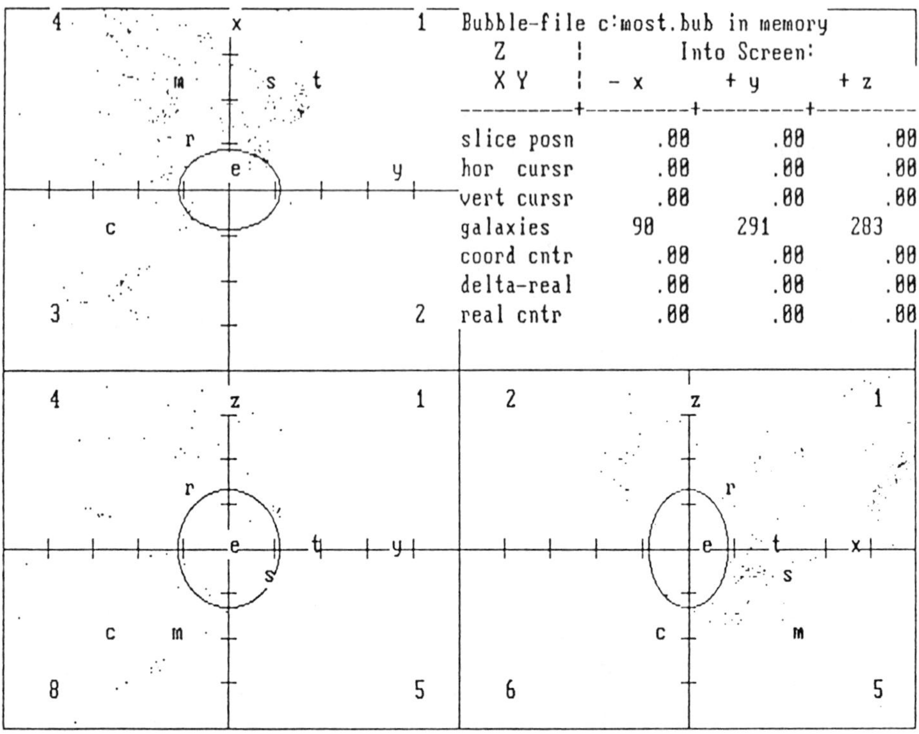

Fig. 1. Cross-sectional slices (thickness 500 km s^{-1}) through an ellipsoidal fit to galaxies surrounding a southern void in Eridanus (see Fairall and Jones 1988). The (x, y, z) coordinates are chosen with respect to the major and minor axes of the fitted ellipsoid. Other labels refer to the visually-selected centroids of voids in Fairall and Jones (1988); Sagittarius (s), Eridanus (both r and e), Microscopium (m), Sculptor (c), and our position (t). Note that our galaxy lies almost in the plane defined by the semi-minor axes of the ellipsoid, that is, the ellipsoid has a major axis nearly perpendicular to the line of sight – in this case 78°.

NAIVE VIEW

We begin by taking a naive view of the late time behavior of voids. They are empty regions in an otherwise homogeneous and isotropic background space-time which we describe by a Friedmann metric. In this view the void is the only perturbation to be considered. (This model is only approximate because we ignore the matching conditions.) We assume that the void has been in existence long enough for the transients left over from an explosion or other creation process to have died out for observational purposes. Under these conditions an argument of Icke (1984) shows that the voids will be spherical because any deviations from sphericity will be smoothed by the interaction between the effective negative mass and the homogeneous background.

For a spherical void the peculiar velocity of the boundary galaxies is given by the Peebles (1980) formula,

$$v = (1/3)\, H\, r\, \Omega^{0.6}\, \delta\rho/\rho, \tag{1}$$

where r is the radius of the void, H is the Hubble constant, ρ is the background density and, for an empty void, $\delta\rho = 0 - \rho = -\rho$, *i.e.*

$$v = -(1/3)\, H\, r\, \Omega^{0.6}. \tag{2}$$

The cause of this peculiar velocity is the same as that which caused the isotropization Icke describes, *i.e.*, the effect of the empty region making up the void.

These peculiar velocities will cause the spherical voids to appear elliptical in redshift space, with the major axis along the line of sight. Table I gives some idea of the magnitude of the effect for a void with radius $H\,r = 1000$ km s^{-1}. It lists the peculiar velocities and resulting ellipticities for various Ω; ϵ is the ratio of the length of the minor axes to the major axis of the resulting ellipsoids in redshift space. Although these predictions of ϵ arise from an highly idealized situation, it is interesting to compare them with the parameters for the best fitting ellipsoids, in redshift space, to some of the southern sky voids. Table II gives the inclination θ to the line of sight of the major axis of a fitted ellipsoid and the ratios ϵ_1^2 and ϵ_1^3 of minor to major axes for three voids identified in Fairall and Jones (1988). From the data it is apparent that θ is very different from zero in all three cases, *i.e.*, that the major axis is pointing well away from the line of sight. This result suggests that the voids are *not distorted* in redshift space *along our line of sight*. This may be significant, because density perturbations should not preferentially distort the voids along or across our line of sight and thus remove, in all three cases, the effect we expect in the high-density case. Thus the distortion in redshift space supports a low-density universe, where the distortion along our line of sight will be smaller, rather than a high-density one. Alternatively, the voids must contain non-luminous matter. More examples are needed to fully test the proposal.

TABLE I Velocities and ellipticities

Ω	v (km s^{-1})	ϵ
1	333	0.75
0.4	193	0.84
0.04	47	0.96

TABLE II Void inclination and axis ratios

Void	θ	ϵ_1^2	ϵ_1^3
Eridanus	78°	0.83	0.76
Sculptor	74°	0.89	0.63
Microscopium	88°	0.92	0.82

The naive view of the late time behavior cannot account for the fact that the voids are in fact ellipsoidal, even if the effect of the homogeneous background velocities is subtracted out. From observation of the distribution of galaxies in the vicinity of the voids it is apparent that it is inhomogeneous at the void scale. To get a measure of this inhomogeneity we have again taken a naive view and have related the ellipsoidal shape to the local shear tensor for the matter flow near the void. Note this need not be the same as the shear for the universe and may not be the same for different voids. However, one may expect it to be of the same order in magnitude for each void.

LESS NAIVE VIEW

Denote the flow vector field for the galaxies in the boundary of the void by $\mathbf{u}^a$. Then, as is well known, the covariant derivative of the $\mathbf{u}^a$ can be decomposed into the shear, rotation, acceleration, and expansion. Projected onto the hypersurface element orthogonal to the fluid flow this gives (Ellis 1971)

$$\dot{\ell}/\ell = \sigma_{ab}\, \mathbf{e}^a\, \mathbf{e}^b + (1/3)\theta, \tag{3}$$

where $\mathbf{e}^a$ are the space-like unit vectors which form a basis in the space orthogonal to $\mathbf{u}^a$, σ_{ab} is the shear tensor, $\theta = \mathbf{u}^a_{;a}$ is the expansion, and ℓ is the relative distance between two nearby flow lines. The vector $\mathbf{X}^a = \ell \mathbf{e}^a$ connects the two neighboring flow lines. For an eigen-direction of the shear with eigen-value λ,

$$\dot{\ell}/\ell = \lambda + (1/3)\theta. \tag{4}$$

Clearly, to integrate this we need to know the time dependence of the right-hand side. We assume that the shear is the result of the re-distribution of the matter which occurred when the void formed, and that the transients have died away. Under these circumstances, for zero pressure and the Friedmann spatial curvature term (k) insignificant, a model for the shear and expansion is provided by Silk (1974) and Johri (1977);

$$\langle \sigma^2 \rangle^{-1/3} \propto t^{-1/3}, \tag{5}$$

$$\theta = 3\, \dot{a}/a = 2t^{-1}, \tag{6}$$

where $a(t)$ is the scale factor of the Friedmann metric. If we assume that the eigenvalues of σ have the same time dependence as the root mean square (5), and that the void creation process leads initially to spherical voids, then equation (4) gives

$$\ell_j = \ell_0\, t^{2/3} \exp[(3/2)\, \gamma_j\, t^{2/3}], \tag{7}$$

where $j = 1, 2, 3$ and γ_j are related to the shear eigenvalues by

$$\sigma_j^j = \gamma_j\, t^{-1/3}, \tag{8}$$

no summation on the j. From equation (7) at time t_0,

$$\epsilon_k^j = \ell_j/\ell_k = \exp[(3/2)\, (\gamma_j - \gamma_k)\, t_0^{2/3}]. \tag{9}$$

Since, from equation (6), $t_0^{-1} = \theta/2 = 3H/2$, this equation may be rewritten in the form

$$\epsilon_k^j = \exp[(1/H)\, (\sigma_j^j - \sigma_k^k)]. \tag{10}$$

The predicted values now are given by

$$\sigma_j^{\ j} - \sigma_k^{\ k} = H \ln (\epsilon_k^{j}), \tag{11}$$

where the ϵ_k^j for the three observed voids are given above. Also, $\sigma_1^{\ 1} + \sigma_2^{\ 2} + \sigma_3^{\ 3} = 0$, and hence one can evaluate the shear eigenvalues for each of the southern voids. Table III gives the shear eigenvalues in multiples of H for the 3 voids in Table II, calculated from equation (11) and the assumption that the shear tensor is trace-free.

These values are consistent with those found by Peebles (1971) from cluster redshift data and show that in the neighborhood of a void the inhomogeneity is not distinctive.

TABLE III Shear Eigenvalues

Void	$\sigma_1^{\ 1}$	$\sigma_2^{\ 2}$	$\sigma_3^{\ 3}$
Eridanus	0.15	-0.03	-0.12
Sculptor	0.19	0.08	-0.27
Microscopium	0.09	0.01	0.10

REFERENCES

Ellis, G. F. R. 1971, *Relativistic Cosmology*, in *General Relativity and Cosmology*, ed R. K. Sacks, (London: Academic Press).

Fairall, A. P. and Jones, A. 1988, *Pub. Astr. Dept. Univ. Cape Town*, Number 10.

Gott, J. R., III, Miller, J., Thuan, T. X., Schneider, S. E., Weinberg, D. H., Gammie, C., Polk, K., Vogeley, M., Scott, J. S., Bhavsar, S. P., Melott, A. L., Giovanelli, R., Hayes, M. P., Tully, R. B., and Hamilton, A. J. S. 1989, *Ap. J.* **340**, 625.

Icke, V. 1984, *M.N.R.A.S.*, **206**, 1P.

Johri, V. B. 1977, *M.N.R.A.S.*, **178**, 395.

Peebles, P. J. E. 1971, *Physical Cosmology*, (Princeton: Princeton University).

Peebles, P. J. E. 1980, *The Large-Scale Structure of the Universe*, (Princeton: Princeton University).

Silk, J. 1974, *Ap. J.*, **194**, 215.

DISCUSSION

E. Bertschinger: It is somewhat surprising that all of your voids seem to be elongated perpendicular to the line of sight. As you point out, expansion in comoving coordinates should lead to elongation parallel to the line of sight. I am concerned that your ellipsoid fitting might be biased by virialized motions. Large voids such as Boötes are surrounded by virialized clusters. If your voids are bounded even partially by clusters, then their 'Fingers of God' would poke into the void, compressing it in the radial (redshift) direction but not in the transverse direction.

A. Maurellis: Agreed. In fact, the presence of large velocity dispersions of clusters may distort the real nature of any features that are usually perceived in the distributions. This probably means that any categorization of structural entities in the observed redshift distribution requires pre-checking of, say, cluster catalogs, and then perhaps a 'collapse' of the 'Finger of God' to a 'fingernail' *before* fitting ellipsoids, or searching for filaments or walls, *etc.*

CLUSTER-CLUSTER CORRELATIONS

JOEL R. PRIMACK AND SCOT OLIVIER
Santa Cruz Institute for Particle Physics
University of California, Santa Cruz, CA 95064

GEORGE R. BLUMENTHAL
University of California Observatories/Lick Observatory, Board of Studies in Astronomy and Astrophysics, University of California, Santa Cruz, CA 95064

AVISHAI DEKEL
Racah Institute of Physics, The Hebrew University of Jerusalem
Jerusalem 91904, Israel

ABSTRACT How much can we learn from correlations between clusters of galaxies? Using the Abell northern and Abell-Corwin-Olowin (ACO) southern catalogs of rich clusters of galaxies, we have studied cluster-cluster correlations in detail in two subsets of these data: (a) the 3-dimensional correlation function $\xi_{cc}(r)$ of the 102 nearby ($D \leq 4$) rich ($R \geq 1$) Abell northern clusters, for which complete redshift data are available; and (b) the 2-dimensional angular correlation function $\omega(\theta)$ of all 2004 $R \geq 1$ clusters in both the Abell and the ACO catalogs. We have attempted to correct these correlation functions for the effects of projection contamination by constructing 'decontaminated' catalogs of richness $R \geq 1$ clusters, using a model in which the galaxy number density is a superposition of a uniform field plus spherical halos centered on each cluster center. The halo profile is derived from observed cluster-galaxy correlation functions, $\xi_{cg}(r)$. This procedure eliminates 9-14% of the clusters, depending on the ξ_{cg} used. But, despite the relatively small number of clusters affected, the effect on the correlation functions is dramatic on small distance or angular scales.

After correction for projection contamination, $\xi_{cc}(r)$ from the nearby sample provides no useful constraints on theory. At small angles, $\theta < 1°$, decontamination of the full catalog reduces the amplitude of ω_{cc} by a factor of two or more, and the large (Poisson) errors indicate that ω_{cc} is positive only at the 2σ level; thus it also cannot provide useful constraints on theory. In the range $1 \leq \theta \leq 3°$, however, ω_{cc} has a zone of statistically significant positivity. Its amplitude is reduced by decontamination only by about 30% at most, it is insensitive to the use of local versus global background normalization, and it is positive at the 5σ level in both the north and the south, indicating spatial correlations on scales of order 10 - 30 h^{-1} Mpc. These data are in significant conflict with the predictions of the 'standard' Cold Dark Matter (CDM) model, but consistent with other models such as a Hybrid of CDM and baryons in an open universe, while an Explosion scenario somewhat overestimates the cluster correlations.

On large angles, $\theta > 3°$, the effect of decontamination is negligible. However, on these scales the north and the south differ significantly; while in the south ω_{cc} is consistent with zero at the 2σ level, it stays significantly positive in the north out to 40°. This inconsistency and other problems with the northern sample argue against drawing any conclusions from the results beyond 3°.

INTRODUCTION

The superclustering of rich clusters of galaxies has been regarded as providing one of the strongest observational constraints on the structure in the universe on very large scales. In a recent letter (Dekel, Blumenthal, Primack and Olivier 1989, hereafter DBPO), we described a robust method for 'decontaminating' cluster catalogs that suffer from projection effects which can produce spurious superclustering and lead to misleading theoretical interpretations (Sutherland 1988). By applying this method to a statistically complete subsample of nearby $D \leq 4$, $R \geq 1$ Abell clusters for which redshifts were available, we found that the spatial cluster-cluster auto-correlation function of the decontaminated sample, $\xi_{cc}(r)$, could not provide statistically significant constraints on any standard theory. In a more recent paper (Olivier, Blumenthal, Dekel, Primack, and Stanhill 1989, hereafter OBDPS) we have applied the decontamination procedure to the entire Abell catalog of northern clusters (Abell 1958) and to the new catalog of southern clusters (Abell, Corwin and Olowin 1989, hereafter ACO) and investigated the constraints provided by the angular correlation function $\omega(\theta)$ of this much larger sample. Here we review this work, and also add some details regarding the Abell $D \leq 4$, $R \geq 1$ sample, including the effects of projection of $D = 5,6$, $R \geq 1$ clusters onto it.

The uncorrected two-point auto-correlation function of the 102 nearby Abell $R \geq 1$ clusters out to distance class D = 4 ($\sim 300\ h^{-1}$ Mpc) has an amplitude about twenty times larger than that of galaxies. It seems to remain positive out to 50 h^{-1} Mpc or more (Bahcall and Soneira 1983, BS; Klypin and Kopylov 1983; see Bahcall 1988 for a review). There is evidence that the angular auto-correlation function, $\omega_{cc}(\theta)$, of the entire Abell catalog out to $D = 6$ in the northern sky, and the complementary southern ACO catalog, scales as expected in accordance with the effective depth of the sample (Bahcall and Soneira 1983; Bahcall, Batuski and Olowin 1988; Batuski *et al.* 1989). These strong correlations are in clear conflict with the traditional scenarios that assume scale-invariant Gaussian initial density fluctuations in an $\Omega = 1$ universe (Barnes *et al.* 1985). In particular, the 'standard' $\Omega = 1$ cold dark matter (CDM) scenario (Blumenthal *et al.* 1984, Davis *et al.* 1985), when normalized to fit either the distribution of galaxies or the isotropy of the microwave background, fails to reproduce the observed ξ_{cc} (Bardeen, Bond and Efstathiou 1987; White *et al.* 1987; Blumenthal, Dekel and Primack 1988). Standard $\Omega = 1$ CDM predicts that ξ_{cc} should become negative beyond $\sim 20\ h^{-2}$ Mpc because if clusters form at high density peaks then $\xi_{cc}(r) \propto \xi(r)$ at large separations (Kaiser 1984, Bardeen *et al.* 1986), and the matter two-point correlation function, $\xi(r)$, is predicted to go negative at $\sim 20(\Omega h^2)^{-1}$ Mpc in a CDM universe (Davis *et al.* 1985). On the other hand, a high ξ_{cc} arises in open cosmological models (Dekel 1984; Blumenthal, Dekel, and Primack 1988; Peebles 1987) or in non-Gaussian models such as the Explosion scenario (Weinberg, Ostriker, and Dekel 1989).

Cluster catalogs inevitably include some clusters that would not have been rich enough for inclusion had their apparent richness not been enhanced by projection effects (Lucey 1983, Frenk *et al.* 1990). Projection effects can be potentially troublesome for measuring ξ_{cc} and ω_{cc}, since contamination of a cluster by galaxies associated with a foreground or background cluster produces a spurious enhancement of the number of cluster pairs at small separations, especially in regions of high density of clusters - in superclusters. Sutherland (1988) and Kaiser and Lahav (1988) have suggested that the high correlations are mostly due to this effect, which they also blame for the anisotropy detected by Bahcall, Soneira and Burgett (1986) in the distribution of cluster-pair separations in redshift space.

We developed in DBPO a simple and direct method to decontaminate a catalog suffering from the contamination effects discussed above. We used a model for the distribution of galaxies around clusters that is constrained by the observed cluster-galaxy cross-correlation function, $\xi_{cg}(r)$. Omitting the clusters whose counts were enhanced to the richness class $R = 1$ by projection of galaxies associated with other clusters, we constructed a new sample corrected for projection contamination, and calculated the correlation function for this sample. In DBPO we applied this method to the $D \leq 4$, $R \geq 1$ northern Abell complete redshift sample. Although only about 10% of the clusters are removed by this decontamination procedure, we found that $\xi_{cc}(r)$ is reduced in amplitude by about a factor of two. The reason why eliminating only a few clusters can have such a dramatic effect on the correlation function is that removing a cluster from a dense region (supercluster) eliminates many cluster pairs with separations of order 20 h^{-1} Mpc.

While this analysis does not exclude a correlation function of high amplitude (as in BS), we found that $\xi_{cc}(r)$ does not necessarily fit a power law, no meaningful correlation length can be determined, the deviations from zero are less than 1.7σ on any scale, and $\xi_{cc}(r)$ is not different from zero at the 1σ level for $r \geq 40$ h^{-1} Mpc. (Here, σ refers to one standard deviation *assuming Poisson statistics*. Because of clustering, this presumably underestimates the actual statistical errors. We discuss this further in OBDPS.) We therefore concluded that the correlation function of this small sample cannot be regarded as a difficulty for any of the standard cosmological scenarios. We also found that the distribution of cluster pairs is much more isotropic in the decontaminated catalog, so there is no significant evidence for cluster peculiar velocities in excess of ~ 1000 km s^{-1}.

Can $\omega_{cc}(\theta)$ provide more significant constraints? On the one hand, the entire Abell catalog, including the new southern ACO catalog, contains about twenty times more clusters of richness $R \geq 1$, so both the Poisson errors and the sample-to-sample scatter should be much smaller. On the other hand, lacking redshifts for the complete sample, we can calculate only the angular correlation function which is a smeared version of $\xi_{cc}(r)$. Our answer is that $\omega(\theta)$ for θ between about 1° and 3° does provide statistically significant constraints on theory.

PROJECTION DECONTAMINATION METHOD

In order to decontaminate a cluster sample, without knowing the actual galaxy positions, we must *assume* a specific model for the spatial distribution of galaxies. Following Seldner and Peebles (1977, SP) and Bahcall (1986), we assume that the galaxy number density $n(\mathbf{r})$ in a large volume V can be approximated as the superposition of two components: (1) spherical halos consisting of galaxies with a universal profile $\rho(r)$ around each of the N_c cluster centers, $\mathbf{r}_i$, and (2) an unclustered component, $(1-f)\langle n\rangle$, where $\langle n\rangle$ is the mean number density of galaxies and f is the fraction of all galaxies in clusters. This reads

$$n(\mathbf{r}) = \sum \rho(|\mathbf{r} - \mathbf{r}_i|) + (1-f)\langle n\rangle, \quad (i = 1,\ldots,N_c) \tag{1}$$

Given the true $\xi_{cc}(r)$ and the cluster-galaxy cross-correlation function, $\xi_{cg}(r)$, the halo profile can be evaluated from the approximate relation

$$\langle n\rangle \xi_{cg}(r) = \rho(r) + \langle n_c\rangle \int_V \rho(x)\, \xi_{cc}(y)\, \delta(|\mathbf{x} + \mathbf{y}| - r)\, d^3x\, d^3y, \tag{2}$$

where $\langle n_c\rangle$ is the mean number density of clusters and δ is the Dirac delta function. SP have applied a similar analysis in two dimensions to their observed angular cross-correlation function, $\omega_{cg}(\theta)$, of Abell clusters of richness $R = 1$ and galaxies from the Lick counts. They obtained a fit with the function

$$\rho_{SP}(r)/\langle n\rangle = (r/8.4\ h^{-1}\ \mathrm{Mpc})^{-2.4}, \quad r \le 15\ h^{-1}\ \mathrm{Mpc}. \tag{3}$$

Alternatively, we use the recent ξ_{cg} of Lilje and Efstathiou (1988, LE), whose amplitude is lower because of the different galaxy luminosity function adopted. By solving equation (2) numerically, we obtain a fit with

$$\rho_{LE}(r)/\langle n\rangle = (r/8.8\ h^{-1}\ \mathrm{Mpc})^{-2.2} - (r/4.9\ h^{-1}\ \mathrm{Mpc})^{-1.5}, \tag{4}$$

where again $\rho(r) = 0$ for $r > 15\ h^{-1}$ Mpc. The first term reflects LE's ξ_{cg} at small r, and the second term accounts for the double counting due to overlapping cluster halos.

The contribution to the Abell count in cluster i by galaxies associated with cluster j is calculated by integrating the projected surface density of the galaxy halo of cluster j over the Abell circle of cluster i. The details are given in OBDPS. The key points are as follows:
- We use either $\rho_{SP}(r)$ or $\rho_{LE}(r)$ as universal halo profiles around all $R \ge 1$ clusters, and we regard the differences between the two resulting correlation functions as a measure of the uncertainty inherent in this decontamination procedure.
- We only include projected galaxies that would fall within the Abell luminosity strip (from the magnitude m_3 of the third brightest cluster galaxy to $m_3 + 2$). In order to calculate the correction due to the shift in the Abell luminosity strip with distance we assume a universal Schechter luminosity function

$$\Phi(\mathrm{L}) = \Phi_* \ (\mathrm{L}/\mathrm{L}_*)^{\alpha} \exp(-\mathrm{L}/\mathrm{L}_*) \tag{5}$$

with $\alpha = -1.07$ and $\Phi_* = 0.0156\ h^3\ \mathrm{Mpc}^{-3}$ (Efstathiou, Ellis and Peterson 1988).
- For purposes of estimating projection contamination, the distances to clusters are needed. These are obtained from redshifts when available, and otherwise calculated from the magnitude m_{10} of the tenth brightest cluster galaxy, following Abell (1958) (but with our fit to clusters with at least four measured redshifts, and a richness - m_{10} correction) and ACO.
- Certain sample biases can be ameliorated by including the same biases in the Poisson sample used in calculating the correlation function. For the Abell catalog a galactic latitude completeness correction was used by Bahcall and Soneira (1983), and for the ACO catalog both a galactic latitude and a declination completeness correction was used by Bahcall, Batuski, and Olowin (1988); we refer to these as the standard correction functions. We also tried other correction procedures.

For every cluster i in the sample, we have used the above method to calculate the contamination due to all other clusters in the sample. The number counts were corrected accordingly, and we iterated until convergence. The clusters that end up with less then 50 members were deleted, leaving behind an 'uncontaminated' sample. Once the sample has been decontaminated, the spatial or angular correlation functions can be calculated and compared to those of the original sample and to theory.

RESULTS

We focused in DBPO on the nearby sample of 102 $D \le 4$, $R \ge 1$ clusters because this sample was the source of the strongest claims regarding evidence for clustering of clusters and for cluster peculiar velocities (*e.g.*, see Bahcall 1988).

In DBPO we showed that decontamination using halo functions ρ_{LE} and ρ_{SP} produced similar results. In Figure 1 we plot $\xi_{cc}(r)$ for all 102 clusters in the original catalog, and for comparison the results in two decontaminated catalogs created using ρ_{LE}. If, as in DBPO,

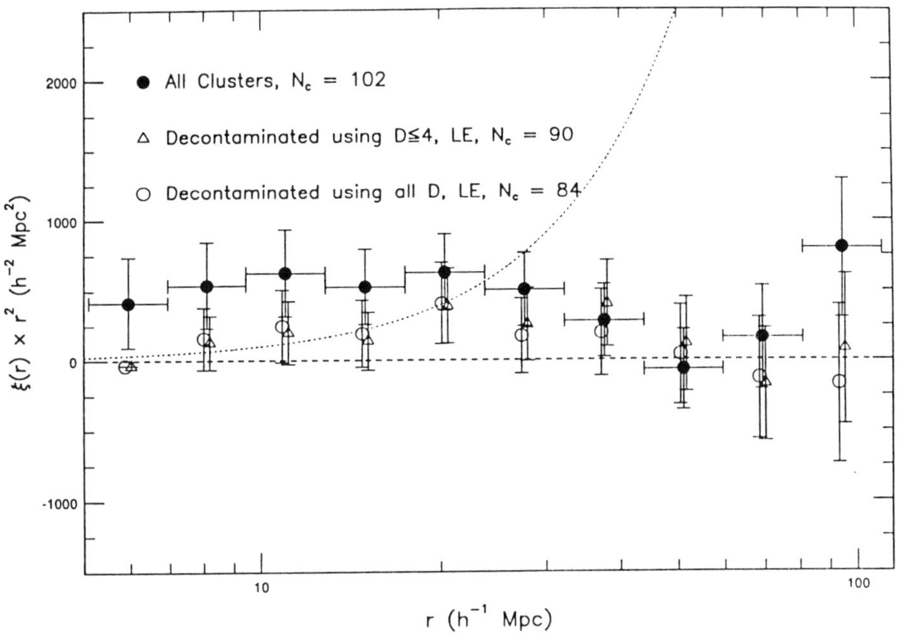

Fig. 1. Spatial correlation function of Abell Clusters ($D \leq 4$)

we include projection effects only from other $R \geq 1$ clusters in the same set of distance classes ($D \leq 4$), 12 clusters are omitted – but $\xi_{cc}(r)$ is decreased by a factor of two or three below about 30 h^{-1} Mpc. Including projection effects from D = 5,6 clusters removes an additional 6 clusters, but as we expected this has very little effect on $\xi_{cc}(r)$ since these more distant clusters just remove a few nearby clusters essentially at random. We discuss this further below.

The point we want to stress is that the projection effects of cluster halos on nearby clusters are capable of causing drastic effects on cluster correlation functions on scales of a few tens of Mpc, corresponding to supercluster sizes. This makes the amplitude of $\xi_{cc}(r)$ uncertain on such scales. It also calls into question the claims of Bahcall, Soneira and Burgett (1986) regarding evidence from the nearby cluster sample for the existence of large cluster peculiar velocities (Sutherland 1988, DBPO).

In OBDPS we discuss our new results on angular correlations using the full Abell and ACO catalogs, but here we will mainly present the ACO results. Figure 2 shows $\omega_{cc}(\theta)$ before and after decontamination, using ρ_{LE} and ρ_{SP} for all clusters of richness $R \geq 1$ and $17.1 < m_{10} \leq 19.3$ (essentially equivalent to distance classes D = 5 and 6) in the ACO catalog. The standard correction functions were used.

There are three distinct regimes: small angles, $\theta < 1°$; intermediate angles, $1° < \theta < 3°$; and large angles, $\theta > 3°$. At small angles the effect of decontamination is strong. The amplitude of $\omega_{cc}(\theta)$ typically decreases by a factor of two when ξ_{cg} of LE is used, and even more for SP. At intermediate angles the effect is moderate; $\omega_{cc}(\theta)$ typically decreases by 30% when SP is used, and it is even less affected for LE. At angles larger than 3° the effect of decontamination is negligible. This general trend is not surprising. With an

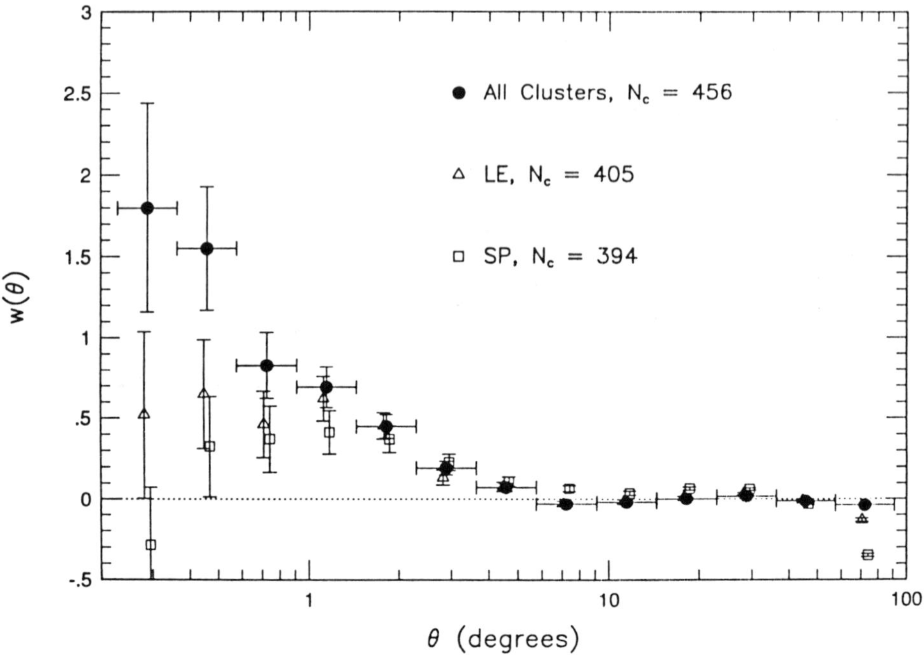

Fig. 2. Angular correlation function of ACO clusters ($17.1 < m_{10} \leq 19.3$, $R \geq 1$)

effective depth of $\sim 600\ h^{-1}$ Mpc, 1° roughly corresponds to 10 h^{-1} Mpc. The projection contamination is thus strong on scales of 10 h^{-1} Mpc and below, and it is still apparent out to scales of $\sim 30\ h^{-1}$ Mpc - the typical size of superclusters. These results are in agreement with the claim of DBPO that most of the contamination occurs inside superclusters where the deletion of one cluster subtracts many pairs on these scales.

The angular correlation function is similar in the north (Abell) and the south (ACO) at small and intermediate angles, both before and after decontamination. They are very different, however, at large angles. While in ACO $\omega_{cc}(\theta)$ practically vanishes beyond 3°, it stays positive in the north out to $\sim 40°$ at a very high σ level. We discuss below several tests that we have carried out trying to understand the source of this difference.

The ultimate question, of course, is whether one can extract any useful constraints on theory from $\omega_{cc}(\theta)$. In Figure 3, we compare the ACO decontaminated ω_{cc} with the Limber conversion (with the selection function of the ACO catalog) of $\xi_{cc}(r)$ predicted by several theoretical scenarios: (a) 'standard' Cold Dark Matter, with $\Omega = 1$, $h = 0.5$ and biasing factor b = 2.5, taken from Figure 5 of White *et al.* (1987), (b) a Hybrid of equal average density of CDM and Baryons with total $\Omega = 0.4$ taken from Blumenthal, Dekel and Primack (1988) (BDP), and (c) the most successful model of superclustering in the Explosion scenario, taken from Figure 6 of Weinberg, Ostriker and Dekel (1989).

We see a significant discrepancy between the observed $\omega(\theta)$ at intermediate angles and the predictions of standard CDM at the 4σ level for the ACO sample. (It is at the 6σ level for the Abell sample; see OBDPS.) The only way out for standard CDM is if the real uncertainties are much larger than the indicated Poisson error-bars. The open Hybrid model is consistent with the data predicting ω_{cc} slightly larger than observed; this could be adjusted

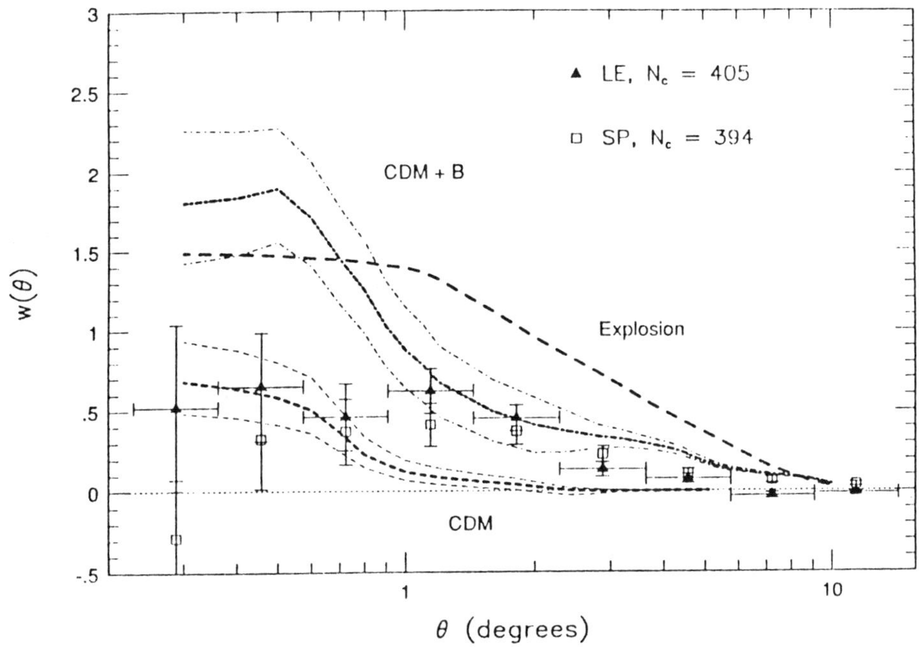

Fig. 3. Angular correlation function of ACO clusters ($17.1 < m_{10} \leq 19.3$, $R \geq 1$)

in the model by modifying the total Ω or the baryonic fraction (although this model is at best marginally consistent with the upper limits on the cosmic microwave background asymmetries for any of these values). The Explosion model predictions are too high.

How significant is $\omega(\theta)$ between one and three degrees? How significant is the discrepancy between these data and the predictions of CDM? The real errors must be larger than the Poisson errors quoted above because the correlation between the objects effectively reduces the number of independent pairs. Following Peebles (1973), one can estimate this effect using a toy model in which the objects are all in 'clusters' (superclusters in our case) whose centers are distributed at random. (We give the details in OBDPS.) Our result is that we expect the superclustering to increase the errors, but by less than a factor of two. The discrepancy between $\omega(\theta)$ in the 1 - 3° zone and CDM is therefore still at the level of at least 3σ. However, our error analysis is only approximate and ignores potentially important effects such as higher order correlations and other sampling errors. Unfortunately, we still lack a general, accurate error analysis for correlation functions.

UNCERTAINTY ON LARGE ANGLES

For angles larger than about 3°, the angular correlation functions $\omega_{cc}(\theta)$ determined from the northern (Abell) and southern (ACO) data disagree significantly. The main difference is that $\omega_{cc}(\theta)$ from ACO is consistent with zero within statistical errors beyond about 3°, while $\omega_{cc}(\theta)$ from the decontaminated Abell catalog remains positive to about 40°. However, as we discuss in detail in OBDPS, various subsamples of the northern data give very different results for $\omega_{cc}(\theta)$ on large angles. Also, the results are very sensitive to the background normalization - local versus global - used in the computation of $\omega(\theta)$. We therefore conclude that the non-vanishing angular correlations in the Abell data beyond about 3° probably represent a systematic sampling error or some other artifact.

A simple test gives some insight into the very inhomogeneous nature of the large scale distribution of Abell clusters and the effects of this distribution on the large angle correlations. We have noticed that by simply cutting the sky in half along certain lines of constant galactic longitude, two equal area subsamples of the Abell catalog are created that have very different number densities and different large scale correlations. Subsample 1 arbitrarily consists of Abell clusters with $b < 0°$, $-60° < \ell < 120°$ and $b > 0°$, $90° < \ell < 270°$ while subsample 2 consists of Abell clusters in the other half of the sky. When the previously described galactic latitude and declination cuts are made, we find that subsample 1 covers 50.8% of the area covered by the full sample, and subsample 2 thus covers 49.2%. However, we also find that subsample 1 contains 971 of the 1548 original $D = 5{,}6$ Abell clusters while subsample 2 contains only 577. Thus, one half of the sky contains over 68% more clusters than the other half. Moreover, the angular correlation functions of these two subsamples are remarkably different on large scales as shown in Figure 4.

If the differences between these two subsamples were accurate representations of mass density perturbations in the universe, virtually all standard cosmological models would be ruled out! However, it seems more likely that these differences are somehow caused by spurious observational effects. This conclusion is supported by the fact that the ACO catalog does not exhibit any such large scale differences. This is particularly evident in the overlap region between the Abell and ACO catalogs where the Abell catalog contains 94 ($R \geq 1$) clusters within the subsample 1 boundaries and only 35 clusters in subsample 2, while the ACO catalog contains 78 clusters in subsample 1 and 67 in subsample 2.

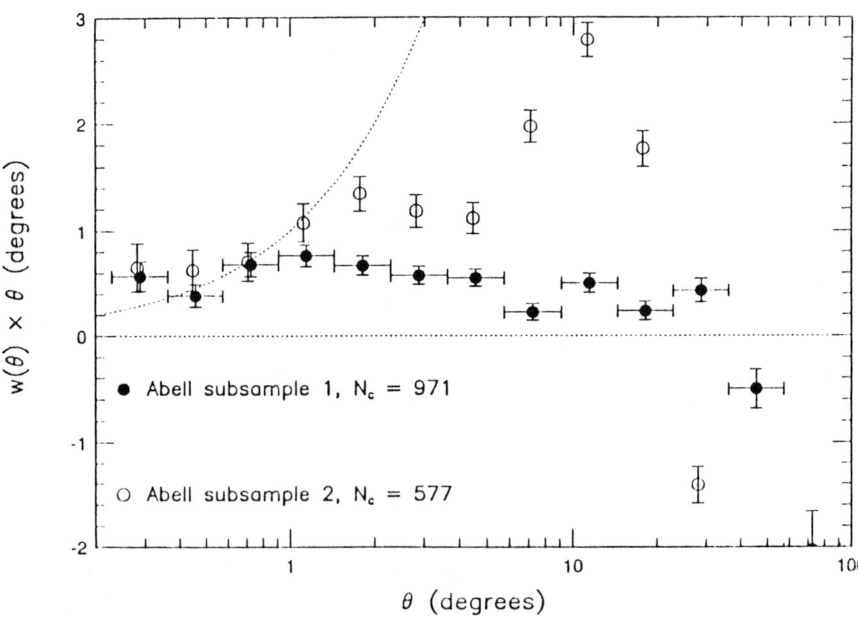

Fig. 4. Angular correlation function of Abell clusters ($4 < D \leq 6$, $R \geq 1$)

DISCUSSION

One might worry about the possibility that Abell's procedure of background subtraction has already, at least partially, corrected for projection effects. All that is said in Abell (1958) is that for each cluster, the background subtracted was the number of galaxies in the appropriate luminosity strip in a circle of the same size in a region free of clusters on the same plate. Since Abell (1958) did not record the actual location of his comparison region(s) on each plate, or his counts there, we cannot be certain to what extent the background subtractions affect the contamination - but it seems unlikely to be a significant effect for the following reason. Even if the cluster-free comparison region were as close as 2° away from the cluster on the 6° × 6° plates, that would correspond to a projected separation of 7 h^{-1} Mpc already at a distance of 200 h^{-1} Mpc. But almost all the contamination that we found in DBPO came from clusters nearer than 6 h^{-1} Mpc in projected separation, and our assumed cluster halo profiles $\rho(r)$ fall off fairly rapidly with r. Since ACO used a global, rather than local, subtraction procedure, our decontamination method certainly does not doubly correct for projection contamination of the ACO catalog. Unfortunately, there are not yet enough redshifts available for ACO clusters to allow a statistically meaningful computation of $\xi_{cc}(r)$. But the fact that $\omega_{cc}(\theta)$ of the decontaminated ACO catalog is generally lower than that of the decontaminated Abell catalog suggests that double-correction for projection is not a significant problem in our decontamination of the Abell catalog.

Our results do depend on the assumed luminosity function, mostly through the mean number density of galaxies $\langle n \rangle$; the projection effect on the cluster correlations is roughly proportional to $\langle n \rangle$. The present results are based on the most recent and comprehensive measurements available, as analyzed by Efstathiou, Ellis and Peterson (1988). Our results do not depend strongly on the cluster density $\langle n_c \rangle$, which is fortunate since $\langle n_c \rangle$ is different for the Abell and ACO catalogs, and even for various subsamples of the Abell catalog. The main dependence on $\langle n_c \rangle$ is in the second term in equation 2, which corrects for double-counting. For small r, the halo profile $\rho(r)/\langle n \rangle$ essentially equals $\xi_{cg}(r)$; the value of $\langle n_c \rangle$ affects the r at which $\rho(r)/\langle n \rangle$ drops significantly below $\xi_{cg}(r)$. Most of our projection correction is for projected separations less than about 6 h^{-1} Mpc, while the effect on $\rho(r)$ is important only for r greater than about 20 h^{-1} Mpc.

Szalay, Hollósi and Tóth (1989), who found a negative angular cross-correlation between $D \leq 4$ and $D = 6$ clusters in the Abell catalog, claim that this argues against the decontamination procedure applied in BDPO to $\xi_{cc}(r)$. We cannot see any relevance of their result to the contamination effects we worry about. First of all, in DBPO, we considered only the effects on ξ_{cc} of projection of galaxies associated with $D \leq 4$ clusters on other $D \leq 4$ clusters. When more distant clusters are also considered, there is some cross-talk between $D \leq 4$ and $D = 6$ clusters, but it is weak. Although the Schechter luminosity function, which we assume, has many low-luminosity galaxies in $D \leq 4$ clusters that could in principle project onto $D = 6$ clusters inside the Abell luminosity strip, only a few would actually fall inside the Abell radius of the distant cluster. On the other hand, galaxies associated with $D = 6$ clusters hardly fall inside the Abell luminosity strip of a $D \leq 4$ cluster, so contamination is not significant in this direction either. Moreover, even if such contamination does occur in some cases, it is unlikely to affect $\xi_{cc}(r)$. As we explained in DBPO, our picture is that the main effect of projection contamination on the auto-correlation functions arises from the spurious cluster pairs in the *same* supercluster, which arise from the inclusion in the $R \geq 1$ catalog of clusters whose richness was enhanced by projection of galaxies associated with the other clusters in the same supercluster. It is a 3-body effect! Therefore, any mutual contamination between $D = 6$ and $D \leq 4$ clusters would not affect the correlation function preferentially at small separations. We have tested this by calculating the effect of the $R \geq 1$, $D = 5,6$ clusters on the $\xi_{cc}(r)$ of the $R \geq 1$, $D \leq 4$ Abell clusters, as discussed above. The result is that, while accounting for projection contamination by the numerous distant clusters removes about another 10% of the $D \leq 4$ clusters from the $R = 1$ richness class, this has essentially no effect on $\xi_{cc}(r)$ except to increase the errors a little - as anticipated.

An independent way to estimate the importance of several potential problems - *e.g.*, nonspherical and sub-clustered cluster halos, or projection contamination by lower-richness clusters - would be to 'observe' clusters in appropriate simulations. Frenk *et al.* (1990) have recently compared the clusters identified in projection to the clusters identified in three-dimensions in the same simulations of Cold Dark Matter. They did not address the effect of projection contamination on the cluster correlations, but it is interesting for us that about a third of their '$R \geq 1$' clusters identified by an Abell-type procedure would not have achieved this richness without projection contamination. It would be very interesting to see the effect of applying our decontamination procedure to the Frenk *et al.* simulated cluster catalogs.

We believe that the main uncertainty in our correlation function results is associated with possible problems specific to the cluster catalogs used, and not with our decontamination procedure. We have discussed some problems with the Abell catalog. Our main hope for progress is to obtain cluster catalogs that are more useful for cosmological purposes. The catalog(s) being prepared by APM machine measurement of the UK Schmidt (southern) plates (see Sutherland, Maddox, and Efstathiou this volume) have the virtue of careful plate calibration and unbiased measurements, and they will allow a better understanding of the effects on cluster correlations of different cluster selection criteria. Because both the ACO and APM work concerns southern clusters, it is especially urgent to obtain sufficient cluster redshifts to allow computation of ξ_{cc}.

When new catalogs become available, it will be necessary to decontaminate them using similar procedures to those discussed here before meaningful cluster correlations can be calculated. Because cluster correlations on small distance or angular scales are so sensitive to the effects of projection contamination and correlations on larger scales are sensitive to sampling inhomogeneities and other problems with the catalogs, both problems should be taken into account in any attempt to measure 2-point as well as 3-point or higher-order cluster correlations, such as that by Tóth, Hollósi and Szalay (1989).

X-ray cluster catalogs avoid projection contamination because of their small X-ray emitting regions. Results on the correlation function of X-ray clusters from the limited data that are presently available were recently reported by Lahav *et al.* (1989). The correlation function of these clusters seems to be similar to that found by BS, but, with the very nonuniform sample of about 50 X-ray clusters used, the statistical uncertainties are very large. It will be very interesting to see the results from the more extensive volume-limited X-ray cluster survey that may be possible with the ROSAT satellite, scheduled to be launched in 1990, and that will certainly be possible with the future AXAF satellite.

ACKNOWLEDGEMENTS

We thank Harold Corwin and Ron Olowin for giving us their raw galaxy counts, Will Sutherland for sharing his unpublished work with us, and all of them for helpful discussions. We have also benefitted from discussions with Neta Bahcall, Alan Dressler, and Anatoli Klypin. This work was supported in part by US - Israel Binational Science Foundation grant 86-00190, NSF grant PHY8800801, a UCSC Faculty Research Grant, and a California Space Institute grant.

REFERENCES

Abell, G. O. 1958, *Ap. J. Suppl.*, **3**, 211.
Abell, G. O., Corwin, H. G., and Olowin, R. P. 1989, *Ap. J. Suppl.*, **70**, 1 (ACO).
Bahcall, N. A. 1986, *Ap. J. (Letters)*, **302**, L41.
Bahcall, N. A. 1988, *Ann. Rev. Astr. Ap.*, **26**, 631.
Bahcall, N. A., and Soneira, R. M. 1983, *Ap. J.*, **270**, 20 (BS).
Bahcall, N. A., Soneira, R. M., and Burgett, W. S. 1986, *Ap. J.*, **311**, 15.
Bardeen, W. A., Bond, J. R., Kaiser, N., and Szalay, A. S. 1986, *Ap. J.*, **304**, 15.
Barnes, J., Dekel, A., Efstathiou, G., and Frenk, C. S. 1985, *Ap. J.*, **295**, 368.
Blumenthal, G. R., Faber, S., Primack, J. R., and Rees, M. J., *Nature*, **311**, 517.
Blumenthal, G. R., Dekel, A., and Primack, J. R. 1988, *Ap. J.*, **326**, 539 (BDP).
Davis, M., Efstathiou, G., Frenk, C. S., and White, S. D. M. 1985, *Ap. J.*, **292**, 371.
Dekel, A. 1984, *Ap. J.*, **284**, 445.
Dekel, A. 1988, in *Large-Scale Motions in the Universe*, ed. V. C. Rubin and G. V. Coyne, S. J., (Princeton: Princeton University), p. 466.
Dekel, A., Blumenthal, G. R., Primack, J. R., and Olivier, S. 1989, *Ap. J. (Letters)*, **338**, L5 (DBPO).
Efstathiou, G., Ellis, R. S., and Peterson, B. A. 1988, *M.N.R.A.S.*, **232**, 431.
Frenk, C. S., White, S. D. M., Efstathiou, G., and Davis, M. 1990, **351**, 10.
Hewett, P. C. 1982, *M.N.R.A.S.*, **201**, 867.
Kaiser, N. 1984, *Ap. J. (Letters)*, **284**, L9.
Kaiser, N. 1986, *M.N.R.A.S.*, **219**, 785.
Kaiser, N., and Lahav, O. 1988, in *Large-Scale Motions in the Universe*, ed. V. C. Rubin and G. V. Coyne, S. J., (Princeton: Princeton University), p. 339.
Klypin, A. A., and Kopylov, A. I. 1983, *Sov. Astr. Letters*, **9**, 41.
Lahav, O., Edge, A. C., Fabian, A. C., and Putney, A. 1989, *M.N.R.A.S.*, **238**, 881.
Lilje, P. B., and Efstathiou, G. 1988, *M.N.R.A.S.*, **231**, 635; (LE).
Olivier, S., Blumenthal, G. R., Dekel, A., Primack, J. R., and Stanhill, D. 1990, *Ap. J.*, **356**, 1 (OBDPS).
Peebles, P. J. E. 1973, *Ap. J.*, **185**, 413.
Peebles, P. J. E. 1987, *Ap. J. (Letters)*, **315**, L73.
Seldner, M., and Peebles, P. J. E. 1977, *Ap. J.*, **215**, 703 (SP).
Sutherland, W.J. 1988, *M.N.R.A.S.*, **234**, 159.
Sutherland, W. J., Maddox, S. J., and Efstathiou, G. this volume, p. 241.
Tóth, G., Hollósi, J., and Szalay, A. S. 1989, *Ap. J.*, **344**, 75.
Weinberg, D., Ostriker, J. P., and Dekel, A. 1989, *Ap. J.*, **336**, 9.
White, S. D. M., Frenk, C., Davis, M., and Efstathiou, G. 1987, *Ap. J.* **313**, 505.

DISCUSSION

J. Lucey: Some real clusters can be lost from catalogs because foreground contamination can bias m_3 to brighter magnitudes. Was this effect considered for your modelling?

J. Primack: We mentioned, in our *Ap. J. Letter*, that this was a potential problem (with a reference to your Monte Carlo simulation paper, which discussed m_3 projection contamination). But we could not think of a good way to correct for this. Unfortunately, there seems to be no record of Abell's m_3 measurements. We hope that in most cases Abell, Corwin, and Olowin disregarded foreground galaxies.

N. Kaiser: I think the most damning result for $\omega(\theta)$ is Will Sutherland's comparison of inter-plate and intra-plate estimates. A coverall hypothesis to explain this is that Abell went further down the richness distribution on the plates which gave high abundance. I have tried to test this against the Kowalski *et al.* x-ray flukes - the prediction is that a smaller fraction of clusters in apparently dense regions should have been detected. If I bin the plates in galactic latitude I see this effect strongly - so it looks like people were right in taking this trend out. However, apart from this I do not see the effect, so I'm a bit puzzled.

D. Burstein: Locally we do find clusters that 1) are near and overlapping, and 2) tend to align along a very long chain (*e.g.* Perseus-Pisces, the Local Supercluster, the Great Attractor, the Shapley/Scaramella *et al.* Clusters). These clusters are probably all physically related (certainly the first set of three are). If we would observe our region from $z \sim 0.1$, we would see small angular separations of clusters over a wide viewing angle. To what extent is the method of Bahcall *et al.* (count all rich clusters) and that of your group (decontaminate the nearest clusters in angle) both consistency checks of the data rather than proofs of one view or the other?

J. Primack: Our procedure tends to remove only a few marginally-rich clusters in a supercluster; many rich clusters remain, even at small angular separations. Some sort of correction for projection contamination will be required to compare two-dimensional observations with three-dimensional predictions; we tried to invent a reasonable and simple one. It would certainly be interesting to see what results other procedures will give.

A. Szalay: Do you have a histogram of the galaxy counts within the Abell circle? If one lowered the threshold of admitting clusters into the catalog, the pair count would come back, and about 40% more new uncorrelated clusters would have to enter to dilute it by a factor of two, if that is the real correlation amplitude. I estimate a mass function slope of –5 to –6 to achieve this effect. Does this agree with your histogram?

J. Primack: Such a histogram appears as Figure 3.2 [reproduced on the next page] in the very preliminary account of our work in Dekel (1988). The number count slope of the $D \leq 4$, $R = 1$ Abell clusters is indeed steep, about –4.5. (The $R = 0$ clusters have a lower slope, but that presumably reflects the incompleteness of the catalog.)

D. Lambas: We have found that $D = 6$ clusters in the Abell Catalog have very little dependence on galactic latitude in comparison with $D = 1$ to 4 or $D = 5$. The $D = 6$ clusters in the ACO catalog behave more in accordance with what is expected. This gives additional evidence that $D = 6$ Abell clusters might be seriously biased. Also, $\omega(\theta)$ in large samples is very sensitive to gradient corrections. They must be carefully estimated for each sample.

J. Primack: I appreciate your comments and agree that one must be careful in making gradient corrections. We tried many methods of doing this for the Abell (north) catalog, but found that the large-angle correlations remained fairly strong, whichever way we corrected for extinction. As I remarked, however, they were very different in different subsets of the northern data.

A. Dressler: Can we return to Neta Bahcall's counter-argument that Abell implicitly corrected for the contamination by determining background *locally* (thus, she says, you've corrected the data a second time)? Ron Olowin has just told us that the ACO Catalog has been constructed with a global background correction. If her explanation were correct, wouldn't this result in an even higher amplitude for the cluster-cluster correlation function for the ACO Catalog? Is this observed?

J. Primack: No, just the opposite is observed: $\omega(\theta)$ is smaller at virtually all angles for ACO, both before and after our decontamination procedure. (Thanks for the question!) When we have enough redshifts of nearby ACO clusters to determine their spatial correlation function, ξ_r, it will be interesting to compare it to that of the (northern) Abell clusters.

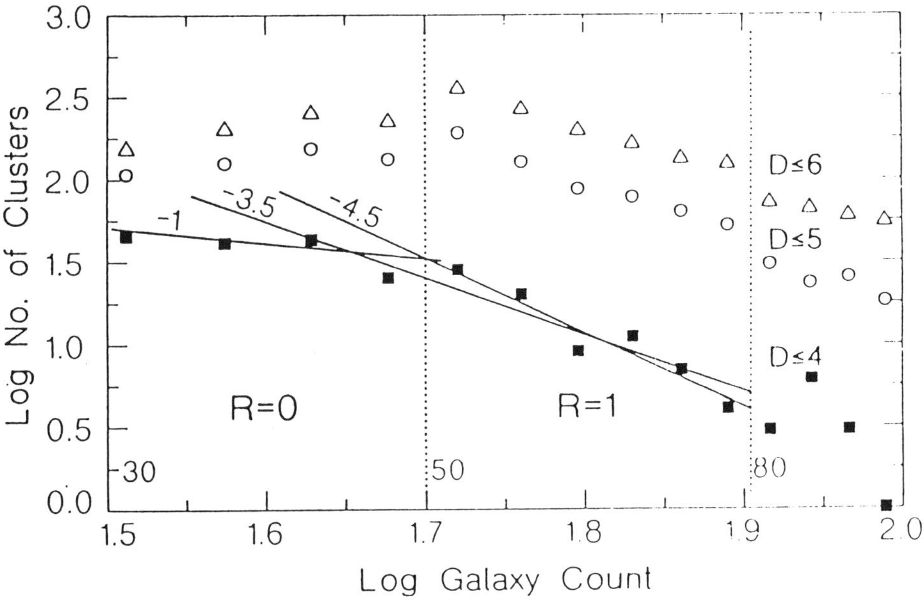

Fig. 3.2. Multiplicity function of Abell clusters in bins of 5 galaxies (from Dekel 1988).

N-POINT CORRELATION AMPLITUDES FROM THE LICK CATALOG

I. SZAPUDI
Loránd Eötvös University, Kun Béla tér 2, H-1083 Budapest, Hungary

A. S. SZALAY
Loránd Eötvös University, Kun Béla tér 2, H-1083 Budapest, Hungary and
Department of Physics and Astronomy, The Johns Hopkins University, Baltimore, MD

P. BOSCHAN
Loránd Eötvös University, Kun Béla tér 2, H-1083 Budapest, Hungary and
University of Duisburg

ABSTRACT We present a novel method for determining the amplitudes of the irreducible N-point correlation functions in the gravitational clustering hierarchy. We use the corrected Lick catalog to obtain values of Q_N up to $N = 8$, considerably beyond previous attempts. The angular amplitudes are relatively close to unity, the spatial amplitudes seem to be slowly decreasing.

INTRODUCTION

Over the years a lot of effort has been spent on determining correlation functions of extragalactic objects. First the emphasis was on the two-point correlations (for a comprehensive review see Peebles 1980, Fall 1979). It soon became clear that clustering does not stop at the two-point level, but it is more non-linear: even the three-, the four-point and the higher order functions are quite strong (White 1979). Their experimental determination was successful only for the three-point (Peebles and Groth 1975, Groth and Peebles 1977) and the four-point functions (Fry and Peebles 1978, Sharp *et al.* 1984).

Theoretically it became clear (Davis and Peebles 1977) that the closure of the BBKGY equations is satisfied with a scaling ansatz, setting the 3-point function to be proportional to the symmetric quadratic product of the 2-point correlation function. This assumption was also extended to the 4-point function. Later, extensive work by Fry (1984, 1986), and Schaeffer (1984) generalized this to an infinite clustering hierarchy with general Q_N coefficients, where the irreducible N-point correlation function can be expanded in terms of all $(N-1)^{\text{th}}$-order symmetric products of the 2-point,

$$\xi^{(N)} = Q_N \sum_{\text{trees}} \xi(r_1) \ . \ . \ . \ \xi(r_{N-1}). \tag{1}$$

There are some constraints for this N-point hierarchy, mostly from the requirement of non-negative total N-point probabilities (Fry 1984, Schaeffer 1984, Balian and Schaeffer 1989). However, these constraints allow extremely different distributions. From the currently known two- and three-point amplitudes it is hard to get any constraint on these models. Balian and Schaeffer (1989) suggest a hierarchical form

$$Q_N = 2^{(N-1)\nu}/N^{\nu}, \text{ where } \nu \approx 1 \pm 1, \tag{2}$$

which includes the somewhat natural choice $Q_N = 1$, for $\nu = 0$. Fry (1984) suggested another alternative,

$$Q_N = 1/2\ (4Q_3/N)^{N-2}\ [N/(N-1)] \tag{3}$$

There were some attempts, using the void probability function (Bouchet and Lachieze-Rey 1986, Fry *et al.* 1989) to constrain the hierarchies, since that function contains integrals over all orders of N; the limits however are not very strong.

Here we present a new method to go beyond previous attempts, and calculate amplitudes directly up to $N = 8$, which already sets some constraints on the possible hierarchies.

CORRELATION OF DENSITY MOMENTS

We consider a distribution of projected densities $\tilde{N}_i$ in cells on the sky, as the data is in the Lick catalog. The usual expression of the angular correlation function $\omega(\theta_{12})$ is

$$\omega(\theta_{12}) = (\langle \tilde{N}_1\ \tilde{N}_2 \rangle - \langle \tilde{N} \rangle^2)/\langle \tilde{N} \rangle^2 \tag{4}$$

This is the normalized expectation value of the product of the two densities at cells V_1 and V_2, with the uncorrelated part subtracted. Here below we generalize this formula and calculate the same 2-point correlation function for the nth and m^{th} moments of the two densities at the distance θ_{12}, using $N = \langle \tilde{N} \rangle$,

$$\omega_{nm}(\theta_{12}) = (\langle \tilde{N}_1^n\ \tilde{N}_2^m \rangle - \langle \tilde{N}^n \rangle \langle \tilde{N}^m \rangle)/N^{n+m} \tag{5}$$

All these functions will go to zero at infinity, since we subtracted the appropriate asymptotics. We also use the nth moments of the cell counts, denoted as ω_{n0},

$$\omega_{n0} = \langle \tilde{N}^n \rangle / N^n, \tag{6}$$

where there is no subtraction. This latter quantity has been used before by Sharp (1981), Sharp *et al.* (1984) to determine higher order correlation coefficients.

In order to fit various powers of the correlation functions to the raw data there are several things to watch. To distinguish between various negative power laws a large dynamic range is needed. This is the principal difficulty in using the moments of cells only. In case of an N-tuple, there are many different pairs, each with a different 2-point function. The irreducible correlation function will be a scalar over $N(N-1)/2$ variables, hard to visualize and handle. Also, a general extended N-tuple will have an irreducible N-point function very close to zero, due to the smallness of the 2-point function at larger separations. This should be fitted to products of the 2-point functions, also near zero, but with a large systematic error due to the uncertainty in the background density.

When doing a similar project to determine the 3-point function of Abell clusters (Tóth *et al.* 1989), we found that the most significant contribution to our fit came from those triangles which were nearly degenerate, *i.e.* the shortest side was almost zero. In that case one of the 2-point functions is quite high, which makes those configurations more probable. In turn this takes them above the noise, which prevents the observation of the 3-point function elsewhere, but due to the two long sides, practically equal, we still retain the large dynamic range for the fit. This is the reason why we decided to use such degenerate configurations to determine the Q_N coefficients.

ANALYTIC CALCULATION OF THE MOMENT CORRELATIONS

We calculate now analytically these ω_{nm}'s, which are *not* irreducible, but rather they contain a variety of other terms as well. During the data reduction we first calculate the expectation values of the density moments. The difficulty is to estimate the Poisson contribution to the correlations. In order to do this we follow the method outlined in Peebles (1980). We break up the cells into infinitesimal ones, where the individual Poisson processes are ν_i in V_1, and μ_j in V_2, at an angular distance θ_{12}. Since these are infinitesimal, the possible count is either zero or 1, so the expectation value of their moments is equal:

$$\langle \nu_i \rangle = \langle \nu_i^2 \rangle = \langle \nu_i^3 \rangle = \ldots \tag{7}$$

because $\nu_i^m = \nu_i$ if $\nu_i = 0,1$. The total count in the cell V_1 is $\tilde{N}_1 = \sum_i \nu_i$. We define the quantity a_{nm} to be the sum of the expectation values

$$N^{n+m} a_{nm} = \sum \langle \nu_1 \ldots \nu_n \mu_1 \ldots \mu_m \rangle, \tag{8}$$

where the sum is over all possible disjoint infinitesimal cells, when we take n from V_1 and m from V_2. This is similar to the irreducible N-point correlation function, degenerated to these two cells, defined by

$$N^{n+m} r_{nm} = \langle \langle \delta\nu_1 \ldots \delta\nu_n \, \delta\mu_1 \ldots \delta\mu_m \rangle \rangle, \tag{9}$$

where $\delta\nu_i = \nu_i - \langle \nu \rangle$, and the symbol $\langle \langle \ldots \rangle \rangle$ here denotes both the ensemble average and the mean over the different possible disjoint configurations of the infinitesimal cells.

Adding a new small cell with ν_{n+1} we can write down the recursion for a_{nm}.

$$a_{n+1,m} = a_{nm} + [\tbinom{n}{1} r_{20} a_{n-1,m} + \tbinom{m}{1} r_{11} a_{n,m-1}] + \tag{10}$$

$$[\tbinom{n}{2} r_{30} a_{n-2,m} + \tbinom{n}{1}\tbinom{m}{1} r_{21} a_{n-1,m-1} + \tbinom{m}{2} r_{12} a_{n,m-2}] + \ldots$$

The first term describes the contribution of the uncorrelated part of ν_{n+1}, the rest of the terms contain the irreducible correlations between ν_{n+1} and a subset of the other cells. The second term comes from all possible 2-point configurations which contain ν_{n+1}, whereas the third term is the sum of all possible irreducible 3-point configurations, etc. One can write down a closed expression

$$a_{n+1,m} = a_{nm} + \sum_{\ell=1}^{n+m} \sum_{k=0}^{\ell} \tbinom{n}{k} \tbinom{m}{\ell-k} r_{k+1,\ell-k} \, a_{n-k,m-\ell+k} \tag{11}$$

This recursion starts at $a_{00} = a_{10} = a_{01} = 1$, and it is symmetric, thus $a_{nm} = a_{mn}$.

Now we would like to express the full probability using the infinitesimal Poisson processes ν_i and μ_j. This is more difficult, since so far we only counted the sums with disjoint configurations. Now we have to count also those when two or more of the infinitesimal processes fall into the same cell. Here we use the property

$$\langle \nu_i \rangle = \langle \nu_i^2 \rangle = \ldots . \tag{12}$$

Counting these cases reduces to the combinatorial problem of placing n objects into k non-empty boxes. This is given by the Stirling numbers of the second kind (*e.g.* Abramowitz and Stegun 1964), which we will denote as $[{}^{n}_{k}]$. With these symbols we can write the moments of cell counts as

$$\langle \tilde{N}_1^n \rangle = \sum_{k=1}^{n} [{}^{n}_{k}]\, N^k\, a_{k0} \tag{13}$$

and for the correlated moments in the two cells

$$\langle \tilde{N}_1^n, \tilde{N}_2^m \rangle = \sum_{k=1}^{n} \sum_{\ell=1}^{m} [{}^{n}_{k}]\, [{}^{m}_{\ell}]\, N^{k+\ell}\, a_{k\ell} \tag{14}$$

These formulas agree with results given by Peebles (1980) for the central moments up to the rank of 4. We calculated these recursions and sums with a computer algebra program analytically. Now we turn to the quantities ω_{nm} and ω_{n0} as defined above by subtracting the appropriate asymptotics, and normalizing by powers of N.

THE MODEL FOR THE IRREDUCIBLE CORRELATIONS

For angular correlations the hierarchy of the correlations is preserved, only in the small angle approximation the spatial coefficients Q_N are replaced by the angular q_N, using Limber's formula to relate the two. Therefore we write the correlation hierarchy for the angular correlations as

$$r_{nm} = q_{n+m} \sum_{\text{trees}} \prod^{n+m-1} \omega(\theta_{ij}) \tag{15}$$

where there are two kinds of possible distances between our cells. If the two points are within the same cell, we use ω_s, the mean of ω over a cell, when they are in different cells, $\omega_{ij} = \omega_\ell = \omega(\theta_{12}) << \omega_s$. With this notation, our expression for the 3-point function becomes

$$r_{12} = r_{21} = q_3\, [2\, \omega_\ell\, \omega_s + \omega_\ell^2] \tag{16}$$

The coefficients increase rapidly with N, there are N^{N-2} possible graphs for a the N-point function. One can already see that the terms will not be equal; the first term is larger by ω_s/ω_ℓ than the second. In a similar manner, in all these functions the leading order will always be linear in ω_ℓ; there must be at least one edge connecting the two cells! Expanding a general case we obtain

$$r_{nm} = \omega_\ell\, q_{n+m},\, (n\, \omega_s)^{n-1}\, (m\, \omega_s)^{m-1}\, \{\, 1 + [(nm-1)/2]\, [\omega_\ell/\omega_s] + \dots \} \tag{17}$$

In this paper we make the following approximations: since we are only dealing with relatively low values of n and m, we ignore the terms higher than linear in ω_ℓ. It can be seen from Figure 1 that this a surprisingly good approximation, nearly all ω_{nm} curves are parallel to ω_{11}. Our other approximation is that we do not distinguish between the different graph types, rather a single mean amplitude q_N is adopted for all correlations of a given order. Thus we use the following expression for the irreducible correlations

$$r_{nm} = \omega_\ell\, q_{n+m}\, (n\, \omega_s)^{n-1}\, (m\, \omega_s)^{m-1} \tag{18}$$

This is substituted into the computer algebra program, and we calculate ω_{nm}, also up to linear order in ω_ℓ. Note that those are not the irreducible functions, only the minimum asymptotics has been subtracted for reasons we will discuss below. As an illustration we present some of the lower order expressions here, which also contain the Poisson noise. First we show the plain moments of the cell counts for n = 2, 3, 4. We use $b = 1/N$ for the Poisson noise term.

$$\omega_{20} = 1+b+\omega_s$$

$$\omega_{30} = 1+3b+b^2+3\omega_s(1+b)+3\omega_s^2 q_3$$

$$\omega_{40} = 1+6b+7b^2+b^3+\omega_s(6+18b+7b^2)+3\omega_s^2+6\omega_s^2 q_3(2+3b)+16\omega_s^3 q_4 \tag{19}$$

For the connected correlations we only show the leading term, the linear coefficient of ω_ℓ.

$$\omega_{21}/\omega_\ell = 2+b+2\omega_s q_3$$

$$\omega_{31}/\omega_\ell = 3+6b+b^2+3\omega_s+6\omega_s q_3(1+b)+9\omega_s^2 q_4$$

$$\omega_{41}/\omega_\ell = 4+18b+14b^2+b^3+6\omega_s(2+3b)+2\omega_s q_3(6+18b+7b^2+12\omega_s)$$
$$+18\omega_s^2 q_4(2+3b)+64\omega_s^3 q_5$$

$$\omega_{22}/\omega_\ell = 4+4b+b^2+4\omega_s q_3(2+b)+4\omega_s^2 q_4$$

$$\omega_{32}/\omega_\ell = 6+15b+8b^2+b^3+3\omega_s(2+b)+2\omega_s q_3(9+15b+4b^2+3\omega_s)$$
$$+3\omega_s^2 q_4(10+7b)+18\omega_s^3 q_5$$

$$\omega_{42}/\omega_\ell = 8+40b+46b^2+16b^3+b^4+6\omega_s(2+b)(2+3b)+24\omega_s^3 q_3^2$$
$$+4\omega_s q_3(8+30b+23b^2+4b^3+18\omega_s+15b\omega_s)$$
$$+2\omega_s^2 q_4(48+108b+41b^2+12\omega_s)+4\omega_s^3 q_5(50+43b)+128\omega_s^4 q_6 \tag{20}$$

We use the ω_{nm} for fitting as appear here, rather than trying to subtract from every term the lower order contributions. This method has considerable advantages. For example, the coefficient of q_3 in ω_{42} is over 100, and in ω_{44} it is over 1000, whereas in those terms q_6 or q_8 have a relatively little weight. If we subtracted a poorly determined q_3 with these high coefficients, the result would have been noise, which prevented previous attempts to go beyond 4-point amplitude. Here all terms depend on q_3, and nearly all depend on q_4. These amplitudes will be determined very well, for the rest our fits will be more uncertain due to the smaller weights, as it was later confirmed by the actual fitting. This way there is no error propagation from lower to higher-order coefficients.

RESULTS

We used the Lick catalog corrected by Peebles and Hauser (1974), considering only the plates with $|b_c| \geq 40°$, where b_c is the galactic latitude of the plate center. Peebles and

Groth (1975) divided this range into four zones because of the large-scale gradient due to the galactic absorption. This was good enough for the ω_{11} function, but the gradient in the densities was still too large for the higher order functions. After having studied the average cell counts as a function of the galactic latitude we chose the twelve zones summarized in Table I to be reasonably uniform in sin b_c.

TABLE I Distribution of Lick plates

Zone	b_{min}	b_{max}	Plates	N
N1	40	45	58	1.17
N2	45	50	50	1.29
N3	50	55	48	1.30
N4	55	62	60	1.53
N5	62	70	51	1.54
N6	70	90	54	1.51
S1	-45	-40	31	1.08
S2	-50	-45	23	1.26
S3	-55	-50	23	1.19
S4	-62	-55	28	1.42
S5	-70	-62	23	1.56
S6	-90	-70	18	1.63
Sum			467	

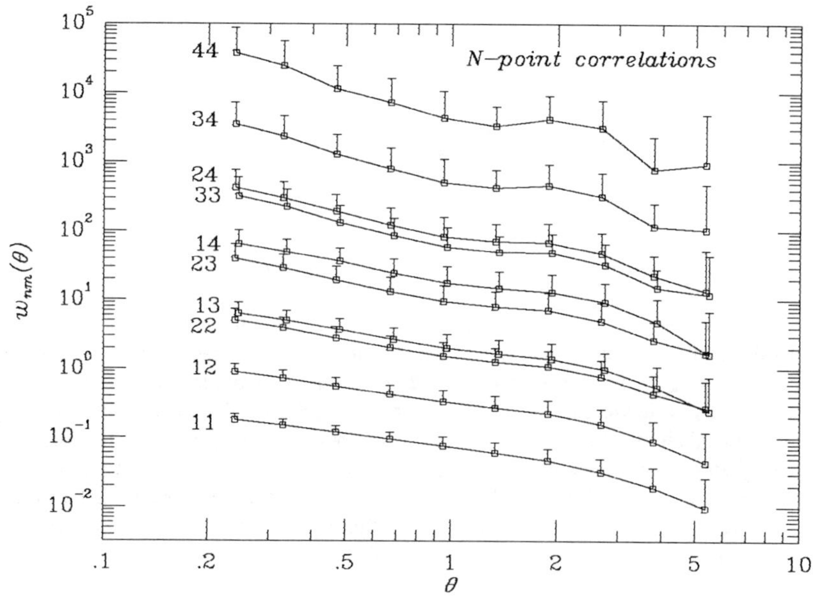

Fig. 1. The degenerate angular N-point correlation functions ω_{nm} are shown *versus* the angular separation of the two cells θ. The labels nm denote the powers of the density in the two cells. Note how similar are the shapes of the curves to $\omega_\ell = \omega_{11}$, the usual angular 2-point functions, indicating that the leading term proportional to ω_ℓ is the dominant one. Also, note the large amplitude of the higher order functions.

We compiled the bin counts N_{nm} with both $n,m = 0, \dots 4$, where

$$N_{nm}(\theta_{12}) = \sum \tilde{N}_1^{\,n}\, \tilde{N}_2^{\,m} \tag{21}$$

This means that N_{00} is the number of cell pairs at a given separation, a form factor. This had to be counted explicitly because the catalog had some bad cells marked with −5, although these were mostly on the edges, which we rejected. Our bin counts are intra-plate values, we have never crossed the boundaries of a plate. The plates contain a 36 × 36 array of cells, covering 6° × 6°. Plate centers are placed in every 5° in declination from δ = −20° to 90°, and within each declination zone the spacing in right ascension is such that each plate overlaps its neighbors by at least 1°. To avoid double counting these regions we used only the central 5° × 5° of each plate (see Peebles and Groth 1975, Groth and Peebles 1977).

The ω_{nm} were calculated from the N_{nm} by dividing and subtracting the appropriate moments. Their values are shown in Figure 1. It is interesting to note how large our N-point functions become, due to the degenerate configurations. The ω_{44} is about 10^5, while $\omega_{11} \sim 0.1$. In an ordinary case ω_{44} would be many orders of magnitude less, buried in the noise. Another interesting fact is that the ω_{nm} corresponding to the same rank, like 14 and 23, or 24 and 33 are of the same magnitude.

We used the formulas derived above to fit the higher order correlation function amplitudes q_3 - q_8. First we determined the value of ω_s directly from ω_{11}. Next we used a maximum likelihood method to fit the q_N parameters. For the probability density we chose

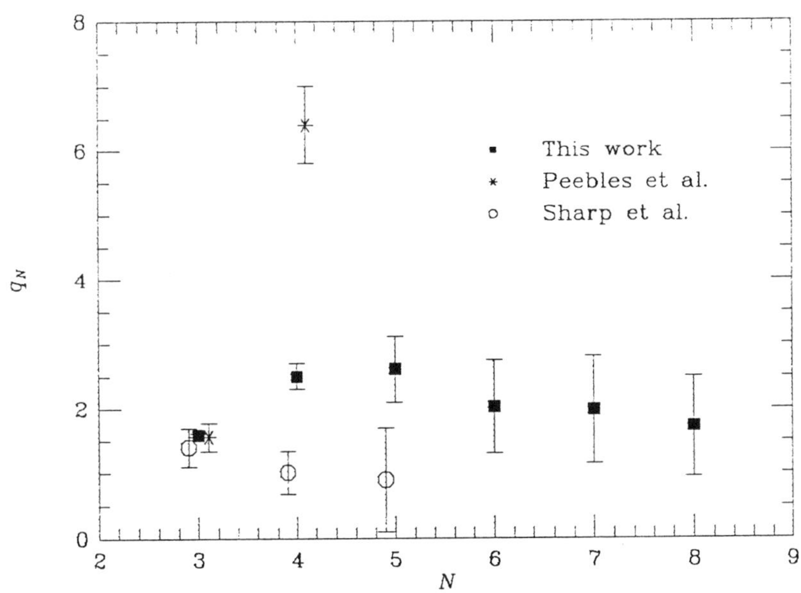

Fig. 2. The known angular amplitudes q_N for the N-point clustering hierarchy. The solid symbols are this paper, empty squares are Peebles and Groth (1977), Fry and Peebles (1978), empty circles are Sharp (1981), and Sharp *et al.* (1984). Note that these are the angular amplitudes, directly determined from the data. The spatial amplitudes connect to these via high moments of the selection function.

the Cauchy distribution instead of the Gaussian, because it handles the outlying points better (Press *et al.* 1986). We built up the likelihood function with the expressions derived by computer algebra, using the leading order amplitudes of ω_ℓ in each term, and the cell moments ω_{n0}. To estimate the errors an iterative procedure was used. First we set the initial parameter values for $q_3 - q_8$, then calculated our errors assuming a 25% relative error with these; later we recalculated the error estimates with the best previous parameter values.

We maximized the likelihood function by using the subroutine AMOEBA from Press et al. (1986). The fits to the higher order parameters were somewhat sensitive to the initial values. We ran AMOEBA 443 times with random initial values between 1.00 and 3.00 for each of the parameters. Results were filtered in the sense that we threw away values lying outside the interval [0.00,4.00]. We present below the averages and the standard deviations of the accepted runs. These standard deviations are the estimators for the real uncertainties in the fitted values, at least for the higher order parameters. Also shown are the conversion coefficients $R_N = q_N/Q_N$, from angular to spatial amplitudes, for various choices (A, B, C) of the luminosity-function parameters. These are quite sensitive to the choice of the luminosity function. We used a Schechter function, with $\alpha = -1.1$, $M_* = -19.68$. We take $\Omega_0 = 1$, and a K-correction $\kappa = 3 - 5$. Model A corresponds to $h = 0.5$, $\kappa = 3$, model B is $h = 1$ and $\kappa = 3$, model C is $h = 1$ and $\kappa = 5$.

TABLE II *N*-point correlation amplitudes and errors

N	3	4	5	6	7	8
q_N	1.59	2.50	2.61	2.03	1.98	1.72
Δq_N	0.030	0.205	0.507	0.724	0.835	0.779
$R_N(A)$	1.05	1.11	1.18	1.25	1.34	1.44
$R_N(B)$	1.10	1.28	1.54	1.91	2.42	3.12
$R_N(C)$	1.15	1.44	1.90	2.58	3.54	4.89

In conclusion, we calculated the angular *N*-point amplitudes from the Lick catalog, up to 8-th order. In future catalogs with better calibrations one can probably go further, using our method. Due to the uncertainty of the luminosity function it is hard to have a strong statement on the spatial hierarchy, although the spatial coefficients are close to unity, and decreasing slowly.

The authors would like to acknowledge a useful conversation with Jim Peebles. This research has been supported by the NSF in the US, and by OTKA in Hungary.

REFERENCES

Abramowitz, M., and Stegun, I. 1964, *Mathematical Functions*, Dover.
Balian, R., and Schaeffer, R. 1989, *Astr. Ap.*, **220**, 1.
Bouchet, F., and Lachieze-Rey, M. 1986, *Ap. J. (Letters)*, **302**, L37.
Davis, M., and Peebles, P. J. E. 1977, *Ap. J. Suppl.*, **34**, 425.
Fall, S. M. 1979, *Rev. Mod. Phys.*, **51**, 21.
Fry, J. N. 1984, *Ap. J. (Letters)*, **277**, L5.
Fry, J. N. 1986, *Ap. J.*, **306**, 358.

Fry, J. N., and Peebles, P. J. E. 1978, *Ap. J.*, **221**, 19.
Fry, J. N., Giovanelli, R., Haynes, M. P., Melott, A. L., and Scherrer, R. J. 1989, *Ap. J.*, **340**, 11.
Groth, E. J., and Peebles, P. J. E. 1977, *Ap. J.*, **217**, 385.
Hamilton, A. J. S., and Gott, J. R. 1988, *Ap. J.*, **331**, 641.
Peebles, P. J. E. 1980, *The Large Scale Structure of the Universe*, (Princeton, Princeton University).
Peebles, P. J. E., and Groth, E. J. 1975, *Ap. J.*, **196**, 1.
Peebles, P. J. E., and Hauser, M. G. 1974, *Ap. J. Suppl.*, **28**, 19.
Press, W. H., Flannary, B. P., Teukolsky, S. A., and Vetterling, V. T. 1986, *Numerical Recipes*, (Cambridge, Cambridge University).
Schaeffer, R. 1984, *Astr. Ap. Lett.*, **134**, L15.
Sharp, N. A. 1981, *M.N.R.A.S.*, **191**, 857.
Sharp, N. A., Bonometto, S. A., and Lucchin, F. 1984, *Astr. Ap.*, **130**, 79.
Tóth, G., Hollósi, J., and Szalay, A. S. 1989, *Ap. J.*, **344**, 75.
White, S. D. M. 1979, *M.N.R.A.S.*, **186**, 145.

DISCUSSION

A. Dekel: What is the physical basis for the hierarchical model?

A. Szalay: The stable closure of the BBKGY hierarchy equations implies (Davis and Peebles 1979) functional form; it is indeed physical.

A. Dekel: Do I understand correctly that you have ignored projection contamination effects? If so, let's recalculate your ω_{11} and ω_{111} in our decontaminated Abell and ACO catalogs. My speculation is that projection tends to increase ω_{11} and ω_{111} because we find that the contamination is a two-cluster effect - most of the time it involves only one significant contaminator, but this is to be tested.

A. Szalay: This can easily be tested, and I am sure it will influence the results somewhat, although we have not used data within 1°, but mostly between 1° and 3° (by our optimal weighting scheme) where you didn't find a very large change, I think.

E. Bertschinger: Did I understand properly that *any* local biasing of a Gaussian field necessarily produces a cubic term in the reduced three-point function with coefficient Q^3 for a Q^2 coefficient of the quadratic terms? If so, this implies that biasing is not responsible for the hierarchical form of correlations and in particular cannot explain the enhanced clustering of Abell clusters. Rather, nonlinear evolution must be responsible for these effects. Does this diminish the role of bias?

A. Szalay: My present feeling is that this is precisely the case. It would be interesting to check in CDM N-body simulations, what Q would they give for the clusters (if any) and whether they possess a cubic term, and how does that change with time.

J. Primack: As I discussed this morning, it is possible to get very different large-angle correlations with different subsets of the Abell (north) catalog - which casts doubt on any such results. Also, as Olowin discussed, the large number of new clusters found by Corwin and Olowin in the region where their catalog overlapped Abell's suggests that Abell's catalog may be seriously incomplete, perhaps in a patchy way. It would therefore be very interesting to see what 3-point correlations result from different subsets, and in particular from the ACO (south) catalog. Did you check this?

A. Szalay: We did take subsets of the Abell 5+6 sample, and the 2-point correlation agreed fairly well. There was a difference of about a factor of two between north and south. Also, some differences appeared in the 3-point function, but testing different weighting schemes implied that this is a result of the large-scale gradients, which are different in the southern part, as I'm sure you will agree.

D. Lambas: Could you comment if it is a bad method to estimate ζ from $\langle N^2 \rangle$?

A. Szalay: One should do the $\langle N^2 \rangle$ in cells on a ring around an object, that is a measure of the 3-point function. One should just take out the Poisson terms carefully.

ON THE KINEMATICAL BEHAVIOR OF GALAXIES IN CLUSTERS

LAERTE SODRÉ, JR., HUGO V. CAPELATO, AND JOÃO E. STEINER
Instituto de Pesquisas Espaciais, Departmento de Astrofísica
Caixa Postal 515, 12201 São José dos Campos (SP), Brazil

ALAIN MAZURE
Université des Sciences et Techniques du Languedoc, C.N.R.S. and
Laboratoire d'Astronomie, 34060 Montpellier Cedex, France

ABSTRACT From an analysis of a sample of 15 galaxy clusters we have found a highly significant trend for spiral and irregular galaxies to have greater velocity dispersion than the ellipticals and lenticulars. This is interpreted as evidence that late type galaxies are infalling towards the virialized cluster cores.

INTRODUCTION

The possibility that spiral galaxies in clusters may display a kinematical behavior different from ellipticals and lenticulars was first discussed by Moss and Dickens (1977). They studied a sample of five clusters and concluded that in four of them the velocity dispersions of the spiral galaxies were in positive excess relative to the early-type galaxies. Although these excesses were not statistically significant, they interpreted this trend as evidence for a shell-like structure infalling over the cores of the clusters.

Models describing the infall over the Virgo Cluster of the Virgo South Extension, a group of late-type galaxies, were discussed by Tully and Shaya (1984). These authors suggest that the spirals now observed in the Virgo core arrived there very recently (<4 Gyr) as a result of the merger of a large cloud of galaxies. Using a sample of 327 galaxies from the CfA Redshift Survey, Huchra (1985) was able to show that the velocity dispersion of the Virgo spirals is a factor 1.4 times that of the ellipticals. He notes that whereas the elliptical galaxies tend to concentrate around M87, the spirals are distributed in several condensations avoiding the core, suggesting a scenario in which the spiral galaxies are still infalling onto the virialized core of the cluster. This is essentially the same conclusion that Binggeli, Tammann, and Sandage (1987) reach in their detailed study of Virgo's morphological and kinematical structure.

Tammann and Binggeli (1988) have discussed a sample of seven nearby clusters, and they also conclude that spirals are late cluster arrivals which have not yet had much dynamical interaction with the older early-type cluster population.

We have studied the kinematical behavior of galaxies of various morphological types using a sample of 15 galaxy clusters from Dressler and Shectman (1988). The details of this study were published elsewhere (Sodré *et al.* 1989), and here we only summarize our main results.

RESULTS

For each cluster of the Dressler and Shectman (1988) sample we have calculated the total mean heliocentric velocity and the total velocity dispersion using the iterative procedure of Yahil and Vidal (1977) and retaining as members of the clusters only those galaxies whose radial velocities were within 3σ of the mean. Among the 15 clusters of the sample, 12 (80%) have $\sigma_{S+I} > \sigma_{E+S0}$, but the individually observed differences are far too small. It is possible, however, to ascertain the statistical significance of this trend by studying the distribution of normalized quantities, such as the ratio of the velocity dispersion of a morphological group to the total velocity dispersion of the cluster. Applying the Kolmogorov-Smirnov test we find a probability of less than 0.05% for the null hypothesis that both the σ_{S+I}/σ_T and σ_{E+S0}/σ_T distributions are drawn from the same population. The median value of $\sigma_{S+I}/\sigma_{E+S0}$ is 1.14 for the clusters of the sample.

One effect which may significantly affect these results is the eventual contamination of this sample by field galaxies which have not been eliminated by the procedure of Yahil and Vidal (1977). The unwanted inclusion of field galaxies in the sample constitutes an important bias due to its high content of spirals. We have tested this possibility by excluding all the galaxies outside a circular region of radius equal to the median radius of the distribution of the (S+I) galaxies and centered on the barycenter of the cluster. This procedure reduces the relative number of field galaxies eventually counted as a cluster galaxy. The distributions of velocity dispersion were analyzed as described before, and no significant changes in the results were observed.

We also tried to see if kinematical differences exist between the elliptical and lenticular galaxies as well. Of the 15 clusters, 11 have $\sigma_{S0} > \sigma_E$, with a median value for all clusters of $\sigma_{S0}/\sigma_E = 1.04$. Comparing the distribution of $\sigma_{S0}/\sigma_{E+S0}$ to that of σ_E/σ_{E+S0}, we find a probability of 18% under the K-S test that these two distributions are drawn from the same population. We conclude that the late-type galaxies in clusters have velocity dispersions which are higher than that of the early-type galaxies by a median factor of 1.14. We have shown that this result is highly significant, with a probability of less than 0.05% that it results from fluctuations in our sample of clusters. Lenticular galaxies in clusters also show a tendency to have higher velocity dispersion than the ellipticals, although this result is not statistically significant.

DISCUSSION

These results obviously raise questions concerning the origin and global dynamics of late-type galaxies in clusters. It is well known that late-type galaxies tend to have broader spatial distributions than the elliptical and lenticulars, which tend to concentrate towards the core of the clusters. This is related to the correlation between the frequency of morphological types and the local density of galaxies in clusters (Dressler 1980). If one supposes that the elliptical and lenticular galaxies in a cluster form an already virialized system ($T/W = 0.5$) and that the spirals are subjected to the same gravitational potential as the virialized component, then $(T/W)_{S+I} \approx 0.7$. This suggests that there is no appreciable dynamical mixing of the spiral and irregular galaxies with the virialized system.

It seems plausible that the distinct kinematical behavior of cluster spiral and irregular galaxies is due to the fact that these galaxies have been accreted by the cluster more recently, after the collapse and violent relaxation of the initial galaxy population which now constitute the cluster core.

In fact, galaxy clusters cannot be considered isolated systems. As they are immersed inside greater structures - the superclusters - a relaxed cluster can capture several groups of galaxies during its life. While these groups do not dissolve, they should be observed like a substructure inside the cluster. This scenario is consistent with the observations of the large scale peculiar velocity fields that are the subject of this Workshop.

The process of accretion of field galaxies by the virialized cluster core may last for the entire cluster life as long as it corresponds to the collapse of less dense bound shells with characteristic turn-around times longer than that of the core of the cluster itself (Gunn and Gott 1972).

What is the fate of these galaxies that are plunging towards the central regions of the clusters? The answer to this question is one of the most controversial of contemporary astronomy. As suggested by Tully and Shaya (1984), the study of galactic populations in distant galaxy clusters allows a natural approach to this question. If late-type galaxies fall continuously into the cluster and if they retain their morphological type, the morphological segregation that we observe today must have been yet stronger in the past. If, on the contrary, they are transmuted into lenticulars or ellipticals, then the relation between morphology and local density may not have substantially changed. Finally, we point out that velocity dispersions of elliptical and lenticular galaxies yield a more reliable estimate of the velocity dispersion of the virialized component than a sample that also includes late-type galaxies. The virial mass of a cluster is proportional to $\sigma^2 R$. The mean radius for the spiral distribution is greater than that for the early-type galaxies, and our analysis shows that spirals have velocity dispersions in excess of that of the elliptical and lenticular galaxies. Thus, the masses of rich clusters obtained by the application of the virial theorem to a sample that includes galaxies of all morphological types are slightly overestimated.

ACKNOWLEDGEMENTS

This work has been partially supported by a CNPq-CNRS protocol. Laerte Sodré acknowledges a fellowship from FAPESP (86/2600-6).

REFERENCES

Binggeli, B., Tammann, G. A., and Sandage, A. 1987, *A.J.*, **94**, 251.

Dressler, A. 1980, *Ap. J*, **236**, 351.

Dressler, A., and Shectman, S. A. 1988, *A.J.*, **95**, 284.

Gunn, J. E., and Gott, J. R. 1972, *Ap. J*, **176**, 1.

Huchra, J. P. 1985, in *The Virgo Cluster of Galaxies*, ed. O.-G. Richter, and B. Binggeli, (Garching: European Southern Observatory), p.181.

Moss, C., and Dickens, R. J. 1977, *M.N.R.A.S.*, **178**, 701.

Sodré, L., Capelato, H. V., Steiner, J. E., and Mazure, A., 1989, *A.J.*, **97**, 1279.

Tammann, G. A., and Binggeli, B. 1988, in *High Redshift and Primeval Galaxies*, ed. J. Bergeron, D. Kunth, B. RoccaVolmerange and J. Tran Thanh Van, (Paris: Editions Frontieres), p. 149.

Tully, R. B., and Shaya, E. J. 1984, *Ap. J*, **281**, 31.

Yahil, A., and Vidal, N. V. 1977, *Ap. J*, **214**, 347.

DISCUSSION

D. Burstein: Could not a foreground/background distribution of spiral galaxies (*i.e.* spirals not occupying the cluster core) also lead to larger velocity dispersion? Spirals are seen in the Local Supercluster distributed over ~2000 km s^{-1} in distance, ellipticals, and S0s are concentrated in the cores of Virgo and Ursa Major and a few isolated groups.

Capelato: The fact that the spiral-galaxy density profiles always peak towards the core of these clusters seems to be evidence that spirals are indeed dynamically bound to clusters. The velocity distributions of spirals were found to be consistent with normal distributions under various statistical tests. We have anyway tried to avoid contamination by field or supercluster galaxies by discarding from our analysis the galaxies which were out from the median distance of the spiral's spatial distribution.

REDSHIFTS OF A THOUSAND ABELL CLUSTERS

HEINZ ANDERNACH
Instituto de Pesquisas Espaciais, Departmento de Astrofísica
Caixa Postal 515, 12201 São José dos Campos (SP), Brazil

ABSTRACT The all-sky ACO catalog of 4073 rich clusters of galaxies and 1175 southern poor or distant S-clusters has been searched for published redshifts. Data for 1059 of them were found and classified into various quality classes, *e.g.* to reduce the problem of foreground contamination of redshifts. Taking the ACO selection criteria for redshifts, a total of 992 entries remain, 21% more than ACO. Redshifts for rich clusters are now virtually complete out to z = 0.05 in the north and 0.04 in the south. In the north the magnitude-redshift (m_{10} - z) relation agrees with that of Kalinkov *et al.* For the southern rich clusters minor adjustments to the m_{10} - z relation of ACO are suggested, while for the S-clusters the redshifts are ~30% lower than estimated.

INTRODUCTION

The recent completion of the extension of Abell's (1958) catalog of northern rich clusters of galaxies to cover the entire sky (Abell, Corwin, and Olowin 1989, hereafter ACO; Olowin this volume) stimulated us to merge and update their available redshifts. An appreciable number of redshifts can be obtained from various publications apparently unrelated to rich clusters. We describe such a literature search and the necessary classification of encountered redshifts. We then analyze the completeness of the present redshift data as function of distance, and possible adjustments to the magnitude-redshift relation. Some by-products of our analysis are mentioned in the final section.

SEARCH FOR PUBLISHED REDSHIFT DATA

The Finder List of Clusters

The ACO catalog is comprised of four different lists of clusters: the 'old' northern clusters, the improved sky-survey data for the latter clusters in the overlap zone ($-17° < \delta < -27°$), the southern rich clusters, and finally the S-clusters, which are too poor or too distant to fulfil the Abell criteria. First we used these lists to search for redshift data from previously-published compilations for northern clusters (Struble and Rood 1987a) and those with southern entries (*e.g.* Schmidt 1986) which have been identified here with new ACO clusters. A single list was prepared of all 5250 clusters, sorted in R.A. (1950). For the clusters in the overlap zone, only the new southern data were used. For each cluster the Abell radius R_A (= 3 h^{-1} Mpc with H_0 = 50 h km s^{-1} Mpc^{-1}) was computed either from the measured redshift or the estimated one according to previously published best fits to the relation between the magnitude m_{10} of the tenth brightest cluster galaxy and the cluster redshift. For the clusters with m_{10} measured from the southern sky survey (m_{10}S) the relation (11) of ACO was used, while for the rest (*i.e.* the clusters north of δ = $-17°$) the m_{10}N - z relations numbers 10, 12, and 13 of Kalinkov *et al.* (1985) were used in their respective magnitude ranges. A value of $q_0 = 0.1$ and no corrections to the m_{10} - z relations of the type proposed by Sarazin *et al.* (1982) were assumed at this stage. The resulting master catalog of ACO clusters with positions, redshifts, and Abell radii served as a finder list for the literature survey.

The Literature Search

The July 1988 diskette version of the catalog of galaxy redshifts south of $\delta = -17°$ (Fairall and Jones 1988) with 6229 entries and 5565 redshifts served as the primary source of new cluster redshifts. A search program was written that checked each galaxy for its location within one or more adjacent Abell circles, and computed the projected distance ratio (in units of R_A) and its logarithmic redshift offset $\Delta z \equiv \log (z_{gal} / z_{cl})$ from the estimated value z_{cl}. To suppress redundant information, only the overlaps with clusters lacking redshifts and cases of $|\Delta z| > 0.3$ for the clusters with measured redshifts were inspected. A total of 88 new cluster redshifts resulted.

Then the list of references given by Fairall and Jones (1988) was searched for galaxies north of $\delta = -17°$. Finally, recent papers on galaxy redshifts were surveyed, applying the above procedure 'by hand', occasionally cross-correlating *three* catalogs (*e.g.* for the papers quoting only the UGC/NGC/Mkn designation of the galaxies and no positions). The major sources of redshifts (sorted by the number of clusters) are listed in Table I, giving a total of 1059 cluster redshifts. Contrary to ACO, this includes discordant redshifts (*i.e.* those far from the m_{10} estimate) which have been flagged as described below.

TABLE I Sources of Cluster Redshifts

Reference	N_z
Abell, Corwin, and Olowin 1989	760
Fairall and Jones 1988	88
Struble and Rood 1987a	34
Schmidt 1986	30
Rhee and Katgert 1988	29
Colless and Hewett 1987	14
da Costa *et al.* 1989	13
Dressler and Shectman 1988	12
Owen *et al.* 1988	11
32 other papers	67

The Catalog and the Redshift Classification

The full catalog will be published elsewhere and contains cluster name, center position, richness, distance, Bautz-Morgan classification, m_{10} (N or S), redshift and source, and one of the following quality classifications:

' ' z measured is within a factor of 2 of the m_{10} estimate,

':' z measured is within a factor of 2 to 4 of the m_{10} estimate,

'*' z measured is outside the above range *i.e.* probably due to foreground or background objects,

'f' same as '*', but for a cluster that has foreground galaxies projected onto it, according to notes by Struble and Rood (1987b) and/or ACO,

'!' listed redshift is the average of very discordant ones, indicating possible line-of-sight projections,

'?' listed redshift is from a single peripheral object, concordant within a factor of 2 of the estimate,

and finally the Abell radius, using $q_0 = 0.1$.

Table II shows the distribution of redshift classes over the different cluster samples (N = northern clusters at $\delta > -17°$; N_{ov} = northern clusters in the overlap zone; R = southern rich, and S = southern poor/distant clusters). The number of clusters with concordant redshifts (992) has been increased by 170 (21%) as compared to ACO.

TABLE II Numbers of Redshifts by Class

z-Class	N	N_{ov}	R	S	Total
' '	578	56	117	142	893
':'	30	1	32	36	99
'*'	6	0	11	7	24
'f'	17	3	2	2	24
'!'	1	1	3	5	10
'?'	2	0	2	5	9
All	634	61	167	197	1059

COMPLETENESS OF CURRENT REDSHIFT DATA

The total fraction of clusters without measured redshifts (76% of the northern, 89% of the southern rich clusters, and 81% of the entire ACO catalog) still appears overwhelming. More edifying, however, is a look at the redshift completeness as function of distance, as displayed in Figure 1. Redshifts classified as 'f' were counted as estimates. Redshifts are

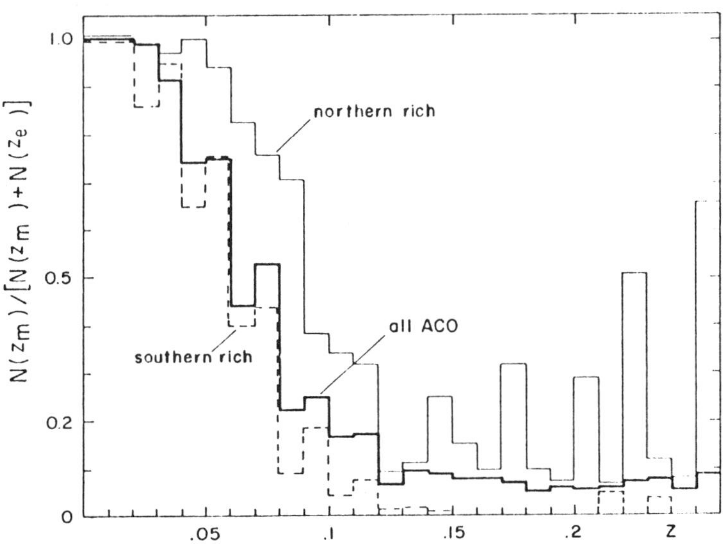

Fig. 1. Fraction of clusters with measured redshifts as a function of distance for different cluster subsamples.

available for 99% (82 out of 83) of the northern (Abell 1958) sample out to $z = 0.05$ and for 93% (43 out of 46) of the southern rich ACO clusters out to $z = 0.04$. The faster decline in available redshifts for the southern clusters is largely due to the delay of 30 years between publication of the northern and southern catalogs, which is the main reason also for the virtual absence of redshift measurements of distant ($z \geq 0.15$) southern clusters.

THE MAGNITUDE-REDSHIFT RELATION

As pointed out by ACO the magnitude scales for the tenth brightest galaxies are substantially different for the northern and southern sky surveys. Since these authors recommend the use of the southern magnitudes wherever available, the clusters were split into three subgroups to analyze the m_{10} - z relation:

-the clusters north of $-17°$ declination (using m_{10}N)

-the southern rich clusters including the 'northern' ones in the overlap zone (using m_{10}S)

-the poor/distant S-clusters (again using m_{10}S).

Instead of trying an improvement of the photometric redshift calibration, only the histogram of logarithmic deviations Δz from the previously known calibration curves has been constructed to look for systematic trends. For the northern clusters this histogram looks normally distributed around $\Delta z = 0$ with little excess for $\Delta z < -0.3$ expected due to contamination by foreground redshifts. The data are thus in good agreement with the calibration of Kalinkov *et al.* (1985). The dispersion in Δz is $\sigma = 0.12$, only slightly higher than the value obtained with the sophistications applied by Sarazin *et al.* (1982).

However, the distributions for the southern rich and southern poor/distant S-clusters (see Figure 2) show marked deviations from the redshift calibration proposed by ACO. For the rich clusters the histogram is again basically a normal distribution around zero with dispersion $\sigma = 0.12$. The peak itself, however, is shifted to $\Delta z \approx 0.05$. An inspection of the m_{10}-z plot reveals that this is due to the ACO calibration being too low and too flat for $14 < m_{10}S < 16$. Considering that the ACO calibration was based on only 30% of the

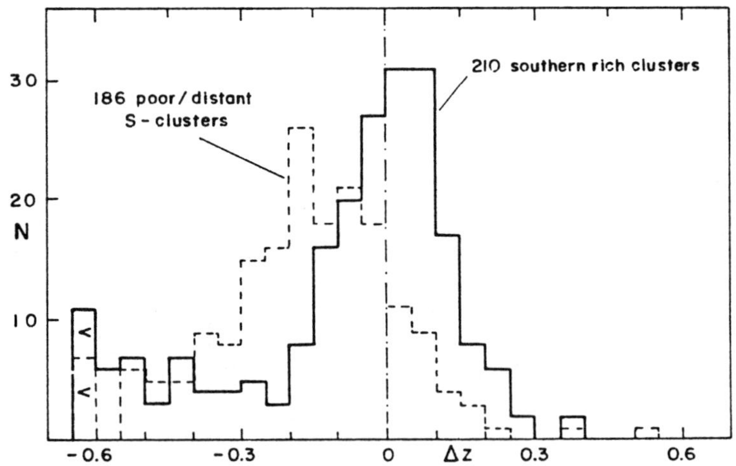

Fig. 2 Histogram of logarithmic deviations of measured redshifts (excluding types *,!,?,f) from ACO's m_{10}S calibration for two different subsamples of clusters.

present data, some adjustments were to be expected. More surprising is that the measured redshifts for the S-clusters are on average 30% lower than the ACO estimate (assumed to be the same as for the rich clusters), without showing significantly larger dispersion ($\sigma \sim 0.13$). It is yet unclear whether this discrepancy is due to an excess foreground contamination of the redshifts or a systematic overestimate of m_{10}S of the clusters caused by background contamination. A hint for the latter is the fact that 98% of the S-clusters of Figure 2 have very low corrected galaxy counts of $C \leq 30$ and thus belong to the poorest of the sample.

At $\Delta z < -0.3$ both distributions show a much more pronounced tail than the northern clusters. This is very likely due to the use of a 'complete' redshift catalog (Fairall and Jones 1988) only for the southern clusters, naturally yielding more foreground objects in the south. Moreover, Table II shows that 20 clusters (= 77%) out of 26 northern ones with $|\Delta z| > 0.6$ are noted by Struble and Rood (1987b) to have background galaxies superimposed, while this is true for only 4 (= 18%) out of 22 southern clusters from the ACO notes. The latter authors thus seem to be much more cautious in claiming objects as foreground (*e.g.*, *cf.* the notes on A14 in both papers). Extreme care is thus needed to establish improved m_{10} - z relations based on the present data. The fit procedure applied by Sarazin *et al.* (1982) might possibly be improved by weighting z with the number of galaxies contributing to each z.

MISCELLANEOUS

The analysis of the ACO catalog yielded as a by-product certain features presumably yet unmentioned in the literature.

While the original Abell catalog does not have a single entry for $|b| < 10°$, the ACO extension lists one rich cluster (A3627) and two poor ones (S610, S642) in this band.

The measured or estimated Abell radii were used to sum up the solid angle Ω subtended by all Abell circles in the different cluster samples. Table III lists Ω (square degrees) for each sample, and the fraction (f_{surv}, listed as a percent) of sky surveyed to find the clusters (subtracting the zone $|b| < 10°$), as well as the fractional area $f_A = \Omega/(4\pi f_{surv})$ subtended by all clusters of a subsample. N is the surface density of the respective cluster subsample, per steradian.

These numbers must be regarded as upper limits due to (a) the overestimate of Abell radii for foreground-contaminated redshifts and (b) the non-correction for occasional overlap of clusters. There is, however, a significant difference in f_A and N between the northern and southern rich samples, probably due to both the deficiency of the northern catalog in its southern-most zones (see ACO) and the superior quality of the southern sky survey plates.

TABLE III Total Area of Abell Circles

Subsample	Ω	f_{surv}	f_A	N
northern rich	834	66.9	3.0	323
southern rich	575	28.6	4.9	456
southern poor	1005	28.6	8.5	327
all	2414	82.6	7.1	506

ACKNOWLEDGEMENTS

Drs. H. J. Corwin, Jr., and A. P. Fairall kindly provided their diskette catalogs of rich clusters and southern galaxy redshifts prior to publication. I thank Dr. H.V . Capelato for letting me share the time for his presentation during the conference and Dr. J. E. Steiner for a critical reading of the text. I am grateful to Dr. D. Nunes for generous help in computer access and use.

REFERENCES

Abell, G. O. 1958, *Ap. J. Suppl.*, **3**, 211.
Abell, G. O., Corwin, H. G., and Olowin, R. P. 1989, *Ap. J. Suppl.*, **70**, 1.
Ciardullo, R., Ford, H., and Harms, R. 1985, *Ap. J.*, **293**, 69.
Colless, M., and Hewett, P. 1987, *M.N.R.A.S.*, **224**, 453
da Costa L. N., Pellegrini, P. S., Willmer, C., de Carvalho, R., and Maia, M. A. G. 1989, *A.J.*, **97**, 315.
Dressler, A., and Shectman, S. A. 1988, *A.J.*, **95**, 985.
Fairall, A. P., and Jones, A. 1988, *Publ. Dept. Astr. Univ. Cape Town* no. 10.
Kalinkov, M., Stavrev, K., and Kuneva, I. 1985, *Astr. Nachr.*, **306**, 283.
Owen, F. N., White, R. A., and Thronson, H. A. 1988, *A.J.*, **95**, 1.
Rhee, G. F. R. N., and Katgert, P. 1988, *Astr. Ap. Suppl.*, **72**, 243.
Sarazin, C. L., Rood, H. J., and Struble, M. F. 1982, *Astr. Ap.*, **108**, L7.
Schmidt, K. -H. 1986, *Astr. Nachr.*, **307**, 69.
Struble, M. F., and Rood, H. J. 1987a, *Ap. J. Suppl.*, **63**, 543.
Struble, M. F., and Rood, H. J. 1987b, *Ap. J. Suppl.*, **63**, 555

DISCUSSION

J. Primack: A comment first: Thanks! I'm sure that your compilation will be helpful to many of us. How can I get a copy?

H. Andernach: A few paper copies of the preliminary version are supplied at this meeting. A version of 5.25-in diskette is available on request. You are welcome to have the spare copy on diskette that I have with me.

REAL AND IMAGINARY CLUSTERS

SIMON D. M. WHITE
Steward Observatory, University of Arizona, Tucson, AZ 85721

ABSTRACT Galaxy clusters are the largest known quasi-equilibrium structures. If they formed by gravitational amplification of linear perturbations, their abundance can be used to determine the amplitude of the primordial fluctuation spectrum. In addition their mass-to-light ratios are an important measure of the extent to which galaxy formation may be biased with respect to the underlying distribution of mass. The observed abundance of clusters is difficult to confront with theory. Small systematic biases in the observables used to define 'rich' clusters (*i.e.* luminosity, velocity dispersion, or X-ray temperature) can translate into large variations in predicted abundance. Numerical simulations illustrate how such biases can enter luminosity and velocity-dispersion measurements. In the interpretation given here, cluster abundances seem well fit by the standard Cold Dark Matter model with a moderately large bias (b ~ 2). However, the uncertainties are large. Within this model, cluster mass-to-light ratios require a large bias.

INTRODUCTION

Galaxy clusters were the first systems in which an apparent need for large amounts of dark matter was identified (Zwicky 1933). They are also the largest known systems in equilibrium under the action of gravity, and so the largest objects for which a mass can be inferred directly from Newton's laws. (Additional assumptions are required before a mass can be derived from the observed kinematics of systems like the Local Supercluster.) Rich and well-observed clusters like Coma may contain about $2 \times 10^{15}\ M_{\odot}$ within the virialized region ($\sim$ 3 Mpc). (Throughout this article I will adopt H_0 = 50 km s^{-1} Mpc.) However, this number is quite uncertain because of the unknown relationship between the distributions of dark matter and of galaxies; it is an underestimate if the galaxies are more concentrated than the mass, as suggested by theories of biased galaxy formation (The and White 1986; Merritt 1987). A randomly placed sphere contains this much mass if its diameter is 2360 $\Omega^{-1/3}$ km s^{-1}. Thus, in gravitational instability theories, clusters arise from initial fluctuations with wavelength of order 5000 km s^{-1}. Cluster properties constrain the power spectrum of linear density fluctuations on relatively large scales.

In order to investigate such constraints in detail it is necessary to model the dynamics which link the observable properties of rich clusters to those of the small-amplitude fluctuations from which they form. Many investigators have dealt with this problem by identifying clusters with sufficiently high peaks of a suitably filtered version of the linear density field (Kaiser 1984; Peacock and Heavens 1985; Bardeen *et al.* 1986; Peebles *et al.* 1989) or by using the analytic theory developed by Press and Schechter (1974) (Narayan and White 1988; Evrard 1989; Cole and Kaiser 1989). There has been relatively little work to test the validity of these approaches. Frenk *et al.* (1988) found that, with a careful choice of parameters, the high peak model did quite well at locating galaxy halos in a Cold Dark Matter (CDM) universe. The Press-Schechter theory can also give moderately accurate results for a wide range of initial conditions (Efstathiou *et al.* 1988). The former fit the galaxy distribution and the latter fit the halo distribution in the experiments of Carlberg and Couchman (1989). Unfortunately, the cluster abundance predicted by these theories depends exponentially on uncertain parameters (filter form, smoothing length, peak threshold). Such predictions are therefore of little use without more detailed dynamical study.

This sensitivity is a direct consequence of working with 'rare events' - objects in the tail of an exponential distribution. It is exacerbated by uncertainties in converting from theoretically accessible quantities (mass, CDM velocity dispersion) to observables (luminosity, galaxy velocity dispersion, X-ray temperature). This conversion is uncertain because all cluster samples to date have been identified as apparent number density enhancements on photographic plates. I argue below that this 2-dimensional selection procedure systematically biases the distribution of cluster properties. Elsewhere in this volume J. Primack and W. Sutherland argue that it also strongly biases their apparent clustering. In addition, the predicted properties of clusters will depend on the theoretical model which is used to relate galaxies and mass. A wide variety of such models are possible (*e.g.* Dekel and Rees 1987). Only the 'natural bias' pointed out by White *et al.* (1987a) and further studied by Carlberg and Couchman (1989) has a clear physical basis and has been investigated in any detail. However, its predictions for rich clusters are still uncertain. The results presented in this paper are based on the less well motivated 'high peak model' of Davis *et al.* (1985; DEFW) and White *et al.* (1987b; WFDE). Agreements and disagreements found in comparing such results to observation must therefore be interpreted as confirming or rejecting CDM plus a specific bias model, rather than CDM itself.

The remainder of this paper presents results for the abundance of clusters in biased CDM universes as a function of the bias strength and also as a function of cluster luminosity and velocity dispersion. Clusters are identified using mass and luminosity criteria in 3 dimensions, and also using 2-dimensional criteria analogous to those of Abell (1958). There are substantial discrepancies between samples defined in these various ways. Although substantial uncertainties remain, I find reasonable fits to observation for models with bias parameters (defined here as the inverse of the *rms* relative mass fluctuation in a sphere of radius 16 Mpc) in the range b $\sim$ 2. Much of this work is discussed in Frenk (1989) and Frenk *et al.* (1990).

SIMULATION DATA

The results of this paper are based on an ensemble of 5 *N*-body simulations evolved from different statistical realizations of initial conditions with the primordial power spectrum expected in a CDM universe. Each simulation represents a comoving cube of present size 360 Mpc containing 262,144 particles each of mass $1.2\times10^{13}\ M_\odot$. The simulations were performed with the P^3M code using a 128^3 mesh and softening parameter $\eta = 0.2$. The numerical techniques and the procedure for laying down initial conditions are described in detail by Efstathiou *et al.* (1985). The simulations were continued past the epoch which we have identified with the present in our previous work (DEFW; Frenk *et al.* 1985, 1988; White *et al.* 1987a, WFDE) and which corresponds to a power spectrum normalization given by b = 2.5. With the chosen initial amplitude for these simulations, this epoch occurs after expansion by a factor $a = 3.0$ from the initial conditions. Identifying a different epoch in the simulations with the present corresponds to adopting a normalization given by b = 2.5 (3.0/a). Results were analyzed for the values of b shown in the second column of Table I. Note that the final time chosen corresponds to a model in which the galaxies trace the mass. The power spectra and other global properties of our models at $a = 3.0$ are very similar to those of model E36 in WFDE.

In the discussion that follows we shall need to associate some particles in the simulations with 'galaxies' and to identify clumps of particles as clusters. To locate galaxies we implemented the 'high peak model' as described by WFDE. Briefly, we filter the large-scale density field and employ the analytic machinery developed by Bardeen *et al.* (1986) to calculate the non-linear function of the smooth large-scale field which gives the number of peaks above some threshold. The smoothing parameters r_s and r_b required for this prescription and the appropriate values for this application are discussed by WFDE. The expected number of peaks in the initial volume occupied by each particle is the 'peak number', which we take to be proportional to the total luminosity associated with that particle. The proportionality constant is set by requiring the mean luminosity density in the simulation to equal the observed value, 0.0019 $L_* \,\mathrm{Mpc}^{-3}$ with $L_* = 3.9\times10^{10}\, M_\odot$, given by Felten (1985). The threshold, ν, in units of σ_s, the *rms* density fluctuation of the smooth density field, is given in the third column of Table I. These values were chosen so that the luminosity-weighted (*i.e.* peak-peak) auto-correlation function is close to the observed galaxy auto-correlation function.

TABLE I Simulation parameters

a	b	ν
2.26	3.3	2.5
3.02	2.5	2.3
3.80	2.0	1.8
4.74	1.6	1.5
5.96	1.3	0.05
7.52	1.0	---

Possible centers of clusters were identified in the mass distribution using a 'friends-of-friends' algorithm to pick out high density groupings of particles (DEFW). The region within one Abell radius (*i.e.* 3 Mpc) of each center was then examined to see if it satisfied the necessary criteria to be considered a rich cluster (*i.e.* sufficiently high mass, luminosity, or velocity dispersion). In cases of overlapping clusters the lower mass object was eliminated. This is the procedure which WFDE used (with a luminosity threshold) to identify the cluster sample analyzed in their paper. Clusters of 'galaxies' were identified in a similar way in projected galaxy catalogs constructed as follows. The luminosity density field of the models was sampled at a frequency corresponding to the mean number density of the (intrinsically) faintest galaxies counted by Abell when defining rich clusters ($R > 0$). The resulting sample of galaxies was then projected along each of the principal axes of the computational cube to produce three 2-dimensional catalogs. Cluster richness was defined as in Abell (1958), using the galaxy count in excess of background within one Abell radius.

THE DISTRIBUTION OF VELOCITY DISPERSIONS

The potential wells of galaxy clusters are entirely dominated by their dark matter content. The velocity dispersion of a population of galaxies moving within such a cluster depends not at all on their abundance, and only weakly on the shape of their distribution. Thus the abundance of rich clusters as a function of velocity dispersion would seem to be almost independent of the details of galaxy formation and so a particularly robust constraint on the CDM model. The abundance of high velocity-dispersion systems is predicted to be a very sensitive function of the amplitude of the CDM spectrum, so one might hope that observations could fix this amplitude with little ambiguity. In their study of rich clusters WFDE found the predicted abundance of rich clusters to agree well with that found by Abell when cluster abundances were compared at fixed luminosity. However, at a given luminosity the velocity dispersions of their model clusters appeared significantly lower than those observed. A possible explanation for this discrepancy might be that the bias, $b = 2.5$, of WFDE's standard model is too large, and that a larger amplitude for the CDM power spectrum is, in fact, required (Kaiser and Lahav 1989). However, we will argue below that the discrepancy can be attributed to the fact that WFDE defined their clusters and measured their velocity dispersions in the 3-dimensional model data, whereas observers have only two spatial dimensions to work with.

An estimate of the observed velocity-dispersion distribution of rich clusters is given in Figure 1a. This histogram contains data for 78 clusters for which at least 10 members have measured redshifts. Of this sample, 49 clusters come from Struble and Rood (1987), 5 from J. Lucey (private communication), 14 from Colless and Hewitt (1987), and 10 from Teague (1987). A number of these clusters, particularly those from the last two sources, have 50 or more redshifts. A comparison of the Abell count distribution for this sample with that of the Abell catalog as a whole (Figure 1b) indicates that the sample is biased towards rich clusters. Clusters with Abell richness class of 2 or greater are over-represented by about a factor of two, while clusters of lower richness are under-represented by about 25%. While bearing this in mind, we will treat the distribution of Figure 1a as representative of Abell clusters with richness class 1 and greater. The errors this introduces are almost certainly small compared to those caused by the projection effects which we discuss below.

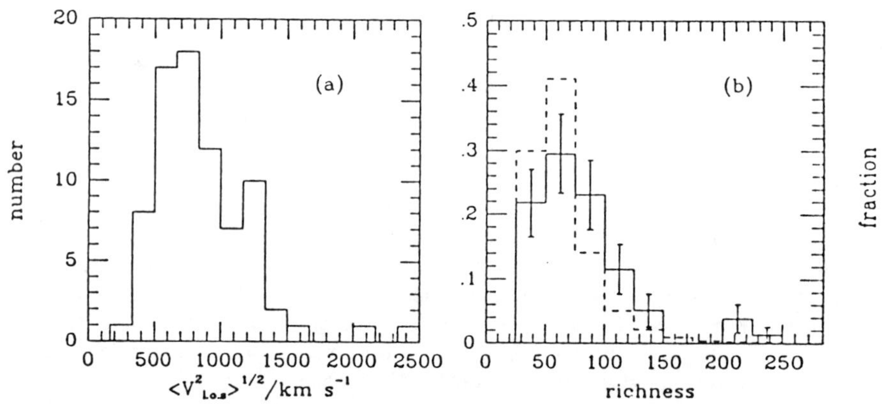

Fig. 1. (a) Distribution of velocity dispersions for our sample of Abell clusters with good velocity-dispersion measurements (see text). (b) Distribution of Abell richness for the entire Abell catalog (dashed lines) and for the sample used in (a) (solid line). The error bars show 1-σ Poisson errors.

The abundance of dark-matter clusters as a function of velocity dispersion is shown in Figure 2a for each of our model output times (corresponding to the values of b shown). The abundance scale is normalized to the observed abundance of Abell clusters of richness class greater than zero, 7.5×10^{-7} Mpc^{-3} (Bahcall and Soneira 1983). Note that the velocity

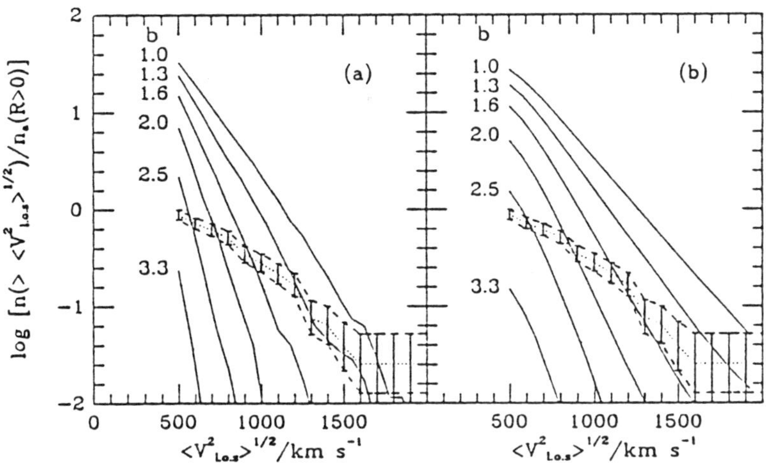

Fig. 2. (a) Cumulative distributions of line-of-sight velocity dispersions of galaxy clusters. The solid lines show the distributions for clusters identified in three dimensions in *N*-body simulations of a CDM universe with the values of the biasing parameter, b, shown. The abundances are normalized to 7.5×10^{-7} Mpc^{-3}, the observed mean density of Abell clusters of richness class $R > 0$. The dotted line shows the cumulative distribution of a sample of Abell clusters with good measurements of $\langle v^2_{los} \rangle^{1/2}$, selected as described in the text. The dashed lines enclose the 1σ range defined by the Poisson statistics of the counts. (b) The theoretical distributions of (a) convolved with a Gaussian of width 0.2 $\langle v^2_{los} \rangle^{1/2}$; the observed distribution is the same as in (a).

dispersions are defined as the 1-dimensional dispersion of the dark-matter particles within one Abell radius of the cluster center. These model predictions thus make no use whatsoever of the assumed distribution of galaxies in the model. They show the expected very steep dependence of predicted abundance both on velocity dispersion and on b. These curves are quite a different shape from the observed distribution derived from Figure 1, and no value of b gives acceptable agreement. In Figure 2b we broaden the predictions with a Gaussian of width 20% to model the effect of Poisson errors in the observed dispersions. This significantly changes the predictions but not nearly enough to get agreement with the observations. The high velocity-dispersion tail of the observed distribution in Figure 2 can be fit for b ~ 1.4, but one then needs to argue that more than 95% of dark-matter clumps with dispersions exceeding 500 km s^{-1} are insufficiently luminous to have Abell richness 1. A factor only slightly smaller than this is, in fact, predicted by our biasing models. (Compare the number of high velocity-dispersion clusters from Figure 2 with the number of high luminosity clusters in Table II below.) It is clearly quite misleading to compare the observed velocity-dispersion distribution to that predicted by CDM theory without accounting for the effective luminosity threshold of the observed sample. This coupling is unfortunate because it reintroduces the need for explicit modelling of the galaxy distribution in order to get predictions for comparison with observation. This dependence is lessened if one concentrates on the high-velocity tail of the distribution, but the observational data are then much more sensitive to the contamination problems which we discuss next.

Real clusters are identified on pictures of the sky and their velocity dispersions are estimated from redshift histograms for all objects projected on the cluster field. On the other hand, the model data for Figure 2 were obtained by identifying clusters as spatial density enhancements and measuring the true velocity-dispersion of material within one Abell radius. These differing procedures lead to significantly different results. In Table II we list the abundances of clusters found in our simulations using the 3-dimensional criterion of WFDE and the 2-dimensional criterion of 50 or more galaxies in excess of background projected within one Abell radius. The observed abundance of richness class 1 and greater clusters corresponds to about 35 objects in one of our simulation volumes. The number of clusters found using the two methods agrees reasonably well and increases with decreasing b. Within the uncertainties, only the largest and smallest values of b give abundances in significant disagreement with observation. Thus, with our biasing model the abundance of Abell clusters does not strongly constrain the amplitude of the CDM spectrum once the bias strength is set to match the galaxy correlation function.

TABLE II Abundance of clusters

b	$N_{2\text{-D}}$	$N_{3\text{-D}}$	f_{true}
3.3	18.1	12.6	0.41
2.5	37.2	32.6	0.55
2.0	35.2	30.2	0.52
1.6	50.1	49.8	0.63
1.3	52.7	49.4	0.65
1.0	65.2	68.8	0.77

The final column of Table II gives the fraction of the clusters identified in projection, which are actually associated with one of the rich systems found in 3-D. Given the near equality of the numbers in columns 2 and 3, this fraction is surprisingly small. What has happened is that many of the projected clusters are identified as rich because a foreground group is superposed on them and so they have a higher than average 'background'. This effect allows some of the more numerous clusters with true richness somewhat below threshold to be boosted above the critical Abell count. The statistics of this effect are complicated by the fact that the background counts are not Poisson distributed. In fact, it is easy to show that if the large-N tail of the distribution of galaxy counts in 3-D cells falls off faster than e^{-N}, then the large-N tail of the distribution of projected counts in cells will be dominated by superpositions of several clusters, rather than by single very rich clusters. Empirical data suggest that the real galaxy distribution is on the boundary of this regime. Superposition effects are often found to be significant, but only rarely to be dominant, when substantial amounts of redshift data are obtained in cluster fields (*e.g.* Dressler and Shectman 1988). The additional clusters found in projection because of this effect are almost exactly balanced by true 3-D clusters which are missed because their foreground contamination is unusually low; the mean background correction then puts their apparent Abell count below threshold.

Although the uncertainties introduced by 2-D selection do not seem to have any substantial systematic effects on the apparent abundances of clusters in these CDM models, the same is not true for their apparent velocity dispersions. Velocity dispersions can be estimated from redshift histograms of apparent cluster members following the membership assignment methods used by observers for real clusters. (We use the procedures of Colless (1987); basically the elimination of all objects more than 4000 km s^{-1} from the apparent centroid of the cluster, followed by a '3-sigma clipping' algorithm). We can also measure the true 1-D dispersion within one Abell radius for the largest clump associated with each projected cluster. Figure 3 is a scatter plot of these two dispersions for a typical 'projected' cluster catalog in our ensemble. Although for many objects the two dispersions agree reasonably well, there is also a substantial population for which the apparent dispersion significantly exceeds the true dispersion. Thus nearly half the clusters in the figure have apparent dispersions exceeding 1000 km s^{-1} (17 clusters out of 39), while only two systems have true dispersions (slightly) in excess of this value. Thus at 1000 km s^{-1} the projected dispersion data overestimate the abundance of high dispersion clusters by an order of magnitude.

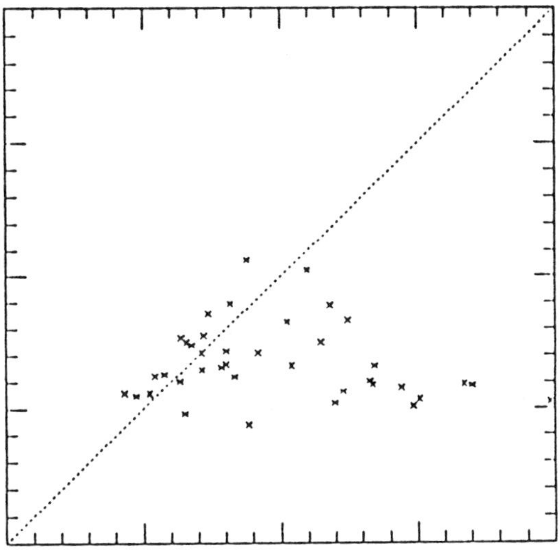

Fig. 3. Scatter plot of the apparent velocity dispersion (horizontally) of clusters selected from 2-D data against the true velocity dispersion of the largest associated real cluster (vertically). Tick marks are at 100 km s^{-1} intervals.

The reasons for this problem are most easily understood by examining in detail the simulation data for individual clusters. Figure 4a plots histograms of recession velocity for six typical clusters found in projection in our models; the bins are 500 km s^{-1} wide. Velocities for all galaxies projected within one Abell radius are used. In several cases, significant groups of galaxies are seen projected onto the cluster but at significantly different redshift. In several cases also the main cluster peak appears significantly asymmetric or to have some substructure, hinting that complex structure may be present. The first effect does not influence velocity-dispersion determinations, but the second can be a severe problem. This is clarified by Figure 4b which plots true distance histograms for the same galaxies in the same six clusters. While in one case the apparent structure of the redshift histogram turns out to be a statistical fluke, in several others there are spatially separated components contributing to the main cluster peak. This substructure usually takes the form of a few relatively small groups within 1000 to 3000 km s^{-1} of the main concentration. Quite similar structures are seen in the real data of Dressler and Shectman (1988); one of their 14 fields actually turned out to be a superposition of two clusters of comparable richness. A small satellite subgroup can substantially bias a velocity dispersion determination, and can be very

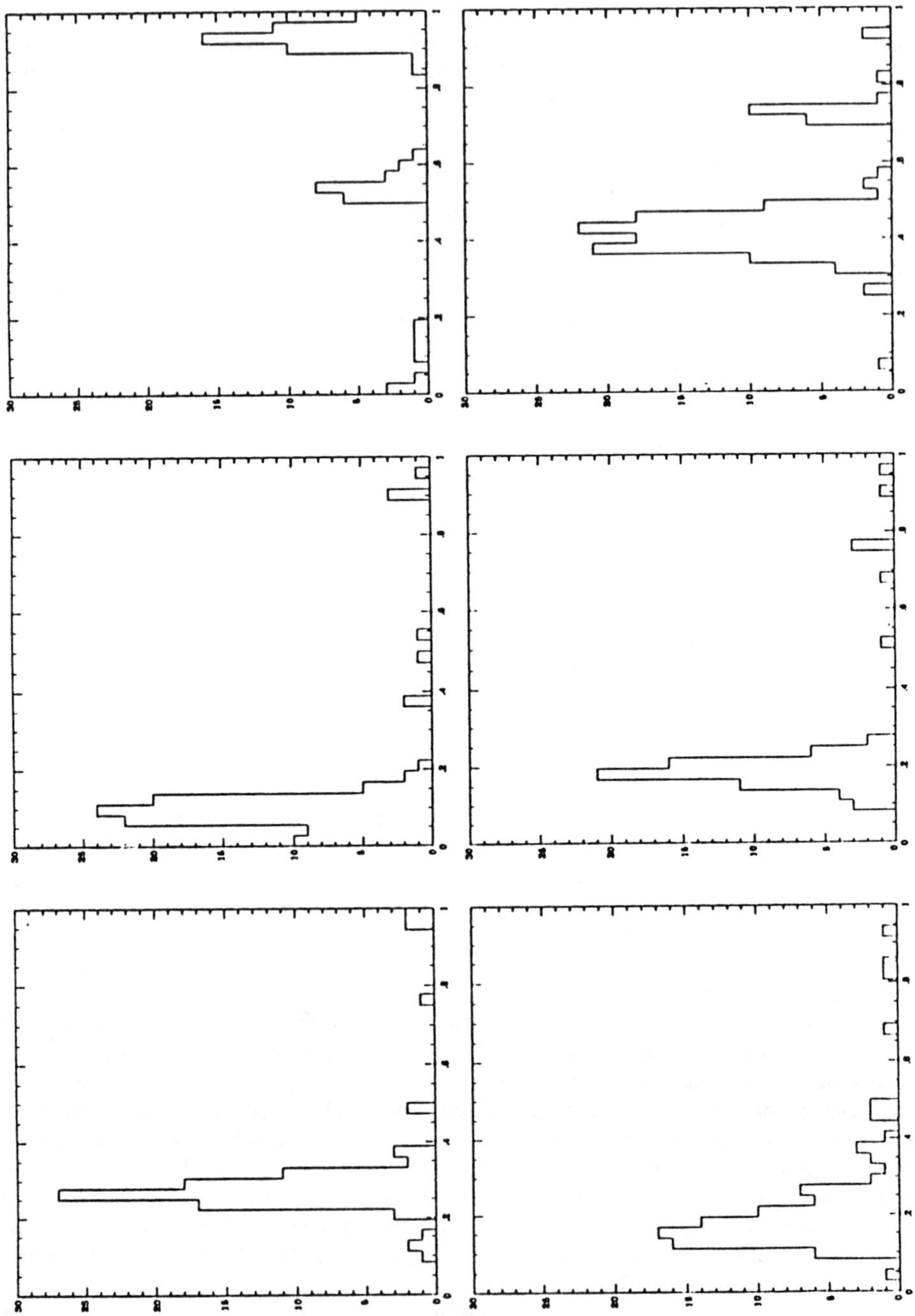

Fig. 4a. Histograms of observed velocity for the galaxies in 6 clusters identified using an 'Abell' criterion on 2-D data. The entire range of velocities is 18000 km s^{-1} and each individual bin is 500 km s^{-1} wide.

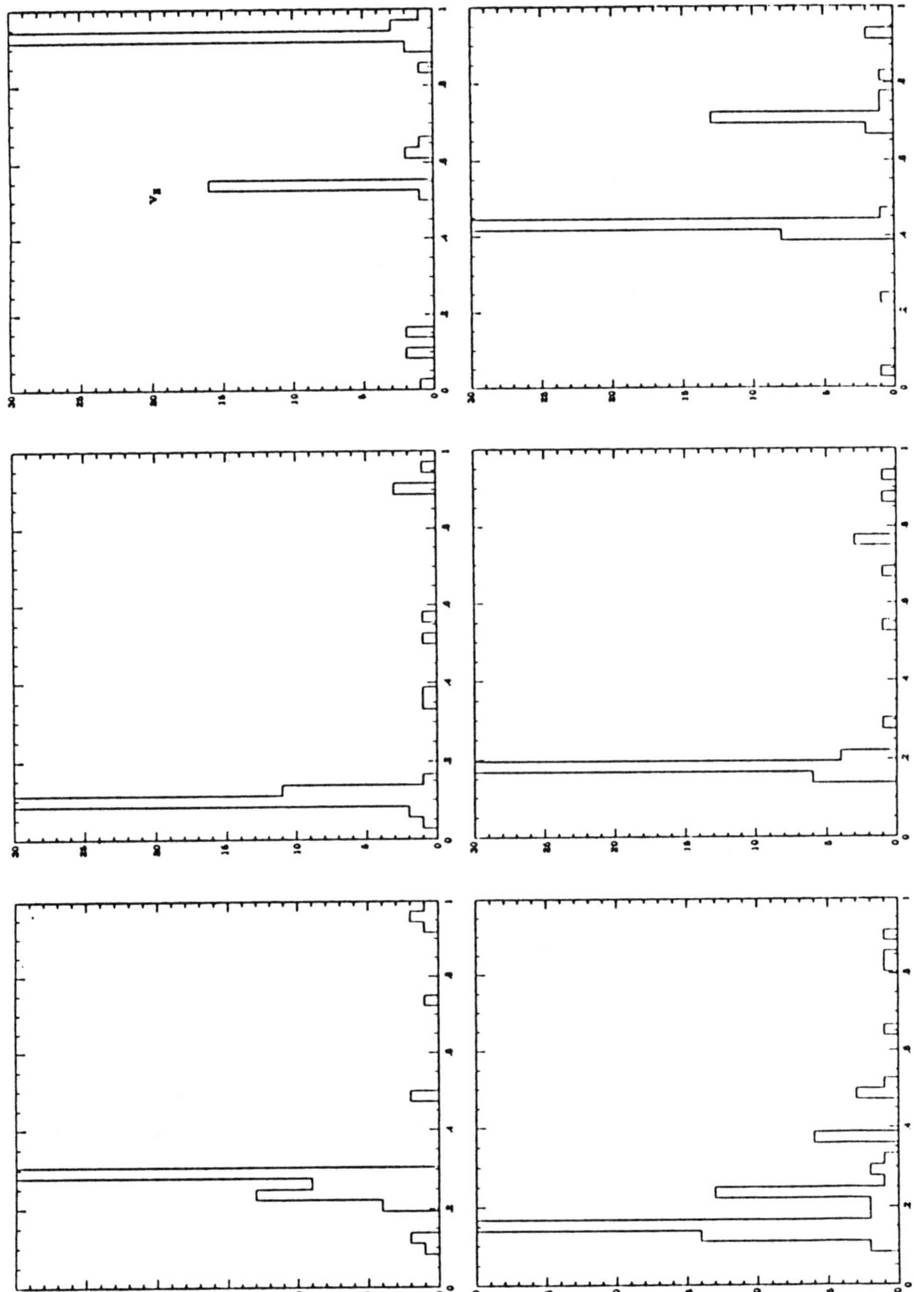

Fig. 4b. Histograms of true distance for the same 6 clusters.

hard to detect from the observationally accessible data, even with ~100 redshifts. Velocity dispersions for most clusters are, of course, determined from far fewer redshifts.

When 'observationally' determined velocity dispersions for 'observationally' selected clusters are used to estimate the abundance-dispersion relation, a very different picture emerges from that shown in Figure 2. The distributions of Figure 5a differ from those plotted earlier as a result of two major effects. For weakly biased models the abundance of low velocity-dispersion clusters is greatly reduced. Most such clusters are not luminous enough to satisfy Abell's criteria, and so are not included in the sample used to make the figure. In addition, the contamination effects just discussed introduce a significant high dispersion tail. The general shape of the distributions is now much closer to that observed, and we find reasonably good agreement for values of b around 2. In Figure 5b we compare the apparent dispersion distribution for b = 2 with the true dispersion distribution for the same clusters, and with the apparent distribution found if only galaxies within 1.5 Mpc are used to estimate the apparent dispersion. This reduces the contamination problem significantly, but is far from eliminating it.

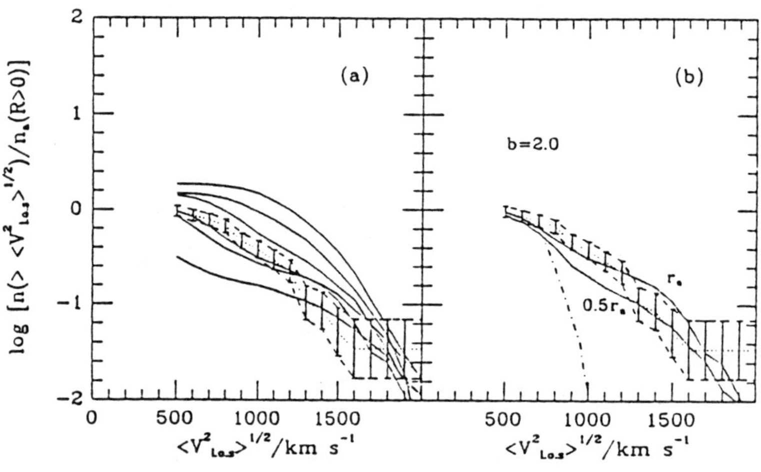

Fig. 5. (a) Cumulative distributions of line-of-sight velocity dispersions for clusters found in projection. From top to bottom the solid curves correspond to CDM models with values of the biasing parameter, b, of 1.0, 1.3, 1.6, 2.0, 2.5, and 3.3 respectively (*cf.* Figure 1). Dotted and dashed lines represent data for real clusters as in Figure 1. (b) Predicted distributions of line-of-sight velocity dispersions for b = 2, estimated using galaxies within the Abell radius (upper solid curve) or only within half this radius (lower solid curve). The dot-dashed line shows the distribution of clusters identified in three dimensions satisfying Abell's $R > 0$ richness criterion. The data for real clusters (dotted and dashed lines) are as in (a).

DISCUSSION

It seems from the above results that the abundance of clusters as a function of velocity dispersion is very difficult to obtain reliably from the observations. At low velocity dispersions the observed samples may be very incomplete because of the natural bias towards observing rich systems. At high velocity dispersions the statistics may be dominated by contaminated clusters where the observed dispersions are inflated by the presence of unvirialized sub-clumps projected onto the main cluster body. Our simulations demonstrate that these effects can conspire to reproduce the velocity-dispersion properties of the observed cluster sample in a CDM universe with b = 2. However, because of the large uncertainties, and their interaction with the specific biasing model employed, it seems best to regard this result as a consistency check, rather than as a determination of the amplitude of the fluctuation spectrum.

In the long term X-ray observations may offer a better solution to this problem. Flux limited cluster samples will be obtained from the ROSAT survey, and imaging and spectroscopic follow-up with AXAF will allow the potential wells of the largest dark-matter distributions to be studied in detail. Their abundance should then give a strong constraint on the initial amplitude of large scale fluctuations. Presently-available X-ray data are incomplete and far from adequate for this task, however, and require a detailed theoretical understanding of the relation between the hot gas and the dark matter before they can be used in this context. The resulting uncertainties seem to preclude any firm conclusion (see Frenk (1989) and Frenk *et al.* (1990) for further discussion).

The observed mass-to-light ratios of rich clusters can be compared to that of the universe as a whole to get a direct measure of the extent of biasing. White *et al.* (1987b) showed that their standard biased CDM model produced rich clusters with M/L about 20% of the universal value, and Frenk *et al.* (1989) showed this number to be very sensitive to b. This result appears to demonstrate that a strongly biased model (with correspondingly weak mass fluctuations) is required for consistency with the cluster data. However, it is important to remember that it hinges critically on the specific 'high peak' bias model used, and that additional physical effects, such as the dynamical biasing investigated by Barnes (1985) and West and Richstone (1988), may act to reduce the observed mass-to-light ratios of clusters. At present it appears that, although a strongly biased model is consistent with all the observed cluster data, it would be premature to rule out a larger amplitude for the initial CDM power spectrum.

ACKNOWLEDGEMENTS

The work reported in this paper was carried out in collaboration with Carlos Frenk, George Efstathiou and Marc Davis. It was supported in part by NSF grant AST-8822297 and by NATO travel grant RG 0882/87.

REFERENCES

Abell, G. 1958, *Ap. J. Suppl.*, **3**, 211.
Bardeen, J. M., Bond, J. R., Kaiser, N., and Szalay, A. S. 1986, *Ap. J.*, **270**, 20.
Barnes, J. 1985, *M.N.R.A.S.*, **215**, 517.
Carlberg, R., and Couchman, H. 1989, *Ap. J.*, **340**, 47.
Cole, S., and Kaiser, N. 1989, *M.N.R.A.S.*, **237**, 1127.
Colless, M. M., and Hewitt, P. 1987, *M.N.R.A.S.*, **224**, 453.
Colless, M. M. 1987, Ph.D. Thesis, Cambridge University.
Davis, M., Efstathiou, G., Frenk, C. S., and White, S. D. M. 1985, *Ap. J.*, **292**, 371; (DEFW).
Dekel, A., and Rees, M. J. 1987, *Nature*, **326**, 455.
Dressler, A., and Shectman, S. 1988, *A. J.*, **95**, 284.

Efstathiou, G., Frenk, C. S., and White, S. D. M. 1988, *M.N.R.A.S.*, **235**, 715.
Evrard, A. E. 1989, *Ap. J. (Letters)*, **341**, L71.
Frenk, C. S. 1989, in *The Epoch of Galaxy Formation*, ed. C. S. Frenk, R. S. Ellis, T. Shanks, A. Heavens and J. Peacock 1989 (Dordrecht: Kluwer Acad. Publ.), p. 257.
Frenk, C. S., White, S. D. M., Efstathiou, G., and Davis, M. 1985, *Nature*, **317**, 595.
Frenk, C. S., White, S. D. M., Efstathiou, G., and Davis, M. 1988, *Ap. J.*, **327**, 507.
Frenk, C. S., White, S. D. M., Efstathiou, G., and Davis, M. 1990, *Ap. J.*, **351**, 10.
Kaiser, N. 1984, *Ap. J. Lett.*, **284**, L9.
Kaiser, N., and Lahav, O. 1989, *M.N.R.A.S.*, **237**, 129.
Merritt, D. 1987, *Ap. J.*, **313**, 121.
Narayan, R., and White, S. D. M. 1988, *M.N.R.A.S.*, **231**, 97P.
Peacock, J. A., and Heavens, A. F. 1985, *M.N.R.A.S.*, **217**, 805.
Peebles, P. J. E., Daly, R. A., and Juszkiewicz, R. 1989, *Ap. J.*, **347**, 563.
Press, W. H., and Schechter, P. L. 1974, *Ap. J.*, **187**, 425.
Struble, M. F., and Rood, H. J. 1987, *Ap. J. Suppl.*, **63**, 543.
Teague, P. 1988, Ph.D. Thesis, Australian National University, Canberra.
The, L. S., and White, S. D. M. 1986, *A. J.*, **92**, 1248.
West, M. J., and Richstone, D. O. 1988, *Ap. J.*, **335**, 532.
White, S. D. M., Frenk, C. S., Davis, M., and Efstathiou, G., 1987a, *Nature*, **330**, 451.
White, S. D. M., Frenk, C. S., Davis, M., and Efstathiou, G. 1987b, *Ap. J.*, **313**, 505 (WFDE).
Zwicky, F. 1933, *Helv. Phys. Acta*, 6, 110.

DISCUSSION

M. Strauss: Could you comment on Jim Peeble's claim (Peebles, Daly, and Juszkiewicz, in press) that the large number of rich clusters at high redshifts put severe constraints on the bias parameter?

S. White: This claim depends critically on the velocity dispersions of these high redshift clusters. I am not sure how reliable these dispersions are in view of the problems I have been discussing. We would predict the true velocity-dispersion distribution of clusters to evolve relatively slowly, but because of its steepness the abundance of clusters above, say, 1000 km s^{-1} nevertheless changes very substantially from $z = 0$ to 0.5.

R. Gott: Your histograms reminded me very much of the results Ed Turner and I got when we studied finding groups of galaxies from the 2-dimensional Zwicky Catalog. Our *N*-body simulations showed that one made mistakes of inclusion and exclusion just remarkably similar to what you have found in the case of Abell clusters. Do you get the right number of 'Abell' clusters in your CDM simulation, and what is ξ_{cc}?

S. White: Yes, the effects are very similar. In fact, the problems are considerably worse in an absolute sense, for relatively poor groups. It is quite easy to 'find' groups which have no real counterparts, whereas 'Abell clusters' in our models almost always correspond to one dominant peak along the line of sight. However, systematic biases of relatively small amplitude (*e.g.* a 20 or 30% overestimate of velocity dispersions) can radically change the interpretation of cluster abundances because of the very steep nature of the theoretical predictions. The cluster abundances we obtain in these models (as a function of richness and luminosity) are in agreement both with those we got earlier and with observation.

A. Szalay: Have you tried to tag particles in the clusters and go back to the initial Lagrangian distribution of those?

S. White: No.

J. Primack: My colleagues and I found that about 10 to 14% of $R = 1$ clusters are so contaminated by projection of galaxies associated with other clusters, that they are really $R = 0$, at least using our projection decontamination procedure. Is this consistent with your simulations?

S. White: I am not sure. The contamination I have been discussing is mostly due to smaller groups, either in the immediate environment of the primary cluster, or in foreground or background objects un-associated with any very rich cluster. I'd be surprised if much more than 10% of the cases of contamination could be assigned to halos of other rich clusters.

N. Kaiser: Do you think the discrepancy between the x-ray temperature abundance and that derived from velocity dispersion may be resolved if clusters have a broad range of L_x at a given velocity dispersion or richness?

S. White: I'm not sure I see how this would work. The x-ray data seem to show that the scatter in the observed L_x *versus* velocity dispersion relation is quite large, whereas the L_x *versus* temperature relation is tighter. Since we expect the temperature *versus* true velocity dispersion relation to show little scatter, it seems to me that the most consistent interpretation is that L_x, T_x, and true velocity dispersion are all tightly related, but that the observed velocity dispersions are seriously contaminated.

P. Shapiro: Your results indicate that, after a considerable amount of effort to bring CDM model theoretical predictions into line with the cluster number density, velocity dispersion, and x-ray temperature data, you are able to get satisfactory agreement only for b greater than about 2, while the large-scale velocity-field data, we were told this week, are only consistent with a CDM model which has b equal to about unity or less. This discrepancy appears to imply that the CDM model is inconsistent with the observational data. Do you agree? If not, *why* not?

S. White: This is true at the level that we have analyzed the model. However, the discrepancy is not large and is probably insufficient to rule out CDM as a whole. Recall that our results for cluster mass-to-light values and small-scale velocity dispersions depend on our particular prescription for putting down the galaxies. This does not have strong physical justification. Indeed, it misses the physical effects expected from the internal degrees of freedom of galaxies and galaxy halos. As shown by Barnes, Evrard, Carlberg, and Couchman, West and Richstone, and others, these effects can cause some dynamically induced bias on small scales. In view of this and of our general ignorance of how galaxies would form in a CDM universe, the difference between the b = 1.3 to 1.6 which Nick Kaiser claims will fit large-scale flows and the b = 2 to 2.5 which we prefer for small-scale statistics does not seem very great.

A. Dekel: Is it indeed true that you get the correct ξ_{gg} for any value of b in the range 1 to 3? Both amplitude and slope? Isn't this different from what you claimed in your original paper on CDM?

S. White: This is at least approximately true! It seems that this particular biasing scheme leads to a galaxy correlation function of similar slope for a wide range of mass fluctuation amplitudes. (We carefully chose parameters to make the amplitude come out right).

J. Primack: Since we're all finding that a little more large-scale power is what's needed to fix up CDM, what's *your* favorite way of getting it? (The ones I investigated, with Blumenthal and Dekel, are more baryons and/or lower Ω.)

R. Gott: One way to get more power at large scales with the CDM spectrum is to simply lower the value of Ωh. One first pushes h as low as possible, which is why one uses h = 0.5 (H_0 = 50 km s^{-1} Mpc^{-1}) and from there one can lower Ω below 1. A CDM model with Ω = 1 and h = 0.5, for example.

M. Davis: George Efstathiou is currently pushing H_0 = 40, Ω_B = 0.22 (Ω_T = 1) in order to get more power on large scales to match the observed $\omega(\theta)$ in the APM survey. Can someone explain the physics of the L_x *versus* T_x relationship? In particular, why is the scatter small given that the x-ray emission is so sensitive to the core parameters?

S. White: I agree this is a puzzle. It seems to require that the galaxy clusters should form a one-parameter family with respect to the overall structure of their inner regions. I don't understand why this should be so.

N. Kaiser: Don't expect the theory to explain everything.

QSO's METAL ABSORPTION LINES

S. M. VIEGAS-ALDROVANDI AND R. B. GRUENWALD
Instituto Astronômico e Geofísico, Av. Miguel Stefano, 4200
01000 São Paulo, (SP) Brazil

ABSTRACT In this paper the possibility that the metal absorption lines from QSOs are produced in H II Regions of intervening galaxies is analyzed. The consequences for cosmology are briefly discussed.

INTRODUCTION

The general properties of the absorption lines from QSOs have been pointed out by several large-scale surveys (Foltz *et al.* 1986, Lanzetta *et al.* 1987, Sargent *et al.* 1980, 1988, Weyman *et al.* 1978, Young *et al.* 1982).

It is generally assumed that absorbing systems with relative velocity to the QSO greater than $0.1c$ are not ejected from the QSO, and are associated to intervening objects (Weyman *et al.* 1979). Thus, the study of these absorbing systems could provide information relevant to cosmology (see, for instance, Sargent 1987, 1988).

Two types of intervening systems are considered: (a) Lyman-α systems - which show only hydrogen lines and no metal lines, and are associated with intergalactic clouds; (b) Metal systems - which contain hydrogen lines as well as metal lines, produced by several elements in different ionization stages, and are associated with intervening galaxies.

It is also generally assumed that the absorbing systems are photo-ionized by the radiation field due to the integrated contribution of the QSOs (Bechtold *et al.* 1987, Bergeron and Stasinska 1986). However, high-resolution data suggest that metal absorption lines could originate in star-forming regions of intervening galaxies, in analogy with absorptions lines in gas-rich dwarf galaxies (York *et al.* 1986, 1988). In addition, an analysis of the metal content of damped Lyman-α systems suggests the same thing (Pettini *et al.* 1988). This point of view is analyzed in this paper by comparing observed metal absorption lines to those calculated from photo-ionization models of H II regions.

H II REGIONS AS ABSORBERS

In order to calculate the absorption lines produced by H II regions, a grid of photo-ionization models has been built using a computer code where the calculation of the column density of different ions has been included. A spherical symmetry is adopted for the gas composed of H, He, C, N, O, Ne, Mg, Si, S, Cl, Ar and Fe. The ionizing radiation spectrum is a black body with temperatures characteristic of O4, O6, O8 and B0 stars. Models with hydrogen density in the range 10^0 to 10^3 cm^{-3}, and chemical abundances from the solar value $Z_\odot$ (Grevesse and Anders 1988) to $Z_\odot/100$, were built. For each model, the column densities of the ions and neutral atoms have been calculated for different lines of sight, corresponding to different distances from the central star.

The method used to analyze the absorption lines of QSOs has been briefly described in Viegas-Aldrovandi and Gruenwald (1989). For a given absorption system we assume that the high ionization absorption lines (C IV, Si IV, Si III, N V) are only produced in an H II region of an intervening galaxy, whereas the low ionization lines (C II, Si II, Fe II, Mg II, O I) are also produced by the neutral interstellar medium. The characteristics of the H II region are obtained using a diagram of the column densities *N*(C IV) versus *N*(Si IV). In this diagram the results corresponding to a given ionizing star define a 'theoretical' region and the column densities obtained from the observed lines, for different values of the parameter *b* (related to the velocity dispersion), define a curve. The superposition of the curve on a 'theoretical' region gives the ionizing star, the range of chemical abundances and the range of hydrogen densities. Then, similar diagrams $N(X_i)$ versus *N*(C IV) are used to test the agreement between the other observed lines and the calculated values obtained from the model previously defined.

The method has been used to study several absorption systems observed by Sargent *et al.* (1988). This survey provides a uniform sample of QSO absorbing systems. The spectra, in a large wavelength range, have a uniform signal-to-noise ratio and high spectral resolution. The method has been tested with the absorption system $z_a = 1.6724$ of the QSO 0237-233. The absorption lines can be produced by an H II region ionized by an O4 star, assuming a density less than 10 cm^{-3} and chemical composition in the range $Z_\odot/30$ to $Z_\odot/100$. Two lines, Si II 1304 Å and O I 1302 Å, should have been observed but they have not been identified by Sargent *et al.* However, the calculated column densities of Si II and O I lead to observed equivalent widths in agreement with the equivalent widths of the observed lines 20 and 21 listed by Sargent *et al.* in the QSO spectrum (Viegas-Aldrovandi and Gruenwald 1989).

We analyzed 20 absorption line systems which show at least C II and Si II absorption lines in addition to C IV and Si IV lines. About 15 systems can be explained by an H II region producing the absorption lines, with density less than 10 cm^{-3} and chemical composition between $Z_\odot/10$ and $Z_\odot/100$, and an ionizing O4 star. Another system indicates the presence of a colder ionizing star. The remaining four systems could not be fitted by the models. The observed Si IV column density is too low for $b \leq 100$ km s^{-1}. A possible explanation is that these systems are dusty and the Si is depleted. A detailed discussion of all these systems will appear elsewhere.

FINAL REMARKS

Our results indicate that at least part of the observed absorption line systems originate in star forming regions of intervening galaxies. The consequences on cosmology are promising.

Abundance determinations from H II regions have been largely used in our galaxy and in nearby galaxies. In addition, photo-ionization models for H II regions have been largely tested and are accurate enough to provide a good indication of the chemical composition of these regions. Thus, our method seems to be a powerful tool to study the chemical evolution of galaxies, if a larger sample of absorbers is analyzed and their chemical abundances is obtained as a function of the redshift.

Regarding the number of absorbing systems $N(z)$ per unit redshift range, observations indicate that for Lyman-α systems $N(z)$ varies as $(1+z)^2$, whereas for metal systems it varies as $(1+z)^{-1.2}$ (Sargent 1987, 1988). It seems that our results could explain this dichotomy. Both types of system should be related to star-forming regions, the difference in the distribution coming from a different chemical enrichment. We expect that Lyman-α systems (corresponding to lower abundances) should dominate at larger redshifts, whereas metal systems (with larger metal enrichment) should decrease as a function of the redshift. It is also plausible that systems showing low H I column density are produced in galaxies with a large rate of star formation which maintains a high ionization rate. In this scenario, the absorbing systems should follow the overall distribution of mass at high redshift.

Finally, since the absorbing systems give the large scale structure of the Universe at high redshift, maybe in the future, when a larger number of systems is analyzed, the dynamical structure of the galaxies could be coupled to the 'chemical structure' providing a better understanding of the Universe.

ACKNOWLEDGEMENTS

We are thankful to D. York for fruitful discussions and to D. Dultzin-Haycian and B. Pagel for some useful references.

REFERENCES

Bechtold, J. P., Green, R. F., and York, D. G. 1987, *Ap. J.*, **312**, 50.

Bergeron, J., and Stasinska, G. 1986, *Astr. Ap.*, **169**, 1.

Foltz, C. B., Weyman, R. J., Peterson, B. M., Sun, L., Malkan, M. H., and Chaffee, F. H., Jr. 1986, *Ap. J.*, **307**, 504.

Grevesse, N., and Anders, E. 1988, *Symposium on Cosmic Abundances of Matter*, ed. C. J. Waddington (AIP).

Lanzetta, K. M., Turnshek, D. A., and Wolfe, A. M. 1987, *Ap. J.*, **322**, 739.

Pettini, M., Boksenberg, A., and Hunstead, R. W. 1988, *Proceedings of the NATO Conference on The Epoch of Galaxy Formation.*

Sargent, W. L. W. 1987, in *IAU Symposium no. 124, Observational Cosmology*, ed. A. Hewitt, G. Burbidge, and L. Z. Fang, (Dordrecht: Reidel), p. 777.

Sargent, W. L. W. 1988, in *IAU Symposium no 130 Large Scale Structure of the Universe*, ed. J. Audouze, M-Ch. Peleton, and A. Szalay (Dordrecht: Kluwer), p. 333.

Sargent, W. L. W., Boksenberg, A., and Steidel, C. C. 1988, *Ap. J. Suppl.*, **68**, 539.

Sargent, W. L. W., Young, P. J., Boksenberg, A., and Tytler, D. 1980, *Ap. J. Suppl.*, **42**, 41.

Viegas-Aldrovandi, S. M., and Gruenwald, R. B. 1989, in *Proceedings of the II Texas-Mexico Conference on Star Forming Regions* (in press).

Weyman, R. J., Williams, R. E., Peterson, B. M., and Turnshek, D. A. 1979, *Ap. J.*, **234**, 33.

York, D. G., Caulet, A., Rybskii, P., Gallagher, J., Blades, J. C., Morton, D. C., and Wamsteker, W. 1988, *Ap. J.*, submitted.

York, D. G., Dopita, M., Green, R., Bechtold, J. 1986, *Ap. J.*, **311**, 610.

Young, P., Sargent, W. L. W., Boksenberg, A. 1982, *Ap. J. Suppl.*, **48**, 455.

DISCUSSION

S. White: I think the striking difference in the apparent clustering properties of the Lyman-α-forest clouds and the metal-line clouds makes it difficult to see how they could reflect aspects of the same population. If the metal lines do reflect H II regions surrounding young stars, what does the observed abundance of such systems imply for the prevalence of star formation at early times in comparison with the present?

S. Viegas-Aldrovandi: The difference in the apparent clustering properties should reflect the difference in chemical evolution of galaxies and proto-galaxies. In order to have column densities in agreement with the observations it seems that the rate of star formation should be higher at early times.

P. Shapiro: Perhaps your photoionized-cloud models would also give a successful fit to the data if, instead of H II regions *inside* a galactic disk, you considered halo clouds illuminated by the photo-ionizing radiation from many galactic early-type stars. After all, the key parameter is not the flux, F, or the gas density, n, but rather the ratio of F to n. Hence, even a lower flux for the starlight in the halo would be adequate if the cloud density were also proportionately lower. Furthermore, the star-formation rate may have been higher at larger z, implying a higher net flux.

S. Viegas-Aldrovandi: It seems to me that it would not be enough. However, we can not exclude that possibility, and some absorption systems could correspond to halo clouds.

M. Strauss: A clear way to test your ideas is to do narrow-band imaging in redshifted H II region emission lines of the quasars. Can you comment? Do the results of Bergeron and others differentiate between HII region and halo models?

S. Viegas-Aldrovandi: In fact, both Jacqueline Bergeron and Don York have observed O II emission line at the same redshift of absorption systems.

SEGREGATION, PROTO-GALAXY AND PECULIAR MOTIONS

RICCARDO GIOVANELLI
National Astronomy and Ionosphere Center
Arecibo Observatory, P.O. Box 995, Arecibo, PR 00613

ABSTRACT A brief review is given of work on the segregation properties of the galaxian population with local density. Recent work based on the Pisces-Perseus Supercluster redshift survey is discussed. Special attention is given to the report of the discovery of an isolated intergalactic cloud, 200 kpc in diameter and total mass near 1.5×10^{10} $M_{\odot}$, likely to be the first example of a proto-galaxy at $z = 0$. Finally, a brief report of ongoing projects to determine the distances and peculiar motions of several samples of spirals, for the study of localized and large-scale deviations from Hubble flow, is included.

PROLOGUE

One of the tasks assigned for this presentation was that of providing a review of the work on segregation of galaxian properties with local density, complemented with recent results derived from the Pisces-Perseus supercluster redshift survey. A discussion of those topics is given in the next two sections. The discovery of a new kind of object, an intergalactic cloud likely to be a proto-galaxy at $z = 0$, receives special attention after that. Its existence raises interesting questions related to the process of galaxy formation - which presumably lurks behind segregation phenomena. Finally, a status report is given of several ongoing observational projects of wide scope, involving the determination of redshift-independent distances and peculiar motions of well over 2000 spiral galaxies. The discussion of the topics in the fourth and fifth sections is kept fairly independent from that in the second and third sections, so that the selective reader may skip ahead without the risk of encountering excessive incoherence.

SEGREGATION: PAST WORK

It is well established that cluster cores exhibit a strong degree of segregation, particularly manifest when the population ratios of different morphological types are considered *vis-a-vis* either the local projected density or the projected distance from the cluster center. This was noticed as early as 1931 by Hubble and Humason, and corroborated more notably by the work of Oemler (1974), Melnick and Sargent (1977) and Dressler (1980). Many mechanisms may have secularly contributed to the observed segregation patterns, well after the 'main season' of galaxy formation. Thus segregation in clusters is unlikely to provide an un-muddled picture of the possible existence of biasing in the galaxy-formation process. We shall concentrate here on the evidence that segregation may exist on larger scales - those ranging between that of a cluster core and that of observed large-scale inhomogeneities in the galaxian distribution (voids, sheets and filaments).

A variety of techniques has been utilized to probe the galaxian distribution. Unavailability of complete redshift samples forced the application of techniques that relied on the projected distribution, *e.g.* the two-point angular correlation function and the measurement of projected morphological type population fractions. With the advent of large redshift data bases, the study of the three-dimensional distribution of galaxies became possible, allowing for the estimation of the spatial correlation functions, luminosity functions, spatial population fractions and the application of more novel methods such as those based on void probability statistics. Table I lists references to work that has been influential in shaping this author's view of the subject. The table also describes in summary the technique used in each case, and an indication on whether each set of authors identified large-scale segregation in the galaxian properties. The identification of the method utilized is as follows: $\omega(\theta)$ and $\xi(r)$ are the 2-point angular and spatial correlation functions, respectively; $T(\theta)$ and $T(\rho)$ are the population fractions of galaxies of type T, respectively projected on the sky and in 3-d space; LSS refers to studies where the space distribution of different types of galaxies is qualitatively compared by looking at wedge plots or redshift histograms; VPF utilizes void probability statistics techniques; $L(\rho)$ and $Mass(\rho)$ study variations of the luminosity function and of the mass function of galaxies with local space density.

Table I Work on Segregation over Large Scales

Authors	Method	Segregation No	Segregation But	Segregation Yes
Davis and Geller 1976	$\omega(\theta)$			√
Postman and Geller 1984	$T(\rho)$		√	
Davis and Djorgovski 1985	$\omega(\theta)$			√
Giovanelli, Haynes, & Chincarini 1986	$\omega(\theta)$, $T(\rho)$			√
Xia, Deng and Zhou 1987	$\xi(r)$			√
Bothun *et al.* 1986	LSS	√		
Thuan *et al.* 1987	LSS	√		
Maurogordato and Lachieze-Rey 1987	VPF, ξ(r)			√
Davis *et al.* 1988	ξ(r))			√
Hamilton 1988	ξ(r)			√
Tully 1988	$T(\rho)$, $L(\rho)$			√
White, Tully and Davis 1988	$Mass(\rho)$			√
Giovanelli and Haynes 1988	$L(\rho)$			√
Hollósi and Efstathiou 1988	ξ(r)			√
Iovino, Melnick, and Shaver 1988	ξ(r)			√
Eder *et al.* 1989	LSS, ξ(r)	√		
Dominguez-Tenreiro and Martinez 1989	ξ(r)		√	
Salzer, Murray and Gavazzi 1989	LSS, ξ(r)			√

Most studies identify rather clearly a monotonic change in average galaxian distribution properties with type, mass, surface brightness and luminosity. With increasing local density, galaxies get brighter, more massive and tend to be of earlier types. Yet, from several studies it appears that the large-scale structures described by galaxies of different types, or by galaxies of different luminosities brighter than about −18, are pretty much the same, once allowance is made for differences in the selection functions of the various subsamples. It is possible that the contrast between these two apparently conflicting views may be one of emphasis, rather than substance. In fact, galaxies of different subsamples can exhibit the same large-scale general topological characteristics and yet do so with different degrees of density contrast. In other words, clustering properties can be monotonically varying with, say, type or luminosity, and yet outline the same voids, sheets and filaments.

Binggeli (1987) has proposed to replace the concept of a *universal* luminosity function (LF) with one that depends on the morphological make-up of the local galaxian population. We have obtained synthetic LF's from a variety of assumed population fractions, using the individual type LF's as given by Sandage, Binggeli and Tammann (1985). The results are mildly surprising, in that the bright end of the global LF is very resilient; only minute changes occur even for drastically different relative proportions of early to late type galaxies (Figure 1). The characteristics of the LF are more vulnerable to population changes in the faint end, which is dominated by dwarf - especially dE galaxies. Unfortunately, large-scale studies fail to sample reliably the faint end of the LF. Thus, we suggest that if a true paucity of bright galaxies exists in low density environments, that may not just be the result of a different morphological type mix: the bright end of, say, the lenticular and spiral LF may be truncated in those regions.

It is sometimes argued that the segregation pattern that several workers find may be heavily weighted by clusters and their peripheries. Once clusters and dense groups - which largely influence estimators of clustering properties - are removed, the 'rural' and 'suburban' galaxy population may look pretty much the same everywhere, and the LSS on scales larger than clusters would thus be roughly invariant with galaxian properties. We shall confront some of these suggestions with data, in the next section.

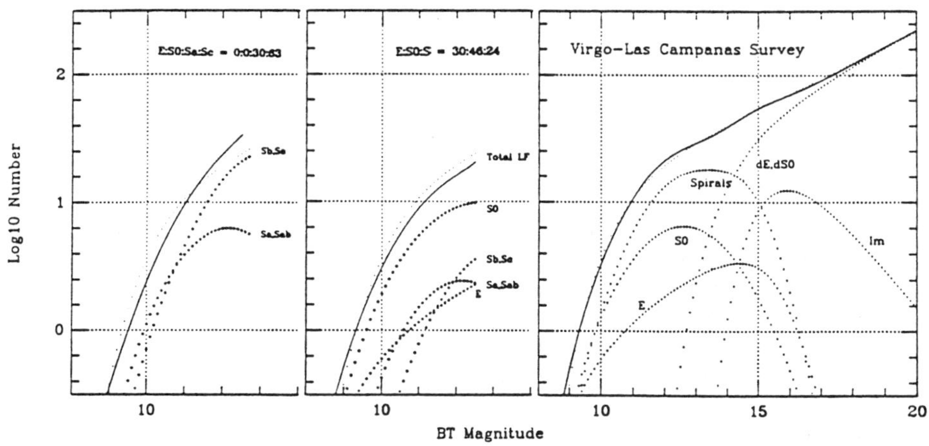

Fig. 1. Three synthesized LF's, obtained by adding the single type LF's, in different proportions. The single-type LF's are those of Sandage, Binggeli and Tammann (1985). In panel (c) the type mix is the same as in the Virgo cluster; the resulting LF is plotted as a heavy line; the abscissa corresponds to B_T magnitudes as measured in Virgo. Panel (a) displays a mix of purely spirals (30% Sa's and Sb's; 63% Sbc's and Sc's; the rest are dwarfs and do not affect the LF in the plotted range of magnitudes); the resulting LF is the solid line, while the light dotted one is the global Virgo LF for comparison. Panel (b) displays the LF typical of a cluster core, with 30% E's, 46% S0's and 24% spirals. Notice that the bright end the total LF is barely distinguishable in all three cases. The truncation of the LF at B_T = 12.5 mag in (a) and (b) is equivalent to a limiting magnitude near 15.5 at ~ 5000 km s^{-1}.

RESULTS FROM THE PISCES-PERSEUS SURVEY

Current Status

About 40% of the data obtained in the course of this survey has been published so far. Scope and general characteristics are described in Giovanelli and Haynes (1988), Haynes and Giovanelli (1988) and references therein. The total sample for the region between 22^h to 04^h, $0°$ to $38°$, consists of about 5000 galaxies brighter than $m = 15.7$, plus an additional 1000 or so which are fainter than $m = 15.7$ but larger than $1'$. The current completion rate is 80%. Completion drops at the faint end ($m > 15.5$) and at declinations at which the gain of the 305-m telescope in Arecibo is degraded ($<5°$ and $>33°$), as illustrated in Figure 2. Current efforts towards completion are carried out with both the 305-m telescope in Arecibo and the 2.4-m telescope at McGraw Hill Observatory, in collaboration with M. Haynes and G. Wegner.

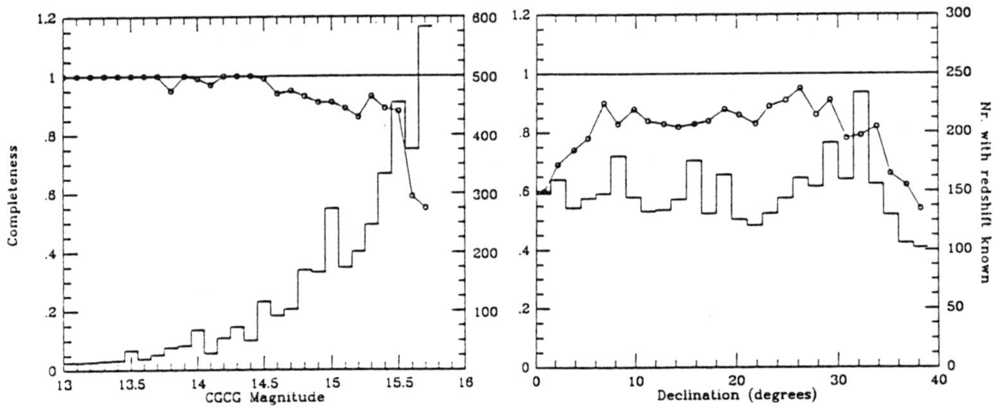

Fig. 2. Pisces-Perseus redshift survey completeness. (left) The completeness (unfilled circles) *versus* apparent magnitude; the number of galaxies with known redshift in each magnitude bin is plotted as a histogram. (right) All galaxies with m(CGCG) $\leq$ 15.7; completeness is plotted versus declination, binned every 1.5°. Redshifts have also been measured North of 39.5°, East of 04^h and West of 22^h as part of the survey, but with lower completeness.

The characteristics of the survey - its depth and wide-angle coverage in both coordinates - make it an excellent test-bench for the study of LSS (see for example Gott's contribution in these proceedings). Figure 3 illustrates a few examples of cartesian slices of a complete, volume-limited subsample drawn from the survey data base.

Segregation

The local density of galaxies can be estimated from volume-limited samples such as the one that was used to produce Figure 3, or by means of other techniques that use more extensively the data base, and take into account the diminishing range of luminosities sampled with increasing distance, as done for example by de Souza, Vettolani and Chincarini (1985). We have analyzed the mean characteristics of the galaxian population as a function of local density ρ, as obtained using several estimators of ρ. The results reported here are method-invariant.

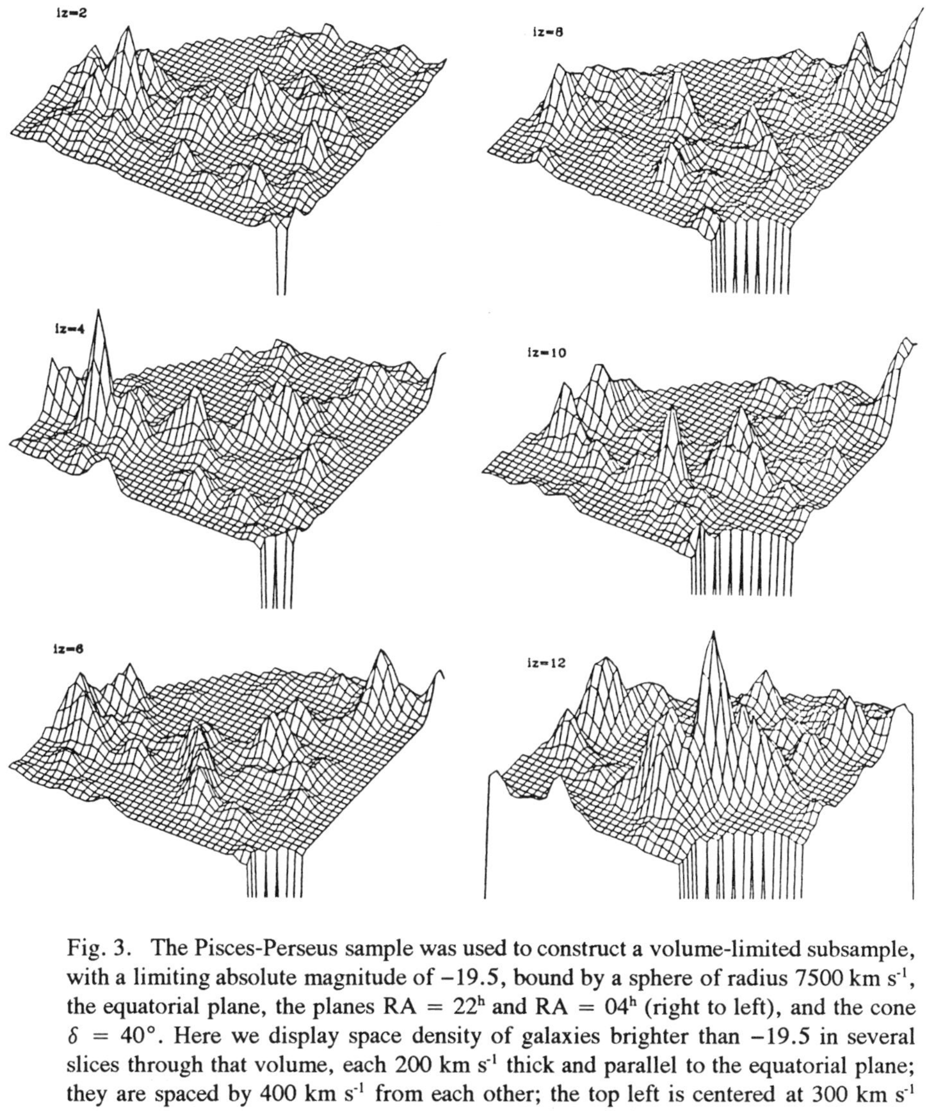

Fig. 3. The Pisces-Perseus sample was used to construct a volume-limited subsample, with a limiting absolute magnitude of −19.5, bound by a sphere of radius 7500 km s^{-1}, the equatorial plane, the planes RA = 22^h and RA = 04^h (right to left), and the cone $\delta = 40°$. Here we display space density of galaxies brighter than −19.5 in several slices through that volume, each 200 km s^{-1} thick and parallel to the equatorial plane; they are spaced by 400 km s^{-1} from each other; the top left is centered at 300 km s^{-1} above the equatorial plane, the bottom right at 2300 km s^{-1} above that plane. Densities are smoothed by a Gaussian of dispersion equal to 200 km s^{-1}.

Figure 4 illustrates that morphological-type segregation exists over the whole range of local densities sampled. While the ratio between the counts of early and late galaxies is greatly enhanced in the cluster regions, a less steep but well marked monotonic change with density occurs also in regions well remote from cluster peripheries. Similarly, the mean luminosity of galaxies changes monotonically with local density, and it does so over density regimes well distinguished from cluster environments, as shown in Figure 5. The density dependences are not strong, less so in fact than anticipated in high-bias scenarios, but they are consistent with a mild genetic bias.

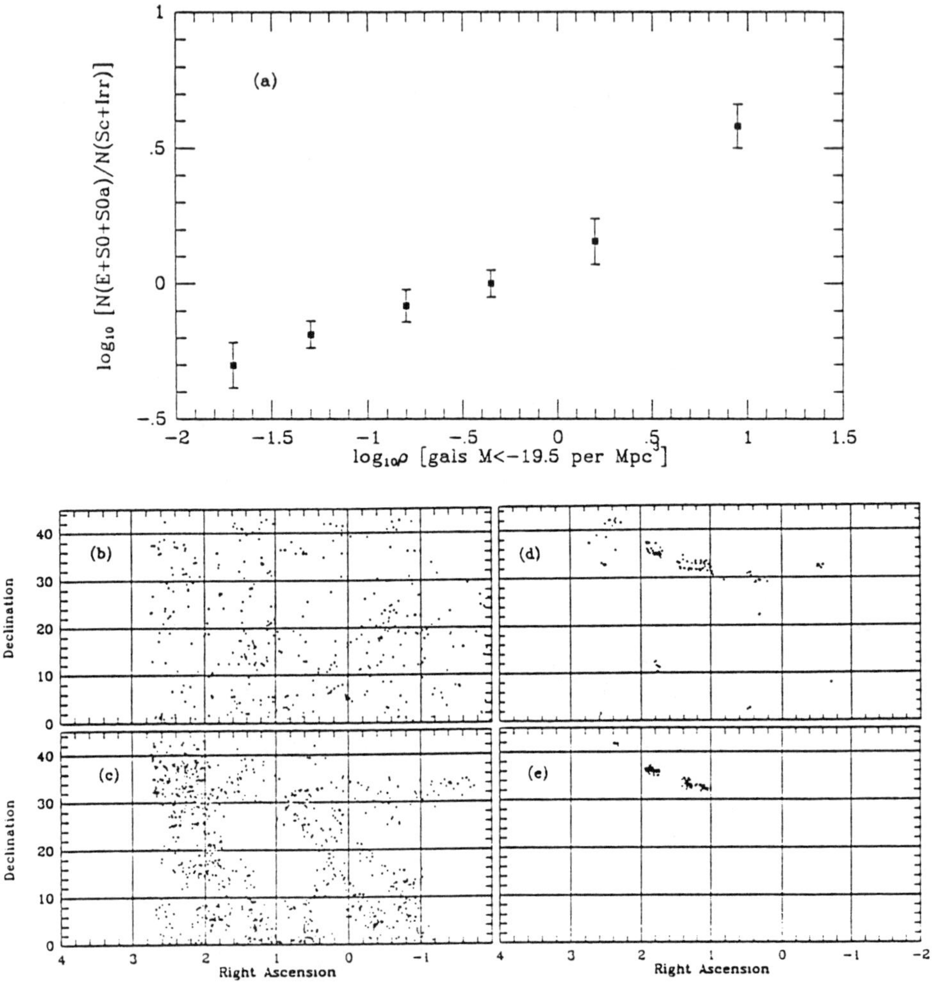

Fig. 4. Panel (a) displays the ratio in counts of galaxies of types E, S0 and S0a to that of late spirals and irregulars, plotted *versus* the local density of galaxies brighter than $M = -19.5$. Panels (b) to (e) show the sky distribution of galaxies in different local density windows, as an aid to visualize the density scale. Panel (b) refers to densities identified with the left-most data point in (a); panel (c) plots galaxies in densities as in the third point from left in (a); panels (d) and (e) plot galaxies in the highest density regime (clusters and their immediate peripheries); the combination of both sets leads to the right-most point in (a).

It could be argued that the change in the mean luminosity with local density is influenced by the fact that our samples do not include dE's, which are faint and known to have similar clustering properties to those of other early-type galaxies (Binggeli, Tammann and Sandage 1987; Ferguson 1989). These objects have magnitudes below the catalog cutoff at the characteristic depth of our sample, and also tend to be of small size, thus missing the

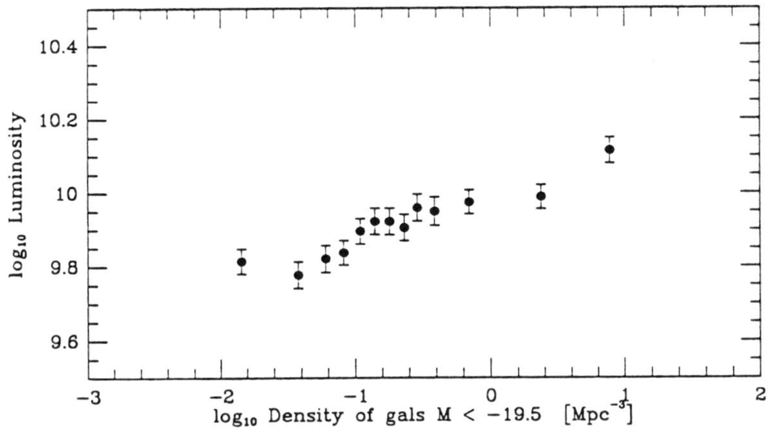

Fig. 5. Mean photographic luminosity (in solar units) of galaxies between 4000 and 5500 km s^{-1} (the redshift interval is narrow so that a differential Malmquist bias between high and low density regions will not contaminate the result), plotted versus local density of galaxies brighter than $M = -19.5$. Note that at this distance range, the survey samples only the bright end of the LF.

angular size cutoff. Their inclusion in a deep sample would undoubtedly change somewhat the characteristics of the picture that we described. Ideally, one would like to have access to a sample that reached as deep as that of Sandage, Binggeli and Tammann (1985) and spanned not only the local supercluster but over and beyond a neighboring giant void. Nonetheless, within the limitations of currently available samples, there is an indication that much of the change in mean luminosity with ρ derives from the bright end of the LF: relative to the galaxies near the knee of the LF, there appears to be a paucity of bright galaxies in the low ρ regions (see Figure 5 in Giovanelli and Haynes 1988).

We conclude that fair evidence exists pointing in the direction that segregation of galaxian properties extends beyond the peripheries of rich clusters, and that changes in the LF are not just a consequence of morphological segregation, but result from a relative paucity of bright galaxies of all types in low density regions. This is consistent with a mild bias in the galaxy formation processes.

LEFT-OVER BARYONS: A PROTO-GALAXY AT $z = 0$?

Background

If the galaxy formation process were not 100% efficient, material may still be slowly collapsing about smaller amplitude perturbations than those that gave rise to present galaxies. Searches for such objects have been made, more recently in LSS voids, resulting in interesting upper limits on their space density rather than clear-cut detections. Most frequently, those searches sought to detect the 21-cm emission of the putative clouds.

A number of intergalactic gaseous features have been detected in the vicinity of nearby galaxies (see Haynes, Giovanelli and Chincarini 1984 for a review); however, they can usually be understood as shreds of loosely-bound disk material, torn from a parent galaxy by tidal forces originating with a companion galaxy. The most interesting feature previously reported is probably that discovered by Schneider *et al.* (1983) in the Leo group of galaxies, which is shaped as a ring around NGC 3384 and M 105. The apparently stable orbital configuration of the ring and the absence of an associated stellar component suggest that the cloud is not the result of a recent tidal event, and have led the authors to suggest that the Leo ring may be a primordial feature. The close vicinity of the cloud to the galaxies in the group and the fact that its dynamics is clearly controlled by the group members do however raise some caution in the adoption of such a view.

An Isolated Intergalactic Cloud

In the course of routine 21-cm observations of spiral galaxies with the 305-m telescope in Arecibo early in 1989, R. Giovanelli and M. Haynes serendipitously discovered emission from an optically blank patch of sky. The signal was later confirmed and the emission region was summarily mapped. Figure 6 illustrates the observations carried out to mid-1989. The boundaries of the cloud are relatively well defined, appearing to extend over more than 200 kpc. The approximate parameters of the object are given in Table II; a distance of 20 Mpc was assumed for all the calculations mentioned here.

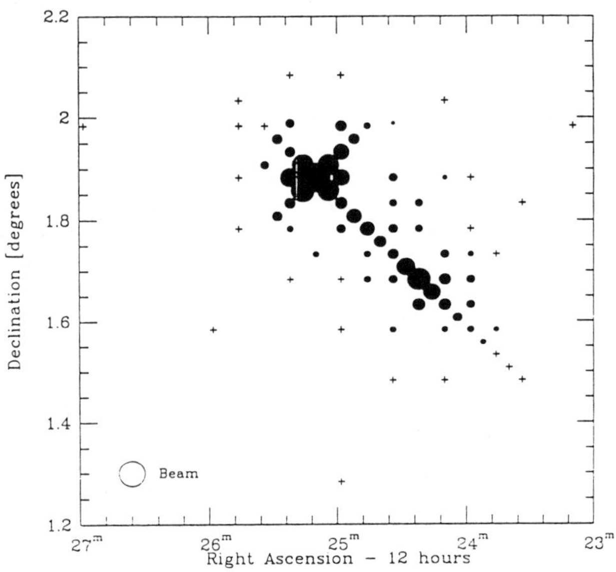

Fig. 6. Sampled points in the cloud field by mid-1989. Crosses identify non-detections, at the rms level of 3 mJy. Filled circles represent HI line detections, with the area of the circle proportional to the HI column density. The peak column density measured is 1.9×10^{20} cm^{-2}. The solid angle subtended by the telescope beam is displayed as an unfilled circle.

Rotation Curve

After the first few spectra were obtained, it was apparent that the direction of maximum velocity gradient is roughly along a position angle of about 45°. Sampling at 2.1′ spacing was thus obtained along that direction, which also appears to define the major axis of the cloud. Figure 7 displays the locus of the peak velocity along the major axis; a well-defined rotation curve reaches a maximum V_{max} = 30 km s^{-1} and then appears to flatten and remain constant to the last detectable point. The rotation curve can be used to estimate the run of total mass with radius, *M(r)*, with the main uncertainty in the determination resulting from not knowing the inclination of the HI disk and the geometrical shape of a possible dark halo. To the last detected point, the total mass is about $1.5\times10^{10}\ M_\odot$.

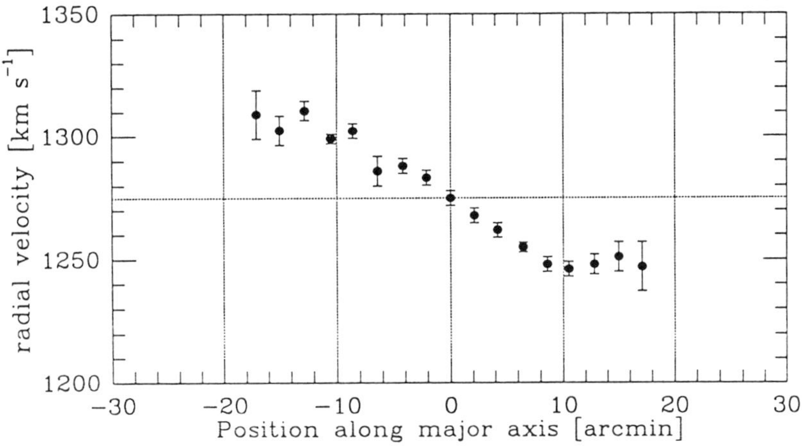

Fig. 7. Rotation curve of the cloud, uncorrected for inclination of the gas disk. Each point plotted is at the velocity of the peak each line profile along the major axis of the cloud, assumed to be the line at PA = 45° through the nominal center listed in Table II. Northeast is to the left.

TABLE II Cloud Parameters

R.A.(1950) center	$12^h\ 24.7^m$
Dec.(1950) center	01° 46′
Systemic $v_\odot$	1275 km s^{-1}
Assumed distance	20 Mpc
Major diameter	204 kpc
Axial ratio	~ 2
HI mass	$\geq 4\times10^9\ M_\odot$
V_{max}	$\geq$ 35 km s^{-1}
Total mass	$\geq 1.5\times10^{10}\ M_\odot$

Discussion

The cloud appears quite isolated in space. Inspection of the Palomar Sky Survey plates yields no indication of optical emission in the field. Figure 8 illustrates the position of the cloud, relative to that of galaxies with known velocity of less than 2500 km s^{-1} and of 8 galaxies of unknown redshift but likely to be nearby objects. It is highly unlikely that the material in the cloud may have been swept from any of the projected neighbors. Its isolation over scales on the order of the galaxy-galaxy separation should on the other hand not distract attention from the fact that the general region in which the cloud is located is characterized by a higher-than-average density of galaxies, at the margins of the Virgo cluster and quite near the supergalactic plane.

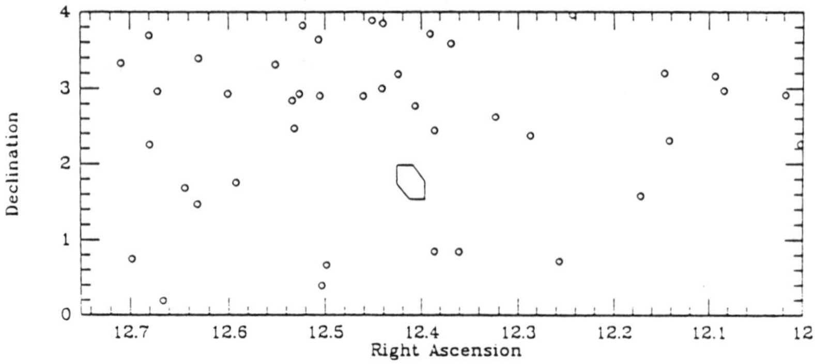

Fig. 8. Sky distribution of galaxies with known radial velocities smaller than 2500 km s^{-1} and of 8 galaxies of unknown redshift (but likely to be in the quoted velocity interval), around the cloud, represented by the polygon.

The total matter density distribution within the cloud is roughly constant in the inner 50 kpc or so, and falls as r^{-2} outside that radius. The flat inner regime of the radial density profile is consistent with the cloud still being in a relatively preliminary phase of collapse. The cloud represents a modest enhancement with respect to the critical density $\langle \rho_{cloud} \rangle / \langle \rho_{crit} \rangle \sim 25$ for $h = 0.7$ - and must have originated from an extremely shallow perturbation in the early universe. It is thus surprising that, if it is indeed primordial, the cloud has maintained its individuality in the relatively crowded environment of the supergalactic plane, where the 'scavenging radius' (*i.e.* the comoving radius within which gas is falling into the potential well) of larger amplitude perturbations - the present normal galaxies - may approach the inter-galaxy separation.

Although only coarsely mapped, it is apparent that the neutral gas in the cloud is not centrally clumped. The two large concentrations along the major axis appear to have peak column densities 3 to 5 times higher than those found in the central parts; a simple model of an edge-on torus does not fit well the observed major-axis column-density profile. The data are most suggestive of two main clumps, in the process of merging. A flat total mass density profile, *i.e.* a surface density profile that is centrally peaked, contrasts with the centrally depressed neutral gas surface density profile. Given the low mean neutral gas density, it is very unlikely that gas be present in the central part of the cloud in molecular form. It is also unlikely that the ratio of ionized to neutral gas density grow toward the center, as no obvious localized ionizing source is present in the cloud. It appears more probable that the dynamics of the cloud be dictated by a dark matter component that makes up for about two-thirds of its total mass.

What is the space density of objects of this kind? Where will they be preferentially found?

DISTANCES AND PECULIAR MOTIONS OF SPIRALS: ONGOING PROJECTS

All Sky Survey

In 1987 a project was initiated to obtain redshift-independent distances of a sample of 2000 *Sbc* and *Sc* galaxies, evenly distributed over the whole sky. This project is currently being carried out (North to South) by G. Wegner (Dartmouth), M. Haynes (Cornell), R. Giovanelli and J. Salzer (Arecibo), L. N. da Costa (Rio), A. Savage (Coonabarabran), D. Mathewson and V. Ford (Mt. Stromlo). The goal is to obtain a large-scale flow solution for galaxies out to a redshift of 6500 km s^{-1}, using 21-cm line widths and I-band CCD photometry. The distance error of the method is 15%, as shown by preliminary data (see Figure 9). The

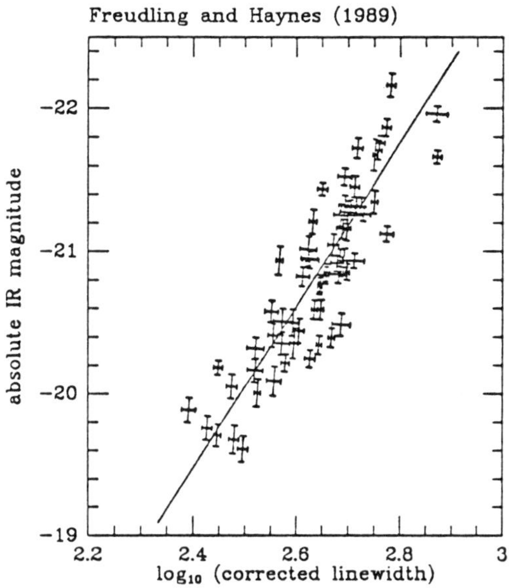

Fig. 9. Infrared magnitude *versus* line width relation for a subsample of about 80 Sc galaxies, produced by Freudling and Haynes (1989), which illustrates the small scatter of the relation (0.31 mags), used for the determination of redshift-independent distances.

TABLE III Distances of Spirals to Measure Global Flows

Goal:	Measure distances and motions of 2000 galaxies over whole sky
Method:	Infrared Tully-Fisher ($\Delta m \sim 0.31$ mag on single galaxy)
Sample:	Sbc and Sc galaxies; $i > 40°$, v < 6500 km s^{-1}, $n(z)\Delta z \sim$ const Δz:
	$4.5' \geq a \geq 2.5'$ for v < 2500
	$a \geq 2.0'$ $2500 \leq v < 4000$
	$a \geq 1.6'$ $4000 \leq v < 5500$
	$a \geq 1.3'$ $5500 \leq v < 6500$
Where:	I-band CCD imaging at: McGraw Hill, Kitt Peak, Siding Spring
	21 cm at: Arecibo, Green Bank, Parkes, Bonn, Nançay
Who:	da Costa, Ford, Giovanelli, Haynes, Mathewson, Salzer, Savage, Wegner
Status:	80% complete at 21 cm; 65% complete at I-band

sample was chosen in such a way that each spherical shell of thickness Δz contains the same number of galaxies as the preceding shell, by forcing the sample to have an approximately flat redshift histogram. The collection of the data is about 80% complete for the radio spectroscopy, and about 65% complete for the I-band photometry. The collapse of the 300-foot telescope dealt a major blow to the completion of the radio work; anticipated use of European facilities in Bonn and Nançay should provide the instrumental capacity necessary to complete the project, later this year. The program is summarized in Table III.

Localized Targets

The study of peculiar motions with a higher degree of detail than that mentioned in the previous subsection - investigations around voids in the LSS - is currently being carried out by several groups. Here we briefly report on ongoing work in two such regions, as detailed in Table IV. Methods are the same as described in the previous paragraph. Figure 9 shows a magnitude *versus* line width plot for a preliminary sample of about 80 galaxies in the Hercules region, obtained by Freudling and Haynes (1989). The magnitude dispersion about the mean relation is 0.31 mag, which translates into a distance error of 15%.

Table IV Distances of Spirals to Measure Peculiar Flows

Goal:	Motions and Distances in a void in Perseus-Pisces	Motions and Distances around the Hercules void
Sample:	350 Sbc's and Sc's, i >40 $a \geq 0.7'$, v < 6000 km s^{-1}	350 Sbc's and Sc's, i >40 $a \geq 0.7'$, v < 12,000 km s^{-1}
Where:	Arecibo, KPNO, McGraw Hill	Arecibo, 300-foot, KPNO
Who:	Giovanelli, Haynes, Wegner	Freudling, Haynes
Status:	21cm: 100%; I-band: 90%	21cm: 100%; I-band → 100%

ACKNOWLEDGEMENTS

This work was made possible by the support of the National Astronomy and Ionosphere Center, which is operated by Cornell University under a management agreement with the National Science Foundation. The author also wishes to thank J. Salzer for critically reading the manuscript, the Brazilian National Research Council (CNPq) for travel support to Rio and, in particular, L. N. and R. da Costa for exceptional organizational efforts, including this conference and the conference party.

REFERENCES

Binggeli, B. 1987, in *Nearly Normal Galaxies*, ed. S. Faber, (New York; Springer Verlag), p. 195.

Binggeli, B., Tammann, G. A., and Sandage, A. 1987, *A. J.*, **94**, 251.

Bothun, G. D., Beers, T. C., Mould, J. R., and Huchra, J. P. 1986, *Ap. J.* **308**, 510.

Davis, M., and Geller, M. J. 1976, *Ap. J.*, **208**, 13.

Davis, M., and Djorgovski, S. 1985, *Ap. J.*, **299**, 15.

Davis, M., Meiksin, A., Strauss, M. A., da Costa, L. N., and Yahil, A. 1988, *Ap. J. (Letters)* **333**, L9.

de Souza, Vettolani, P., and Chincarini, G. L. 1985, *Astr. Ap.* **143**, 143.
Dominguez-Tenreiro, R., and Martinez, V. J. 1989, *Ap. J. (Letters)* **339**, L9.
Dressler, A. 1980, *Ap. J.*, **236**, 351.
Eder, J., Schombert, J. M., Dekel, A., and Oemler, A., Jr. 1989, *Ap. J.*, **340**, 29.
Ferguson, H. C. 1989, Ph. D. Thesis, Johns Hopkins University.
Giovanelli, R., and Haynes, M. P. 1988, in *IAU Symposium No. 130, Large Scale Structures in the Universe*, ed. Audouze, J., Pelletan, M.-C., and Szalay, A., (Dordrecht: Kluwer), p. 113.
Giovanelli, R., and Haynes, M. P. 1989, *A. J.*, **97**, 633.
Giovanelli, R., Haynes, M. P., and Chincarini, G. L. 1986, *Ap. J.*, **300**, 77.
Hamilton, A. J. S. 1988, *Ap. J. (Letters)*, **331**, L59.
Haynes, M. P., and Giovanelli, R. 1988, in *Large Scale Motions in the Universe*, ed. V. C. Rubin and G. V. Coyne, S. J., (Princeton: Princeton University), p. 31.
Haynes, M. P., Giovanelli, R., and Chincarini, G. L. 1984, *Ann. Rev. Astr. Ap.*, **22**, 445.
Hollósi, J., and Efstathiou, G. 1988, in *Large-Scale Structures in the Universe. Observational and Analytical Methods.*, ed. W. C. Seitter, H. W. Duerbeck, and M. Tacke, (Berlin: Springer-Verlag), p. 106.
Hubble, E., and Humason, M. 1931, *Ap. J.*, **74**, 43.
Iovino, A., Melnick, J., and Shaver, P. 1988, *Ap. J. (Letters)*, **330**, L17.
Maurogordato, S., and Lachieze-Rey, M. 1987, *Ap. J.*, **320**, 13.
Melnick, J., and Sargent, W. L. W. 1977, *Ap. J.*, **215**, 401.
Oemler, A. 1974, *Ap. J.*, **194**, 1.
Postman, M., and Geller, M. J. 1984, *Ap. J.*, **281**, 95.
Salzer, J. J., Murray Hanson, M., and Gavazzi, G., 1990, *Ap. J.*, **353**, 39
Sandage, A., Binggeli, B., and Tammann, G. A. 1985, *A. J.*, **90**, 1759.
Schneider, S. E., Helou, G., Salpeter, E. E., and Terzian, Y. 1983, *Ap. J. (Letters)*, **273**, 1.
Thuan, T. X., Gott, J. R., III and Schneider, S. E. 1987, *Ap. J. (Letters)*, **315**, L93.
Tully, R. B. 1988, *A. J.*, **96**, 73.
White, S. D., Tully, R. B., and Davis, M. 1988, *Ap. J. (Letters)*, **333**, L45.
Xia, X. Y., Deng, Z. G., and Zhou, Y. Y. 1987, in *IAU Symposium No. 124, Observational Cosmology*, ed. by A. Hewitt, G. Burbidge, and L. Z. Fang, (Dordrecht: Reidel), p. 363

DISCUSSION

M. Strauss: There are interesting papers by Alimi and Blanchard (*Astr. Ap. Lett.* 1988, *Nature*) which show that the correlation functions of CfA galaxies fainter than about −20 show absolutely no dependence on luminosity. Why is the scatter you measure in the Tully-Fisher diagram so much smaller than what, *e.g.*, Faber and Burstein quote (0.31 *versus* 0.45 mag)?

R. Giovanelli: Thanks for pointing out the reference to the work of Alimi and Blanchard, which I was unaware of. As for the dispersion in the Tully-Fisher relation, I pointed out that the number quoted by Freudling and Haynes is based on preliminary data for 80 galaxies, and I'd rather not speculate on the difference compared to other results until the final numbers are available.

M. Strauss: Can you explain in a little more detail how you construct a volume-limited subsample from your catalog?

R. Giovanelli: One chooses an absolute magnitude limit and samples galaxies brighter than the limit out to a distance for which satisfactory completeness is available to that limit.

M. Davis: I was intrigued to see your results showing luminosity segregation in the Perseus-Pisces region. Certainly such segregation is expected in our models, but the easiest parameterization for comparison to models is to study the segregation *versus* circular velocities of the galaxies, or depth of the potential well. Have you sorted the data in this manner?

R. Giovanelli: You refer to work similar to White, Tully and Davis (*Ap. J. (Letters)*, **333**, L45, 1988), which I must apologize for having had time to only briefly mention. We have looked at the data in the same way you did, but we have not cleaned the sample from biases deriving from the way the HI subsample is selected, which is why I chose to keep a low profile on the topic.

J. Lucey: How many spirals do you have in the Great Attractor region?

R. Giovanelli: It depends on what we call the Great Attractor region. If we take the currently most favored view that it subtends a circle of 60° diameter, and allowing for some overdensity in that region of our mainly field spirals, then I would estimate that we'll end up with about 300 galaxies there.

POSSIBLE LARGE-SCALE STRUCTURES IN THE UNIVERSE

R. TRIAY
Centre de Physique Théorique, Laboratoire Propre, LP7061 and
Université de Provence, Université d'Aix-Marseille I
CNRS, case 907 Luminy, F-13288 Marseille Cedex 9, France

ABSTRACT Two large-scale dipole anisotropies are reported: a singular zone in the 3-dimensional distribution of quasars and a common orientation in the sky distribution of large-scale H I extents of galaxies. These anisotropies are within 6° to each other, the mean pole is at equatorial coordinates (1950) RA $\approx 5^h\,33^m$, Dec $\approx +60°\,20'$. The statistical significance, as well as this coincidence, of both anisotropies are discussed. It turns out that the existence of the zone must be checked with on-purpose observations, while the anisotropy found in the 21-cm position-angle distribution of galaxies seems to be probable. A speculative scenario is proposed if these anisotropies are real.

INTRODUCTION

Fliche *et al.* (1982) pointed out that the 3-dimensional distribution of quasars shows a singular zone about 200 Mpc wide and covering half the sky. This zone was utilized to determine a Friedmann-Lemaître model with eternal expansion defined by $q_0 \approx -1.1$ and $0.05 \leq \Omega_0 \leq 0.2$, giving an age of 13-20 h_0^{-1} Gyr. The persistence of this zone is checked with twice as much data since the previous publication, and its statistical significance is discussed. If this zone is real, a cosmic dipole anisotropy is expected and oriented nearby to the Super Galactic Pole. The description of this anisotropy is given. This work might have a renewal of interest because of the large scale anisotropy in the sky distribution of extragalactic radio sources found by Shaver and Pierre (1988), Shaver (1989). Recently, Fliche and Souriau (1990) pointed out an anisotropy in the sky distribution of the large-scale H I extents of galaxies showing the same pole; we summarize their result.

THE SPACE DISTRIBUTION OF QUASARS

The analysis of the 3-dimensional distribution of quasars as a whole, instead of investigating the distribution of redshifts and of positions in the sky separately, is not so usual as that. The reason is that such an approach is (as a matter of fact) model dependent because of the large redshift extent and that a non-Euclidian geometry is not trivial to handle. On the other hand, 2-dimensional plots are of great use in such an investigation because the eyes are powerful tools for checking whether structures are produced by selection effects. Herein, we define a part of the approach which makes this analysis easier, further details are given in Triay (1981). This investigation is developed in Friedmann-Lemaître models without assuming (*a priori*) a zero cosmological constant Λ, which might seem to be a problem. Indeed, it is generally believed that Λ is a free parameter of Einstein's equations, which was introduced artificially to obtain a static description of the Universe at the detriment of the simplicity of the model (the Poisson equation is modified and reads $div\,(g) = -4\pi G\rho + \Lambda$, hence the vacuum is repulsive if $\Lambda > 0$ and the related equilibrium density reads $\Lambda/(4\pi G)$. To show the arbitrariness of the usual approach, Souriau (1985) pointed out that the Λ term appears in the field equations of the gravitation, as obtained from the general theory of relativity, as well as Newton's constant G; hence, the previous arguments for removing Λ might be used also to remove G. On the other hand, observational evidence suggests that $\Lambda < 2\times10^{-55}$ cm^{-2} in our galactic neighborhood (Peach 1970), but it must be noted that the related equilibrium density $< 2\times10^{-28}$ g cm^{-3} shows that Λ may not be negligible at scales of clusters; Fliche *et al.* (1982) estimated a value of $\Lambda \approx 4\times10^{-56}\,h_0^2$ cm^{-2}.

Cosmographic Tools

The Friedmann-Lemaître models are defined by the Hubble constant $H_0 = 100h_0$, the deceleration parameter q_0 and the density parameter Ω_0; the reduced cosmological constant reads $\lambda_0 = \Omega_0/2 - q_0$. The quasars have fixed positions in the comoving space V_3, which has a constant curvature $k_0 = 3/2\Omega_0 - q_0 - 1$. The world lines of photons are projected on V_3-geodesics, and the tangent space at the observer's location is the celestial sphere (see Souriau 1974). Hereafter, we assume that $k_0 > 0$ (while this approach can be easily extended to $k_0 \leq 0$), V_3 is a 3-sphere of radius $k_0^{-1/2}$. By embedding V_3 in 4-dimensional Euclidian space, and with a particular choice of coordinates, the position of a quasar of redshift z seen toward the line of sight $\mathbf{x}$ is given by the 4-vector $k_0^{-1/2}\,\Xi(\mathbf{x},z)$, where

$$\Xi(\mathbf{x},z) = \begin{bmatrix} \sin(\tau\, k_0^{1/2}\,\mathbf{x}) \\ \cos(\tau\, k_0^{1/2}) \end{bmatrix} \tag{1}$$

is a 4-vector localizing a point of S^3; $\tau = \tau(z)$ is the *comoving distance*, and $\tau\, k_0^{1/2}$ denotes the *angular distance*. The distribution in V_3 of a quasars catalog depends on the values of the parameters q_0 and Ω_0. The calculation of the comoving distance d between two quasars $(\mathbf{x}_1, z_1)$ and $(\mathbf{x}_2, z_2)$ is obvious in this frame-work, it is given by the following equation

$$\cos(d\, k_0^{1/2}) = « \Xi(\mathbf{x}_1, z_1), \Xi(\mathbf{x}_2, z_2) » \tag{2}$$

where the symbol «,» denotes the scalar product (in 4-D as well as in 3-D). Note that the statistical investigation of the distribution of quasars is more convenient on S_3 rather than on V_3 because it is normalized. With equation (1), let $\mathbf{P}$ be a point of S_3 defined by the line of sight $\mathbf{p}$ and the angular distance π and $\mathbf{T}$ the earth's location on V_3 (obtained with $z = 0$). The variables

$$\begin{aligned} u &= \sin(\tau\, k_0^{1/2})\, « \mathbf{x}, \mathbf{p} » \\ v &= \cos(\tau\, k_0^{1/2}) \end{aligned} \tag{3}$$

define a model-dependent 2-D diagram which is the orthogonal projection of S_3 on the plane generated by the 4-vectors $\mathbf{P}$ and $\mathbf{T}$. Therefore, the whole Universe is projected on a disk. The interesting feature is that, for the true values of q_0 and Ω_0, a uniform distribution of quasars in the Universe appears still uniform through this projection. Indeed, the comoving volume element reads $dV = \sin^2(\tau)\, d\tau\, d\Omega$, where the solid angle element can be written as $d\Omega = d(«\mathbf{x},\mathbf{p}»)\, d\varphi$; hence with (3) we have $dV = 2\pi\, du\, dv$.

A Singular Zone In The Space Distribution Of Quasars

The distribution of 2566 quasars of secure redshift given in the Veron-Cetty and Veron

(1987) catalog is plotted on Figure 1(a) with the world model defined by $q_0 = -1.12$ and $\Omega_0 = 0.03$; the (1950) equatorial coordinates of the direction **p** are RA $= 5^h\ 42^m$ and Dec $= +4°\ 40'$. The distribution covers more than half of the disk, which corresponds to the percent of the observed Universe through the distribution of known quasars. The v axis is the vector radius which points **T** at the border of the disk ($u = 0$, $v = 1$). In accordance with equation (3), the region of the Universe at redshift z is projected on a chord which is orthogonal to the v axis. Note on Figure 1(a) that the number density decreases with v (from **T**) until reaching a bump and hence decreases again. This bump corresponds to a redshift $z \approx 1.7$ where the Lyman-α emission line enters in the observable wavelength range (which favors the quasar detection by optical techniques). In accordance with equation (3), deep surveys produce trails about ellipses having the v axis as the major axis.

The anomalies due to selection effects can be seen more clearly by looking sideways at Figure 1(a). Figure 1(b) is obtained by compressing Figure 1(a) *versus* the horizontal axis. The symbol (S) indicates a lack of quasars within a zone $\approx 65\ h_0^{-1}$ Mpc wide. Figure 1(c) is an enlargement of this region; the expected number of missing objects is $\langle n_{\rm missing} \rangle =$ 26 (= 28 − 2) as given by the two adjacent bands of same width which contain 21 + 35 objects (the values of $q_0 = -1.12$ and $\Omega_0 = -0.03$ were obtained by maximizing this number). By going back to Figure 1(a) one sees that this zone is located about a diameter of the disk, which is transverse to anomalies produced by selection effects. It is obvious that (S) is not clearly prominent, while the probability that it is produced by a statistical fluctuation is $\exp(-\langle n_{\rm missing} \rangle) \approx 10^{-12}$ (as obtained by assuming a local Poisson distribution), and the real statistical significance of this lack of objects must be determined with on-purpose observations. An indication of the likelihood of (S) is that the expected number of missing objects is larger than the previous estimate $\langle n_{\rm missing} \rangle \approx 18$ of Fliche *et al.* (1982), while the width is much smaller. It must be noted that the width does account of the statistical significance but intervenes for checking the physical scenario which can be proposed for explaining such an absence of quasars. This approach is not the aim of this paper, and further details can be found in Desert and Schatzman (1986). What is the physical situation of (S)? It turns out that any region in the Universe which is projected on a diameter in Figure (1a) is a plane for the Riemannian geometry. Thus (S), as defined by the neighboring quasars, is a plane which covers half the sky and extends to redshift bounds of the quasars catalog. Its nearest point is toward **−p** at a redshift $z \approx 0.85$, and the Earth shows no peculiar location with respect to this plane.

THE LARGE-SCALE H I EXTENTS OF GALAXIES

A review of the main properties of the large-scale H I density distribution in the outer parts of spiral galaxies and in interacting systems can be found in Sancisi (1987). These H I structures are seen through 21-cm maps, they show tails and bridges of various shapes and extending up to 200 kpc with masses up to $10^{10}\ M_\odot$. It is general believed that these features are of tidal origin, especially when a bright companion is found in the optical image. However, peculiar H I structures which show some resemblance with these few clearly interacting systems are observed around apparently isolated galaxies. It is believed that these puzzling examples may be the result of recent gas accretion or mergers.

A Common Orientation Of 21-cm Position Angles

The sample utilized by Fliche and Souriau (1989), which is given in Table I, consists of 83 galaxies with published 21-cm maps showing large-scale H I density distributions. The

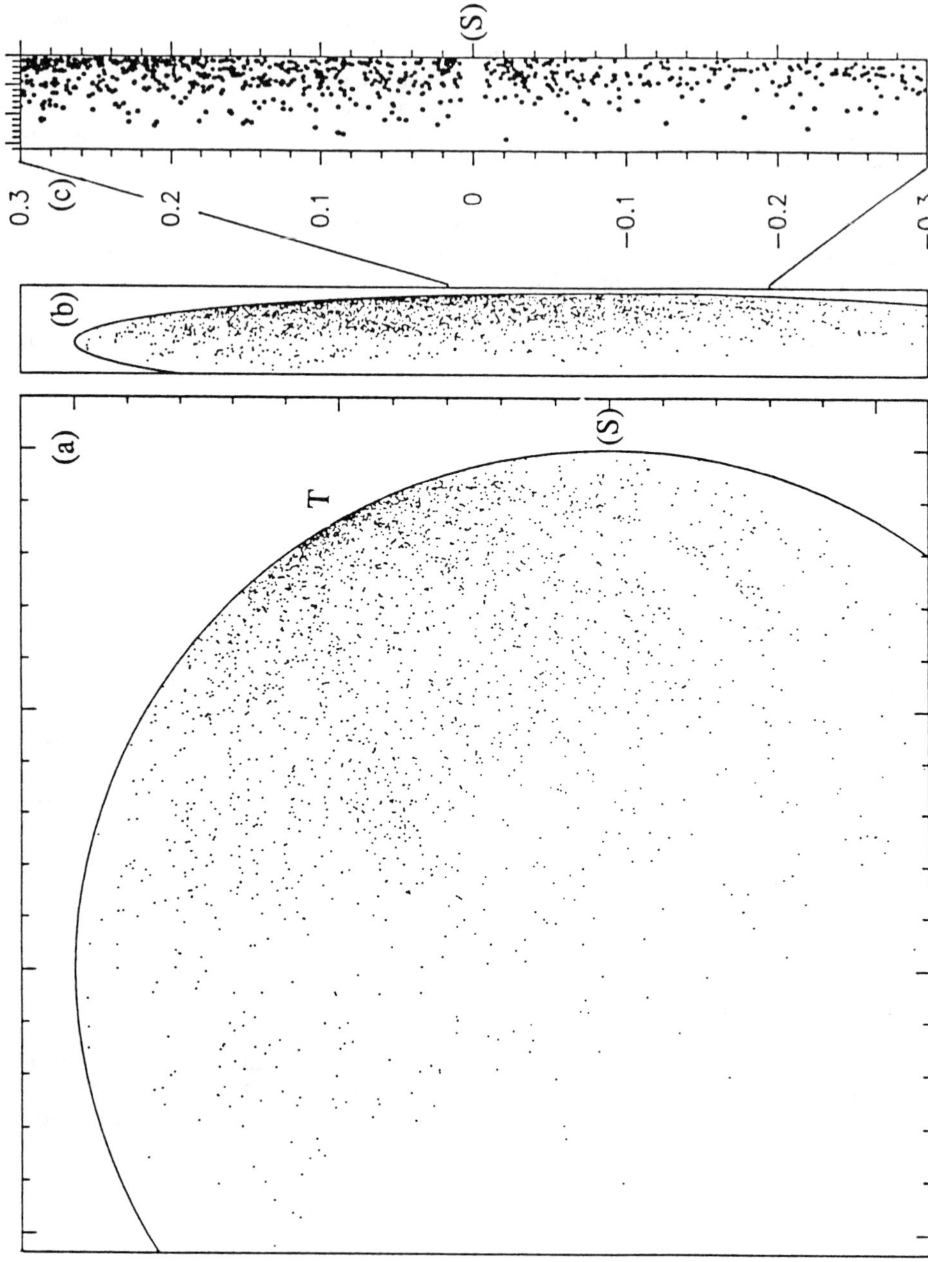

Fig. 1. A map of the Universe. The distribution in the Universe of 2566 quasars of secure redshift given in the Veron-Cetty and Veron (1987) catalog. The high number density region corresponds to the Earth's location. The symbol (*S*) locates the singular zone.

position angle *pa* gives the orientation of the H I extent with respect to the North pole. The evidence of anisotropy in the distribution of position angles is obvious: the Kolmogorov-Smirnov test gives a probability of $\approx 5 \times 10^{-7}$ that the measured deviation from uniformity is due to a random fluctuation. However, this result depends on the pole (the North pole) which is used to define the position angle. Hence, one might ask whether the distribution shows a higher anisotropy by changing this pole. Then, letting $\alpha(\mathbf{p})$ be the position angle with respect to a given pole $\mathbf{p}$, one has $pa = \alpha(\mathbf{p})$ if $\mathbf{p}$ is the North pole. The first step of this analysis is to note that two possible situations might account for this anisotropy: either the HI extents are elongated toward a common pole $\mathbf{p}$ or orthogonal to that direction, and it turns out that the second scenario is clearly preferred. The significance of this results is investigated by means of the following statistic,

$$\varsigma(\mathbf{p}) = (N^{1/2}/0.31) \{ (1/N) \sum_{k=1}^{N} |\cos [\alpha(\mathbf{p})_k]| - 0.64\}, \tag{4}$$

which is normal random variable if the distribution $\{\alpha(\mathbf{p})_k\}_{k=1,N}$ is uniform. Then, let t_P be the value of (4) obtained from the sample when choosing a given pole $\mathbf{p}$ and let

$$\pi(\mathbf{p}) = Prob\,(|\varsigma(\mathbf{p})| > |t_P|) = 1 - erfc\,(|t_P|)$$

be the probability that the random variable $\varsigma(\mathbf{p})$ is more distant from its mean 0 than t_P. Then $\pi(\mathbf{p})$ gives the probability that the measured deviation from uniformity is due to a random fluctuation (*i.e.* it is as smaller as the orientation of H I extents of galaxies is not uniform with respect to this pole). The pole of largest anisotropy was searched all over the sky.

Figure 2 shows the iso-probability contours $\pi(\mathbf{p})$ in a 'wide-angle' view of the sky. The outer border of black regions corresponds to a probability of $\pi(\mathbf{p}) = (1/2)10^{-n}$ with $n = 0$ (for the outer contour) to $n = 7$ (for the central contour). One can note that the

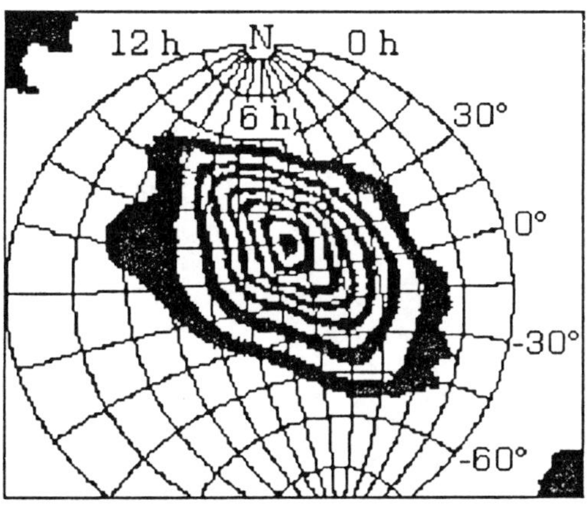

Fig. 2 Level of anisotropy in the sky. The outer border of black regions correspond to a probability of $(1/2)10^{-n}$ with $n = 0$ (for the outer contour) to $n = 7$ (for the central contour) that the measured deviation from uniformity of the distribution of position angles with respect to a given pole chosen in the map is due to a random fluctuation. The black regions at the edges of the diagram are antipodal to the outermost contour.

TABLE I The orientation of the H I extent of 83 galaxies

Name	pa	δ	Obs	Name	pa	δ	Obs
NGC2841	16	0	7	NGC6503	12	28	5
NGC4038/9	5	0	7	NGC2903	26	29	3
NGC7331	16	1	7	NGC4747	46	29	7
NGC3953	10	1	5	NGC4244	45	30	3
NGC2655	14	2	3	NGC4236	16	32	6
NGC3729	9	2	7	NGC3521	15	33	5
NGC5236, M83	172	2	3	IC10	43	35	4
NGC2403	125	2	6	NGC5301	174	38	7,3
NGC300	162	3	6	NGC3198	32	38	7
NGC3359	172	3	5	IC1727	150	39	5
NGC5457, M101	35	3	3	NGC4559	157	39	3
NGC891	23	4	7	NGC5194, M51	170	39	7
NGC1961	90	4	7	A0813+70	175	39	2
NGC2712	162	4	3	NGC4395	154	41	5
NGC3998	15	4	7	NGC4490/85	155	42	3
UGC2885	44	5	7	NGC1058	165	43	7
NGC4151	19	5	7	NGC4258, M106	152	43	7
NGC3718	13	7	7	NGC4214	150	44	3
NGC7640	165	8	5	NGC5033	160	44	7
NGC3031, M81	152	9	7	NGC1560	25	50	5
NGC3938	20	9	7	NGC2683	35	51	5
A235915, DDO221	0	10	5	NGC5383, Mrk281	85	52	7
NGC4535	0	11	1	NGC772	130	55	5
NGC247	176	11	3	NGC672	65	55	5
NGC598, M33	21	11	6	IC2574, DDO81	45	56	5
NGC784	0	13	5	NGC2146	160	57	5
NGC4096	0	13	5	IC239	147	60	2
NGC3368, M96	170	16	5	A0355	125	60	5
NGC2715	163	16	3	NGC2336	0	60	5
NGC4618/25	35	16	3	NGC6384	30	67	1
NGC5364	30	16	1	NGC253	51	68	6
NGC4262	29	17	7	NGC4631	86	68	7
NGC5832	71	17	3	NGC1530	0	72	5
NGC4725	35	18	7,4	NGC3109, DDO236	93	76	3
NGC660	160	19	5	NGC6946	62	77	6
DDO154	39	21	3	NGC4826, M64	115	82	5
IC342	39	23	2	NGC3338	90	85	5
NGC2805	130	24	7	NGC925	115	86	5
NGC628, M74	25	25	7	NGC4736, M94	114	87	7
NGC2541	170	25	5	NGC2366, DDO42	35	89	5
NGC224, M31	33	26	2	NGC5055, M63	116	89	7
NGC4449	45	28	7				

Observatory key: 1 = Arecibo, 2 = Cambridge, 3 = Effelsberg, 4 = Jodrell Bank, 5 = NRAO, 6 = Owens Valley, 7 = Westerbork highest anisotropy is given by a single pole located in the central contour. A more precise estimate, which takes into account that the determination of a pole of anisotropy is less accurate as the galaxy involved in the calculation is nearer to this pole, locates the pole of anisotropy at RA $\approx 5^h\ 24^m$, Dec $\approx +8°$. Let $\delta = |pa - \alpha(\mathbf{p})|$ be the gap between the orientation of H I extents of galaxies and the one which is expected by the anisotropy. The galaxies are listed in Table I with increasing δ and grouped within steps of 15°. The decrease of the number of galaxies in the groups with δ shows the magnitude of the anisotropy. It is interesting to note that a χ^2 statistic is not suitable to evaluate this result since a permutation of these groups will not change the statistical significance at the detriment of the geometric interpretation. Since the observatories where the galaxies were observed are distributed at random in Table I, the possibility that this anisotropy is a bias due to selection effect in observation is excluded. In addition, the effect of elongation of the beamwidth was also checked.

DISCUSSION

If the dipole anisotropy found in the distribution of 21-cm position angles of galaxies is real, it has to be explained by a non-geocentric scenario. The simplest situation (*i.e.* the one of lowest symmetry) is that the large H I extents of galaxies are parallel sheets in space. Since the position angles of optical regions of galaxies do not show a significant anisotropy, (MacGillivray *et al.* 1982), a warping is required to combine these two results. This effect (which, sometimes, extends far out beyond the optical image) was checked on the sample and it turns out that the H I external regions show in general the predicted orientation.

For the galaxy NGC4013 (Botema *et al.* 1987, Sancisi 1987), which is not in the sample, the expected position angle $pa \approx 13°$ coincides better with the H I regions beyond the elbows than with the ones in the inner part (near the optical disk). This suggests that the measured position angle will be as close to the predicted one as the H I extent is large. Conversely, the edge-on galaxy NGC891, (Sancisi and Allen 1979, Sancisi 1987), is not warped and its H I structures are elongated toward the expected orientation (see Table I). This scenario might explain puzzling systems without (tidal) companions. Two striking examples are the cases of the SO galaxy NGC1023 (Sancisi *et al.* 1984) and of the Seyfert galaxy Mkn348 (Heckman *et al.* 1982, Sancisi 1987). These galaxies are not in the sample, and in both cases the expected position angle ($pa \approx 52°$ for NGC1023 and 26° for Mkn348) coincides with the orientation of their H I structures. Moreover, the galaxies with largest H I extent (NGC2841, M83, M101, DDO154, NGC628, NGC4449) show a pa within 30° of the expected orientation (see Table I). Nevertheless, this is not as clear cut as that, since to this category of isolated systems belong also the galaxies NGC4631 (Weliachev *et al.* 1978) and NGC2146 (Fisher and Tully 1976), which show a 21-cm position angle more than 45° off the expected orientation (the gap is $\delta = 68°$ for NGC4631 and 57° for NGC2146, see Table I). On the other hand, if this effect is real, as the tidal origin was proposed to account for these structures, the same process might disrupt the predicted symmetry of these systems. This should produce irregular and asymmetric H I distributions and make the galaxy H I deficient, as it is the case for galaxies close to cluster centers.

In this scenario, the hypothesis of primordial gas remnants is preferred. A remarkable candidate is the H I ring of NGC4262 (Krumm *et al.* 1985), apparently in circular rotation about the galaxy center. It was proposed as a remnant of a primordial proto-galaxy cloud, while the capture gas hypothesis was used to explain the large relative inclination of the gas ring and optical disk. On the other hand, this scenario does not account for the orientation of the giant intergalactic H I ring found by Schneider (1985), its position angle is 46° off the expected orientation.

Since most of the galaxies in the sample belong to the Local Super Cluster (LSC), one may ask whether the origin of such an anisotropy is not located into it. However, the flat distribution of the LSC has a pole (*RA* $\approx 6^h\ 52^m$, *Dec* $\approx -16°$) which is more than 30° off, unless (if the scenario is real!) the LSC pole might require a more accurate determination. On the other hand, the dipole anisotropy found in the distribution of quasars is within 6°. Then the singular zone (S) found in the distribution of quasars is plane parallel to large-scale H I sheets of galaxies.

If this is not an accidental coincidence, a common origin for these anisotropies must be found, and one might speculate the following. As Fliche *et al.* (1982) suggested, since (S) is visible at high redshift, the related anisotropy has a primordial origin. Therefore, the appearance (at large scale) of the medium where the galaxies were born was stratified as a 'mille feuille' (pastry) and hence partially disrupted (at small scale) by the galaxies and/or clusters formation.

Some observations in the Galactic neighborhood coincide with this scenario: the Local Cloud (de Vaucouleurs 1975), the system of the Galaxy with its close satellites and the high velocity H I clouds (a warp of $\approx 80°$, Fliche *et al.* 1982), the system of M31 (Brinks and Burton 1984), and its major satellites (see Fliche *et al.* 1983), show such an anisotropy. A possible prediction is that, if the mass of the H I sheet is much more important than observed (not seen because ionized), this formation process would spin the galaxies as tops (because of the angular momentum conservation of the part of the primordial H I sheet which collapsed).

CONCLUSION

The anisotropy in the sky of the distribution of large-scale H I extents of galaxies seems to have a real existence, while the singular zone found in the 3-dimensional distribution of quasars still has to be confirmed with on-purpose observations. The world model which shows the higher anisotropy ($q_0 = -1.12$ and $\Omega_0 = 0.03$) in the quasar distribution is spatially closed with an eternal expansion. The main problem is that its age is $\approx 25\ h_0^{-1}$ Gyrs, which is really too old. However, the zone is still (but less) statistically significant with $\Omega_0 = 0.1$-0.2 giving an age of $\approx$ 13-16 h_0^{-1} Gyrs, which still favors the high value of H_0. If the singular zone is real, the coincidence of these anisotropies suggests that the medium where the galaxies were formed was stratified as a 'mille feuille'.

ACKNOWLEDGEMENTS

I thank H. H. Fliche and J. M. Souriau for helpful discussions and for providing me with details of their unpublished work.

REFERENCES

Botema, R., Shostak, G. S., and van der Kruit, P. C. 1987, *Nature*, **328**, 401.
Brinks, E., and Burton, W. B. 1984, *Astr. Ap.*, **141**, 195.
de Vaucouleurs, G. 1975, *Ap. J.*, **202**, 319.
Desert, F. X., and Schatzman, E. 1986, *Astr. Ap.*, **158**, 135.

Fisher, J. R., and Tully, R. B. 1976, *Astr. Ap.*, **53**, 397.
Fliche, H. H., and Souriau, J. M. 1990, *Astr. Ap.* **233**, 317.
Fliche, H. H., Souriau, J. M., and Triay, R. 1982, *Astr. Ap.* **108**, 256.
Fliche, H. H., Souriau, J. M., and Triay, R. 1983, in proceedings of *Journées Relativistes*, ed. S. Benenti, M. Ferreris, and M. Francaviglia (Turin).
Heckman, T. M., Sancisi, R., Sullivan, III, W. T., and Balick, B. 1982, *M.N.R.A.S.*, **199**, 425.
Krumm, N., van Driel, W., and van Woerden, H. 1985, *Astr. Ap.* **144**, 202.
MacGillivray, H. T., Dodd, R. J., MacNally, B. V., Corwin, H. G. 1982, *M.N.R.A.S.*, **198**, 605.
Peach, J. V. 1970, *Ap. J.* 159, 753.
Sancisi, R. 1987, in *QSO Absorption Lines: Probing the Universe*, ed. J. C. Blades, D. A. Turnshek, and C. A. Norman, (Baltimore: Space Telescope Science Institute), p. 241.
Sancisi, R., and Allen, R. J. 1979, *Astr. Ap.* **74**, 73.
Sancisi, R., van Woerden, H., Davies, R. D., and Hart, L. 1984, *M.N.R.A.S.*, **210**, 497.
Schneider, E. S. 1985, *Ap. J. (Letters)*, **288**, L33.
Shaver, P. A. 1989, in *The quest for the Fundamental Constants in Cosmology*, 24th Rencontres de Moriond, in press.
Shaver, P. A., and Pierre, M. 1988, *Astr. Ap.* submitted.
Souriau, J. M. 1974, in *Coll. Internationaux CNRS*, **237**, 59.
Souriau, J. M. 1985, *C. R. Académie des Sciences*, serie gén. **2**, 213.
Triay, R. 1981, thesis.
Veron-Cetty, M. P., and Veron, P. 1987, *Catalog of Quasars and Active Nuclei, 3rd Edition*, (Garching: European Southern Observatory).
Weliachev, L., Sancisi, R., and Guelin, M. 1978, *Astr. Ap.*, **65**, 37.

DISCUSSION

J. Primack: To what extent could selection biases be responsible for the effects you discussed?

R. Triay: The situation of this empty (of quasars) zone, as described in the diagrams, shows that selection effects cannot be responsible for the lack of objects. A line of sight is projected onto an ellipse, and a sphere of redshift is projected onto a chord of the disk; these two effects are transverse to the singular zone (the redshift of the zone depends on the line of sight where it is observed in the sky).

D. Lambas: The position angles of classical double sources seem to show strong anisotropy (P. Birch, *Nature*, **298**, 451, 1982). The pole is at right ascension $13^h\ 20^m$ and declination $-40°$, which is far away from your two poles. Could you comment about this?

R. Triay: I do not know this reference. The scenario which is suggested is that the large H I extents are (statistically) flat and parallel. If the orientation of these double sources is related to this anisotropy, then it is either orthogonal or parallel to the direction of the anisotropy. In the first case it is not possible to detect such an anisotropy from the distribution of position angles, while in the second case it might be detectable.

DOMAIN WALLS AND LARGE-SCALE STRUCTURE

BARBARA S. RYDEN
Harvard-Smithsonian Center for Astrophysics
60 Garden Street, Cambridge, MA 02138

ABSTRACT The presence of a scalar field with two degenerate discrete minima leads to the formation of domain walls. Numerical simulations of the evolution of domain walls in an expanding universe show that there will be approximately one wall per horizon at any given time. Closed walls within the horizon collapse rapidly; thus, there is not a hierarchical distribution of domain wall bags of different radii. The *rms* velocity of the walls is approximately half the speed of light. If domain walls survive to the present, they cannot, without violating the observed constraints on the quadrupole anisotropy of the microwave background, have generated the large-scale structure of the universe.

INTRODUCTION

Topological defects arising from spontaneous symmetry breaking, as reviewed by Vilenkin (1985), have been examined as a possible source of large-scale structure. More intellectual effort has been expended on strings (linear defects) than on domain walls (two-dimensional defects). The neglect of walls has resulted from the realization of Zel'dovich *et al.* (1975) that high energy walls, forming before recombination, rapidly dominate the energy density of the universe, incidentally making hash out of the cosmic microwave background. More recently, however, Hill, Schramm, and Fry (1989) have proposed a late phase transition, at a redshift $z \sim 10$ - 100, as a means of producing domain walls which are not cosmologically disastrous. It has been suggested (Hill, Schramm, and Fry 1989, Stebbins and Turner 1989) that domain walls can contribute to large-scale structure and streaming motions in the universe. The purpose of this paper is to review recent numerical work by Press, Ryden, and Spergel (1989) on the dynamics of domain walls in an expanding universe, and to point out why, in the simplest models, domain walls are unlikely to affect the large-scale structure of the universe.

In the simplest case giving rise to domain walls, a scalar field ϕ has a potential $V(\phi)$ with two degenerate discrete minima. The numerical results which follow assume a standard ϕ^4 potential,

$$V(\phi) = V_0(\phi^2/\phi_0^2 - 1)^2 \tag{1}$$

with potential minima at $+\phi_0$ and $-\phi_0$ and a barrier of height V_0. The precise shape of the potential, however, does not critically affect the dynamics of the walls. Future work (Ryden, Press, and Spergel 1989) will examine the evolution of hybrid string-wall networks arising from complex fields with multiple degenerate minima. This work deals with the simplest domain scenario.

In an expanding universe with expansion factor $a(t)$, it is useful, for computational purposes, to define a conformal time η through the relation $d\eta = dt/a$. The equation of motion for a ϕ field with potential V then takes the form of the wave equation

$$(\partial^2\phi/\partial\eta^2) + 2\,(d \ln a \,/\, d \ln \eta)\,(1/\eta)\,(\partial\phi/\partial\eta) - \nabla^2\phi = -a^2\,(\partial V/\partial\phi) \tag{2}$$

Assuming a ϕ^4 potential, as given in equation (1), an adiabatic solution for the equations of motion (exact in the limit of a non-expanding universe) is

$$\phi = \phi_0 \tanh\,[(2V_0)^{1/2}\, a\, \gamma\, (z - vt)/\phi_0], \tag{3}$$

where

$$\gamma = (1 - a^2\, v^2)^{-1/2}. \tag{4}$$

This particular solution represents a planar wall moving with a velocity v along the z axis. A domain wall at rest will have a characteristic thickness, in comoving units, of

$$w \propto \phi_0/V_0^{1/2}. \tag{5}$$

The domain wall will have a surface density (energy per unit comoving area) $\sigma \propto wV_0 \propto \phi_0 V_0^{1/2}$, and a surface tension equal to $-\sigma$.

CALCULATIONS

Press, Ryden, and Spergel (1989) followed the dynamics of domain walls by numerically integrating the equations of motion for the ϕ field, both on a $200 \times 200 \times 200$ cubical grid and on a 1024×1024 square grid with slab symmetry imposed in the third dimension. Integration on the two-dimensional grid allows integration over a longer conformal time before the finite grid size affects the results. The equation of motion used is

$$(\partial^2\phi/\partial\eta^2) + \alpha\,(d \ln a \,/\, d \ln \eta)\,(1/\eta)\,(\partial\phi/\partial\eta) - \nabla^2\phi = -a^{\beta}(\partial V/\partial\phi) \tag{6}$$

The constants α and β which have crept into the above equation are $\alpha = \beta = 2$ in the case of the true equation of motion (equation 2). The factors α and β are introduced to surmount a severe numerical difficulty: the grid used is a constant comoving size, but the walls are of constant physical thickness. Thus, in comoving units, the wall thickness is inversely proportional to the expansion factor a. Once the comoving thickness of a wall is less than five grid zones or so, its dynamics are no longer accurately represented by the code. To prevent the thickness of our walls from dwindling to nothing, we found the dynamics of walls with constant comoving thickness, by setting $\beta = 0$ in equation (6). The factor α in the Hubble damping term is then adjusted to preserve various physical properties. It can be shown that the scalar field damps to a minimum at the rate

$$(\phi - \phi_0)_{rms} \propto a^{-\alpha/2 - \beta/4} \tag{7}$$

and that moving walls slow down at the rate

$$\gamma v \propto a^{-\alpha - \beta/2}. \tag{8}$$

Thus, the computations of Press, Ryden, and Spergel (1989) use $\alpha = 3$, $\beta = 0$ instead of the true values of $\alpha = \beta = 2$. In the limit that the wall thickness is much less than the radius of curvature of the walls, the dynamics of the wall network are not affected by the wall thickness. The errors of using a constant comoving wall thickness are seen only in the last stages of the collapse of a closed domain wall bag, when the radius of the bag becomes comparable to the comoving wall thickness.

The simulations were started at a conformal time $\eta = 1$. The initial value of ϕ at each grid point was randomly chosen between $-\phi_0$ and $+\phi_0$, and the time derivative of ϕ set equal to zero everywhere. The wall thickness is equal to five grid lengths. The universe is assumed to have a density $\Omega = 1$, with the bulk of the density being in the form of matter (the density in walls is assumed to be negligible.) After a conformal time $\eta \sim w$, the field settles down into domains where the ϕ lies in one or the other potential minimum, separated by walls of thickness w, where the value of ϕ has a large gradient.

Figure 1 shows snapshots of a two-dimensional simulation taken at times $\eta = 32$, when the box size is 32 times the horizon size (left panel) and $\eta = 93$, when the box size is 11 times the horizon size (right panel). The walls are under tension, so that walls which stretch from horizon to horizon tend to straighten themselves out, while closed wall 'bags' within the horizon tend to collapse, at velocities comparable to that of light, and dissipate their energy in the form of oscillations about the potential minima. At a given time, then, the typical radius of curvature of the walls is approximately equal to the horizon size, and there is roughly one wall within the horizon at any time.

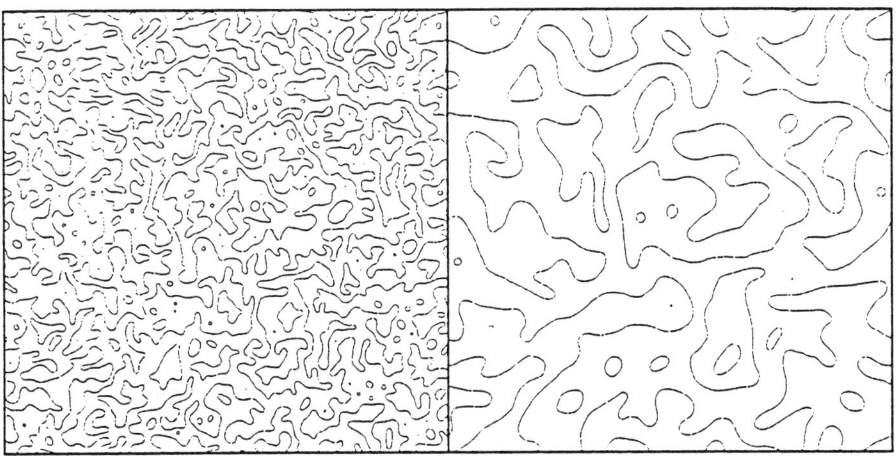

Fig. 1. Pictures of the domain wall network in a 1024 × 1024 numerical simulation. The left picture is taken at a conformal time $\eta = 32$, and the right picture is taken at $\eta = 93$. The wall thickness is $w = 5$.

If there were strictly one wall per horizon, then the comoving wall area A per comoving volume V would have the time dependence $A/V \propto \eta^{-1}$. Fitting the relation $A/V \propto \eta^{\nu}$ for ten separate two-dimensional simulations yields a mean and sample deviation of $\nu = -0.92 \pm 0.06$. The deviation of A/V from a $\nu = -1$ can also be fit by a logarithmic correction term, as shown in Figure 2, which plots $\eta A/V$ versus the logarithm of the conformal time. The best-fitting straight line, shown as the dashed line in Figure 2, gives

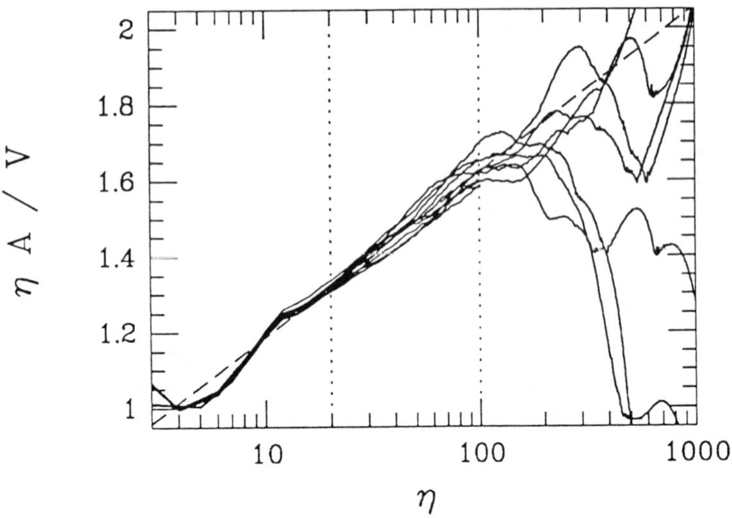

Fig. 2. The comoving wall area per comoving unit volume, multiplied by conformal time, for ten 1024 × 1024 simulations. The dashed line is the best fitting relation (given in equation 9) for the interval 20 < η < 100.

a relation

$$\eta A/V = 1.06 + 0.19 \ln(\eta/w), \tag{9}$$

where the domain wall thickness is $w = 5$. A comparison of the total kinetic energy of the ϕ field with the rest energy of the domain wall network gives an rms wall velocity of v ~ 0.4 c. Full three-dimensional simulations, run on a 200 × 200 × 200 grid, without the artificially imposed slab symmetry of the two-dimensional simulations, give results for A/V which are not in disagreement with the two-dimensional results. However, the cubical simulations have a limited dynamic range in time over which the results are valid. The walls and domains are not well formed until η is equal to a few times the wall thickness. Moreover, once η starts to approach the grid length, the domain dynamics are affected by the periodic boundary conditions of the grid.

CONCLUSIONS

It is possible to make an envelope-back estimation of the matter perturbations caused by the domain wall network, and the distortions in the cosmic microwave background which result. If there is one domain wall within the horizon at any given time, which the numerical results indicate is true to within a logarithmic factor, then at the present time t_0, the matter perturbations on a scale λ will have an amplitude

$$(\delta \rho / \rho) \sim \Omega_{\rm walls}(t_0)\,(\lambda / c t_0). \tag{10}$$

On the present horizon scale ($\lambda = ct_0$), the domain wall, moving at a velocity comparable to that of light, will produce a quadrupole distortion in the microwave background with amplitude $\delta T/T \sim \Omega_{walls}$. The observed upper limits on the quadrupole anisotropy then put an upper limit on the present wall density of roughly $\Omega_{walls} < 10^{-4}$, while in the past $\Omega_{walls} \propto (1 + z)^{-3/2}$. If domain walls behave in the manner indicated by the numerical simulations, with only one wall or so per horizon, moving about at a fair fraction of the speed of light, then domain walls cannot contribute significantly to large-scale structure without violating the microwave background constraints.

ACKNOWLEDGEMENTS

I would like to thank my collaborators Dave Spergel and Bill Press for letting me present our results at this meeting. The work presented in this talk was supported by the NSF and NASA.

REFERENCES

Hill, C. W., Schramm, D. N., and Fry, J. N. 1989, *Comments Nucl. Part. Phys.*, **19**, 25.
Press, W. H., Ryden, B. S., and Spergel, D. N. 1989, *Ap. J.*, **347**, 590.
Press, W. H., and Spergel, D. N. 1989, in preparation.
Stebbins, A., and Turner, M. S. 1989, *Ap. J. (Letters)*, **339**, L13.
Vilenkin, A. 1985, *Phys. Reports*, **121**, 263.
Zel'dovich, Ya. B., Kobzarev, I. Yu., and Okun, L. B. 1975, *Sov. Phys. JETP*, **40**, 1.

DISCUSSION

S. Shandarin: Did you study the geometry of your pictures? They look very similar to positive and negative regions of smoothed random Gaussian fields. Is this impression correct?

B. Ryden: This is correct. For our initial values, we assign a random value of ϕ to each grid point. At a later time t, the domain walls can be thought of, in an approximate sense, as the zero contours of a Gaussian field smoothed on the horizon size ct.

E. Bertschinger: You concluded that large-scale structure cannot be made by presently existing walls. However, the comoving horizon size at the transition to a matter-dominated universe ($z \sim 10^4$) was about the desired size, so that perturbations imprinted then might lead to large-scale structure. Is this consistent with microwave-background wall limits on the wall tension?

B. Ryden: Since Ω_{wall} varies as $(1+z)^{-3/2}$, to within a logarithmic correction factor, the present limit of $\Omega_{wall} < 10^{-4}$ imposed by the microwave-anisotropy limits implies that the domain walls had $\Omega_{wall} < 10^{-10}$ at the end of the radiation-dominated epoch, if they existed at that epoch and survived until the present.

STRING PRODUCTION AS A RESULT OF THERMAL FLUCTUATIONS

G. C. MARQUES
Instituto de Física, Universidade de São Paulo, São Paulo (SP), Brazil

ABSTRACT We study the production of cosmic strings as a result of thermal fluctuations. Within the dilute gas approximation we get explicit expressions for the length of strings as well as the contrast density in terms of the Free Energy per unit length of the string. In the high temperature limit one gets simple expressions for these quantities. The contrast density is scale free and, for the SO(10) model, has the right order of magnitude.

In this paper we will be concerned with string formation in the early Universe. Our analysis differs from previous ones in the fact that we have studied string formation as a result of thermal fluctuations. Thermal fluctuations induces the production of a large number of strings. As a matter of fact, above a critical temperature (T_c) even infinite strings, of the Nielsen-Olesen type (Nielsen and Olesen 1973), can be produced. Under these circumstances the system goes to a new phase for $T > T_c$. In this phase there is condensation of strings. The conclusion is that thermal fluctuations are extremely relevant in any string-driven structure formation (Bazeia et al. 1985, 1987).

The thermodynamical argument on which we have based our analysis is the so-called Kosterlitz-Thouless picture of phase transition. As a matter of fact, in their classical paper Kosterlitz and Thouless (1973) analyzed spin configurations called vortices which are configurations analogous to infinite strings studied here. In the case of strings, the argument of Kosterlitz and Touless can be stated in the following way: the Free Energy per unit length associated with a given string can be written as

$$F_{\text{string}}(T) = M - TS(T), \tag{1}$$

where M is the mass per unit length at zero temperature and $S(T)$, is an entropy term. Assuming that M is positive (as in the case at classical level) there is no chance, at low temperatures, for the appearance of strings in the system, since the energy cost for the introduction of a string in the system is infinite. However, as the temperature increases, the entropy term takes over the mass term, so that for a critical temperature we will have

$$F_{\text{string}}(T_c) = 0. \tag{2}$$

In the high-temperature limit we can write, on general grounds (Bazeia *et al.* 1985, Guerra *et al.* 1989, Carvalho *et al.* 1986):

$$F_s(T) = M_s - BT^2 \equiv M_s[1 - T^2/T_c^2]. \tag{3}$$

That is, at the critical temperature the cost in energy for introducing an infinite string is zero. In this way for temperatures high enough there is a condensation of strings. The system goes to a phase in which there is a condensate of strings. This entails a new phase of the system. Below the critical temperature only finite strings are allowed to exist in the system. In the following we will consider the formation of finite strings of length L, and, by making simple hypothesis, we will compute their lengths as well as their distribution.

We will make only two hypothesis. The first one is that the strings that are formed below the phase transition do not differ (except for the finite length) from infinite strings. More explicitly we assume that the Mass (M_s), Free Energy, and other relevant physical quantities associated with a single string can be obtained from analogous quantities defined per unit length, for the infinite string (M_s, F_s, ...), by just multiplying these quantities by the length of the string, L. That is:

$$\begin{aligned} f_{\text{string}} &= LF_s, \\ M_{\text{string}} &= LM_s. \end{aligned} \tag{4}$$

The length of the string should depend on the temperature. One can expect that as one increases the temperature the length of the string increases. That is, thermal fluctuations induces the creation of strings of larger and larger length as the temperature increases. As a matter of fact, we have seen that at the critical temperature even strings of infinite length can be produced. This means that at this temperature one should expect that

$$L(T_c) = \infty. \tag{5}$$

The other hypothesis that we will make is that the gas of strings is a dilute one. Under this hypothesis one can write, for the partition function Z of the dilute gas,

$$Z/Z_V = \exp\,(Z_s/Z_V) = \exp\{1/(2\pi)^3 \int d^3\mathbf{x} \int d^3\mathbf{p}\ \exp\,(-\beta E_{\text{string}})\}, \tag{6}$$

where Z_s is the partition function associated with one string.

In the non-relativistic limit one can write

$$Z_s/Z_V = V \exp\{-(1/T)\, f_{\text{string}}(T)\}\ [TM_{\text{string}}/2\pi]^{3/2}. \tag{7}$$

It follows from equations (6) and (7) that the Free Energy of a dilute gas of strings ($VF = -T \ln(Z/Z_V)$) will be given by

$$F = -T \exp\{-(1/T)\, f_{\text{string}}(T)\}\ [TM_{\text{string}}/2\pi]^{3/2}. \tag{8}$$

Although we have made use of the non-relativistic approximation one can show that the relevant term $TM^{3/2}_{\text{string}}$ can be obtained without resorting to this approximation. In Field Theory it follows by just taking into account the zero modes. This has been shown by Guerra *et al.* (1989).

By making use of our first hypothesis one can state that strings of length L, in the dilute gas approximation, have a free energy whose expression is

$$F = -T^{5/2} \exp\,\{-LF_s(T)/T\}\ [LM_s/2\pi]^{3/2}. \tag{9}$$

The length L is a parameter as yet unknown. However, L can be determined by remembering that, for each temperature, there will be a length L favored by statistical arguments. This length L is the one that minimizes the Free Energy. The system produces strings whose size can be determined from the condition

$$dF/dL|_{L=\langle L\rangle} = 0 \quad \text{and} \quad d^2F/dL^2|_{L=\langle L\rangle} > 0. \tag{10}$$

It follows then from equations (9) and (10), that the length of a string as a function of temperature will be

$$\langle L(T)\rangle = (3/2)\,[T/F_s(T)]. \tag{11}$$

At this point we would like to comment on our expression (11) for the length of strings produced as a result of thermal fluctuations. In the first place the length of these strings tends to zero as T goes to zero. This follows from (11) and expression (1). This is also expected on physical grounds. Finally, one can see from this expression that at the critical temperature, for which $F_s(T_c) = 0$, strings of infinite length are favored. In our scheme the critical temperature T_c defined by (2) is the same as the critical temperature defined by (5). Both schemes are, then, totally compatible.

Let us turn to the computation of the contrast density due to string production as a result of thermal fluctuations. The contrast density due to strings is defined, as usual, by

$$\delta\rho/\rho = \rho_{\text{string}}/(\rho_{\text{string}} + \rho_{\text{elem.part.}}). \tag{12}$$

The energy density associated with strings whose average number is $N(T)$ and whose mass is M_{string} is

$$\rho_{\text{string}} = \langle N(T)\rangle\, M_{\text{string}}/V. \tag{13}$$

Since $N(T)$ is given by (*e.g.* see Gross *et al.* 1981)

$$\langle N(T)\rangle = Z_s/Z_V. \tag{14}$$

One can write, from equations (13) and (14), for a dilute gas of strings of length L:

$$\rho_{\text{string}} = \exp\{-L(T)\,F_s(T)/T\}\,[T\,L(T)\,M_s/2\pi]^{3/2}\,L(T)\,M_s. \tag{15}$$

For the most favored strings, those obeying (10), one gets

$$\rho_{\text{string}} = [T/2\pi e]^{3/2}\,\{[3M_s/2]\,[T/F_s(T)]\}^{5/2}. \tag{16}$$

Expressions (11) and (16) are the main results of this paper.

For $F_S(T)$ given by (3) one gets

$$L(T) = (3/2)\,T/[M_s(1 - T^2/T_c^2)], \tag{17}$$

with T_c defined in (2) and

$$\rho_{\text{string}} = (3/2)\,(3/4\pi e)^{3/2}\,T^4/(1 - T^2/T_c^2)^{5/2}. \tag{18}$$

Since the contribution of the elementary particles can be written in terms of the number of degrees of freedom Fermionic (N_F) and Bosonic (N_B) as

$$\rho_{\text{elem.part.}} = (\pi^2/30)\,[N_B + (7/8)\,N_F]T^4, \tag{19}$$

the contrast density will be given as

$$\delta\rho/\rho = 1/\{1 + (2\pi^2/90)\,[N_B + (7/8)\,N_F]\,[1 - T^2/T_c^2]^{5/2}\,[4\pi e/3]^{3/2}\}. \tag{20}$$

For T below T_c and for $N_B + (7/8)N_F >> 1$ the contrast density is small and can be approximated by

$$\delta\rho/\rho = 1/(2\pi^2/90)\,[N_B + (7/8)\,N_F]\,[1 - T^2/T_c^2]^{5/2}\,[4\pi e/3]^{3/2}. \tag{21}$$

For the SO(10) model one can write (for $T < T_c$), taking the effective degrees of freedom of the minimal SU(5) model,

$$N_B + (7/8)N_F = 160.75. \tag{22}$$

So that, for $T < T_c$, one gets

$$\delta\rho/\rho \sim 6\times 10^{-4}. \tag{23}$$

This result is compatible with the bounds imposed by the anisotropy of the background radiation.

Another interesting result of this paper is that the contrast density is almost independent of temperature for $T < T_c$ and can be approximated by

$$\delta\rho/\rho \sim 1/\{9[N_B + (7/8)\,N_F]\}. \tag{24}$$

In this way we showed that the density contrast is the scale-invariant (Zel'dovich) spectrum.

REFERENCES

Bazeia, D., Carvalho, C. A., Eboli, O. J. P., and Marques, G. C. 1985, *Phys. Rev.*, **D32**, 3256.

Bazeia, D., Eboli, O. J. P., Guerra, J. M., and Marques, G. C. 1987, *Phys. Rev.*, **D39**, 1838.

Carvalho, C. A., Marques, G. C., Silva, A. J., and Ventura, I. 1986, *Nucl. Phys.*, **B265**, 45.

Guerra, J. M., Marques, G. C., and Rodrigues, S. J. 1989, preprint.

Gross, D. J., Pisarski, R. D., and Yaffe, L. G. 1981, *Rev. Mod. Phys.*, **53**, 43.

Kosterlitz, J. M., and Thouless, D. J. 1973, *J. Phys.*, **C6**, 1181.

Nielsen, H. K., and Olesen, P. 1973, *Nucl. Phys.*, **B61**, 45.

DISCUSSION

E. Bertschinger: Your application of the Kosterlitz-Thorless methods to cosmic strings is very nice and is the correct treatment for thermal equilibrium. However, cosmological phase transitions are non-equilibrium. This is most easily shown by the requirement that the correlation length not exceed the causal horizon (Kibble 1976). The evolution of the topological defects is then determined mainly by dynamical evolution rather than by thermodynamic equilibrium. At low temperatures the abundance of defects can exceed the prediction of thermal equilibrium by orders of magnitude.

D. Lynden-Bell: You have taken an infinite string and assumed this gives free energy per unit length. This neglects any free energy associated with the end points, at which points any energy density per unit length has, *e.g.*, an infinite gravity field.

GENERATION OF NON-FLAT AND NON-GAUSSIAN PERTURBATIONS FROM INFLATION

LEV KOFMAN
Tartu Astrophysical Observatory, 202444 Toravere, Estonia, USSR

GEORGE R. BLUMENTHAL
University of California Observatories/Lick Observatory, Board of Studies in Astronomy and Astrophysics, University of California, Santa Cruz, CA 95064

HARDY HODGES AND JOEL R. PRIMACK
Santa Cruz Institute for Particle Physics
University of California, Santa Cruz, CA 95064

ABSTRACT We begin by discussing motivations for considering non-flat (*i.e.*, non-Zel'dovich) or non-Gaussian primordial fluctuations, and briefly review standard cosmological inflation and the generation of a flat spectrum of Gaussian adiabatic fluctuations. Then we discuss recent work on the inflationary generation of non-flat and non-Gaussian fluctuations in models with multiple scalar fields (including a possible difficulty in achieving the necessary fine-tuned initial conditions), and non-flat fluctuations in a model with a single scalar inflaton with a polynomial potential including quartic, cubic, and quadratic terms. We conclude by briefly discussing what needs to be done to characterize non-Gaussian fluctuations.

MOTIVATION FOR NON-STANDARD INITIAL PERTURBATIONS

Standard initial fluctuations are Gaussian random fluctuations with a scale-free (flat, Harrison-Zel'dovich) spectrum. There are three basic reasons for studying non-standard – *i.e.* non-flat Gaussian or non-Gaussian – initial fluctuations: (1) non-standard fluctuations may be needed to reconcile theories such as cold dark matter (CDM) with observations; (2) non-standard fluctuations may arise in theoretical models for the origin of fluctuations; and (3) an understanding of the consequences of non-standard fluctuations may help in analyzing observations. We will discuss each of these motivations briefly in turn.

CDM is remarkably successful as a theory of galaxy and cluster formation, but it may predict less structure on very large scales than is observed. Although there are several known physical mechanisms that can add additional large-scale power to the CDM fluctuation spectrum in order to improve the agreement with the observed galaxy and cluster angular correlations, bulk motion, and very large scale structures such as voids, filaments, and 'Great Walls', these mechanisms generally increase the predicted amplitude of the microwave background anisotropies. For example, within the context of standard primordial perturbations, increasing the baryonic fraction or lowering the total cosmological density Ω below unity each increase the power in the fluctuation spectrum at wavelengths of tens of Mpc – but at the cost of possible incompatibility with the current $\Delta T/T$ constraints (Bardeen, Bond and Efstathiou 1987; Blumenthal, Dekel and Primack 1988, Holtzman 1989).

Any model, such as cosmic strings, in which the primordial fluctuations arise in connection with extended physical structures necessarily gives rise to non-Gaussian fluctuations (hereafter abbreviated as nGf), since there must be phase correlations between different wave numbers. We are more interested here in exploring the possibility that nGf may arise during a period of cosmic inflation in the very early universe, and in the nature of nGf that may be produced. We will discuss in this lecture as much as is known about the generation of non-standard fluctuations from inflation in models with one or two scalar fields.

Finally, even if the fluctuations that gave rise to galaxies and large-scale structures in the universe actually were perfectly Gaussian, it would still be useful to understand the consequences of nGf in analyzing the relevant observational data. It is difficult to test a theoretical idea unless it can be put in the context of alternative ideas, specified by one or more parameters – for example, General Relativity as a limit of the Brans-Dicke theory. Then one can see how to test the theory by measuring the relevant parameters.

A particular problem with nGf is that, while Gaussian fluctuations are fully specified merely by giving the power spectrum $P(k)$, additional information is in principle needed to specify nGf fully. In order to clarify the implications of non-standard perturbations for cosmology it will be necessary to address several questions, which we list below:

i) What physically plausible models give non-standard fluctuations?

ii) What sorts of nGf arise, and how can these fluctuations be most usefully characterized?

iii) How can observers test for nGf, and distinguish primordial nGf from the non-Gaussian effects that will inevitably arise from nonlinear gravitational evolution at late times? In the present paper, we can only begin to address these difficult questions.

INFLATIONARY COSMOLOGY

Let us remember the main ideas of the standard inflationary cosmology (for a recent review, see Olive, 1989). The evolution of the universe corresponds to the hot Friedmann scenario beginning not from the initial singularity $t = 0$, but from $t_{\rm reheat} \geq 10^{-35}$ sec, when the temperature is $T \leq 10^{14}$ GeV. After reheating, the equation of state of matter is that of radiation $p = \rho/3$ until the epoch when the density of dust-like matter dominates over radiation. At the present time, the dust-like equation of state is $p = 0$. If we consider the far past, before the hot Friedmann scenario, then the equation of state changes at the energy scale where all fundamental interactions are expected to be unified. Among all fields (φ, ψ, A_μ, $h_{\mu\nu}$) which are present at these energy scales, the main contribution to the gravity of the universe comes from scalar fields φ, or from condensates which look like them. More formally, in the fundamental Lagrangian $\mathcal{L}(R, \varphi, \psi, A_\mu, \mathrm{h}_{\mu\nu})$ for large energy density, we can retain for cosmological applications only Einstein gravity and scalar field(s):

$$\mathcal{L} = -(M^2_{\rm Planck}/16\pi)R + \tfrac{1}{2}\varphi_{,\alpha}\varphi^{,\alpha} - V(\varphi) \tag{1}$$

where R is the space-time curvature and $V(\varphi)$ is the potential of the scalar field φ. For a Friedmann universe with scale factor $a(t)$, the evolution of the background metric is described by the Einstein equation for the Hubble parameter

$$H^2 \equiv (\dot{a}/a)^2 = (8\pi/3M^2_{\rm Planck})\,[\tfrac{1}{2}\dot{\varphi}^2 + V(\varphi)]. \tag{2}$$

The Lagrangian (1) leads to the Klein-Gordon equation for a scalar field

$$\ddot{\varphi} + 3H\dot{\varphi} + V_{,\varphi} = 0. \tag{3}$$

The system of equations (2)-(3) is relatively simple. For realistic potentials leading to chaotic inflation (Linde 1983), and for an initial value φ of order several $M_{\rm Planck}$, the friction term, $3H\dot{\varphi}$, in (3) dominates over $\ddot{\varphi}$. This leads to a slow evolution of $\varphi(t)$ when $V(\varphi)$ is nearly constant. For large φ ($\geq M_{\rm Planck}$), the equation of state for the scalar field is similar to the one for an effective cosmological constant: $p \simeq -\rho \simeq -V(\varphi) \simeq$ const. This yields the (quasi) exponential solution (Inflation) for the scale factor

$$a(t) = a_0 \exp\left[\int dt\, H(t)\right], \qquad |\dot{H}| << H^2. \tag{4}$$

During inflation the geometry of space-time is (quasi-) de Sitter. Inflation is an inevitable intermediate stage of the dynamical evolution of the Universe for a wide range of initial conditions with various potentials. Later, when $\varphi(t)$ reaches a small value and friction in (3) also becomes small, rapid oscillations near the minimum of $V(\varphi)$ occur. The interaction of the scalar field with other particles of the fundamental Lagrangian (which we remember again) leads to the creation and reheating of the matter filling the universe. All scalar field energy converts into the heating of ultra-relativistic particles, and the equation of state becomes $p = \rho/3$.

Let us note that there is another way to mimic an effective cosmological constant in the theory of gravity with higher derivatives, containing an additional R^2 term in the Lagrangian. Such a term appears in Quantum Gravity due to vacuum polarization or in superstring theory. However, for our purposes, such a theory is equivalent to the model with a scalar field.

The most interesting consequences of the inflationary scenario are that inflation produces a flat homogeneous background universe with $\Omega = 1$ (assuming that the cosmological constant $\Lambda = 0$, for reasons not yet understood), it washes out all inhomogeneities which might exist prior to inflation, and with an appropriate choice of parameters in $V(\varphi)$, it leads to the creation of small amplitude cosmological perturbations afterwards on large scales.

GENERATION OF STANDARD PERTURBATIONS FROM INFLATION

Before the invention of inflation the initial perturbations were 'put in by hand' in cosmological models. In the framework of scenarios with an initial inflationary stage, density perturbations $\delta\rho/\rho$ inevitably arise as a consequence of vacuum fluctuations $\delta\varphi$ of the inflaton scalar field during inflation (Guth and Pi 1982; Hawking 1982; Starobinsky 1982; Bardeen, Steinhardt and Turner 1983). These fluctuations $\delta\varphi$ obey an oscillator-like equation, also containing the friction term $3H\delta\dot{\varphi}$. For short waves with wavelength less than the horizon size $\lambda \leq H^{-1}$, this term can be neglected, and fluctuations $\delta\varphi$ oscillate as in ordinary Minkowski space-time. The physical wavelengths of all fluctuations $\delta\varphi$ grow exponentially along with the scale factor. When the wavelength of any particular fluctuation becomes greater than the horizon size, its amplitude freezes because of the large friction term. Let $\Phi(\mathbf{x},t)$ be the initial peculiar gravitational potential which describes the metric perturbations in the longitudinal gauge:

$$ds^2 = (1 + 2\Phi)\, dt^2 - (1 - 2\Phi)\, a^2(t)\, [dx^2 + dy^2 + dz^2]. \tag{5}$$

Fluctuations $\delta\varphi$ generated during inflation lead to perturbations of the metric (adiabatic perturbations of density) after inflation, having Fourier amplitude

$$\Phi_k = C(\varphi,\dot{\varphi})\ \delta\varphi_k. \tag{6}$$

For the long-wave limit $\delta\rho/\rho = -2\Phi$. The main aim of the theory of adiabatic perturbations is to determine the function $C(\varphi,\dot{\varphi})$, which depends on background evolution of the scalar field. In the simplest model containing a single scalar inflaton field φ, there is an analytical solution (Mukhanov 1985):

$$\Phi_k = \{\ (1/\dot{\varphi})\ \frac{d}{dt}\ [(1/a) \int dt\ \ a\ ]\ \}\ \delta\varphi_k. \tag{7}$$

This gives the value of $C(\varphi,\dot{\varphi})$ in (6). Long–wave fluctuations of the massless scalar field at the inflationary stage have a scale-free spectrum $\delta\varphi_k = H/2^{1/2}k^{3/2}$. Using the gauge-invariant variable $\zeta \approx \delta\rho/(\rho+p)$ (Bardeen *et al.* 1983), which is a linear function of Φ_k, $\dot{\Phi}_k$, and which remains constant outside the horizon (even through reheating), one finds that

$$\delta\rho/\rho \approx 16\ (6\pi)^{1/2}\ [V(\varphi)]^{3/2}/(V_{,\varphi}\ M^3_{\rm Planck}). \tag{8}$$

Equation (8) assumes the slow roll conditions, and pressureless equation of state; for length scales that re-enter the horizon in the radiation era, there is an additional factor of 4/3. Here, $\delta\rho/\rho$ refers to the fluctuation amplitude when a given length scale reenters the horizon, and φ is to be evaluated at the time the scale leaves the horizon during inflation. For example, for the potential $V(\varphi) = \lambda\varphi^4/4$ and $\lambda \sim 10^{-14}$, we get the value $k^{3/2}\ \langle\Phi_k\rangle^{1/2} \sim 10^{-4}$, which is consistent with observations. The reason for scale invariance in this model is that φ varies very weakly (logarithmically) with scale.

Let us now discuss briefly the properties of random fields of peculiar (Newtonian) gravitational potential

$$\Phi(t,\mathbf{x}) = 1/(2\pi)^{3/2} \int d^3\mathbf{k}\ \Phi_k(t)\ \exp\ (i\ \mathbf{k}\cdot\mathbf{x}).$$

We are interested in this object because the initial post-inflationary perturbations are calculated directly in term of Φ. Inside the horizon, the matter density $\rho(t,\mathbf{x})$ and Φ are connected by the Poisson equation. Nevertheless, the Φ-field has specific interesting properties of its own. For the simplest case of a single inflaton scalar field, the gravitational potential is a realization of a random Gaussian field, which is a superposition of spatial waves exp $(i\ \mathbf{k}\cdot\mathbf{x})$ with random phases and scale-free (Zel'dovich) spectrum (8). In the linear theory, at the beginning of the matter-dominated era, the amplitude Φ_k is modified by a factor $T(k)$, the transfer function, which is determined by the model of dark matter; afterward, its shape remains constant in time up to the non-linear regime.The statistical properties of the Gaussian random field Φ_k depend more strongly on the cosmological model of dark matter than does the density field (Kofman 1989). The linear gravitational potential Φ plays a crucial role in an approximate theory of Large Scale Structure known as the Adhesion model (Kofman and Shandarin 1988).

Negative peaks in the gravitational potential Φ are associated with stars, galaxies, and so on. In cosmology the gravitational potential has positive as well as negative peaks. Giant voids in the galaxy distribution form around high positive peaks of Φ. Consequently, the statistics of voids directly depends on the statistical properties of the random field Φ.

Now we will consider the following mechanisms for the generation of non-standard perturbations from inflation: adiabatic (curvature) and isothermal (isocurvature) non-standard perturbations for models with multiple scalar fields; and adiabatic non-standard perturbations arising from a single scalar field model.

NON-FLAT AND NON-GAUSSIAN PERTURBATIONS WITH MULTIPLE SCALAR FIELDS

As a rule, modern unified theories of elementary particles contain a number of scalar fields, and the potentials governing the interactions among these fields are fairly arbitrary. In the Multiple Inflation scenario, a number of successive stages of inflation are driven by a number of classical scalar fields φ_i with different effective potentials $V_i(\varphi_i)$, i = 1,2,...,n, without interaction between the fields. One might naively expect that the resulting spectrum could be a direct sum of the separate inflationary stages. This would typically give a non-flat spectrum falling off towards long wavelength. However, this expectation is wrong. Actually, taking into account the whole evolution of all fluctuations of all fields $\delta\varphi_1,\ldots,\delta\varphi_n$ during all successive stages of inflation leads to the inverse result: the spectrum of the adiabatic perturbations after multiple inflation increases with wavelength (Kofman and Pogosyan 1988; Bond, Salopek and Bardeen 1988; Starobinsky 1985).

There are much more interesting ways to get non-standard perturbations in models with multiple interacting scalar fields (Kofman and Linde 1987; Kofman and Pogosyan 1988; Salopek, Bond, and Bardeen 1989). Let us consider the model with two interacting scalar fields with the common potential

$$V(\varphi,\chi) = (\tfrac{1}{2}m_\varphi^{\,2}\varphi^2 + \tfrac{1}{4}\lambda_\varphi\varphi^4) + (-\tfrac{1}{2}m_\chi^{\,2}\chi^2 + \tfrac{1}{4}\lambda_\chi\chi^4) + \tfrac{1}{2}\nu\varphi^2\chi^2. \tag{9}$$

Suppose that the (isoinflaton) scalar field χ is present, but does not drive inflation. Inflation is driven by another inflaton field φ. In this model with a single inflaton but with two interacting scalar fields, we may obtain non-flat as well as non-Gaussian adiabatic perturbations from inflation for an appropriate choice of parameters in (9). The joint evolution of all fluctuations $\delta\chi$, $\delta\varphi$ and metric perturbations during the inflationary stage turns out to be pretty non-trivial. The gauge-invariant equations for the perturbations are obtained by linearization of the Einstein and Klein-Gordon equations for the scalar fields:

$$(3H^2 - a^{-2}\nabla^2)\,\Phi + 3H\dot{\Phi} = (4\pi/M^2_{\rm Planck})\,[\dot{\varphi}\delta\dot{\varphi}+\dot{\chi}\delta\dot{\chi}+V_{,\varphi}\delta\varphi+V_{,\chi}\delta\chi-\Phi(\dot{\varphi}^2+\dot{\chi}^2)]; \tag{10}$$

$$(\dot{\varphi}^2 + \dot{\chi}^2)\,\Phi = \dot{\varphi}^2\, d/dt\,(\delta\varphi/\dot{\varphi}) + \dot{\chi}^2\, d/dt\,(\delta\chi/\dot{\chi}); \tag{11}$$

$$(\Box + V_{,\chi\chi})\,\delta\chi = 4\dot{\Phi}\,\dot{\chi} - 2V_{,\chi}\,\Phi - V_{,\varphi\chi}\delta\varphi; \tag{12}$$

$$(\Box + V_{,\varphi\varphi})\,\delta\varphi = 4\dot{\Phi}\,\dot{\varphi} - 2V_{,\varphi}\,\Phi - V_{,\varphi\chi}\delta\chi, \tag{13}$$

where $\Box = \partial_t^2 + 3H\partial_t - a^{-2}\,\nabla^2$. These equations are not completely independent, as equations (11-13) fully describe the evolution of the perturbed variables.

Initial conditions are defined by vacuum quantum fluctuations of the scalar field during inflation. The essential idea in making the spectrum nonflat involves the possibility that the effective mass-squared of the isoinflaton field $m^2_{\rm eff}(\chi) = V_{,\chi\chi} = -m_{\chi 2} + \nu\varphi^2 + 3\lambda_\chi\chi^2$ depends on time due to the interaction between φ and χ. The solution to equation (12) during inflation in the WKB approximation is

$$\delta\chi_k(t) \sim \exp\,\{(3/2)\int_{t_k}^{t} dt\, H\,[(1 - 4m^2_{\rm eff}/9H^2)^{1/2} - 1]\,\}. \tag{14}$$

The moment t_k corresponds to the transition to the long-wave regime. If the value of the mass $m^2_{\rm eff}(\chi)$ can change from positive to negative, then the time evolution of the long-wave perturbations will change from damping to exponential growth.

In the model defined by equation (9), we may make $m^2_{\rm eff} < 0$ during some time interval during inflation. Although χ does not dominate during inflation, its fluctuations $\delta\chi$ do dominate over $\delta\varphi$ in the metric perturbations Φ in (11) when $m^2_{\rm eff} < 0$ because of exponential growth. Therefore the spectrum of the (adiabatic) perturbations of the metric acquire the properties of the spectrum of the isoinflaton fluctuations for the modes exceeding the horizon in the region $m^2_{\rm eff} < 0$:

$$\Phi_k = C(\varphi,\dot{\varphi};\chi,\dot{\chi})\ \delta\chi_k. \tag{15}$$

It is non-flat, because δ_k is non-flat and contains a maximum. As usual, the function $C(\varphi,\dot{\varphi};\chi,\dot{\chi})$ depends on the background evolution of the scalar fields.

Let us follow the time dependence of $m^2_{\rm eff}(\chi)$ along a typical trajectory L of the background fields in the $\varphi - \chi$ phase diagram (see Figure 1). Initially ($t < t_*$), the mass is positive; then, at the moment t_* the trajectory L crosses the curve B, corresponding to $m^2_{\rm eff} = 0$, and the isoinflaton field attains its minimal value $\chi(t_*) = \chi_*$. After L passes through the interesting region of negative mass, it again crosses the curve B, enters the region of positive mass, and ends up on the inclined part of A corresponding to a minimum of the potential $V(\varphi,\chi)$. In the region below B on the upper half plane (and vice versa on the lower half) $m^2_{\rm eff} < 0$, isoinflaton fluctuations grow exponentially and dominate over inflaton fluctuations with small mass. If the minimal value of χ_* is greater than the fluctuations $\delta\chi$

$$\chi_* \geq \delta\chi, \tag{16}$$

then we can treat the background evolution as classical. Solving (10)-(14), the resulting spectrum of Gaussian adiabatic perturbations Φ_k after inflation has a break (the quantity

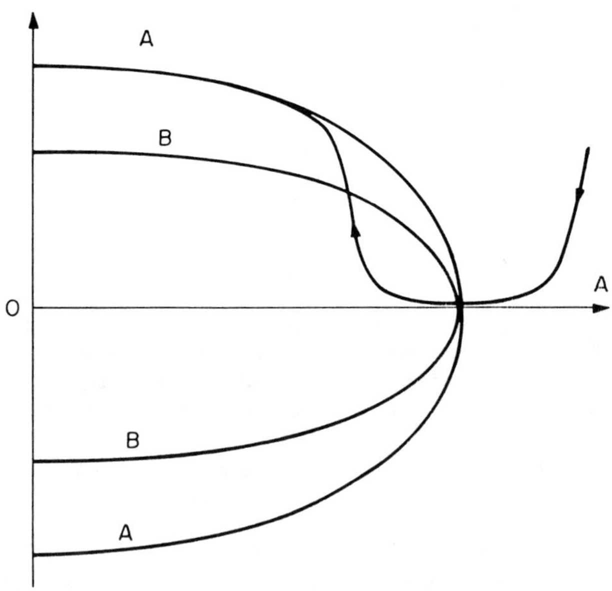

Fig. 1. The $\varphi - \chi$ phase diagram for the potential (9). See text for discussion. (From Kofman and Pogosyan 1988.)

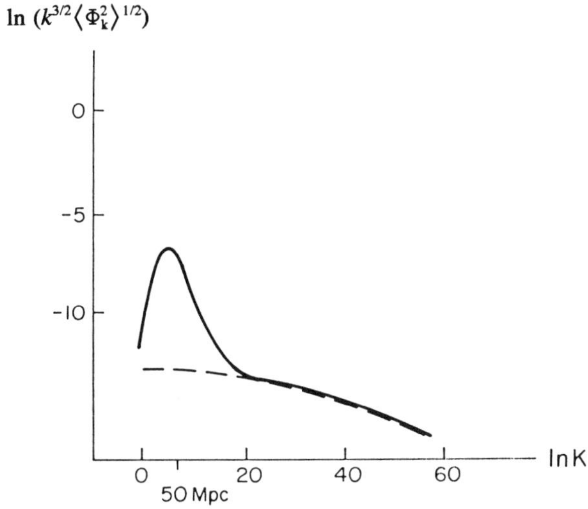

Fig. 2. Two examples of adiabatic fluctuation spectra $k^{3/2}\langle\Phi_k^2\rangle^{1/2}$. The solid line is a spectrum arising from $V(\varphi,\chi)$ given by (9), and the dashed line is a corresponding scale-free spectrum arising from a similar potential with a single inflaton φ.

$k^3\langle\Phi_k^2\rangle$ has a maximum) on the scale $\lambda_* \approx H^{-1} \exp(\pi m_\chi^2/\nu M^2_{\text{Planck}})$; see Figure 2. This mode undergoes the largest amplification due to exponential growth of $\delta\chi(t_*)$. The smaller χ_* is, the larger the amplitude Φ_{k*}. However, when

$$\chi_* \leq \delta\chi, \tag{17}$$

then the fluctuations of $\delta\chi_k$ influence the classical evolution of the background field χ near χ_*. This means that in the model with condition (17), we have non-Gaussian perturbations of the gravitational potential

$$\Phi_k = C(\varphi,\dot{\varphi};\delta\chi_k)\,\delta\chi_k \tag{18}$$

because of the non-linear dependence on the Gaussian fluctuations $\delta\chi$.

In addition, in the model with a one-component isoinflaton field χ satisfying condition (17), there are also space domains of size λ_* with a different sign $+\chi_0$ or $-\chi_0$, separated by gravitating walls. Even if the domains of χ disappear after some phase transition, there can still be remnant negative walls of gravitational potential Φ. If χ is a complex scalar field, then the model also contains strings generated at the moment t_*. If χ is a more complicated field with an appropriate symmetry, we could avoid any topological defects and still have non-Gaussian perturbations from inflation. This is also the idea behind global texture (Turok 1989).

J. Bardeen (1989) has improved the model under consideration, changing the potential to

$$V(\varphi,\chi) = (\tfrac{1}{2}m_\chi^2\chi^2 + \tfrac{1}{4}\lambda_\chi\chi^4) + (\tfrac{1}{2}m_\varphi^2\varphi^2 + \tfrac{1}{4}\lambda_\varphi\varphi^4) + \tfrac{1}{2}\nu\,[(\varphi-\varphi_1)^2 - \varphi_2^2]\,\chi^2,$$

where φ_1 and φ_2 are some constants of order M_{Planck}. In this model there are no defects after inflation, and non-Gaussian adiabatic perturbations of the metric look like the square of Gaussian fluctuations

$$\Phi \sim (\delta\chi)^2. \tag{19}$$

This is a typical dependence for equation (18).

Another possible way to generate non-Gaussian perturbations in multiple scalar fields models is connected with the axion, corresponding to the cold dark matter scenario with isocurvature perturbations. Let χ be an axion field, which is present but not dominant during inflation. The axion field potential is

$$V(\chi) = V_0 \, [1 - \cos(\chi/\chi_0)]. \tag{20}$$

If the dispersion of the fluctuations of the light axion field $(\int d^3\mathbf{k} \, \langle \delta\chi_k{}^2 \rangle)^{1/2}$ is more than χ_0, then space perturbations of the isocurvature axion $V[\chi(\mathbf{x})]$ are non-Gaussian because of the non-linear dependence (20) (Linde 1984, 1985; Kofman 1986). Additional higher derivatives of the axion field in the Lagrangian also lead to non-Gaussian fluctuations (Allen, Grinstein and Wise 1987).

Finally, we mention a possible problem with some of the scenarios described in this section. In the double inflation model, for example, a second field must inflate at a precisely tuned value in order that a break in the power spectrum appears on astrophysical scales of interest. Due to the stochastic evolution of the second field during inflation dominated by the first field, the second field develops a huge dispersion. One can show (Hodges 1989b) that it is very unlikely that the second field will have the right initial value for its period of inflation to produce a break in the spectrum on interesting large scale structure scales if inflation begins near the Planck scale. The same reasoning applies to other models that require fine tuning the value of an auxiliary scalar field.

NON-FLAT AND NON-GAUSSIAN PERTURBATIONS IN A SINGLE SCALAR FIELD MODEL

Quantum fluctuations $\delta\varphi$ generated during inflation may, in principle, affect the background classical evolution even in the simplest model with a single scalar field φ. On a scale which is much greater than the present horizon, according to (8) (note the logarithmic growth of amplitude with growth of wavelengths), these fluctuations make the universe very inhomogeneous. The feedback of fluctuations with wavelength larger than the horizon on the classical evolution has been considered in the framework of the (semiclassical) stochastic approach (Starobinsky 1986; Bardeen and Bublik 1987; Goncharov, Linde, and Mukhanov 1987). We now consider the possibility of obtaining non-flat or non-Gaussian adiabatic perturbations inside the present horizon from inflation with a single scalar inflaton field.

The influence of fluctuations is significant if the change of the field $\delta\varphi$ associated with fluctuations during time interval δt is comparable to its change $\Delta\varphi$ due to classical slow rolling during the same time interval

$$\langle \delta\varphi^2 \rangle^{1/2} / \Delta t \geq \Delta\varphi / \Delta t. \tag{21}$$

For the popular chaotic inflation scenario, equation (21) is valid for values of φ much greater than the value φ_h corresponding the scale of our horizon. Therefore in the simplest chaotic inflationary scenarios, adiabatic perturbations inside the present horizon are Gaussian as well as scale-free.

A more rigorous analysis is based upon the stochastic Langevin-type equation (Starobinsky 1986; recent references are Sasaki, Nambu, and Nakao 1988; Kandrup 1989), which has the slow-roll limit

$$\dot{\varphi} = -(1/3H)\, V_{,\varphi} + (H^{3/2}/2\pi)\, \hat{\eta}(t), \tag{22}$$

where $\hat{\eta}$ is a quantum noise term, assumed to have a Gaussian distribution with $\langle \hat{\eta}(t) \rangle = 0$ and $\langle \hat{\eta}(t)\, \hat{\eta}(t') = \delta(t-t')$. Equation (22) is derived in the 'coarse-grained' approximation, *i.e.* it describes the evolution of spatial regions having size larger than the horizon. This equation leads to a Fokker-Planck equation for the probability distribution $\mathcal{P}(\varphi,t)$ of the coarse-grained field φ at time t. If one chooses an initial $\varphi(t_i)$ at initial time t_i, then one can solve for $\mathcal{P}(\varphi,t_0)$, the distribution function today for an ensemble of universes of typical size H_0^{-1}. Note that the distribution $\mathcal{P}$ will generally be very sensitive to the initial condition $\varphi(t_i)$. Analytical solutions of the Fokker-Planck and Langevin equations as well as numerical simulations (Hodges 1989a, Bond and Salopek 1989) show that for initial conditions corresponding to universes with the size of the present horizon, non-Gaussian effects in $\mathcal{P}(\varphi,t)$ are negligible. This result was checked for a wide class of potentials $V(\varphi)$ including chaotic, new- and power-law inflations. For initial conditions corresponding to an ensemble of universes with sizes much greater than our horizon, non-Gaussian effects could be significant (Linde 1987; Matarrese, Ortolan and Lucchin 1989), according to the qualitative criterion (21). Physically, this means that for a large initial value $\varphi(t_i)$, the duration of inflation is long, and the generated long-wave quantum fluctuations accumulate so that they significantly affect the classical rolling down of the scalar field. On the other hand, for $\varphi(t_i)$ much less than the value at which fluctuations dominate the evolution of φ, as happens in 'eternal inflation' (Linde 1986), there is insufficient time to generate such large fluctuations. Then, non-Gaussian effects are negligible, even on scales well outside our present horizon.

In the large wavelength limit, and in the region of the potential that does not contain a false vacuum, there is a general relationship, in the large wavelength limit, between the amplitude of density fluctuations and the non-Gaussian criterion (21) (Hodges, Blumenthal, Kofman, and Primack 1989)

$$\delta\rho/\rho \simeq (\langle \delta\varphi^2 \rangle^{1/2}/\Delta t)/(\Delta\varphi/\Delta t)\, \{1 + (V_{,\varphi} + 4H\dot{\varphi})/[\ddot{\varphi} - H\dot{\varphi} + \dot{\varphi}a(t)/\textstyle\int a(t)dt]\} \tag{23}$$

Therefore, to satisfy (21) and have a small enough amplitude, $\delta\rho/\rho \leq 10^{-4}$, would require that the factor in brackets become $<<$ 1 at some point in the evolution of φ. This appears very unlikely, as the factor in brackets is typically $>>$ 1. Assuming slow-roll conditions, interesting non-Gaussian fluctuations arising from a single field potential can only be associated with a large fluctuation amplitude, as $\delta\rho/\rho \sim 1$ when (21) is just satisfied. This would indicate that an acceptable spectrum with non-Gaussian fluctuations, if possible at all, should contain a large peak over a rather short range of wavelengths. In this case, non-Gaussian fluctuations are necessarily associated with broken scale-invariance.

We now discuss the possibility of obtaining non-flat power spectra from single scalar field inflation. At present, we are aware of only two models. Power law inflation (Abbott and Wise 1984; Lucchin and Mataresse 1985), resulting from an exponential-type potential, can lead to non-Zel'dovich power-law spectra. The other model is somewhat more interesting (Hodges, Blumenthal, Kofman, and Primack 1989), leading not only to non-Zel'dovich (nearly) power-law spectra over astrophysical scales, but also to the possibility of a dip in the power spectrum. Such spectra arise in the polynomial potential

$$V(\varphi) = A\varphi^4/4 + B\varphi^3/3 + C\varphi^2/2$$

with $A > 0$ for stability, and $B^2 < 4AC$ so that the global minimum of the potential (which is at $\varphi = 0$ for this range of parameters) corresponds to $V(\varphi=0) = 0$. If we consider the case of the field rolling down from the right side of the global minimum of the potential, then a significant feature in the potential, a plateau, can be encountered if $B < 0$ and $B^2 \simeq 4AC$. Spectra, for specific sets of parameters, are shown in Figure 3. The dip in the spectra, if located around 5 Mpc (the normalization scale), may help reconcile CDM with the data as the increase in power (over a pure Zel'dovich spectrum) on small scales could lead to earlier galaxy formation, and the increase on larger scales would lead to more large scale structure.

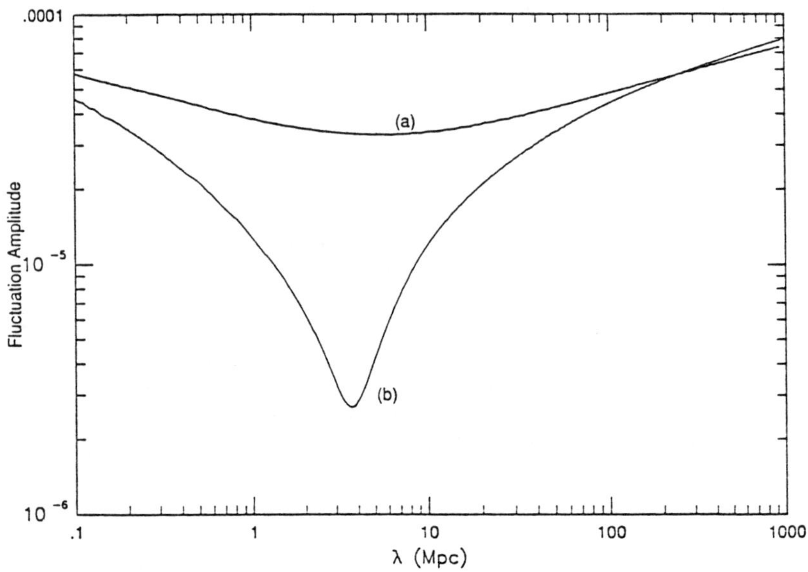

Fig. 3 Spectra with a dip, as a function of scale λ, arising in the polynomial potential models. For curve (a), $B = -2.3\ AM_{\rm Planck}$ and $C = 1.3525\ AM^2_{\rm Planck}$, and for curve (b) $B = -1.3\ AM_{\rm Planck}$ and $C = 0.42345\ AM^2_{\rm Planck}$. The amplitude of the spectra, shown in arbitrary units, can be adjusted by varying A.

CONCLUSION

We have demonstrated that in inflationary models with two scalar fields or even a single scalar field, by specifying the potential appropriately, one may get non-flat Gaussian adiabatic perturbations from inflation. This is possible because the evolution of perturbations during inflation depends in a quite complicated way on the inflaton field φ and mainly on its time derivative; the evolution also depends on the iso-inflaton field $\chi,\dot{\chi}$ (if it exists) even if it does not dominate. The price for this behavior in these models is fine-tuning of the parameters of the potential so that they produce the desired fluctuations. Another problem occurs in models with multiple inflators if the auxiliary scalar fields must be tuned in order to obtain interesting effects on large scales – such tuning may be unlikely (Hodges 1989b).

Models which produce non-flat Gaussian perturbations may also lead to the generation of non-Gaussian perturbations, by an appropriate choice of parameters in the potential. As a rule, the non-Gaussian random field of perturbations of the gravitational potential $\Phi(\mathbf{x})$ resembles a non-linear function of the Gaussian random field of fluctuations $\delta\varphi$:

$$\Phi(\mathbf{x}) = F\,[\delta\varphi(\mathbf{x})] \tag{24}$$

see, for instance, (19).

We propose to call these perturbations a local non-Gaussian field, contrary to the more general case where the non-Gaussian field can contain non-local dependences on spatial derivatives of $\delta\varphi$, $\nabla_i\delta\varphi$, $\nabla_i\nabla_j\delta\varphi$. The strings generated during inflation [in model (9) for a complex isoinflaton field] are an example of a non-local non-Gaussian field.

The statistics of a local non-Gaussian field (24) can be determined from the Gaussian statistics of $\delta\varphi$. Let $P(\delta\varphi)$ be the Gaussian probability density of $\delta\varphi$; then the probability density $Q(\Phi)$ of Φ is given by

$$Q(\Phi) = \textstyle\sum_a P(\delta\varphi_a)/F'(\delta\varphi_a), \tag{25}$$

where $F' = dF/d(\delta\varphi)$, and $\delta\varphi_a$ are all the real roots of equation (25). Let $P(\delta\varphi,\nabla_i\delta\varphi,\nabla_i\nabla_j\delta\varphi)$ be the joint Gaussian probability density of $\delta\varphi$ and its first and second spatial derivatives. Then the joint probability density of a local non-Gaussian field and its first and second spatial derivatives

$$Q(\Phi,\nabla_i\Phi,\nabla_i\nabla_j\Phi) = \textstyle\sum_a P(\delta\varphi_a,\nabla_i\delta\varphi_a,\nabla_i\nabla_j\delta\varphi_a)/\,\mathfrak{J}(\delta\varphi_a,\nabla_i\delta\varphi_a,\nabla_i\nabla_j\delta\varphi_a) \tag{26}$$

where $\mathfrak{J}$ is the Jacobian of the transformation

$$\Phi = F(\delta\varphi)$$

$$\nabla\Phi = F'(\delta\varphi)\nabla\delta\varphi$$

$$\nabla_i\nabla_j\Phi = F''(\delta\varphi)\nabla_i\delta\varphi\nabla_j\delta\varphi + F'(\delta\varphi)\nabla_i\nabla_j\delta\varphi, \tag{27}$$

and $(\delta\varphi_a,\nabla\delta\varphi_a,\nabla_i\nabla_j\delta\varphi_a)$ are all real solutions of equations (27). The interesting problem is to calculate the statistics of peaks of the non-Gaussian field $\Phi(\mathbf{x})$

$$\langle n_P(\nu)\rangle = \int D[\Phi]\, D[\nabla\Phi]\, D[\nabla_i\nabla_j\,\Phi]\ \delta^3(\nabla_i\Phi)\ |\det \nabla_i\nabla_j\ \Phi|\ Q(\Phi,\nabla\Phi,\nabla_i\nabla_j\Phi \tag{28}$$

for simple functions F, where the multiple integrals are over all derivatives and values of the field such that Φ exceeds its *rms* value by a factor of ν. Evaluating the integrals in (28) represents a worthwhile challenge.

ACKNOWLEDGMENTS

We acknowledge discussions with Francesco Lucchin and Sabino Matarrese that have helped us to understand their approach, and with Dick Bond, Jim Bardeen, Andrei Linde, and Alex Starobinsky. L. K. thanks the Canadian Institute for Theoretical Astrophysics and G. B., H. H., and J. P. thank the Aspen Center for Physics for hospitality. This work was supported in part by NSF grant PHY-8800801 and by a UCSC faculty research grant.

REFERENCES

Abbott, L. F. and Wise, M. B. 1984, *Nucl. Phys.* **B244**, 541.

Allen, T., Grinstein, B., and Wise M. 1987, *Phys. Lett.* **197B**, 66.

Bardeen, J., Bond, J., and Efstathiou, G. 1987, *Ap. J.* **321**, 28.

Bardeen, J. 1989, Lectures at Second Guo Shou-jing Summer School on Particle Physics and Cosmology, in *Particle Physics and Cosmology*, ed. A. Zee (Gordon and Breach), in press.

Bardeen, J., and Bublik, G. 1987, *Class. Quant. Gravity*, **4**, 573.

Bardeen, J., Steinhardt, P., and Turner M. 1983, *Phys. Rev.* **28D**, 679.

Blumenthal, G., Dekel, A., and Primack, J. 1988, *Ap. J.* **326**, 539.

Bond, J. and Salopek, D. 1989, in preparation.

Bond, J., Salopek, D., and Bardeen, J. 1988, in *Large Scale Motions in the Universe*, ed. V. C. Rubin and G. V. Coyne, S. J., (Princeton: Princeton University), p. 385.

Goncharov, A., Linde, A., and Mukhanov, V. 1987, *Int. J. Mod. Phys.* **2A**, 561.

Guth, A., and Pi, S.- Y. 1982, *Phys. Rev. Lett.* **49**, 110.

Hawking, S. 1982, *Phys. Lett.* **115B**, 295.

Hodges, H. 1989, *Phys. Rev.* **D39**, 3568.

Hodges, H. 1990, *Phys. Rev Lett.*, **64**, 1080.

Hodges, H., Blumenthal, G., Kofman, L. and Primack, J. 1989, UCSC preprint SCIPP-89/31, *Nucl. Phys.*, in press.

Holtzman, J. 1989, *Ap.J. Suppl.*, **71**, 1.

Kandrup, H. 1989, *Phys. Rev.* **D39**, 2245.

Kofman, L. 1986, *Phys. Lett.* **173B**, 400.

Kofman, L. and Linde, A. 1987, *Nucl. Phys.* **282B**, 555.

Kofman, L. and Pogosyan, D. 1988, *Phys. Lett.* **214B**, 508.

Kofman, L. and Shandarin, S. 1988, *Nature* **334**, 129.

Kofman, L. A. 1989 in *Morphological Cosmology*, ed. P. Flin and H. Duerbeck, *Lecture Notes in Physics*, **332**, (Berlin: Springer), p. 354.

Linde, A. 1983, *Phys. Lett.* **129B**, 177.

Linde, A. 1984, *JETP Lett.* **40**, 1333.

Linde, A. 1985, *Phys. Lett.* **158B**, 375.

Linde, A. 1986, *Phys. Lett.* **175B**, 395.

Lucchin, F. and Matarrese, S. 1985, *Phys. Rev.* **D32**, 1316.

Matarrese, S., Ortolan, A. and Lucchin, F. 1989, *Phys. Rev.* **D40**, 290.

Mukhanov, V. 1985, *JETP Lett.* **41**, 493.

Mukhanov, V., Kofman, L. and Pogosyan, D. 1987, *Phys. Lett.* **157B**, 427.

Mukhanov, V. and Chibisov, G. 1981, *JETP Lett.* **33**, 532.

Olive, K. 1989, "Inflation", University of Minnesota preprint.

Salopek, D., Bond, J., and Bardeen, J. 1989, *Phys. Rev. D.*, **40**, 1753.

Sasaki, M., Nambu, Y. and Nakao, K. 1988, *Nucl. Phys.* **308B**, 868.

Starobinsky, A. 1982, *Phys. Lett.*, **117B**, 175.

Starobinsky, A. 1985, *JETP Lett.* **42**, 152.

Starobinsky, A. A. 1986, in *Field Theory, Quantum Gravity and Strings*, ed. H. J. de Vega and N. Sanchez, *Lecture Notes in Physics*, **246**, (Berlin: Springer), p. 107.

Turok, N. 1989, *Phys. Rev. Lett.*, **63**, 2625.

DISCUSSION

S. Matarrese: It seems that the real discrepancy between our work (Matarrese, Ortolan, and Lucchin 1989) and the results you referred to by Bond and Salopek comes from the initial value of H (or Φ). What you and Bond and Salopek assume corresponds to having no fluctuations at all beyond the horizon size (or a scale somewhat larger than that). This is not necessary. One only needs to demand that $\delta\rho/\rho < 10^{-5}$ on the horizon scale. That of course makes a big difference!

L. Kofman: I agree.

S. Shandarin: You have mentioned that in the Bond-Salopek model the distribution function of the perturbation was Gaussian. Generally speaking, that does not prove that the perturbation field is Gaussian. Has it been shown that phases were statistically independent?

L. Kofman: Yes.

PRIMORDIAL GENERATION OF NON-GAUSSIAN ADIABATIC PERTURBATIONS

S. MATARRESE, F. LUCCHIN
Dipartimento di Fisica *Galileo Galilei*, Università di Padova
Via Marzolo 8, I-35131 Padova, Italy

A. ORTOLAN
Dipartimento di Fisica, Università di Ferrara and *INFN*
Via Paradiso 12, I-44100 Ferrara, Italy

ABSTRACT The stochastic dynamics of *inflaton*, both in chaotic and power-law inflation, determines a *scaling regime* during which adiabatic perturbations originate with non-Gaussian scale-invariant statistics.

INTRODUCTION

It is commonly believed that density fluctuations generated during inflation (Guth 1981, Linde 1982, Albrecht and Steinhardt 1982) have random phases, owing to the flatness of the potential for the scalar field (*inflaton*) driving the accelerated expansion of the universe. Recently, however, it has been shown that adiabatic perturbations can be characterized by non-Gaussian statistics (Ortolan, Lucchin, and Matarrese 1988; Matarrese, Ortolan and Lucchin 1989). Due to the back-reaction of field fluctuations on the background geometry, phase correlations arise during the stochastic evolution of horizon-size inflaton modes. This provides non-Gaussian initial conditions for the linear evolution of adiabatic perturbations. It has been shown that, for a wide class of potentials leading either to Linde's chaotic inflation (Linde 1983), or to power-law inflation (Abbott and Wise 1984, Lucchin and Matarrese 1985), fluctuations in the scalar field are non-Gaussian distributed around the classical trajectory. At the end of inflation the distribution of density perturbations can be highly non-Gaussian. This may have relevant effects on the large-scale matter distribution in the universe (Lucchin, Matarrese, and Ortolan 1989). Different mechanisms for producing non-Gaussian perturbations in inflation have been recently proposed by Allen, Grinstein, and Wise (1987), Peebles (1988), Salopek, Bond, and Bardeen (1988), Kofman and Pogosyan (1989); all these are based on the use of multiple scalar fields during inflation and usually generate non-Gaussian isocurvature perturbations (see also Kofman this volume).

As we shall shortly see, this mechanism for the generation of non-Gaussian adiabatic perturbations is characterized by the absence of intrinsic lengths on cosmologically relevant scales, so that it gives rise to a 'scale-invariant' fluctuation field. The scale-invariance property is reflected by the behavior of the *N*-point connected correlation functions $\xi_R^{(N)}(\mathbf{x}_1,\ldots,\mathbf{x}_N)$ of mass fluctuations (filtered over the scale *R*); these obey the scaling relations (Otto *et al.* 1986; Lucchin and Matarrese 1988)

$$\xi_{\mu R}^{(N)}(\mu\mathbf{x}_1,\ldots,\mu\mathbf{x}_N) \approx \mu^{-N(3+n_p)/2}\, \xi_R^{(N)}(\mathbf{x}_1,\ \ldots,\ \mathbf{x}_N), \tag{1}$$

where n_p is the primordial spectral index. It is easy to show that such statistics have some interesting properties: the functional form of the probability is left invariant by time evolution in the linear regime; the *N*-point correlation functions scale in a simple way at horizon-crossing (for instance, if $n_p = 1$, they are constant – a generalization of the Zel'dovich spectrum); and the two-point correlation function of peaks, evaluated at their mean separation, stays constant at different resolution scales (Lucchin, Matarrese, and Vittorio 1988). Simple toy-model examples of these statistics are: the model proposed by Peebles (1983) (and analyzed in more detail by Lucchin and Matarrese 1988), where the density-fluctuation field is the convolution of two independent scale-free Gaussian processes; and a model where the density field is the absolute value of a Gaussian random process. As we shall see in the following, the non-Gaussian scale-invariant density fluctuations obtained from inflation may be simply related to the latter toy-model.

STOCHASTIC DYNAMICS OF INFLATION

As shown by many authors, the dynamics of the inflaton on scales larger than the comoving Hubble radius $r_H(t) \approx 1/\dot{a}$ (a being the universe scale factor) can be studied by a stochastic approach (see, *e.g.*, Starobinskii 1986, Bardeen and Bublik 1987). One defines a *coarse-grained* variable $\varphi_x(t)$ as the average of the quantum field over a volume of size $\sim r_H{}^3$.

In the chaotic inflation scenario it is assumed that there are initial domains of size $\sim r_H{}^3$ characterized by a nearly homogeneous value φ_* for the scalar field. Provided that φ_* is large enough, a few Hubble times suffice to depress both the kinetic energy and the spatial gradients compared to the potential energy $V(\varphi)$. The resulting coarse-grained dynamics is *friction* dominated and can be described by a Langevin-type equation

$$\dot{\varphi}_x = -[1/3H(\varphi_x)]\,[\partial V(\varphi_x)/\partial\varphi_x] + [H^{3/2}(\varphi_x)/\sigma^{1/2}]\,\eta_x(t) \tag{2}$$

where $H(\varphi_x) \approx [V(\varphi_x)/(3\sigma^2)]^{1/2}$ and $\sigma \equiv 1/(8\pi G)^{1/2}$. In this approach $\mathbf{x}$ labels the coarse-grained variable of different cells. In writing equation (2) we implicitly assumed a perturbed Friedmann line-element in an appropriate synchronous gauge

$$ds^2 \approx dt^2 - a^2(\mathbf{x},t)\, d\ell^2, \tag{3}$$

with a local scale factor $a(\mathbf{x},t) \approx a(\mathbf{x},t_*) \exp[\int_{t_*}^{t} H(\varphi_x)\, dt']$ (as noticed by Goncharov, Linde, and Mukhanov 1987). Non-diagonal scalar perturbations of the metric are quickly depressed, on scales larger than the horizon, by the inflationary expansion. The first term on the right-hand side of equation (2) plays the role of a classical convective force, while the second one represents the diffusion process induced by fine-grained quantum fluctuations. The dependence of H on φ introduces multiplicative effects in the process. The Markovian Gaussian noise $\eta_x(t)$ has zero mean and auto-correlation function

$$\langle \eta_x(t)\, \eta_{x'}(t') \rangle = (\hbar\sigma/4\pi^2)\, j_0\,[\,|\mathbf{x} - \mathbf{x}'|/r_H(t)]\ \delta(t - t'), \tag{4}$$

where j_0 is the spherical Bessel function of order 0. The quantity

$$\wp(\varphi,t)\, d\varphi = \langle \delta(\varphi - \varphi[\eta_x(t)]\,)\rangle_\eta d\varphi,$$

yielding the ('one-particle') probability that, in a randomly chosen point $\mathbf{x}$, φ_x takes a value in the interval φ, $\varphi + d\varphi$, evolves according to a Fokker-Planck equation

$$\partial \wp / \partial t = (\partial / \partial \varphi) \{(2\sigma / 3^{1/2}) (\partial V^{1/2} / \partial \varphi) \wp + (\hbar / 4\pi^2) (V^{3/4} / 3^{3/2} \sigma^3) (\partial / \partial \varphi)(V^{3/4} \wp)\}. \quad (5)$$

Once the functional form of the inflaton potential $V(\varphi)$ has been specified, one looks a for a time-dependent solution with the initial condition (at t_*) $\wp(\varphi, t_*) = \delta(\varphi - \varphi_*)$ that corresponds to a homogeneous configuration φ_* with potential energy $V_* \approx 3\sigma^2 H_*^2$. The requirement that the initial energy density does not exceed the Planck energy implies $H_*/\sigma \leq 8\pi/3^{1/2}$. The Langevin equation (2), together with the η correlation function (4) actually contain more information than the (one-particle) probability $\wp$, for they take into account the whole space-correlation properties of the distribution, that is, they constrain the whole probability-density function.

It has been shown (Matarrese, Ortolan, and Lucchin 1989) that the dynamics described by equation (5) [or equation (2)] present universal properties for a large class of models leading either to chaotic or to power-law inflation. These can be summarized as follows. At early times the diffusion term in equation (5) causes the initial delta function to spread around its maximum which starts moving, due to the convective term. Convection, at this stage, can be approximated by a constant positive force, later on, however, force gradients start to be felt: *Brownian particles* which are closer to the minimum of the potential suffer smaller attraction compared to more distant ones; at the same time the diffusion coefficient becomes smaller and smaller as the minimum is approached and can be neglected. This causes a shrinking of the distribution at late times characterizing the so called *scaling regime* (see, *e.g.*, Suzuki 1981): at the *scaling time* Δt_{sc} the system turns from a diffusion-dominated regime (due to quantum fluctuations) to a convection-dominated regime (due to classical non-linearities), or from a disordered state to a macroscopically ordered state. In this picture inflation is described as the non-equilibrium decay of the system from the unstable state φ_* to the minimum of the potential, with the whole coarse-grained field distribution undergoing slow rolling. During the scaling regime the distribution is strongly peaked around the classical homogeneous configuration [the solution of equation (2) in the $\hbar \to 0$ limit] $\varphi_{cl}(t)$, with fluctuations giving rise to classical adiabatic density perturbations. In this *scaling limit* the solution tends to a self-similar time-independent function $\mathfrak{R}$ of a single *scaling variable* ξ: $\wp(\varphi,t)\, d\varphi \sim \mathfrak{R}(\xi)\, d\xi$. Such a scaling variable can be chosen so that it is, with very good accuracy (or even exactly), Gaussian with zero mean and equal-time two-point function

$$\langle \xi_{x}(t)\, \xi_{x'}(t) \rangle = 1/(2\pi^2) \int_0^\infty dk\, k^2\, W^2(k,t)\, P(k)\, j_0\, (k|\mathbf{x} - \mathbf{x}'|), \quad (6)$$

where $W^2(k,t) \approx \vartheta\, [k - 1/r_H(t)] - \vartheta\, [k - 1/r_H(t_*)]$ is such that only wavelengths that left the horizon during the interval $t_* \to t$ contribute to the integral in equation (4). The ξ power-spectrum has a 'flicker-noise' form $P(k) = P_0\, k^{-3}$. The amplitude P_0 (which may possibly contain a residual logarithmic k-dependence) as well as the non-linear transformation between φ and ξ depend upon the precise form of the inflation potential.

CHAOTIC AND POWER-LAW INFLATION

In this paper we shall restrict our discussion to two simple models: a chaotic inflation based on a quartic potential (Linde 1983) and a power-law inflation based on an exponential potential (Lucchin and Matarrese 1985; see also Ratra 1989 and references therein).

Chaotic inflation can be driven by a quartic potential $V(\varphi) = (\lambda/4)\varphi^4$. The classical solution of the φ dynamics in the slow-rolling down phase is described by

$$\varphi_{cl}(t) = \varphi_* \exp\, [-2(\lambda/3)^{1/2}\, \sigma\, \Delta t],$$

where $\Delta t = t - t_*$. If the expansion is dominated by the classical homogeneous mode, the scale factor is given by

$$a(t) = a_* \exp \{(H_*/4\sigma)(3/\lambda)^{1/2} [1 - \exp(-4(\lambda/3)^{1/2} \sigma \Delta t)]\}.$$

Inflation is expected to end when the kinetic energy equals the potential energy, *i.e.* when $\langle\varphi\rangle = \pm 8^{1/2}\sigma$. The exponential potential $V(\varphi) = V_* \exp[-\lambda(\varphi - \varphi_*)/\sigma]$, leads to power-law inflation $a(t) = a_* [1 + \lambda^2 H_* \Delta t/2]^{2/\lambda^2}$, for any $\lambda \leq 2^{1/2}$. Unless the potential is suitably modified to allow for reheating, inflation lasts forever, the ratio of kinetic and potential energy being constant along the classical trajectory $\varphi_{cl}(t) = \varphi_* + (2\sigma/\lambda) \ln(1+\lambda^2 H_* \Delta t/2)$.

For the quartic potential one can solve exactly the Langevin equation and find, for $\delta\varphi_x(t) = \varphi_x(t) - \varphi_{cl}(t)$ [where $\varphi_{cl}(t)$ generally differs from $\langle\varphi(t)\rangle$ by a small, time-dependent, space-homogeneous quantity which does not affect $\delta\rho$];

$$\delta\varphi_x(t) = \pm\sigma (12H_*^2/\lambda\sigma^2)^{1/4} \exp[-2(\lambda/3)^{1/2} \sigma \Delta t] [|1 + \xi_x(t)|^{-1/2} - 1], \tag{7}$$

where the solution with the plus (minus) sign must be taken if φ_* is positive (negative). In this case

$$P_0 \approx \hbar(\lambda/3)^{1/2} (H_*/\sigma) [1 - 4(\lambda/3)^{1/2} (\sigma/H_*) \ln(k/a_*H_*)].$$

For the exponential potential, appropriate use of the 'scaling approximation' yields

$$\delta\varphi_x(t) = (2\sigma/\lambda) \ln\{[|1 + \xi_x(t)|^{2/3} + \lambda^2 H_* \Delta t/2] / [1 + \lambda^2 H_* \Delta t/2]\} \tag{8}$$

and $P_0 = (9\hbar/16)[\lambda^2/(2-\lambda^2)](H_*/\sigma)^2$. The random-phase approximation in this context would amount to expand the right-hand side of equations (7) and (8) to first order in ξ; however ξ can have a large *rms* value, depending on the value of H_*, so that such an approximation would fail in the general case. In the exponential potential model, for instance,

$$\xi_{rms}(t) = (\hbar/2\pi^2)^{1/2} (3H_*/4\sigma) \ln^{1/2}(1 + \lambda^2 H_* \Delta t/2), \tag{9}$$

which, for times after the onset of the scaling regime $\Delta t_{sc} \sim 2(e^{1/2} - 1)\lambda^{-1} H_*^{-1}$, can be larger than unity. This fact is at the origin of the non-Gaussian behavior of $\delta\varphi$ discussed in Matarrese, Ortolan, and Lucchin (1989). For the quartic potential model, for times much larger than $\Delta t_{sc} \sim (3/\lambda)^{1/2} (1/8\sigma) \ln 3$, due to the large value of $\xi_{rms}(t)$ one can approximate $\varphi_x(t) \propto |\xi_x(t)|^{-1/2}$. Positive moments of φ, being related to negative moments of a semi-Gaussian distribution, are infinite (as for the Cauchy distribution), except for $\langle\varphi\rangle$. One finds that the probability of crossing a level ν times a suitable 'effective dispersion' $\langle\delta\varphi_{eff}^2\rangle^{1/2}$ is far from the Gaussian expectation already for small ν. This result holds, in the scaling regime, for suitably large values of H_*/σ, independently on the value of λ and on time (in disagreement with Hodges [1989], who uses a Gaussian approximation to estimate these crossing probabilities), as follows from the scale-invariance property

$$\wp(\varphi,t) \approx \mu\wp(\mu\varphi, t - (3/4\lambda\sigma^2)^{1/2} \ln\mu). \tag{10}$$

For the exponential potential, in the same way, one finds at large times $\varphi_x(t) \propto |\xi_x(t)|^{2/3}$ and

$$\langle\varphi^N\rangle \sim \pi^{-1/2} \Gamma(1/2 + N/3) \{(2^{7/3}\sigma)/(\lambda^3 H_* t)\}^N \xi_{rms}^{2N/3}(t), \tag{11}$$

with ξ_{rms} given in equation (9). In the scaling regime the N-th connected moments normalized to the $(N/2)$-th power of the variance relax to constant non-zero values: clear evidence for non-Gaussian scale-invariant behavior of the statistics. These dimensionless ratios do depend neither on the value of λ, nor on time (scale), nor on the initial condition through H_*. The role of H_*, however, is more subtle: it must be large enough so that the system has well entered the scaling regime when the cosmologically interesting scales leave the inflationary horizon. It is clear that a low value of H_* (just enough to solve the horizon problem), as that quoted by Kofman (this volume), keeps the system in the diffusive regime, where the fluctuations are practically Gaussian, and implies that our universe is exactly homogeneous on scales larger than the Hubble radius. This property is much stronger than what required by the isotropy of the cosmic microwave background, which only demands (for reasonable fluctuation spectra) that the *rms* density fluctuation on the scale of the present horizon is less than about 10^{-5}. Fluctuations on super-horizon scales only determine the local value of 'average quantities'.

DENSITY FLUCTUATIONS

The coarse-grained fluctuation $\delta\varphi_x(t)$ contains all wavelengths larger than the horizon size at the time t. In order to obtain the density fluctuation $\delta\rho/\rho$, it is necessary to Fourier transform $\delta\varphi$ at each time t keeping only that mode $k \approx 1/r_H(t)$ that crosses the Hubble radius then. We can obtain the fluctuation field at the time $t_H(k)$ when each Fourier mode crosses the Hubble radius during the Friedmann era by using the approximate constancy of the gravitational potential outside the horizon. One has $\delta\rho(\mathbf{k},t_H)/\rho \approx -b\, H_{cl}(t_1)\ \delta\varphi(\mathbf{k},t_1)/\dot\varphi_{cl}(t_1)$, where t_1 is the time when the wave-number k left the horizon during inflation and $b = 1$ if $t_H < t_{eq}$ or $b = 2/5$ if $t_H > t_{eq}$, with t_{eq} the equivalence time (see, *e.g.*, Lyth 1985). One finds

$$\delta\rho/\rho(\mathbf{k},t_H) \approx (b/2)(3H_*^2/\lambda\sigma^2)^{1/2} \exp[-4(\lambda/3)^{1/2}\sigma\, t_1] \int d^3x \exp(-i\,\mathbf{k}\cdot\mathbf{x})[\,|1+\xi_x(t_1)|^{-1/2}-1], \tag{12}$$

for the quartic potential, and

$$\delta\rho/\rho(\mathbf{k},t_H) \approx -4b/(\lambda^4 H_* t_1) \int d^3x \exp(-i\,\mathbf{k}\cdot\mathbf{x})[\,|1+\xi_x(t_1)|^{2/3} - 1], \tag{13}$$

for the exponential potential at times much larger than Δt_{sc}. It is clear that the non-linear transformation from the Gaussian variable ξ to $\delta\varphi$ [equations (7) and (8)] implies that all the Fourier modes of ξ contribute to $\delta\rho(\mathbf{k},t_H)/\rho$. As we have shown before, in the scaling regime one can approximately write $[\,|1 + \xi_x(t)|^\alpha - 1] \sim |\xi_x(t)|^\alpha$ in the right-hand side of equations (12) and (13), where α is respectively $-1/2$ and $2/3$. Note that this is the opposite of the random-phase approximation, valid in the initial diffusion dominated regime, where one writes $[\,|1 + \xi_x(t)|^\alpha - 1] \sim \alpha\, \xi_x(t)$. Of course, if one had linearized from the beginning the evolution equation for $\delta\varphi$, the coarse-grained fluctuations would come out proportional to ξ and then Gaussian (non-Gaussian fluctuations can be obtained in the linear approximation by interacting multiple scalar fields, as in Salopek, Bond, and Bardeen [1988]): the linear approximation for $\delta\varphi$, therefore, does not adequately follow the dynamics during the convective regime. The density fluctuation field $\delta\rho(\mathbf{x},t)/\rho$ at a given time t (before the onset of non-linear evolution; for instance at the recombination epoch) is obtained from equations (12) and (13) after specifying the linear growth factor and multiplying by $k^2T(k)$, with $T(k)$ the transfer function. The non-linear transformation $\xi \to \delta\rho/\rho$ implies that fluctuations of ξ on all scales, even of super-horizon size (where, due to the k^{-3} tail of the spectrum, ξ largely fluctuates) affect the present statistics of sub-horizon density perturbations which will then be strongly non-Gaussian. In summary, fluctuations in the ensemble of universes (which one can only perceive as a local 'monopole' contribution to the energy density) are well coupled to fluctuations inside the present Hubble radius, which makes density fluctuations non-Gaussian on observable scales.

CONCLUSIONS

The stochastic approach allows to study the dynamics of inflation on scales larger than the Hubble radius. It accounts for the generation of large-scale classical fluctuations from quantum oscillations inside the horizon, the effect of non-linearities on the evolution of inflaton perturbations and the back-reaction of matter fluctuations on the background geometry.

A generic feature of models leading to chaotic or power-law inflation is the occurrence of a scaling regime where the coarse-grained distribution sharply peaks around the classical solution. In this regime the distribution is non-Gaussian and scale-invariant, due to the dominance of the inflaton non-linearities over the diffusion caused by quantum fluctuations. The deviation from the Gaussian behavior does not depend on the strength λ of the non-linearity, on time (scale), and on the initial condition, provided the latter permits the system to enter the scaling regime. Because the probability is peaked, its bulk properties may be described by a Gaussian centered on $\varphi_{cl}(t)$ with suitable dispersion, but this approximation would fail in estimating the likelihood of rare events for which the actual distribution is required.

The scale-invariance of the inflation is expected to be reproduced by the density fluctuation, during its linear evolution. As a consequence, the N-point mass correlation functions will obey equation (1), with a spectral index n_p determined by the inflationary parameters; in the exponential potential model, for instance, n_p depends upon the value of λ. The non-Gaussian scaling invariance of the primordial perturbation field will affect the properties of the universe on large scales (see also the discussion by Lucchin *et al.* this volume). In particular the probability for the occurrence of rare events, such as high peculiar velocities, large empty regions and 'great attractors' are related to the statistics of the primordial fluctuation field, even though this has been partially masked by the non-linear action of gravity after recombination.

REFERENCES

Abbott, L. F., and Wise, M. B. 1984, *Nucl. Phys.*, **B244**, 541.
Albrecht, A., and Steinhardt, P. J. 1982, *Phys. Rev. Lett.* **48**, 1220.
Allen, T. J., Grinstein, B., and Wise, M.B. 1987, *Phys. Lett.* **197B**, 66.
Bardeen, J. M., and Bublik, G. J. 1987, *Class. Quantum Grav.*, **4**, 573.
Goncharov, A. S., Linde A. D., and Mukhanov, V. F. 1987, *Int. J. Mod. Phys.*, **A2**, 561.
Guth, A. H. 1981, *Phys. Rev.*, **D23**, 347.
Hodges, H. 1989, *Phys. Rev.* **D40**, 1798.
Kofman, L. K. this volume, p. 339.
Kofman, L. K., and Pogosyan, D. Yu. 1989, *Phys. Lett.*, **214B**, 508.
Linde, A. D. 1982, *Phys. Lett.*, **108B**, 389.
Linde, A. D. 1983, *Phys. Lett.*, **129B**, 177.
Lucchin, F., and Matarrese, S. 1985, *Phys. Rev.*, **D32**, 1316.

Lucchin, F., and Matarrese, S. 1988, *Ap. J.*, **330**, 535.
Lucchin, F., Matarrese, S., Messina, A., and Moscardini, L. this volume, p. 205.
Lucchin, F., Matarrese, S., and Ortolan, A. 1990, in preparation.
Lucchin, F., Matarrese, S., and Vittorio, N. 1988, *Ap. J. (Letters)* **330**, L21.
Lyth, D. H. 1985, *Phys. Rev.*, **D31**, 1792.
Matarrese, S., Ortolan, A., and Lucchin, S. 1989, *Phys. Rev.*, **D40**, 290.
Ortolan, A., Lucchin, F., and Matarrese, S. 1988, *Phys. Rev.*, **D38**, 465.
Otto, S., Politzer, H. D., Preskill, J. and Wise, M.B. 1986, *Ap. J.*, **304**, 62.
Peebles, P. J. E. 1983, *Ap. J.*, **274**, 1.
Peebles, P. J. E. 1988, in *Large-Scale Structure and Motions in the Universe*, ed. G. Giuricin *et al.* (Dordrecht: Kluwer Academic Publishing).
Ratra, B. 1989, preprint.
Salopek, D. S., Bond, J. R. and Bardeen, J. M. 1988, preprint.
Starobinsky, A. 1986, in *Field Theory, Quantum Gravity and Strings*, ed. H. J. de Vega and N. Sanchez (*Lecture Notes in Physics*, **246**) (Berlin: Springer), p. 107.
Suzuki, M. 1981, *Adv. Chem. Phys.*, **46**, 195.

DISCUSSION

L. Kofman: If you start with the initial condition $\wp(\varphi,t_*) = \delta(\varphi-\varphi_*)$, where φ_* corresponds to the size of our universe ($\ell = H^{-1}\exp(\pi\varphi_*^2/M_{\rm Pl}^2)$; in the $V(\varphi) = \lambda\varphi^4/4$ model $\ell = 10^{28}$ cm for $\varphi_* \simeq 5M_{\rm Pl}$), do you get Gaussian perturbations? My point is that it is the only initial condition for our part of universe.

S. Matarrese: I agree that, if one starts with such a low value for φ_* (or what we call H_*, the initial Hubble parameter), the fluctuation distribution is likely to be Gaussian just because the system is still in the diffusive regime when quantum fluctuations are dominating the dynamics. Our main point is that the system must enter the convective ('scaling') regime where non-Gaussian perturbations arise: in order to relate this phase to the cosmologically relevant scales (our universe), one needs a quite large value of φ_*. The onset of such a scaling regime is in any case necessary for the classical slow rolling down picture of inflation to be valid, and for classical perturbations to arise from quantum fluctuations.

A DISC WITH STRING-LIKE PROPERTIES

JOSÉ P. S. LEMOS
Observatório Nacional, Rua General Bruce 586, Rio de Janeiro (RJ), Brazil

ABSTRACT In this communication I will show that a static relativistic disc and a cosmic string have similar properties, namely, that the space-time which both create is flat and has a conical geometry.

Cosmologists and particle physicists have been working on the likelihood of the existence of cosmic strings as these line-like structures may be able to explain the formation of galaxies and clusters of galaxies. In General Relativity cosmic strings produce gravitational fields without Newtonian analogues, the main feature being that the exterior space-time is flat.

To be more specific, the exterior metric of a static infinite string lying along the z-axis is given by (Vilenkin 1987):

$$ds^2 = -dt^2 + dz^2 + dr^2 + (1 - 4\mu)^2 r^2 d\varphi^2 \tag{1}$$

where μ is the inertial mass per unit length and we use geometrical units ($G = c = 1$). If we make the coordinate transformation $\varphi \rightarrow (1 - 4\mu)\varphi$, we see that the space outside the cosmic string is flat. However, since φ is now in the range $0 \leq \varphi \leq (1 - 4\mu) 2\pi$, the space structure of surfaces with constant t and z is not that of a plane sheet but rather of a cone. This conical geometry can be observed by the formation of double images of objects located behind the string (Figure 1).

The energy-momentum tensor $\mathbf{T}_{\mu\nu}$ of an infinitesimally thin string is provided by the Grand Unified Theories and given by $\mathbf{T}^{t}_{t} = \mathbf{T}^{z}_{z} = \mu\,\delta(r)$. This energy-momentum tensor implies that the pressure is negative $p_{zz} = -\mu$ and is therefore a tension. Since the gravitational mass is given by the Tolman formula

$$M_{\rm grav} = \int(-\mathbf{T}^{t}_{t} + \mathbf{T}^{z}_{z})\,(-g)^{1/2}\,d^3x$$

(where g is the determinant of the metric) we have that cosmic strings have zero gravitational mass and can be considered as a missing part of a Riemannian manifold, *i.e.*, a topological defect. Some membranes have analogous properties (Letelier and Oliveira 1988).

Having summarized the properties of cosmic strings that are of interest in our context, we now turn to the disc. This disc is formed by counter-rotating particles moving at the speed of light. Counter-rotation means that an equal number of particles move in each sense of rotation. The disc as a whole does not rotate, rather, it is held by the tangential stresses caused by the two counter-rotating streams of inter-penetrating matter, each moving with the velocity of light (for details see Lynden-Bell and Pineault 1978; Lemos 1988; Lemos 1989).

The disc is assumed to lie on the equatorial plane, $\theta = \pi/2$. We also assume that the disc is infinite. In this case the space-time it creates is self-similar since it is not possible to construct a length scale. The corresponding finite disc is not self-similar but has the property that for sufficiently small radii its solution, within higher order corrections, is the

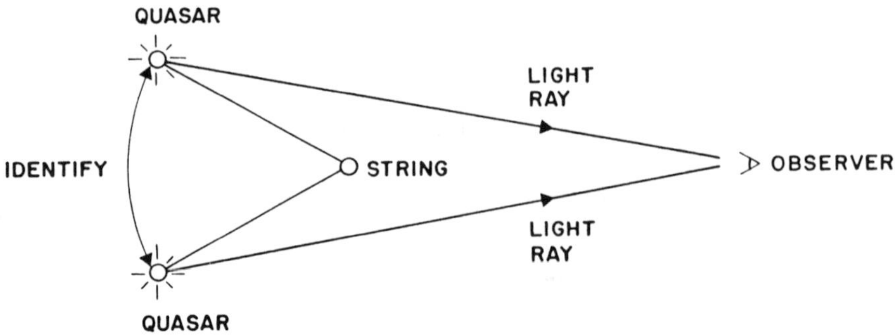

Fig. 1. The observer sees two images of the same quasar due to the conical structure of space. The cone is most easily seen by pulling the point (apex) where the string is located until the two lines from the string to the quasar meet.

same as if the disc were infinite. This ensures that the boundary conditions make sense. The infinite self-similar disc is the limiting case of the finite disc.

The striking property of this disc is that space-time is flat on both sides of the disc but not on the disc itself. To see this in detail we write the metric of a counter-rotating null disc:

$$ds^2 = -r\cos^2(\theta/2)\, d\tau^2 + 1/r\,(dr^2 + r^2\, d\theta^2) + 4r\sin^2(\theta/2)\, d\varphi^2 \tag{2}$$

where $0 < \theta < \pi/2$. In the domain $\pi/2 \leq \theta \leq \pi$ one replaces θ by $\pi - \theta$. If we now change to quasi-cylindrical coordinates $\zeta = 2r^{1/2}\cos(\theta/2)$, $\eta = 2r^{1/2}\sin(\theta/2)$ the metric to the upper half of the disc will be given by

$$ds^2 = -\zeta^2\, d(\tau/2)^2 + d\zeta^2 + d\eta^2 + \eta^2\, d\varphi^2 \tag{3}$$

which is the usual form of the Rindler metric. Here $0 \leq \eta \leq \infty$, and $0 \leq \zeta \leq \infty$. Then, if we transform to inertial frame coordinates by writing $t = \sinh(\tau/2)$, $z = \cosh(\tau/2)$, $x = \eta\cos\varphi$, $y = \eta\cos\varphi$, we obtain for $\zeta > \eta$ (which is $z^2 - t^2 > x^2 - y^2$) a Minkowskian space-time:

$$ds^2 = -dt^2 + dx^2 + dy^2 + dz^2 \tag{4}$$

However, observers above and below the disc have to choose different coordinates to do this. Figures 2, 3 and 4 correspond to the three types of coordinates used in equations (2), (3) and (4), respectively. The equation of the disc for an inertial observer in Minkowski coordinates is the hyperboloid $z = \pm(t^2 + x^2 + y^2)^{1/2}$.The disc itself has a surface layer stress-energy tensor $\mathbf{T}^{\alpha\beta}$ given by $-\mathbf{T}^t_t = \mathbf{T}^\varphi_\varphi = \delta(r\cos\theta)/8\pi$. Recurring to the Tolman formula it has a non-zero gravitational mass. On the other hand its rest mass is zero.

Like a cosmic string, the disc has the property that the exterior space-time is locally flat. However a string produces a deficit angle in the azimuthal direction while the deficit angle produced by the disc is in the polar direction. From Figure 4 it can be easily worked out that multiple images of quasars can be formed if the disc is between the object and the observer. We also point out that, like the cosmic string, the disc is a topological defect since its Riemann tensor is zero outside the matter.

We have compared two space-times with a remarkable property: outside the matter the metric is Minkowskian. Penrose (1967) considered a gravitational wave pulse with a similar property.

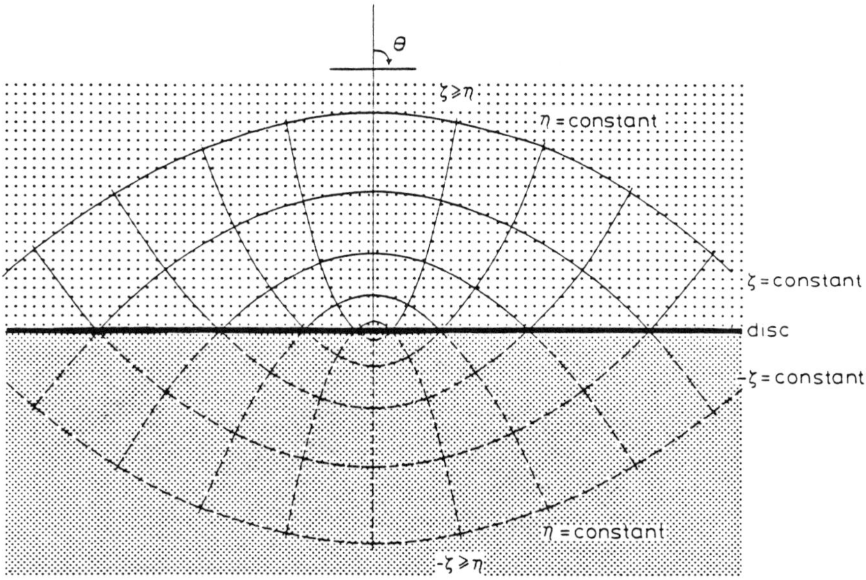

Fig. 2. Diagram for the hypersurface τ = constant in the coordinate frame given by equation (2). In the upper part of the disc, lines ζ = constant are downward opening parabolas and lines η = constant are upwards opening parabolas. In the lower part of the disc this is reversed. The φ coordinate has been suppressed.

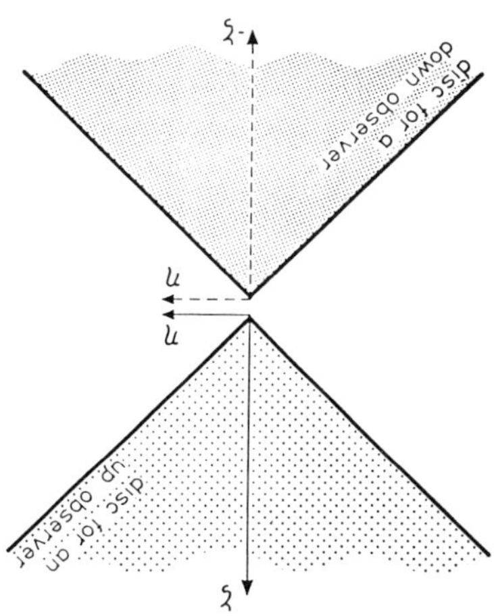

Fig. 3. Diagram for the hypersurface τ = constant in Rindler coordinates. Lines ζ = constant are planes, η = constant are cylinders. Only the interiors and surfaces of these cones represent the physical space, *i.e.*, the boundaries should be identified.

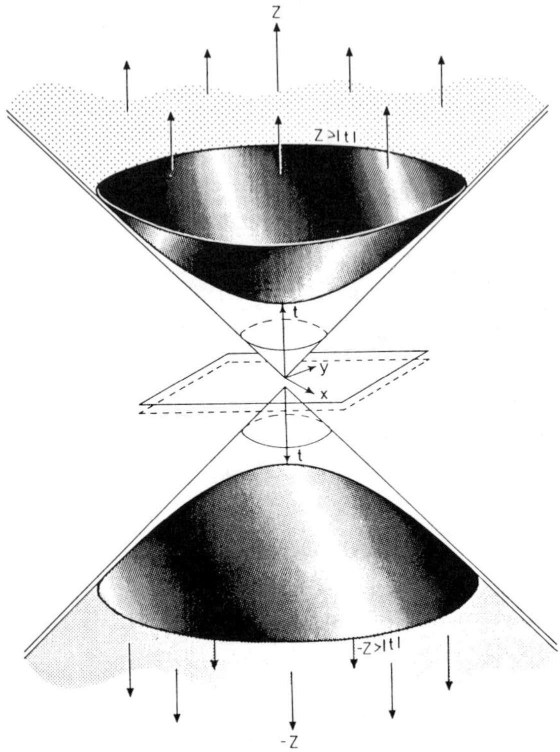

Fig. 4. Diagram showing two hyper-surfaces of t = constant in the Minkowski coordinate system. At $t = 0$ the disc is seen as a cone, while at all other times it is observed as a rectangular hyperboloid. Only the shaded areas represent the physical space.

ACKNOWLEDGEMENTS

I thank conversations with Joseph Katz, Patricio Letelier and Donald Lynden-Bell. I also thank C. Willmer for his comments on the manuscript.

REFERENCES

Lemos, J. P. S., 1988, *M.N.R.A.S.*, **230**, 451

Lemos, J. P. S., 1989, *Class. Quantum Grav.*, submitted.

Letelier, P. S., Oliveira, S. R., 1988, *Class. Quantum Grav.*, **5**, L47.

Lynden-Bell, D., Pineault, S., 1978, *M.N.R.A.S.*, **186**, 679.

Penrose, R., 1968, in *Batelle Rencontres*, ed. C. M. DeWitt and J. A. Wheeler, (New York: Benjamin Inc.)

Vilenkin, A., 1987, in *300 Years of Gravitation*, ed. S. W. Hawking and W. Israel, (Cambridge: Cambridge University).

THE FIRST STARS

J. C. N. DE ARAUJO AND R. OPHER
Instituto Astronômico e Geofísico, Universidade de São Paulo
Caixa Postal 30627, CEP 01051, São Paulo (SP), Brazil

ABSTRACT We study the formation of Population III stars. We also investigate the formation of voids from negative density perturbations.

INTRODUCTION

Several authors have studied the formation and evolution of Population III objects, *e.g.* Hirasawa, Aizu, and Taketani (1969); Matsuda, Sato, and Takeda (1969); Yoneyama (1972); Hutchins (1976); Silk (1977, 1983); Lepp, and Shull (1984); and Lahav (1986). Lahav was the first to use more realistic initial conditions. He did not begin the calculation, however, at the appropriate epoch: the beginning of the recombination era.

We investigate the cooling and collapse of perturbations of mass $M \sim M_J$ and $M << M_J$, taking into account the expansion of the Universe, internal pressure, photon drag, photon cooling, photo-ionization, collisional ionization, and the formation and cooling of hydrogen molecules. We begin the calculation at the recombination era. We use a spectrum of perturbations

$$\delta\rho/\rho = (M/M_0)^{-1/3}\,(1 + z_{rec})^{-1}, \tag{1}$$

where M_0 is the mass scale and z_{rec} is the redshift at recombination (*cf.* Lahav 1986, de Araujo and Opher 1988).

Studies of aspherical collapse have been made by several authors: Mestel (1965a,b); Lin, Mestel, and Shu (1965); Hutchins (1976); Silk (1982); White, and Silk (1979); Palmer (1981). Detailed calculations were not made, however, of aspherical perturbations taking into account the relevant physical processes described above. Detailed effects of rotation and a magnetic field were also not made. We study the influence of non-sphericity, rotation, and magnetic fields in the collapse of clouds of mass $M \sim M_J$, in particular, the influence on the turn-around and collapse redshift, cloud temperature and density for a given eccentricity, rotational energy, and magnetic field.

We study the formation of Population III stars that can be formed by fragmentation from a cloud of mass $M \sim M_J$. We use the fact that a perturbation for $M << M_J$, that can survive with some residual amplitude, can produce the fragmentation of clouds of mass $M \sim M_J$, in the first free-fall time scale of M (see de Araujo and Opher 1989).

Some authors analyzed the possibility of formation of voids from a negative spectrum of perturbations (*e.g.* Hausman, Olson, and Roth 1983; Hoffman, and Shaham 1983; Fillmore, and Goldreich 1984; Bertschinger 1985). We adopt here a spectrum such as given by equation (1), but a negative one, and include all the physical effects described above.

We now discuss the basic equations used and then the calculations made.

BASIC EQUATIONS

The basic equations used for the calculation are:

$$\partial \rho / \partial t + \nabla \cdot (\rho \mathbf{v}) = 0 \tag{2}$$

$$\rho (d\mathbf{v}/dt) + \nabla P + \rho \nabla \Phi + (\sigma b T^4)/(m_p c) [x_e \rho (\mathbf{v} - H_r)] = 0 \tag{3}$$

$$\nabla^2 \Phi = 4\pi G \rho \tag{4}$$

$$P = N \rho K_B T (1 + x_e) \tag{5}$$

$$dU/dt = -L + P/\rho^2 (d\rho/dt) \tag{6}$$

where ρ is the matter density, $\mathbf{v}$ is the fluid velocity, P is the matter pressure, the last term in equation (2) is the photon drag due to the background radiation, x_e is the degree of ionization, U is the internal energy per unit mass, and L is the cooling function.

For the calculation of the degree of ionization (recombination), we use Peebles (1968), who made a detailed calculation taking into account photo-ionization due to background radiation and recombination photons. We take into account also collisional ionization. We include in the cooling function: Compton heating (cooling), molecular hydrogen cooling, Lyman-α cooling, recombination, collisional ionization, and photo-ionization. The cooling due to hydrogen molecules, as used by Lahav (1986) and de Araujo and Opher (1988, 1989), was taken from Lepp and Shull (1983). For Lyman-α cooling, we used Carlberg (1981). To solve the system of equations, we use a square density profile.

CALCULATIONS AND DISCUSSION

We begin our calculations at the beginning of the recombination era ($T_R \sim 4000$ K). For the study of the cooling and collapse of clouds of mass $M \sim M_J$ and $M << M_J$, we study the effects of non-sphericity, rotation, and magnetic fields. We use $M_0 = 10^{15} M_\odot$ (galactic cluster). For the study of the formation of void regions, we examine different values of M_0 and the spectral index α ($\delta \rho / \rho \propto M^{-\alpha}$), *e.g.* $\alpha = 1/3$ and $1/2$.

For the study of the collapse of Population III objects, we assume $\Omega = 0.1$ (where $\Omega = \rho_0 / \rho_{0C}$, ρ_0 is the present density, and ρ_{0C} is the present critical density) and $h = 1.0$ (Hubble constant in units of 100 km s^{-1} Mpc^{-1}). We obtain:

–the first bound objects were $\sim 10^4 M_\odot$;
–the collapse of the clouds was not significantly delayed by the internal pressure of the cloud (since at turn-around the cloud temperature is close to the ambient temperature, due to photon cooling); and
–after turn-around, the temperature does not rise adiabatically.

Including also the effects of non-sphericity (assuming oblate spheroids), rotation, and magnetic fields, we obtain:

–the efficiency of the production and cooling of H_2 molecules decreases with increasing eccentricity, e. For example, for $M = 10^6 M_\odot$ we have $x_{2C} \sim 2 \times 10^{-3}$ to 3×10^{-5} for $e = 0.0$ to 0.9;
–oblate spheroids have larger collapse Jeans masses, M_{JC}, than spherical clouds (collapse defined for oblate spheroids ($e > 0$) as $e_{\text{final}} \sim 1.0$. For $M = 10^6 M_\odot$ we have

$M_{JC}/M_\odot$ = 0.3 to 5 for e = 0.0 to 0.9).
$-M_{JC}$ increases with increasing parent cloud mass (*e.g.* for e = 0.0 we have $M_{JC}/M_\odot$ = 0.3 to 5 for $M_C = 10^6$ to 10^8);
–rotation decreases the efficiency of cooling due to H_2, producing larger M_{JC}. For example, for a cloud initially spherical of mass = $10^6\ M_\odot$ and ratio of rotational energy to gravitational energy β_r = 0.0 to 0.1, we have T_{MC} = 440 to 640 K, $M_{JC}/M_\odot$ = 0.3 to 5; and
–magnetic fields $\mathbf{B} = 10^{-6}$ to 10^{-5} Gauss make an initially spherical cloud into an oblate spheroid, which collapses into a disk.

We evaluated numerically the decay of isothermal density fluctuations of mass $M << M_J$ during the recombination era due to the non-sudden elimination of dynamical coupling with the radiation and the gradual loss of thermal contact with it. We find that δ_f/δ_i (where $\delta = \delta\rho/\rho$) for $M << M_J$ is not negligible in general. For example, for $M/M_J = 10^{-3}$, Ωh^2 = 0.025 to 1.0, and δ_i = 0.01 to 1.0, we have $\delta_f/\delta_i \sim$ 25 to 50%.

We study the formation of proto-stars following the suggestion of de Araujo and Opher (1988) that the fluctuations that persist for $M << M_J$ can fragment the clouds of mass $M_C \sim M_J$. We note that there is no direct collapse for $M_C < 10^4\ M_\odot$. We take the spectrum of perturbations given by equation (1) for M_C ($\sim M_J$) and M_S (proto-stars). We take $M_C = 10^6\ M_\odot$ and M_S = 10 to 500 $M_\odot$. We obtain, for example, that when $n_s \sim 10^8$ to 10^9 cm^{-3} we have $\delta_S = 900$ for $M_S = 500\ M_\odot$, and $\delta_S = 3.2$ for $M_S = 50\ M_\odot$ at $z \sim 180$ (z is the redshift). The minimum mass that can collapse is $M_S \sim 50\ M_\odot$.

In figure 1 we show the behavior of $\mathbf{v}_{1S}/\mathrm{v}_f$ (where $\mathbf{v}_{1S}$ is the collapse velocity of M_S relative to M_C, and $\mathrm{v}_f = \mathbf{v}_{1C}(r_S/r_C) + (\dot{R}/R)r_S$ is the collapse velocity of the M_S cloud if $\delta = 0$, $\mathbf{v}_{1C}$ is the collapse velocity of M_C, r_S (r_C) is the radius of M_S (M_C), and R is the scale factor) *versus* z for z = 200 to 180, for M_S = 500, 50, and 20 $M_\odot$ in order to understand how the fragmentation takes place. When M_S becomes greater than the Jeans mass, the gravitational attraction becomes sufficiently strong to compensate for the outward pressure. When the fast collapse starts at $z \sim 181$, one has $\mathbf{v}_{1S}/\mathrm{v}_f \sim 0.6$ for $M_S = 500\ M_\odot$, which aids in collapsing M_S. The argument is valid also for 50 $M_\odot$. For $M_S = 20\ M_\odot$, $\mathbf{v}_{1S}/\mathrm{v}_f$ and δ_S are too small to cause M_S to collapse appreciably before the actual collapse of M_C.

Finally, we might ask what happens to the rest of the M_C cloud that did not form stars. A possible scenario is that after the first stars of mass M_S formed and started nuclear burning, they produced strong winds (like present day newly formed stars) and dispersed the rest of the M_C cloud.

For the formation of void regions, as mentioned above, we take different values of M_0 (mass scale) and α. In general, we produce underdense regions of present radius, r_p, on the order of several Mpc, appreciably less than the recently discovered voids of tens of Mpc (which may have been produced by hierarchical clustering of smaller voids in conjunction with hierarchical clustering of galaxies, produced by the positive density perturbation spectrum). We find, for example, for $\alpha = 1/2$ and $M_0 = 10^{15}\ M_\odot$ (with the physical process described above), underdense regions 35 to 75% of the present density of the Universe, containing masses $M \sim 10^9$ to $10^{12}\ M_\odot$.

ACKNOWLEDGEMENTS

The authors wish to thank the Brazilian agency CNPq for support.

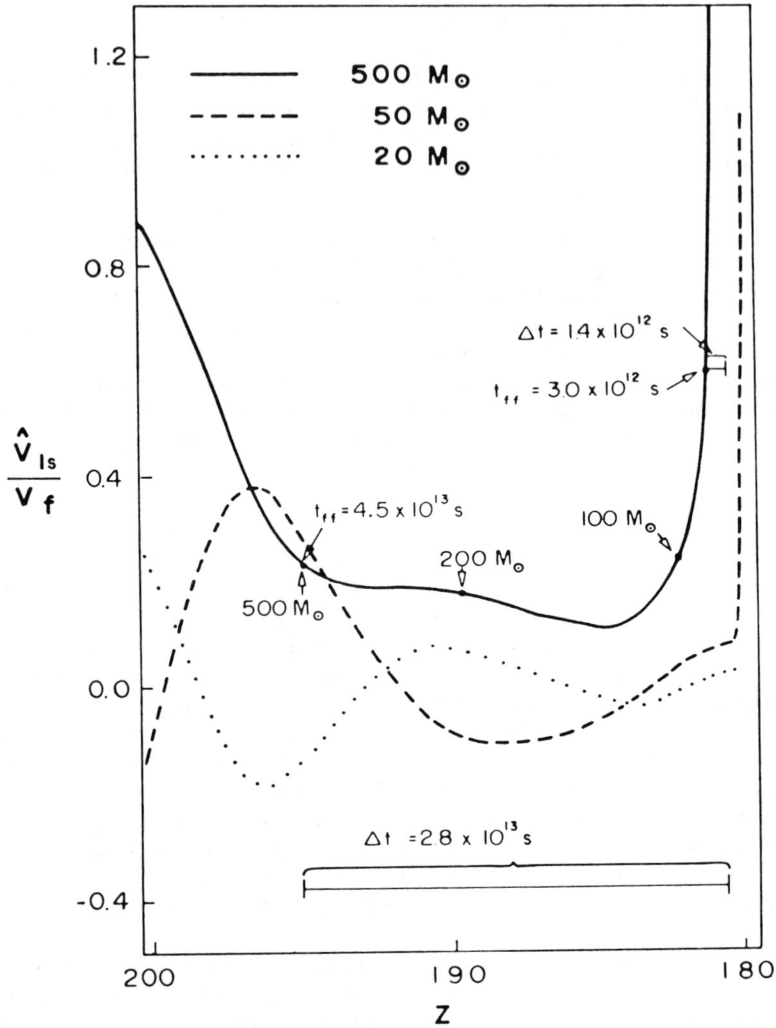

Fig. 1. $\mathbf{v}_{1S}/\mathbf{v}_f$ as a function of redshift z for $M_S = 500$, 50, and 20 $M_\odot$. The Jeans masses are noted on the $M_S = 500\ M_\odot$ curve. The free-fall time (t_{ff}) when $M_S \sim M_J$ is noted on the $M_S = 500\ M_\odot$ curve, as well as the time for M_S to collapse (ΔT). Also noted: t_{ff} and Δt for $\mathbf{v}_{1S}/\mathbf{v}_f \sim 0.6$ and $z \sim 181$.

REFERENCES

Bertschinger, E. 1985, *Ap. J. Suppl.*, **58**, 1.
Carlberg, R. G. 1981, *M.N.R.A.S.*, **197**, 1021.
de Araujo, J. C. N., and Opher, R. 1988, *M.N.R.A.S.*, **231**, 923.
de Araujo, J. C. N., and Opher, R. 1989, *M.N.R.A.S.*, **239**, 371.

Fillmore, J. A., and Goldreich, P. 1984, *Ap. J.*, **281**, 9.
Hausman, M. A., Olson, D. W., and Roth, B. D. 1983, *AP. J.*, **270**, 351.
Hirasawa, T., Aizu, K., and Taketani, M. 1969, *Prog. Theor. Phys.*, **41**, 835.
Hoffmann, Y., and Shaham, J. 1983, in *IAU. Symp. 104, Early Evolution of the Universe and Its Present Structure*, ed. G. O. Abell and G. Chincarini, (Dordrecht: Reidel), p. 203.
Hutchins, J. B. 1976, *Ap. J.*, **205**, 103.
Lahav, O. 1986, *M.N.R.A.S.*, **220**, 259.
Lepp, S., and Shull, J. M. 1983, *Ap. J.*, **270**, 578.
Lepp, S., and Shull, J. M. 1984, *Ap. J.*, **280**, 465.
Lin, C. C., Mestel, L., and Shu, F. H. 1965, *Ap. J.*, **142**, 1431.
Matsuda, T., Sato, H., and Takeda, H. 1969, *Prog. Theor. Phys.*, **42**, 219.
Mestel, L. 1965a, *Quart. J. R.A.S*, **6**, 161.
Mestel, L. 1965b, *Quart. J. R.A.S*, **6**, 265.
Palmer, P. L. 1981, *M.N.R.A.S.*, **197**, 721.
Peebles, P. J. E. 1968, *Ap. J.*, **153**, 1.
Silk, J. 1977, *Ap. J.*, **211**, 638.
Silk, J. 1982, *Ap. J.*, **256**, 514.
Silk, J. 1983, *M.N.R.A.S.*, **205**, 705.
White, S. D. M., and Silk, J. 1979, *Ap. J.*, **231**, 1.
Yoneyama, T. 1972, *P.A.S.J.*, **24**, 87

DISCUSSION

P. Shapiro: How does the calculation by Lepp and Shull of the thermal and chemical evolution of a collapsing, post-recombination Jeans-mass cloud differ from yours? In particular, if you parameterize the temperature and density evolution by the spherical Jeans-mass formula, do you get the same result as they do, that the Jeans mass drops to solar-mass levels or below?

R. Opher: Our calculations are very much different from those of Lepp and Shull (*Ap. J.* **280**, 465, 1984). We examined the evolution of approximately Jeans masses ($M_J \sim 10^6 M_\odot$) and masses $M << M_J$ (*e.g.* $M \sim 10^4 M_\odot$) from the recombination era $T_0 \sim 4000$ K ($z > 1000$), taking into account pressure effects and other physical processes. Lepp and Shull neglected pressure gradients and assumed that the cloud's mass is several times the local Jeans mass. They started the calculation at $T_0 = 64$ K ($z = 50$).

J. Primack: Have you considered power spectra which are flatter (less dependence on M for $M \sim 10^6 M_\odot$) and adiabatic, and included dark matter? Your technique could usefully be applied here.

R. Opher: We confined ourselves to isothermal perturbations. We are examining at the present time the effects of dark matter and a flatter power spectrum.

A MILLION MPC^3 CHUNK OF THE UNIVERSE

J. LOVEDAY
Department of Astrophysics, University of Oxford
Keble Road, Oxford, OX1 3RH, England

B. A. PETERSON
Mt. Stromlo and Siding Spring Observatories
Australian National University, Private Bag, Canberra, Australia

G. EFSTATHIOU AND S. J. MADDOX
Department of Astrophysics, University of Oxford
Keble Road, Oxford, OX1 3RH, England

ABSTRACT In order to study the distribution of galaxies on scales >15 Mpc, a volume of space $\sim 10^6$ Mpc^3 must be sampled. We have recently completed a large solid-angle galaxy redshift survey with a median redshift of 12,000 km s^{-1}, and which samples a volume about four times larger than that of previous surveys.

Existing large redshift surveys such as the CfA survey (Huchra *et al.* 1983) and the Southern Sky Redshift Survey (SSRS; da Costa *et al.* 1988) sample galaxies to a median depth of ≈ 6000 km s^{-1}, but this is too small a volume to reliably determine even the mean density (Davis and Huchra 1982). An additional problem affecting the CfA survey is the unreliability of the Zwicky magnitudes. In the SSRS the galaxies are selected using diameters, so comparisons with the CfA survey are difficult to interpret.

We have constructed a new, deeper wide-angle redshift catalog based on the APM Galaxy Survey (Maddox *et al.* this volume). We visually inspected the non-stellar images brighter than a magnitude limit $b_J \approx 16.5$ over the whole 4300-square-degree area ($\delta < -20°$, $b < -40°$) of the faint survey. This gives a sample of 17,770 galaxies to which we assigned rough morphological types (elliptical, lenticular and spiral). We estimate the sample to be 95% complete. The distribution of these galaxies is shown in Figure 1 in an equal-area projection centered on ($\alpha = 15°$, $\delta = -45°$).

One in 20 of these galaxies were sampled at random to form the basis of a redshift survey. By sparse sampling in this way we have been able to observe a *much* larger volume of the universe than would have been possible for a fully sampled survey. This strategy maximizes the efficiency of telescope time in estimating large-scale statistics of the galaxy distribution (Kaiser 1986).

We have used the Australian National University 2.3-m telescope and dual-beam spectrograph to measure redshifts for 760 out of the 918 galaxies in the random sample. The *rms* velocity errors of our measurements are ≈ 35 km s^{-1}. In addition, 63 of our survey galaxies have had redshifts determined as part of the SSRS. We intend to measure the remaining redshifts this year, and extend the survey to a fainter magnitude limit in the future.

The distribution of galaxies in our current redshift sample is shown in projection on the sky in Figure 2 and in right ascension - redshift space for four declination slices in

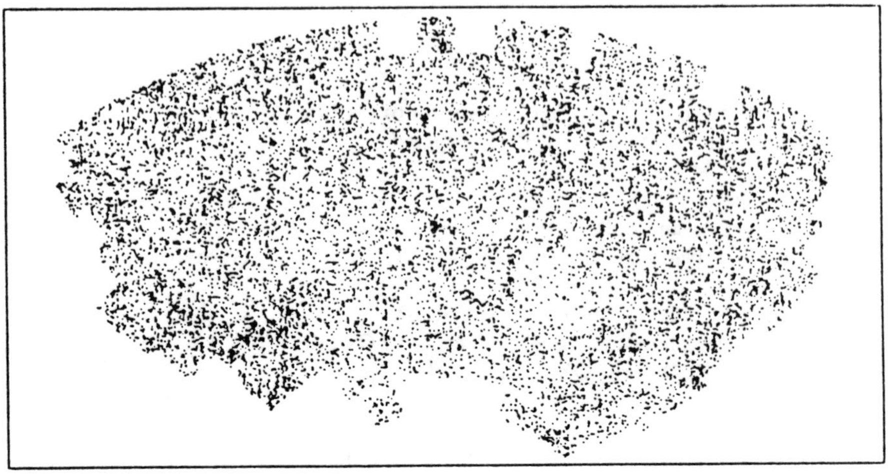

Fig. 1. All 17770 galaxies in the APM bright galaxy survey in an equal-area projection centered on (α = 15°, δ = −45°).

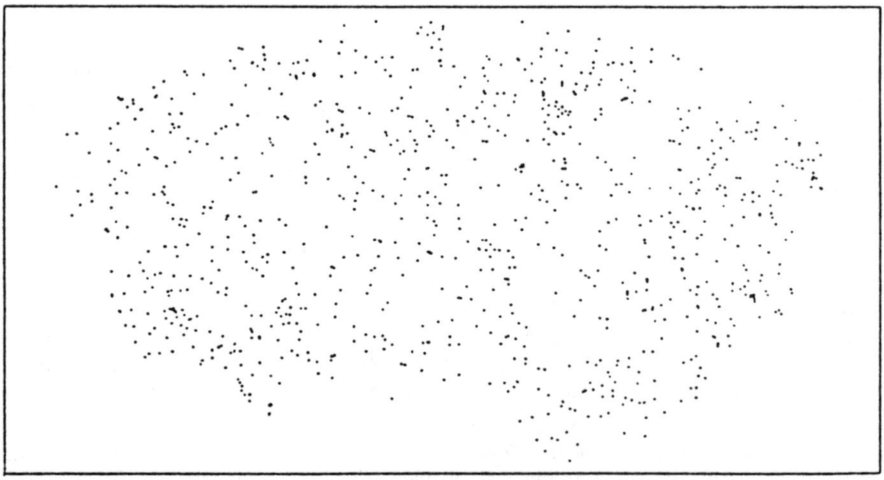

Fig. 2. A random (1:20) subsample of 918 galaxies from the APM bright galaxy survey selected for redshift observations.

Figure 3. The median redshift of our galaxies is ≈ 12,000 km s^{-1}, and so we are probing a volume roughly twice as deep as the fully sampled CfA and SSRS redshift surveys.

We have also obtained two-color CCD photometry for about 100 galaxies from the Siding Spring 40-in and the SAAO 40-in telescopes. These observations enable us to calibrate accurately the APM magnitudes and measure saturation effects for the brighter galaxies. The reliable photometry, together with the large survey volume, $\sim 10^6$ Mpc^3, (assuming a Hubble constant of $H_0 = 100$ km s^{-1} Mpc^{-1}) make this an ideal redshift survey for the study of large-scale clustering, and for the determination of the field-galaxy luminosity function. These results will be presented in due course.

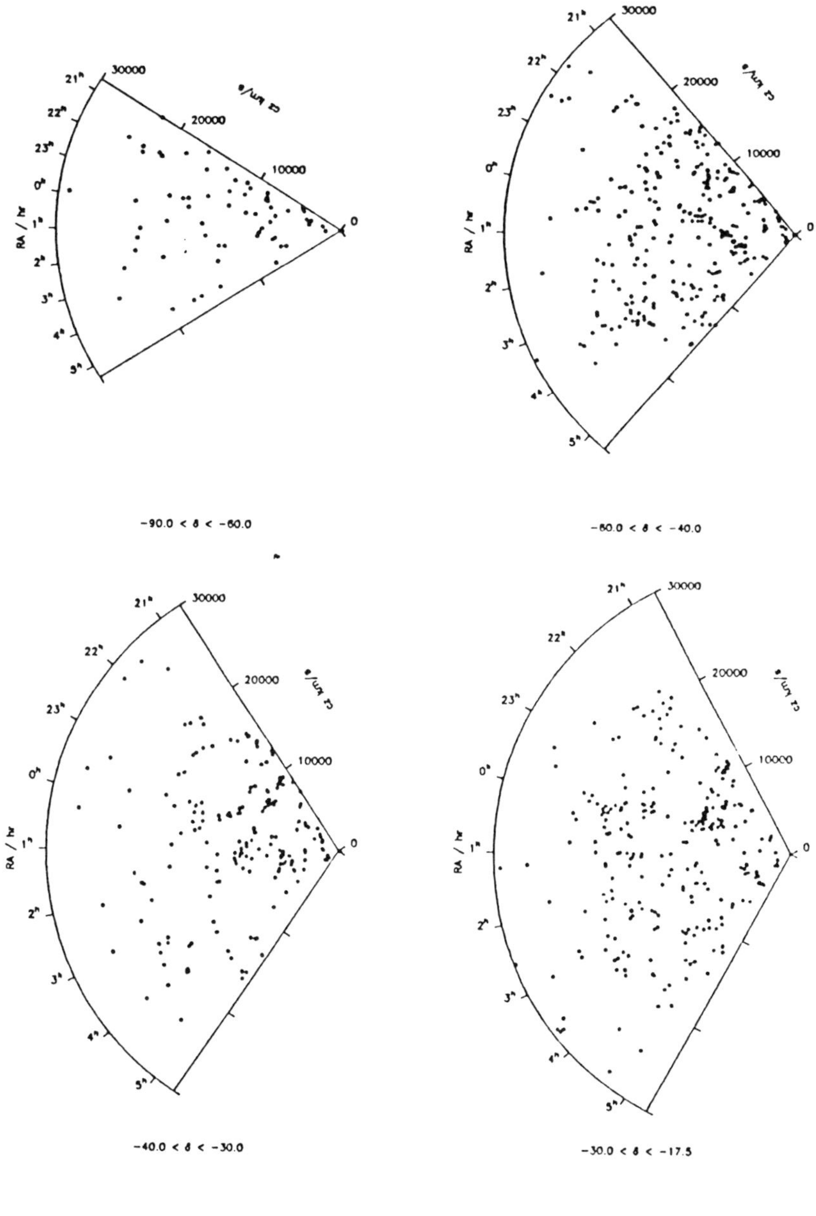

Fig. 3. Distribution of survey galaxies in right ascension - redshift space for four declination ranges.

REFERENCES

da Costa L. N., Pellegrini, P. S., Sargent, W. L. W., Tonry, J., Davis, M., Meiksen, A. Latham, D. W., Menzies, J. W., and Coulson, I. A. 1988, *Ap. J.* **327**, 544.

Davis, M., and Huchra, J. P. 1982, *Ap. J.* **245**, 437.

Huchra, J. P., Davis, M., Latham, D. W., and Tonry, J. 1983, *Ap. J. Suppl.*, **52**, 89.

Kaiser, N. 1986, *M.N.R.A.S.*, **219**, 785.

CATALOG OF RADIAL VELOCITIES OF GALAXIES: THE UPDATED VERSION

G. G. C. PALUMBO
Dipartimento di Astronomia, Università di Bologna, and Istituto Te. S.R.E. CNR, Casella Postale 596, 40100 Bologna, Italy

G. VETTOLANI
Instituto di Radioastronomia CNR, via Irnerio 46, 40126 Bologna, Italy

G. BAIESI-PILLASTRINI
Dipartimento di Astronomia, Università di Bologna, via Zamboni 33, 40126 Bologna, Italy

A. P. FAIRALL
Department of Astronomy, University of Cape Town, Rondebosch, 7700 South Africa

The updated version of the Catalogue of Radial Velocities of Galaxies (CRVG) will include all radial velocities published since 1981, the year in which the first issue ended, to the end of 1987.

At present we have on file about 95 percent of all radial velocities which have appeared in the literature; small refinements on accurate positions, names *etc.* are still necessary. The final version of the CRVG will contain some 30,000 radial velocities taken from about 1700 references.

For easy identification of objects we have greatly expanded the cross identification of galaxies with other catalogues such as UGC, ESO-Uppsala, MCG, *etc.* Furthermore, identification tables provide finding lists for all NGC, UGC, IC, *etc.* galaxies which have an entry in the CRVG. Special attention has been given to clusters of galaxies: objects in clusters will be identified as cluster members (Abell *etc.*) and for those galaxies, for which no coordinates were provided, numbers for finding charts published in the original article will be given.

The new CRVG will again be published (possibly as a two volume set) by Gordon and Breach Science Publishers of New York but with the novel feature that a tape version of the data can be obtained directly from the publisher.

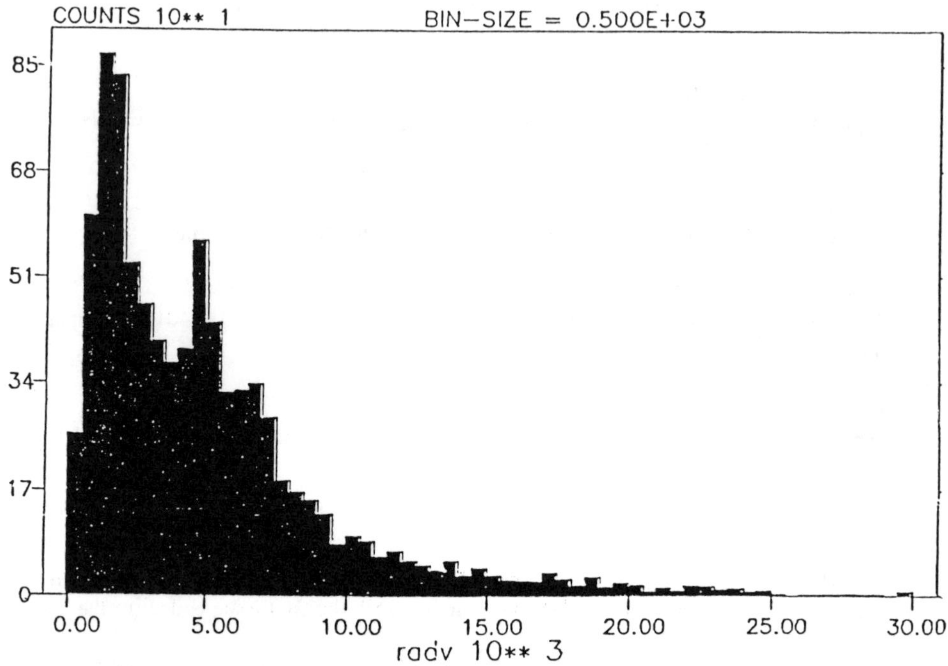

Fig. 1. Diagram showing number of galaxies per redshift interval of the data already collected. The peaks can be identified with the Virgo Cluster (~ 1500 km/s) and the Perseus Cluster (~ 5000 km/s)

SOUTHERN REDSHIFT CATALOGS AND PLOTS

A. P. FAIRALL AND A. JONES
Department of Astronomy, University of Cape Town, Rondebosch, 7700 South Africa

This is an ongoing project (since 1981) to keep track of available redshifts for southern galaxies. It is not intended as a definitive work, but is kept reasonably up to date. The most recent edition (early 1988 - on display at the workshop), has entries for about 6200 galaxies. It is also distributed through the Goddard and Strasbourg data centers. Work has started on the next version. It is hoped to include radio as well as optical measurements and possibly extend the northern Declination limit from $-17.5°$ to the equator.

Readers who did not receive this catalog in the past, and who would like to be added to the mailing list (for when the next edition is produced), should contact the first author at the address above.

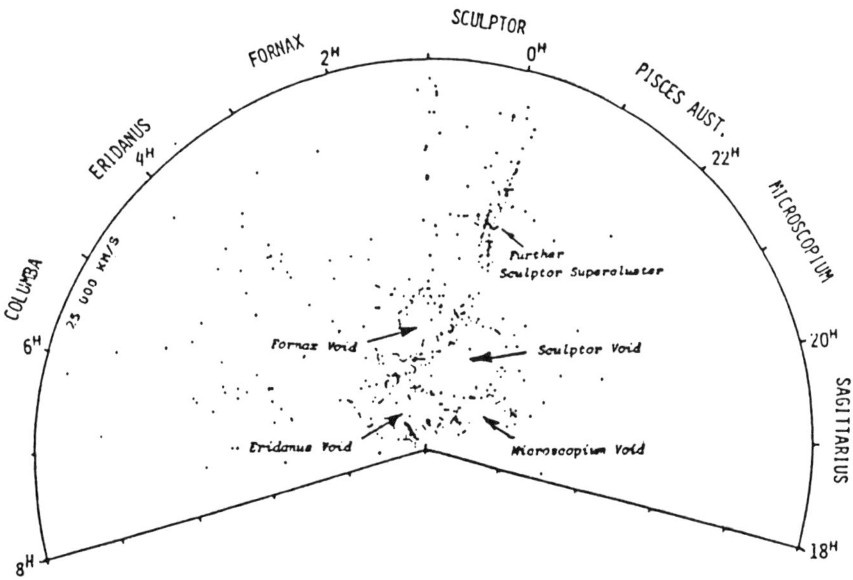

Fig. 1. A sample of one of the many R.A. and Dec. redshift plots in the catalog. This shows galaxies between Dec. $-27.5°$ and $-37.5°$ in the Southern Galactic Hemisphere. Note the large structure running diagonally.

SUMMARY: PROBLEMS AND PROSPECTS FOR LARGE-SCALE STRUCTURE STUDIES

MARC DAVIS
Astronomy Department, University of California, Berkeley, CA 94720

INTRODUCTION

One conclusion from this week is incontrovertible: Rio is a great place for a conference! Nicolaci and all his assistants are to be congratulated for all their effort to organize a very memorable week for all of us.

It was apparent also from the week's discussion that good progress is being made on numerous fronts in large-scale structure studies. I have high expectations for continued rapid progress, both in new high quality observational data, and in new tools of simulation and analysis.

A final general conclusion concerns the degree to which our theoretical models can explain the barrage of new data. In the last few years, the Cold Dark Matter (CDM) model has become the 'standard model' of structure in the Universe, and since this week was an advanced workshop, nobody reviewed the fundamentals. CDM is undoubtedly a good model that provides a natural explanation for a wealth of data concerning structure in the Universe. However, no-one would object if the model provided more power on large scale. Some would say that additional power is essential.

DATA MOST DAMNING TO THE CDM LITANY

Let me present four arguments for more power on large scales than given by standard CDM with $h = 0.5$. I list them in decreasing order of seriousness of challenge to the established orthodoxy.

Angular correlations measured in the APM survey

Steve Maddox gave a clear account of the meticulous care that has gone into the Cambridge APM survey of two million galaxies in one steradian of the Southern sky. Their measurement of $\omega(\theta)$ for the galaxy distribution satisfies scaling reasonably well, and has a gently increasing slope at large angles that appears to be real and unrelated to galactic contamination. The problem is not that CDM fails to predict a break in $\omega(\theta)$, but that the observations are curving at a larger scale than expected in the model, indicating excess power on scales larger than 20 h^{-1} Mpc. $\omega(\theta)$ is simply related to $\xi(r)$, which is the Fourier transform of the power spectrum, and the initial power spectrum is the fundamental underpinning of the model. It is difficult to imagine how the galaxy distribution could have a power spectrum on large scales different in shape from the mass distribution. Furthermore, the shape of the large-scale power spectrum should be undisturbed from the initial conditions at the recombination epoch, so it would seem these new measurements present a serious challenge.

The excess power on large scales is not huge, and simple modifications to CDM could account for the difference. The 'standard' model assumes $H_0 = 50$, but a lower value would simply shift the bend in the power spectrum to larger scales. Raising Ω_b, the baryonic component of the universe, to higher values, say $\Omega_b = 0.2$, would make up most of the excess power. Each of these fixes has its own problems, and it is certainly worthwhile to continue to explore alternatives to CDM that can perhaps provide a clean explanation for excess power. To date no suggested alternatives are particularly compelling.

The Tree Rings of the CfA2 Slices and Other Large Filaments

Although neither of the principals of the ongoing CfA redshift survey were here this week, Ed Bertschinger presented a summary of their recent activity. The data speak for themselves. Large filamentary structures are seen both in the northern and southern Galactic hemispheres, their extent is at least 100 h^{-1} Mpc, with a thickness of less than 5 h^{-1} Mpc. The third dimension is less constrained, but is roughly half the length, although the structures are not continuous sheets. It appears there are substantial gaps in the plane of the structures, so that the overall topology appears spongy.

By now we know that filamentary structures can be produced in many models of large-scale structure, and the relevant question is the coherence length of the most prominent structures. By visual inspection, the CfA filaments *appear* larger and more prominent than in any available CDM *N*-body simulation. The challenge is to proceed beyond the beauty-contest comparison of pictures. Is the power spectrum derived from the CfA survey consistent with CDM? Do the data suggest non-gaussian initial fluctuations? Answers to these quantitative issues are still lacking.

The Amplitude of Large-Scale Flows

We have heard from Alan Dressler, Dave Burstein, Gary Wegner, and Riccardo Giovanelli about continuing efforts to map the deviations from Hubble flow on scales up to 100 Mpc. This difficult work has been very exciting during the last several years, as the quality and quantity of data have steadily improved. The Seven Samurai group (Lynden-Bell *et al.* 1988) and others have shown good evidence for very substantial matter flows on large scale. Interpretation of the results has been evolving and is continuing. Nick Kaiser discussed the implications of the observations for the CDM model. The amplitude of large-scale flows is sensitive to the mass fluctuations, so they can test the normalization of the CDM model. It appears difficult to reconcile the CDM model with *any* of the flow data unless the bias parameter is smaller than 2. To reconcile the large-amplitude flow of the Seven Samurai data with CDM requires an essentially unbiased CDM model, $b = 1$. Yet, on small scales there is reason to suggest the CDM bias parameter is larger than 2 (Frenk *et al.* 1990).

There are two possibilities for resolving this problem. The bias mechanism is not very well defined in the CDM models; perhaps we simply do not understand biasing correctly and galaxies are biased on small scales but not on large scales. This is contrary to the expected behavior of peaks in Gaussian noise, but alternative mechanisms such as dissipation and dynamical friction of the baryonic galaxies do behave in the desired manner. Alternatively, the velocity-field problem would be neatly solved if there were additional power on large scales. The excess power needed to match the APM survey data would be perfectly adequate to explain all the large-scale flow data. I rank this objection to CDM as third rather than first because of the two alternative reconciliations.

The Correlations of Abell Clusters

Joel Primack discussed his work on the rich-cluster correlations, and presented a bewildering list of problems with the data base. In spite of this list, he argued that the angular correlations in the range $1° < \theta < 3°$ were robust and not a result of selection effects, and that the implied correlation strength was sufficiently large to be a problem for CDM. My own feeling is that this might indeed be a problem consistent with the discussion above, but I am not going to get worked up about it. For example, Will Sutherland's talk emphasized the extreme sensitivity of the cluster correlation statistics to the fine selection details in the APM survey. I think we should all pray for the success of the upcoming ROSAT mission, which promises to provide a well-defined cluster sample with a new set of problems.

REDSHIFT SURVEYS

Much of our current understanding of large-scale structure has come from redshift surveys, and the status of current surveys was summarized by Luiz da Costa this week. A fair comment is that the clustering on scales less than 10 h^{-1} Mpc is well measured and is reasonably reproducible. Attention should now be focused on larger-scale surveys, which is why the recent APM results are so significant. This catalog represents an enormous improvement over previously available databases, and is particularly well suited for large-scale studies. Even without redshift information a good catalog can be a gold mine for large-scale structure studies.

The large filaments so apparent in the CfA and Perseus redshift surveys are also visible in the APM and Lick catalogs. Rodney Smith showed that several of the Lick filaments, the reality of which has been a topic of considerable discussion, are reproduced on the Palomar Sky Survey plates.

For the future, we should consider carefully the scientific questions we seek to address before embarking on enormously large and expensive redshift surveys. Rather than do a redshift survey of every galaxy in the APM catalog, would a dilute sample of the brightest galaxies give us equivalent information? What is the optimal strategy for proving whether the galaxy fluctuations are Gaussian on large scales? What is the best method for studying the time evolution of the clustering?

VELOCITY FIELDS- A NEW PROBE OF LARGE-SCALE STRUCTURE

The week's agenda included considerable discussion of large-scale flows, the Great Attractor model, new distances for galaxies in the Centaurus region, and comparison of observed peculiar velocities of galaxies to the velocity-field expectations derived from the *IRAS* density field. The situation is clearly fluid, and is likely to have evolved by the time of the publication of this proceeding. Since the flows are so sensitive to the largest scales of structure, their study provides an important test of any theory, as discussed above.

Several fundamental questions were raised by Alan Dressler, Dave Burstein, and John Lucey. The new data presented by Alan beyond the distance of Centaurus shows some hint of backflow into the Centaurus region. Yet, the Centaurus region itself is remarkably complex, as described by the impressive new data of John Lucey. How many components comprise the Centaurus cluster, and what is their sub-clustering? It is entirely likely that the system is too evolved to be describable by linear perturbation theory, which is perhaps why the *IRAS* iteration procedure does such a poor job at reproducing the observed velocity field in this region.

In two year's time the *IRAS* redshift catalog will be doubled in size, to 5500 galaxies, and other groups will have measured the velocity field with good precision over much of the sky. We can look forward to much more detailed comparisons between predicted and observed velocity fields. The enlarged samples will have greatly reduced shot noise and sampling problems that plague current comparisons. These projects require full sky coverage, so it is natural for many groups to contribute in a truly international effort. With an improved data set, we will have better constraints on the degree of large-scale power affecting the velocity field, and the adequacy of linear theory for predicting the velocity field.

At present the optical and *IRAS*-selected galaxy samples yield slightly inconsistent dipoles, but it is likely that the disagreements will be fully reconciled within a few months. The inferred cosmological density parameter is higher for the *IRAS*-selected galaxies, but this is not surprising since *IRAS* galaxies undercount the cluster centers relative to the optically selected galaxies. The fact that the two samples give very different estimates for Ω is a reflection of the relative bias of the two samples. Neither density estimate should be considered reliable; the important issue is that both samples imply that our motion relative to the comoving frame has been induced by material within 3500 km s^{-1} redshift of us, which is rather local by cosmological standards.

There is no particular reason for the details of the deviations from the Hubble flow to be well fit by any simple parameterized model of several attractors. The flow should be complex, but if described by linear theory it should be irrotational. Therefore it is extremely interesting to use the available noisy data to construct a velocity *field*. We heard discussion of two promising algorithms, by Bertschinger and Dekel and by Kaiser and Stebbins, that use the available velocity data to construct a unique irrotational velocity field, and therefore a density field of perturbations. Much work must yet be done on both of these projects, and we can look forward to continued heated discussions in the coming years.

There was little discussion this week about methods of testing models of large-scale flow fields. The main method used to visualize the data is to assume a calibrated Tully-Fisher or D_n - σ relationship, to use the measured velocity width or dispersion to predict the absolute luminosity or diameter, and therefore distance d, which when compared to the measured redshift cz yields the estimated peculiar velocity $v_p = cz - d + v_0 \cos\theta$, where $v_0 \cos\theta$ is the component of our peculiar velocity directed toward the galaxy in question. This procedure is an essential step to help us visualize the velocity field, but we must beware of Malmquist biases.

A second essential method that has not yet received sufficient attention is to test parameterizations of the flow field by minimizing the scatter in the Tully-Fisher diagram. In particular, assume the flow is parameterized with coefficients a_i. For each galaxy measure cz and apparent diameter or flux. Solve self consistently for

$$d = cz - v_p(\mathbf{r}, a_i) + v_0 \cos\theta$$

and infer the absolute distance or diameter for each object. Note that the solution can be multi-valued around cluster regions. Finally, compute the scatter in the Tully-Fisher or D_n - σ relationship, and minimize with respect to variation in a_i using the velocity width as the dependent variable, in order to minimize Malmquist effects. One advantage of this scheme is that it can be used to compare the 'goodness' of alternative models such as the *IRAS* predicted flow field versus the Faber and Burstein (1988) parameterization of the Great Attractor model, or the Bertschinger-Dekel velocity field versus the Kaiser-Stebbins field. The Faber-Burstein model leads to very small scatter of the AHM spiral galaxies, implying it gives a very good description of the velocity field, better than has been achieved with the *IRAS* data thus far.

NEW METHODS

We heard this week a review by Rich Gott of his large collaborative topological study of large-scale structure. The method is relatively insensitive to the power spectrum of fluctuations but seems more sensitive to the statistics of phase coherence. It therefore appears to be an excellent complement to the standard covariance analysis, which is sensitive *only* to the power spectrum of fluctuations. My basic question is whether it is realistic to expect to construct a statistic that with high confidence can reject a model of large-scale structure for having insufficiently large and/or common filaments and sheets. Are the topology statistics of Gott and Ryden the best for this purpose? Or is a beauty contest, that is a direct visual comparison of model versus data, the best test we can manage? The topological studies have not yet been applied to the new CfA survey data; hopefully this will be done before the publication of these proceedings.

Also on the topic of methods, we heard presentations on new simulation techniques for cosmological studies. Jens Villumsen described his Hierarchical Particle Mesh (HPM) *N*-body code, which features a larger dynamic range than all the competition, and Paul Shapiro described a Smooth Particle Hydro code merged with the HPM gravity code, which looks extremely promising for cosmological dissipational studies at high resolution. My main comment here is that these guys have competition: several groups have now begun serious examination of SPH and sticky particle gas codes merged with PPPM and TREE-CODE *N*-body codes. Over the next several years we will have to do a lot of testing of these codes, but hopefully they can be used to probe some details of the formation and evolution of galaxies and clusters, Lyman-α clouds, and the mechanism of biasing.

In summary, the excitement of this week's workshop parallels the excitement in this field in general. The last decade has been especially productive for the field of large-scale structure. We have witnessed the birth of large redshift surveys, new accurate methods of measuring distance, a new exquisite catalog of several million galaxies, theories of large-scale structure that have some plausibility and modest agreement with observations, and enormous advances in computing hardware that has permitted bigger and better numerical simulations of the evolution of structure. The prospects are equally bright for continued rapid progress during the next few years. The ROSAT mission will lead to well defined samples of thousands of galaxy clusters; the Hubble Space telescope will permit high-resolution studies of the evolution of galaxies to $z = 0.5$; COBE and further Soviet RELICT satellites will give critical constraints on the microwave and infrared background radiation. Further ground-based work and simulation studies will likely lead to results as exciting as these major space missions. I look forward to the next gathering in Rio and expect it to be as productive as this meeting has been.

REFERENCES

Faber, S. M. and Burstein, D. 1988, in *Large-Scale Motions in the Universe: A Vatican Study Week*, ed. V. C. Rubin and G. V. Coyne, S. J. (Princeton: Princeton University), p. 115.

Frenk, C. S., White, S. D. M., Efstathiou, G., and Davis, M. 1990, *Ap. J.*, **351**, 10.

Lynden-Bell, D., Faber, S. M., Burstein, D., Davies, R. L., Dressler, A., Terlevich, R. J., and Wegner, G. 1988, *Ap. J.*, **326**, 19.

DISCUSSION

D. Lynden-Bell: As an old forester I know little about trees but every trunk has its edge. Every redshift survey has its limit too. Perhaps the round shapes seen in the great walls of the redshift survey are enhanced by the excess densities where the redshift surveys reach radii at which detected galaxies are most common.

A. Dressler: I would like to comment on the ordering of Davis' list of observations that challenge the standard CDM model. The extra power in the angular correlation function on large scales found in the APM survey, and the 'Great Wall' seen in the continuing CfA redshift survey, do appear to pose considerable difficulty to the standard picture. In comparison, the observations of large-scale streaming motions might appear on shakier ground, but I would like to emphasize that they are the only true measures of the mass distribution, and thus are not subject to assumptions about the degree to which galaxies trace the dark matter. Let me also recall two points which have not been resolved but which I hope will receive further discussion in the future. First, I wonder whether the extra large-scale power found in the APM correlation function is roughly what is needed to explain the large-scale streaming motions. Second, Kaiser has found that the Great Attractor, as modelled by the Seven Samurai, is not in gross conflict with standard CDM, contrary to the conclusion reached by Bertschinger and Juszkiewicz. Furthermore, Kaiser had little trouble reconciling the velocity correlations in the elliptical galaxy samples with an unbiased CDM model, though Davis and coworkers had reported a fairly strong contradiction. I hope to hear sometime what was the difference in treatment that might account for these different conclusions.

R. Gott: I would like to say a word of thanks to Luiz da Costa. Two years ago in China Luiz told me of his dreams of completing the Southern Sky Redshift Survey and of having a meeting in Brazil on large-scale structure. We have now seen the wonderful way that he has brought these two projects to completion.

I would like to thank Luiz for holding the workshop in this famous Copacabana Palace Hotel. Roosevelt and de Gaulle have stayed here, and Richard Feynman has stayed here, so we are in good company. I would like to thank Luiz for holding the workshop in Rio. I have a book at home called *Wonders of the World*. It has pictures of about 50 man-made and natural wonders. It has pictures of the Grand Canyon, the Great Wall of China, and the Taj Mahal. There is a picture of Rio on the *cover* of that book.

We are fortunate to be able to travel around the world, to have friends from around the world with whom we can work together to help unravel some of the secrets of the universe. Luiz has shown us what a tremendous difference just one person can make.

Aaronson, M., Bothun, G. D., Cornell, M. E., Dawe, J. A., Dickens, R. J., Hall, P. J., Sheng, H. M., Huchra, J. P., Lucey, J. R., Mould, J. R., Murray, J. D., Schommer, R. A., and Wright, A. E. 1989, *Ap. J.*, **338**, 654.

Aaronson, M., Bothun, G. D., Mould, J. R., Huchra, J. P., Schommer, R. A., and Cornell, M. E. 1986, *Ap. J.*, **302**, 536.

Aaronson, M., Huchra, J. P., and Mould, J. R. 1979, *Ap. J.*, **229**, 1.

Aaronson, M., Huchra, J. P., Mould, J. R., Schechter, P. L., and Tully, R. B. 1982, *Ap. J.*, **258**, 64.

Aaronson, M., Huchra, J. P., Mould, J. R., Tully, R. B., Fisher, J. R., van Woerden, H., Goss, W. M., Chamaraux, P., Mebold, U., Siegman, B., Berriman, G., and Persson, S. E. 1982, *Ap. J. Suppl.*, **50**, 241.

Abbott, L. F., and Wise, M. B. 1984, *Nucl. Phys.*, **B244**, 541.

Abell, G. O. 1958, *Ap. J. Suppl.* **3**, 211.

Abramowitz, M., and Stegun, I. 1964, *Mathematical Functions*, Dover.

Abell, G. O., Corwin, H. G., and Olowin, R. P. 1989, *Ap. J. Suppl.* **70**, 1.

Adler, R. J. 1981, *The Geometry of Random Fields*, (New York: Wiley).

Albrecht, A., and Steinhardt, P. J. 1982, *Phys. Rev. Lett.* **48**, 1220.

Allen, T. J., Grinstein, B., and Wise, M. B. 1987, *Phys. Lett.* **197B**, 66.

Argyres, P. C., Groth, E. J., Peebles, P. J. E., and Struble, M. F. 1986, *A.J.*, **91**, 471.

Arnold, V. I., Shandarin, S. F., and Zel'dovich, Ya. B. 1982, *Geophys. Ap. Fluid Dyn.*, **20**, 111.

Bahcall, N. A. 1986, *Ap. J. (Letters)*, **302**, L41.

Bahcall, N. A. 1988, *Ann. Rev. Astr. Ap.*, **26**, 631.

Bahcall, N. A., and Soneira, R. M. 1983, *Ap. J.*, **270**, 20.

Bahcall, N. A., Batuski, D. J., and Olowin, R. P. 1988, *Ap. J. (Letters)*, **333**, L13.

Bahcall, N. A., Soneira, R. M., and Burgett, W. S. 1986, *Ap. J.*, **311**, 15.

Balian, R., and Schaeffer, R. 1989, *Astr. Ap.*, **220**, 1.

Bardeen, J. 1989, Lectures at Second Guo Shou-jing Summer School on Particle Physics and Cosmology, in *Particle Physics and Cosmology*, ed. A. Zee (Gordon and Breach), in press.

Bardeen, J. M., and Bublik, G. J. 1987, *Class. Quantum Grav.*, **4**, 573.

Bardeen, J. M., Bond, J. R., and Efstathiou, G. 1987, *Ap. J.* **321**, 28.

Bardeen, J. M., Bond, J. R., Kaiser, N., and Szalay, A. S. 1986, *Ap. J.*, **304**, 15.

Bardeen, J., Steinhardt, P., and Turner M. 1983, *Phys. Rev.* **28D**, 679.

Barnes, J. 1985, *M.N.R.A.S.*, **215**, 517.

Barnes, J., Dekel, A., Efstathiou, G., and Frenk, C. S. 1985, *Ap. J.*, **295**, 368.

Barrow, J. D., and Bhavsar, S. P. 1987, *Q.J.R.A.S.*, **28**, 109.

Batuski, D. J., and Burns, J. O. 1985, *A.J.*, **90**, 1413.

Batuski, D. J., Bahcall, N. A., Olowin, R. P., and Burns, J. O. 1989, *Ap. J.*, **349**, 599.

Bazeia, D., Carvalho, C. A., Eboli, O. J. P., and Marques, G. C. 1985, *Phys. Rev.*, **D32**, 3256.

Bazeia, D., Eboli, O. J. P., Guerra, J. M., and Marques, G. C. 1987, *Phys. Rev.*, **D39**, 1838.

Bechtold, J. P., Green, R. F., and York, D. G. 1987, *Ap. J.*, **312**, 50.

Bennett, D. P., and Bouchet, F. E. 1988, *Phys. Rev. Lett.*, **60**, 257.

Bergeron, J., and Stasinska, G. 1986, *Astr. Ap.*, **169**, 1.

Bertschinger, E. 1985, *Ap. J. Suppl.*, **58**, 1.

Bertschinger, E., and Dekel, A. 1989, *Ap. J. (Letters)*, **336**. L5.

Bertschinger, E., and Gelb, J. M. 1989, in preparation.

Bertschinger, E., and Juszkiewicz, R. 1988, *Ap. J. (Letters)*, **334**, L59.

Bertschinger, E., Dekel, A., Faber, S. M., Dressler, A., and Burstein, D. 1990, *Ap. J.*, **364**, 370 .

Bhavsar, S. P., and Barrow, J. D. 1983, *M.N.R.A.S.*, **205**, 61.

Bhavsar, S. P., and Ling, E. N. 1988, *Ap. J. (Letters)*, **331**, L63.

Binggeli, B. 1982, *Astr. Ap.*, **107**, 338.

Binggeli, B. 1987, in *Nearly Normal Galaxies*, ed. S. Faber, (Berlin: Springer), p. 195.

Binggeli, B., Tammann, G. A., and Sandage, A. 1987, *A.J.*, **94**, 251.

Blumenthal, G. R., Dekel, A., and Primack, J. R. 1988, *Ap. J.*, **326**, 539.

Blumenthal, G. R., Faber, S., Primack, J. R., and Rees, M. J., *Nature*, **311**, 517.

Bond, J. R., and Salopek, D. 1989, in preparation.

Bond, J. R., Centrella, J. Szalay, A., and Wilson, J. R. 1984, *M.N.R.A.S.*, **210**, 515.

Bond, J. R, Salopek, D., and Bardeen, J. M. 1988, in *Large-Scale Motions in the Universe: A Vatican Study Week*, ed. V. C. Rubin and G. V. Coyne, S. J., (Princeton: Princeton University), p. 385.

Boroson, T. Y. 1988, private communication.

Botema, R., Shostak, G. S., and van der Kruit, P. C. 1987, *Nature*, **328**, 401.

Bothun, G. D., Aaronson, M., Schommer, R. A., Huchra, J. P., and Mould, J. R. 1984, *Ap. J.*, **278**, 475.

Bothun, G. D., Beers, T. C., Mould, J. R., and Huchra, J. P. 1986, *Ap. J.*, **308**, 510.

Bouchet, F., and Lachieze-Rey, M. 1986, *Ap. J. (Letters)*, **302**, L37.

Boughn, S., Cottingham, D., Fixsen, D., and Cheng, E. 1989, private communication.

Brinks, E., and Burton, W. B. 1984, *Astr. Ap.*, **141**, 195.

Burgers, J. M. 1974, *The Non-linear Diffusion Equation*, (Dordrecht: Reidel).

Burstein, D. 1989, in *Les Mondes des Galaxies*, ed. H. C. Corwin and L. Bottinelli, in press.

Burstein, D. 1990, *Rep. Prog. Phys.*, **53**, 421.

Burstein, D., and Heiles, C. 1982, *A. J.*, **87**, 1165.

Burstein, D., and Lebofski, 1986, *Ap. J.*, **301**, 683.

Burstein, D., Davies, R. L., Dressler, A., Faber, S. M., Lynden-Bell, D., Terlevich, R. J., and Wegner, G. 1986, in *Galaxy Distances and Deviations from Universal Expansion*, ed. R. B. Tully and B. F. Madore (Dordrecht: Reidel), p. 123.

Burstein, D., Davies, R. L., Dressler, A., Faber, S. M., Stone, R. I. S., Lynden-Bell, D., Terlevich, R. J., and Wegner, G. 1987, *Ap. J. Suppl.*, **64**, 601.

Burstein, D., Faber, S. M., and Dressler, A. 1990, *Ap. J.*, **354**, 18.

Cameron, L. M. 1989, preprint.

Carlberg, R. G. 1981, *M.N.R.A.S.*, **197**, 1021.

Carlberg, R. G., and Couchman, H. 1989, *Ap. J.*, **340**, 47.

Carruthers, P., and Minh, D.-V. 1983, *Phys. Lett.*, **131B**, 116.

Carter, D., and Metcalfe, N. 1988, *M.N.R.A.S*, 191, 325.

Carvalho, C. A., Marques, G. C., Silva, A. J., and Ventura, I. 1986, *Nucl. Phys.*, **B265**, 45.

Cawson, M. G. M. 1987, GRASP Documentation, (Tucson: University of Arizona).

Centrella, J. M., Gallagher III, J. S., Melott, A. L., and Bushouse, H. A. 1988, *Ap. J.*, **333**, 24.

Chibisov, G. V., and Mukhanov, V. F. 1983, Lebedev Phys. Inst., Moscow, preprint.

Chincarini, G. L., and Rood, H. 1979, *Ap. J.*, **230**, 648.

Ciardullo, R., Ford, H., Harms, R. 1985, *Ap. J.*, **293**, 69.

Colafrancesco, S., Lucchin F., and Matarrese S. 1989, *Ap. J.*, **345**, 3.

Cole, S., and Kaiser, N. 1989, *M.N.R.A.S.*, **237**, 1127.

Coles, P., and Barrow, J. D. 1987, *M.N.R.A.S.*, **228**, 407.

Colless, M. M. 1987, Ph.D. dissertation, University of Cambridge.
Colless, M. M., and Hewitt, P. 1987, *M.N.R.A.S.*, **224**, 453.
Collins, C. A., Heydon-Dumbleton, N. H., and MacGillivray, H. T., 1989, *M.N.R.A.S.*, **236**, 7P.
Collins, C. A., Joseph, R. D., and Robertson, N. A., 1986, *Nature*, **320**, 506.
Conklin, E. K. 1969, *Nature*, **222**, 971.
Conklin, E. K., and Bracewell, R. N. 1967, *Nature*, **216**, 777.
Couchman, H. M. P., McGill, C., and Olowin, R. P. 1989, *M.N.R.A.S.*, **239**, 513.
Cousins, A. W. J. 1976, *Mem. R.A.S.*, **81**, 25.
da Costa, L. N., Pellegrini, P. S., Sargent, W. L. W., Tonry, J., Davis, M., Meiksin, A., Latham, D. W., Menzies, J. W., and Coulson I. A. 1988, *Ap. J.*, **327**, 544.
da Costa, L. N., Pellegrini, P. S., Willmer, C., de Carvalho, R., Maia, M. 1989, *A.J.*, **97**, 315.
da Costa, L. N., Pellegrini, P. S., Willmer, C., and Latham, D. W. 1989, *Ap. J.*, **344**, 20.
Davies, R. L., Burstein, D., Dressler, A., Faber, S. M., Lynden-Bell, D., Terlevich, R. J., and Wegner, G. 1987, *Ap. J. Suppl.*, **64**, 581.
Davis, M., and Djorgovski, S. 1985, *Ap. J.*, **299**, 15.
Davis, M., and Geller, M. J. 1976, *Ap. J.*, **208**, 13.
Davis, M., and Huchra, J. P. 1982, *Ap. J.*, **254**, 425.
Davis, M., and Peebles, P. J. E. 1977, *Ap. J. Suppl.*, **34**, 425.
Davis, M., and Peebles, P. J. E. 1983, *Ann. Rev. Astr. Ap.*, **21**, 109.
Davis, M., Efstathiou, G., Frenk, C. S., and White, S. D. M. 1985, *Ap. J.*, **292**, 371.
Davis, M., Meiksin, A., Strauss, M. A., da Costa, L. N., and Yahil, A. 1988, *Ap. J. (Letters)* **333**, L9.
Dawe, J. A., and Metcalfe, N., 1982, *Proc. Astr. Soc. Aust.*, **4**, 466.
de Araujo, J. C. N., and Opher, R. 1988, *M.N.R.A.S.*, **231**, 923.
de Araujo, J. C. N., and Opher, R. 1989, *M.N.R.A.S.*, **239**, 371.
de Lapparent, V., Geller, M. J., and Huchra, J. P. 1986, *Ap. J. (Letters)*, **302**, L1.
de Lapparent, V., Geller, M. J., and Kurtz, M. J. 1986, *Ap. J.*, **304**, 585.
de Lapparent, V., Geller, M. J., and Huchra, J. P. 1988, Ap. J., **332**, 44.
de Lapparent, V., Geller, M. J., and Huchra, J. P. 1989, *Ap. J.* **343**, 1.
de Souza, Vettolani, P., and Chincarini, G. L. 1985, *Astr. Ap.* **143**, 143.
de Vaucouleurs, G. 1953, *A. J.*, **58**, 30.
de Vaucouleurs, G. 1975, *Ap. J.*, **202**, 319.
de Vaucouleurs, G., and Peters, W. L. 1984, *Ap. J.*, **287**, 1.
Dekel, A. 1983, *Ap. J.*, **264**, 373.
Dekel, A. 1984, *Ap. J.*, **284**, 445.
Dekel, A. 1988, in *Large-Scale Motions in the Universe: A Vatican Study Week*, ed. V. C. Rubin and G. V. Coyne, S. J., (Princeton: Princeton University), p. 465.
Dekel, A. 1988, in *Large-Scale Motions in the Universe: A Vatican Study Week*, ed. V. C. Rubin and G. V. Coyne, S. J., (Princeton: Princeton University), p. 561.
Dekel, A., and Rees, M. J. 1987, *Nature*, **326**, 455.
Dekel, A., Bertschinger, E., and Faber, S. M. 1990, *Ap. J.*, **364**, 370.
Dekel, A., Blumenthal, G. R., Primack, J. R., and Olivier, S. 1989, *Ap. J. (Letters)*, **338**, L5.
Desert, F. X., and Schatzman, E. 1986, *Astr. Ap.*, **158**, 135.
Dey, A., Strauss, M. A., and Huchra, J. P. 1990, *A.J.*, **99**, 463.
Dickens, R. J., Currie, M. J., and Lucey, J. R. 1986, *M.N.R.A.S.*, **220**, 679.
Djorgovski, S. 1987, in *Nearly Normal Galaxies*, ed. S. Faber, (Berlin: Springer), p. 227.
Djorgovski, S., and Davis, M. 1987, *Ap. J.*, **313**, 59.

Djorgovski, S., de Carvalho, R., and Han, M. S. 1989, in *The Extra-Galactic Distance Scale*, ed. S. Van den Bergh and C. J. Pritchet, *A.S.P. Conf. Ser.*, **4**, (Provo: Astronomical Society of the Pacific), p. 329.

Dobrzycki, A. 1989, *M.N.R.A.S.*, **236**, 403.

Dominguez-Tenreiro, R., and Martinez, V. J. 1989, *Ap. J. (Letters)* **339**, L9.

Doroshkevich, A. G. 1970, *Sov. Ap.*, **6,** 320 (translated from *Astrofizika*, **6**, 581).

Doroshkevich, A. G., Kotok, E. V. 1988, Inst. Appl. Math., Moscow, preprint.

Doroshkevich, A. G., Kotok, E. V., Novikov, I. D., Polyudov, A. N., Shandarin, S. F., and Sigov Yu. S. 1980, *M.N.R.A.S.*, **192**, 321.

Dressler, A. 1980, *Ap. J.*, **236**, 351.

Dressler, A. 1987, *Ap. J.*, **317**, 1.

Dressler, A. 1988, *Ap. J.*, **329**, 519.

Dressler, A., and Faber, S. M. 1990, *Ap. J.*, **354**, 13.

Dressler, A., and Shectman, S. A. 1988, *A.J.*, **95**, 284.

Dressler, A., and Shectman, S. A. 1988, *A.J.*, **95**, 985.

Dressler, A., Faber, S. M., Burstein, D., Davies, R. L., LyndenBell, D., Terlevich, R. J., and Wegner, G. 1987, *Ap. J. (Letters)*, **313**, L37.

Dressler, A., Lynden-Bell, D., Burstein, D., Davies, R. L., Faber, S. M., Terlevich, R. J., and Wegner, G. 1987, *Ap. J.*, **313**, 42.

Drinkwater, M. J., 1987, Ph.D. dissertation, University of Cambridge.

Eddington, A. S. 1922, *The Mathematical Theory of Relativity*, (Cambridge: University of Cambridge).

Eder, J., Schombert, J. M., Dekel, A., and Oemler, A., Jr. 1989, *Ap. J.*, **340**, 29.

Efstathiou, G., Ellis, R. S., and Peterson, B. A. 1988, *M.N.R.A.S.*, **232**, 431.

Efstathiou, G., Frenk, C. S., White, S. D. M., and Davis, M. 1988, *M.N.R.A.S.*, **235**, 715.

Einasto, J., Klypin, A. A., Saar, E., and Shandarin, S. F. 1984, *M.N.R.A.S.*, **206**, 529.

Ellis, G. F. R. 1971, *Relativistic Cosmology*, in *General Relativity and Cosmology*, ed R. K. Sacks, (London: Academic).

Ellis, G. F. R., and Fairall, A. P. 1988, *Ap. J.*, **327**, 34.

Evrard, A. E. 1988, *M.N.R.A.S.*, **235**, 911.

Evrard, A. E. 1989, *Ap. J. (Letters)*, **341**, L71.

Faber, S. M., and Burstein, D. 1988, in *Large-Scale Motions in the Universe: A Vatican Study Week*, ed. V. C. Rubin and G. V. Coyne, S. J., (Princeton: Princeton University), p. 115.

Faber, S. M., Wegner, G., Burstein, D., Davies, R. L., Dressler, A., Lynden-Bell, D., and Terlevich, R. J. 1989, *Ap. J. Suppl.*, **69**, 763.

Fairall, A. P. 1987, *M.N.A.S.S.A.*, **46**, 47.

Fairall, A. P. 1988, in *Morphological Cosmology*, ed. H. E. Duerbeck and Flin (Dordrecht: Kluwer) in press.

Fairall, A. P. 1988, *M.N.R.A.S.*, **230**, 69.

Fairall, A. P., and Jones, A. 1988, *Pub. Astr. Dept. Univ. Cape Town*, Number 10.

Feitzinger, J. V., and Stüwe, J. A. 1984, *Astr. Ap. Suppl.*, **58**, 365.

Ferguson, H. C. 1989, Ph.D. dissertation, Johns Hopkins University.

Fillmore, J. A., and Goldreich, P. 1984, *Ap. J.*, **281**, 9.

Fisher, J. R., and Tully, R. B. 1976, *Astr. Ap.*, **53**, 397.

Fliche, H. H., Souriau, J. M. 1990, *Astr. Ap.* **233**, 317.

Fliche, H. H., Souriau, J. M., and Triay, R. 1982, *Astr. Ap.* **108**, 256.

Fliche, H. H., Souriau, J. M., and Triay, R. 1983, in proceedings of *Journées Relativistes*, ed. S. Benenti, M. Ferreris, M. Francaviglia (Turin).

Focardi, P., Marano, B., and Vettolani, G. 1984, *Astr. Ap.*, **136**, 178.

Foltz, C. B., Weyman, R. J., Peterson, B. M., Sun, L., Malkan, M. H., and Chaffee, F. H., Jr., 1986, *Ap. J.*, **307**, 504.
Frenk, C. S. 1989, in *The Epoch of Galaxy Formation*, ed. C. S. Frenk, R. S. Ellis, T. Shanks, A. Heavens, and J. Peacock 1989 (Dordrecht: Kluwer), p. 257.
Frenk, C. S., White, S. D. M., Efstathiou, G., and Davis, M. 1985, *Nature*, **317**, 595.
Frenk, C. S., White, S. D. M., Efstathiou, G., and Davis, M. 1988, *Ap. J.*, **327**, 507.
Frenk, C. S., White, S. D. M., Efstathiou, G., and Davis, M. 1990, *Ap. J.*, **351**, 10.
Fry, J. N. 1984, *Ap. J. (Letters)*, **277**, L5.
Fry, J. N. 1985, *Ap. J.*, **289**, 10.
Fry, J. N. 1986, *Ap. J.*, **306**, 358.
Fry, J. N., Giovanelli, R., Haynes, M. P., Melott, A. L., and Scherrer, R. J. 1989, *Ap. J.*, **340**, 11.
Geller, M. J., de Lapparent, V., and Kurtz, M. J. 1984, *Ap. J. (Letters)*, **287**, L55.
Geller, M. J., and Huchra, J. P. 1988, in *Large Scale Motions in the Universe: A Vatican Study Week*, ed. V. C. Rubin and G. V. Coyne, S. J., (Princeton: Princeton University), p. 3.
Geller, M. J., and Huchra, J. P. 1989, *Science*, **246**, 897.
Giovanelli, R., and Haynes, M. P. 1988, in *IAU Symposium No. 130, Large-Scale Structures in the Universe*, ed. Audouze, J., Pelletan, M.-C., and Szalay, A., (Dordrecht: Kluwer), p. 113.
Giovanelli, R., and Haynes, M. P. 1989, *A. J.*, **97**, 633.
Giovanelli, R., Haynes, M. P., and Chincarini, G. L. 1986, *Ap. J.*, **300**, 77.
Goncharov, A. S., Linde A. D., and Mukhanov, V. F. 1987, *Int. J. Mod. Phys.*, **A2**, 561.
Górski, K., Hoffman, Y., and Bertschinger, E. 1989, preprint.
Górski, K., Davis, M., Strauss, M. A., White, S. D. M., and Yahil, A. 1989, *Ap. J.*, **344** 1.
Gott, J. R., III, and Turner, E. L. 1977, *Ap. J.*, **216**, 357.
Gott, J. R., III, Melott, A. L., and Dickinson, M. 1986, *Ap. J.*, **306**, 341.
Gott, J. R., III, Miller, J., Thuan, T. X., Schneider, S. E., Weinberg, D. H., Gammie, C., Polk, K., Vogeley, M., Scott, J. S., Bhavsar, S. P., Melott, A. L., Giovanelli, R., Haynes M. P., Tully. R. B., and Hamilton, A. J. S. 1989, *Ap. J.*, **340**, 625.
Gott, J. R. III, Weinberg, D. H., Melott, A. L. 1987, *Ap. J.*, **319**, 1.
Gray, P. M. 1986, *Proc. S.P.I.E.*, **627**, 96.
Gregory, S. A., and Thompson, L. A. 1978, *Ap. J.*, **222**, 784.
Grevesse, N., and Anders, E. 1988, *Symposium on Cosmic Abundances of Matter*, ed. C. J. Waddington (AIP).
Gross, D. J., Pisarski, R. D., and Yaffe, L. G. 1981, *Rev. Mod. Phys.*, **53**, 43.
Groth, E. J., and Peebles, P. J. E. 1977, *Ap. J.*, **217**, 385.
Groth, E. J., and Peebles, P. J. E., 1986, *Ap. J.*, **310**, 499.
Groth, E. J., Juszkiewicz, R., and Ostriker, J. P. 1989, *Ap. J.*, **344**, 558.
Gull, S. F., and Daniell, G. J. 1978, *Nature*, **272**, 686.
Guerra, J. M., Marques, G. C., and Rodrigues, S. J. 1989, preprint.
Gunn, J. 1989, in *The Extragalactic Distance Scale*, ed. S. van den Bergh and C. J. Pritchet, *A.S.P. Conf. Ser.*, **4**, (Provo: Astronomical Society of the Pacific), p. 344.
Gunn, J. E., and Gott, J. R. 1972, *Ap. J*, **176**, 1.
Gurbatov, S. N., Saichev, A. I., and Shandarin, S. F. 1985, *Sov. Phys. Doklady*, **30**, 921.
Gurbatov, S. N., Saichev A. I., and Shandarin, S. F. 1989, *M.N.R.A.S.*, **236**, 385.
Guth, A., and Pi, S.- Y. 1982, *Phys. Rev. Lett.* **49**, 110.
Guth, A. H. 1981, *Phys. Rev.*, **D23**, 347.
Hamilton, A. J. S. 1988, *Ap. J. (Letters)*, **331**, L59.

Hamilton, A. J. S., and Gott, J. R., III 1988, *Ap. J.*, **331**, 641.
Hamilton, A. J. S., Gott, J. R., III,, and Weinberg, D. H. 1986, *Ap. J.*, **309**, 1.
Harmon, R. T., Lahav, O., and Meurs, E. J. A. 1987, *M.N.R.A.S.*, **228**, 5.
Hauschildt, M. 1987, *Astr. Ap.*, **184**, 45.
Hausman, M. A., Olson, D. W., and Roth, B. D. 1983, *AP. J.*, **270**, 351.
Hawking, S. 1982, *Phys. Lett.* **115B**, 295.
Haynes, M. P., and Giovanelli, R. 1986, *Ap. J. (Letters)*, **306**, L55.
Haynes, M. P., and Giovanelli, R. 1988, in *Large-Scale Motions in the Universe: A Vatican Study Week*, ed. V. C. Rubin and G. V. Coyne, S. J., (Princeton: Princeton University), p. 31.
Haynes, M. P., Giovanelli, R., and Chincarini, G. L. 1984, *Ann. Rev. Astr. Ap.*, **22**, 445.
Heckman, T. M., Sancisi, R., Sullivan, III, W. T., Balick, B. 1982, *M.N.R.A.S.*, **199**, 425.
Hernquist, L., and Katz, N. 1989, *Ap. J. Suppl.*, **70**, 419.
Hewett, P. C. 1982, *M.N.R.A.S.*, **201**, 867.
Heydon-Dumbleton, N. H., Collins, C. A., and MacGillivray, H. T., 1989, *M.N.R.A.S.*, **238**, 379.
Hill, C. W., Schramm, D. N., and Fry, J. N. 1989, *Comments Nucl. Part. Phys.*, **19**, 25.
Hirasawa, T., Aizu, K., and Taketani, M. 1969, *Prog. Theor. Phys.*, **41**, 835.
Hodges, H. 1989, *Phys. Rev.* **D39**, 3568.
Hodges, H. 1989, *Phys. Rev.* **D40**, 1798.
Hodges, H., Blumenthal, G. R., Kofman, L., and Primack, J. R. 1989, UCSC preprint SCIPP-89/31, *Nucl. Phys.*, submitted.
Hoffmann, Y., and Shaham, J. 1983, in *IAU. Symp. 104, Early Evolution of the Universe and Its Present Structure*, ed. G. O. Abell and G. Chincarini, (Dordrecht: Reidel), p. 203.
Hoffmann, Y., and Zurek, W. H. 1988, *Nature*, **333**, 46.
Hollósi, J., and Efstathiou, G. 1988, in *Large-Scale Structures in the Universe. Observational and Analytical Methods.*, ed. W. C. Seitter, H. W. Duerbeck, and M. Tacke, (Berlin: Springer), p. 106.
Holtzman, J. 1989, *Ap.J. Suppl.*, **71**, 1.
Hopp, U., and Materne, J. 1985, *Astr. Ap. Suppl.*, **61**, 93.
Hubble, E., and Humason, M. 1931, *Ap. J.*, **74**, 43.
Hubble, E. P. 1929, *Proc. Nat. Acad. Sci.*, **15**, 158.
Huchra, J. P. 1985, in *The Virgo Cluster of Galaxies*, ed. O.-G. Richter and B. Binggeli, (Garching: European Southern Observatory), p.181.
Huchra, J. P. 1989, in *The Extragalactic Distance Scale*, ed. S. van den Bergh and C. J. Pritchet, *A.S.P. Conf. Ser.*, **4**, (Provo: Astronomical Society of the Pacific), p. 257.
Huchra, J. P., Davis, M., Latham, D. W., and Tonry, J. 1983, *Ap. J. Suppl.*, **52**, 89.
Huchra, J. P., Geller, M. J., de Lapparent, V., and Corwin, H. E. 1990, *Ap. J. Suppl*, **72**, 433.
Hutchins, J. B. 1976, *Ap. J.*, **205**, 103.
Icke, V. 1984, *M.N.R.A.S.*, **206**, 1P.
Icke, V., and Van der Weygaert, R. 1987, *Astr. Ap.*, **184**, 16.
Iovino, A., Melnick, J., and Shaver, P. 1988, *Ap. J. (Letters)*, **330**, L17.
IRAS Catalogs and Atlases, Explanatory Supplement 1985, ed. C. A. Beichman, G. Neugebauer, H. J. Habing, P. E. Clegg, and T. J. Chester (Washington DC: U.S. Government Printing Office).
IRAS Point Source Catalog 1985, Joint *IRAS* Science Working Group (Washington DC: U.S. Government Printing Office).
Johri, V. B. 1977, *M.N.R.A.S.*, **178**, 395.

Juszkiewicz, R., Vittorio, N., and Wyse, R. J. 1990, *Ap. J.*, **349**, 408.
Kaiser, N. 1983, *Ap. J. (Letters)*, **273**, L17.
Kaiser, N. 1984, *Ap. J. (Letters)*, **284**, L9.
Kaiser, N. 1986, *M.N.R.A.S.*, **219**, 785.
Kaiser, N. 1988, in *Large-Scale Structures and Motions in the Universe*, ICTP Trieste, in press.
Kaiser, N., and Lahav, O. 1988, in *Large-Scale Motions in the Universe*, ed. V. C. Rubin and G. V. Coyne, S. J., (Princeton: Princeton University), p. 339.
Kaiser, N., and Lahav, O. 1989, *M.N.R.A.S.*, **237**, 129.
Kalinkov, M., Stavrev, K., Kuneva, I. 1985, *Astr. Nachr.*, **306**, 283.
Kandrup, H. 1989, *Phys. Rev.* **D39**, 2245.
Kashlinsky, A. 1987, *Ap. J.*, **317**, 19.
Kashlinsky, A. 1989, *Ap. J. (Letters)*, **343**, L5.
Kerr, F. J., and Westerhout, G. 1965, in *Galactic Structure*, (Chicago: University of Chicago), p. 186.
Kibblewhite, E. J., Bridgeland, M. T., Bunclark, P., Cawson, M., and Irwin, M. J. 1983, in *Astronomy with Schmidt-type Telescopes*, ed. M. Capaccioli, (Dordrecht: Reidel), p. 89.
Kibblewhite, E. J., Bridgeland, M. T., Bunclark, P., and Irwin, M. J., 1984, *Astronomical Microdensitometry Conference*, NASA Conf. Pub., **2317**, p. 277.
Kirshner, R. P., Oemler, A., Schechter, P. L., and Shectman, S. A. 1981, *Ap. J. (Letters)*, **248**, L57.
Klypin, A. A., and Kopylov, A. I. 1983, *Sov. Astr. Letters*, **9**, 41.
Klypin, A. A., and Shandarin, S. F. 1983, *M.N.R.A.S.*, **204**, 891.
Kofman, L. A. 1986, *Phys. Lett.* **173B**, 400.
Kofman, L. A. 1989 in *Morphological Cosmology*, ed. P. Flin and H. E. Duerbeck, *Lecture Notes in Physics*, **332**, (Berlin: Springer), p. 354.
Kofman, L. A. 1989, preprint.
Kofman, L. A., and Linde, A. D. 1987, *Nucl. Phys.*, **282B**, 508.
Kofman, L. A., and Linde, A. D. 1987, *Nucl. Phys.*, **282B**, 555.
Kofman, L. A., and Pogosyan, D. Yu. 1988, *Phys. Lett.*, **214B**, 508.
Kofman, L. A., and Shandarin, S. F. 1988, *Nature*, **334**, 129.
Kofman, L. A., Pogosyan, D. Yu., and Shandarin, S. 1990, *M.N.R.A.S.*, **242**, 290.
Kosterlitz, J. M., and Thouless, D. J. 1973, *J. Phys.*, C61181.
Kotok, E. V., and Shandarin, S. F. 1987, *Sov. Astr.*, **64**, 1144.
Kraan-Korteweg, R. C. 1988, in *Large-Scale Structure: Observations and Instrumentation*, ed. C. Balkowski and S. Gordon (Meudon: Observatoire de Paris), p. 155.
Kraan-Korteweg, R. C. 1989, in preparation.
Krumm, N., van Driel, W., van Woerden, H. 1985, *Astr. Ap.* **144**, 202.
Kuhn, J. R., and Uson, J. M. 1982, *Ap. J. (Letters)*, **263**, L47.
Lahav, O. 1986, *M.N.R.A.S.*, **220**, 259.
Lahav, O. 1987, *M.N.R.A.S.*, **225**, 213.
Lahav, O., Edge, A. C., Fabian, A. C., and Putney, A. 1989, *M.N.R.A.S.*, **238**, 881.
Lahav, O., Rowan-Robinson, M., and Lynden-Bell, D. 1988, *M.N.R.A.S.*, **234**, 677.
Lambas, D. G., Groth, E. J., and Peebles, P. J. E. 1988, *A. J.*, **95**, 975.
Lambas, D. G., Groth, E. J., and Peebles, P. J. E. 1988, *A. J.*, **95**, 996.
Landolt, A. U. 1983, *A. J.*, **88**, 439.
Lanzetta, K. M., Turnshek, D. A., and Wolfe, A. M. 1987, *Ap. J.*, **322**, 739.
Lauberts, A. 1982, *The ESO/Uppsala Survey of the ESO(B) Atlas*, (Garching: European Southern Observatory).

Lepp, S., and Shull, J. M. 1983, *Ap. J.*, **270**, 578.
Lepp, S., and Shull, J. M. 1984, *Ap. J.*, **280**, 465.
Lemos, J. P. S., 1988, *M.N.R.A.S.*, **230**, 451.
Lemos, J. P. S., 1989, *Class. Quantum Grav.*, submitted.
Letelier, P. S., Oliveira, S. R., 1988, *Class. Quantum Grav.*, **5**, L47.
Lilje, P. B., and Efstathiou, G. 1988, *M.N.R.A.S.*, **231**, 635.
Lin, C. C., Mestel, L., and Shu, F. H. 1965, *Ap. J.*, **142**, 1431.
Linde, A. D. 1982, *Phys. Lett.*, **108B**, 389.
Linde, A. D. 1983, *Phys. Lett.*, **129B**, 177.
Linde, A. D. 1984, *JETP Lett.*, **40**, 1333.
Linde, A. D. 1985, *Phys. Lett.*, **158B**, 375.
Linde, A. D. 1986, *Phys. Lett.*, **175B**, 395.
Loveday, J., Efstathiou, G., and Peterson, B. A., 1989, in preparation.
Lubin, P. M., and Villela, T. 1986, in *Galaxy Distances and Deviations from Universal Expansion*, ed. B. F. Madore and R. B. Tully, (Dordrecht: Reidel) p. 169.
Lucchin, F., and Matarrese, S. 1985, *Phys. Rev.*, **D32**, 1316.
Lucchin, F., and Matarrese, S. 1988, *Ap. J.*, **330**, 535.
Lucchin, F., Matarrese, S., and Ortolan, A. 1990, in preparation.
Lucchin, F., Matarrese, S., and Vittorio, N. 1988, *Ap. J. (Letters)* **330**, L21.
Lucey, J. R., and Carter, D. 1988, *M.N.R.A.S.*, **235**, 1177.
Lucey, J. R., Currie, M. J., and Dickens, R. J. 1986, *M.N.R.A.S.*, **221**, 453.
Lucey, J. R., Currie, M. J., and Dickens, R. J. 1986, *M.N.R.A.S.*, **222**, 427.
Lucey, J. R., Dickens, R. J., and Dawe, J. D. 1980, *Nature*, **285**, 305.
Lynden-Bell, D. 1987, *Quart. J.R.A.S.*, **28**, 187.
Lynden-Bell, D. 1988, *Observatory*, **108**, 145.
Lynden-Bell, D., and Lahav, O. 1988, in *Large-Scale Motions in the Universe: A Vatican Study Week*, ed. V. C. Rubin and G. V. Coyne, S. J., (Princeton: Princeton University), p. 199.
Lynden-Bell, D., Faber, S. M., Burstein, D., Davies, R. L., Dressler, A., Terlevich, R. J., and Wegner, G. 1988, *Ap. J.*, **326**, 19.
Lynden-Bell, D., Katz, J., and Redmount, I. H. 1989, *M.N.R.A.S.*, **239**, 201.
Lynden-Bell, D., Lahav, O., and Burstein, D. 1989, *M.N.R.A.S.*, **241**, 325.
Lynden-Bell, D., Pineault, S., 1978, *M.N.R.A.S.*, **186**, 679.
Lyth, D. H. 1985, *Phys. Rev.*, **D31**, 1792.
MacGillivray, H. T., Dodd, R. J., MacNally, B. V., Corwin, H. G. 1982, *M.N.R.A.S.*, **198**, 605.
Maddox, S. J. 1988, Ph.D. dissertation, University of Cambridge.
Maddox, S. J., Efstathiou, G., and Loveday J., 1988, in *IAU Symposium 130, Large-Scale Structures in the Universe*, ed. J. Audouze, M.-C. Pelletan, and A. Szalay, (Dordrecht: Kluwer), p. 151.
Maddox, S. J., Sutherland, W. J., Efstathiou, G., and Loveday J., 1990a, *M.N.R.A.S.*, **243**, 692.
Maddox, S. J., Sutherland, W. J., Efstathiou, G., and Loveday J., 1990b, *M.N.R.A.S.*, **246**, 433.
Matarrese, S., Ortolan, A., and Lucchin, F. 1989, *Phys. Rev.*, **D40**, 290.
Matsuda, T., Sato, H., and Takeda, H. 1969, *Prog. Theor. Phys.*, **42**, 219.
Maurellis, A., Fairall, A. P., Matravers, D., and Ellis, G. F. R. 1989, *Astr. Ap.*, submitted.
Maurogordato, S., and Lachieze-Rey, M. 1987, *Ap. J.*, **320**, 13.
Meiksin, A., and Davis, M. 1986, *A.J.*, **91**, 191.
Melnick, J., and Sargent, W. L. W. 1977, *Ap. J.*, **215**, 401.

Melott, A. L., Cohen, A. P., Hamilton, A. J. S., Gott, J. R., III, and Weinberg, D. H. 1989, *Ap. J.*, **345**, 618.
Melott, A. L., Einasto, J., Saar, E., Suisalu, I., Klypin, A. A., Shandarin, S. F. 1983, *Phys. Rev. Lett.*, **51**, 935.
Melott, A. L., and Shandarin, S. F. 1989, *Ap. J.*, **343**, 26.
Melott, A. L., Weinberg, D. H., and Gott, J. R., III 1988, *Ap. J.*, **328**, 560.
Merritt, D. 1987, *Ap. J.*, **313**, 121.
Messina, A., Moscardini, L., Lucchin, F., and Matarrese, S. 1989, in preparation.
Mestel, L. 1965a, *Quart. J. R.A.S*, **6**, 161.
Mestel, L. 1965b, *Quart. J. R.A.S*, **6**, 265.
Meurs, E. J. A., and Harmon, R. T. 1988, *Astr. Ap.*, **206**, 53.
Michel, A. and Huchra, J. P. 1988, *P.A.S.P.*, **100**, 1423.
Monaghan, J. J. 1985, *Comp. Phys. Reports*, **3**, 71.
Moody, J. E., Turner, E. L., and Gott, J. R., III 1983, *Ap. J.* **273**, 16.
Morton, D. C., Krug, P. A., and Tritton, K. P., *M.N.R.A.S.*, **212**, 325.
Moss, C., and Dickens, R. J. 1977, *M.N.R.A.S.*, **178**, 701.
Mukhanov, V. 1985, *JETP Lett.* **41**, 493.
Mukhanov, V., and Chibisov, G. 1981, *JETP Lett.* **33**, 532.
Mukhanov, V., Kofman, L. A., and Pogosyan, D. Yu. 1987, *Phys. Lett.* **157B**, 427.
Narayan, R., and White, S. D. M. 1988, *M.N.R.A.S.*, **231**, 97P.
Neckel, Th., and Klare, G. 1980, *Astr. Ap. Suppl.*, **42**, 251.
Nielsen, H. K., and Olesen, P. 1973, *Nucl. Phys.*, **B61**, 45.
Nilson, P. 1973, *Uppsala General Catalogue of Galaxies*, (Uppsala: University of Uppsala).
Nusser, A., and Dekel A. 1990, *Ap. J.*, **362**, 14.
Oemler, A. 1974, *Ap. J.*, **194**, 1.
Olive, K. 1989, "Inflation", University of Minnesota preprint.
Olivier, S., Blumenthal, G. R., Dekel, A., Primack, J. R., and Stanhill, D. 1990, *Ap. J.* **356**, 1.
Olowin, R. P., Chincarini, G, and Corwin, H. G. 1987, *B.A.A.S.*, Vol. **19**, No. 4.
Ortolan, A., Lucchin, F., and Matarrese, S. 1988, *Phys. Rev.*, **D38**, 465.
Ostriker, J. P., and Cowie L. L. 1981, *Ap. J. (Letters)*, **243**, L127.
Ostriker, J. P., and Suto, S. 1990, **348**, 378.
Otto, S., Politzer, H. D., Preskill, J., and Wise, M.B. 1986, *Ap. J.*, **304**, 62.
Owen, F. N., White, R. A., Thronson, H. A. 1988, *A.J.*, **95**, 1.
Palmer, P. L. 1981, *M.N.R.A.S.*, **197**, 721.
Park, C., Gott, J. R., and Melott, A. L. 1989, in preparation.
Park, C., Gott, J. R., da Costa, L. N., and Melott, A. L. 1989, in preparation.
Peach, J. V. 1970, *Ap. J.* 159, 753.
Peacock, J. A., and Heavens, A. F. 1985, *M.N.R.A.S.*, **217**, 805.
Peebles, P. J. E. 1968, *Ap. J.*, **153**, 1.
Peebles, P. J. E. 1971, *Physical Cosmology*, (Princeton: Princeton University).
Peebles, P. J. E. 1973, *Ap. J.*, **185**, 413.
Peebles, P. J. E. 1976, *Ap. J.*, **205**, 318.
Peebles, P. J. E. 1976, *Astr. Ap.*, **53**, 131.
Peebles, P. J. E. 1980, *The Large-Scale Structure of the Universe*, (Princeton: Princeton University).
Peebles, P. J. E. 1983, *Ap. J.*, **274**, 1.
Peebles, P. J. E. 1986, *Nature*, **321**, 27.
Peebles, P. J. E. 1987, *Ap. J. (Letters)*, **315**, L73.
Peebles, P. J. E. 1987, *Nature*, **327**, 210.

Peebles, P. J. E. 1988, *Ap. J.*, **332**, 17.

Peebles, P. J. E. 1988, in *Large-Scale Motions in the Universe: A Vatican Study Week*, ed V. C. Rubin and G. V. Coyne, S. J., (Princeton: Princeton University), p. 457.

Peebles, P. J. E. 1988, in *Large-Scale Structure and Motions in the Universe*, ed. G. Giuricin *et al.* (Dordrecht: Kluwer).

Peebles, P. J. E., Daly, R. A., and Juszkiewicz, R. 1989, *Ap. J.* **347**, 563.

Peebles, P. J. E., and Groth, E. J. 1975, *Ap. J.*, **196**, 1.

Peebles, P. J. E., and Hauser, M. G. 1974, *Ap. J. Suppl.*, **28**, 19.

Pellegrini, P. S., and da Costa, L. N. 1990, *Ap. J.*, **357**, 408.

Pellegrini, P. S., da Costa, L. N., Huchra, J. P., Latham, D. W., and Willmer, C. 1990, *A.J.* **99**, 751.

Pellegrini, P. S., Willmer, C. N., da Costa, L. N., and Santiago, B. X. 1990, *Ap. J.*, **350**, 95.

Penrose, R., 1968, in *Batelle Rencontres*, ed. C. M. DeWitt and J. A. Wheeler, (New York: Benjamin).

Pettini, M., Boksenberg, A., and Hunstead, R. W. 1988, *Proceedings of the NATO Conference on The Epoch of Galaxy Formation*.

Phillipps, S., Fong, R., Ellis, R. S., Fall, S. M., and MacGillivray, H. T., 1978, *M.N.R.A.S.*, **182**, 673.

Pierce, M. J., and Tully, R. B. 1988, *Ap. J.*, **330**, 579.

Plionis, M. 1988, *M.N.R.A.S.*, **234**, 401.

Pogosyan, D. Yu. 1989, preprint.

Postman, M., and Geller, M. J. 1984, *Ap. J.*, **281**, 95.

Postman, M., Geller, M. J., and Huchra, J. P. 1986, *A.J.*, **91**, 1267.

Press, W. H., Flannary, B. P., Teukolsky, S. A., and Vetterling, V. T. 1986, *Numerical Recipes*, (Cambridge, Cambridge University).

Press, W. H., and Schechter, P. L. 1974, *Ap. J.*, **187**, 425.

Press, W. H., and Spergel, D. N. 1989, in preparation.

Press, W. H., Ryden, B. S., and Spergel, D. N. 1989, *Ap. J.* **347**, 590.

Ramella, M., Geller, M. J., and Huchra, J. P. 1989, *Ap. J.* **344**, 57.

Ramella, M., Geller, M. J., and Huchra, J. P. 1990, *Ap. J.*, **353**, 51.

Ratra, B. 1989, preprint.

Raychaudhury, S. 1989, *Nature*, **342**, 251.

Rhee, G. F. R. N., Katgert, P. 1988, *Astr. Ap. Suppl.*, **72**, 243.

Ritcher, O-G. 1984, *Astr. Ap. Suppl.*, **58**, 131.

Rubin, V. C., and Coyne, G. V., S. J. 1988, *Large-Scale Motions in the Universe: A Vatican Study Week*, (Princeton: Princeton University).

Ryden, B. S. 1988, *Ap. J. (Letters)*, **333**, L41.

Ryden, B. S., Melott, A. L., Craig, D. A., Gott, J. R. III, Weinberg, D. H., Scherrer, R. J., Bhavsar, S. P., and Miller, J. M. 1989, *Ap. J.*, **340**, 647.

Ryden, B. S., Scherrer, R. J., Melott, A. L., Craig, D. A., Gott, J. R., III, Weinberg, D. H., Bhavsar, S. P., and Miller, J. M. 1989, *Ap. J.*, **340**, 647.

Ryden, B. S., and Turner, E. L. 1984, *Ap. J. (Letters)*, **287**, L59.

Salopek, D. S., Bond, J. R., and Bardeen, J. M. 1989, *Phys. Rev. D.* **40**, 1753.

Salzer, J. J., Murray Hanson, M., and Gavazzi, G., 1990, *Ap. J.*, **353**, 39.

Sancisi, R. 1987, in *QSO Absorption Lines: Probing the Universe*, ed. J. C. Blades, D. A. Turnshek, and C. A. Norman, (Baltimore: Space Telescope Science Institute), p. 241.

Sancisi, R., Allen, R. J. 1979, *Astr. Ap.* **74**, 73.

Sancisi, R., van Woerden, H., Davies, R. D., Hart, L. 1984, *M.N.R.A.S.*, **210**, 497.

Sandage, A., and Tammann, G. A. 1984, in *Large-Scale Structure of the Universe,*

Cosmology and Fundamental Physics, First ESO-CERN Symp., ed. G. Setti and L. van Hove, (Garching: European Southern Observatory), p. 127.
Sandage, A., Binggeli, B., and Tammann, G. A. 1985, *A. J.*, **90**, 1759.
Santiago, B. X., and da Costa, L. N. 1990, *Ap. J.*, **362**, 386.
Sarazin, C. L., Rood, H. J., Struble, M. F. 1982, *Astr. Ap.*, **108**, L7.
Sargent, W. L. W. 1987, in *IAU Symposium no. 124, Observational Cosmology*, ed. A. Hewitt, G. Burbidge, and L. Z. Fang, (Dordrecht: Reidel), p. 777.
Sargent, W. L. W. 1988, in *IAU Symposium No. 130, Large-Scale Structures in the Universe*, ed. Audouze, J., Pelletan, M.-C., and Szalay, A., (Dordrecht: Kluwer).
Sargent, W. L. W., Boksenberg, A., and Steidel, C. C. 1988, *Ap. J. Suppl.*, **68**, 539.
Sargent, W. L. W., Young, P. J., Boksenberg, A., and Tytler, D. 1980, *Ap. J. Suppl.*, **42**, 41.
Sasaki, M., Nambu, Y., and Nakao, K. 1988, *Nucl. Phys.* **308B**, 868.
Saslaw, W. C., and Hamilton, A. J. S. 1984, *Ap. J.*, **276**, 13.
Scaramella, R., Baiesi-Pillastrini, G., Chincarini, G., Vettolani, G., and Zamorani, G. 1989, *Nature*, **338**, 562.
Schaeffer, R. 1984, *Astr. Ap. Lett.*, **134**, L15.
Schmidt, K. -H. 1986, *Astr. Nachr.*, **307**, 69.
Schneider, S. E. 1985, *Ap. J. (Letters)*, **288**, L33.
Schneider, S. E., Helou, G., Salpeter, E. E., and Terzian, Y. 1983, *Ap. J. (Letters)*, **273**, 1.
Seldner, M., and Peebles, P. J. E. 1977, *Ap. J.*, **215**, 703.
Seldner, M., Siebers, B., Groth, E. J., and Peebles, P. J. E. 1977, *A.J.*, **82**, 249.
Shandarin, S. F. 1988, in *IAU Symposium No. 130, Large-Scale Structures in the Universe*, ed. J. Audouze, M.-C., Pelletan, and A. Szalay, (Dordrecht: Kluwer), p. 273.
Shandarin S. F., and Zel'dovich, Ya. B. 1989, *Rev. Mod. Phys.*, **61**, 185.
Shane, C. D., and Wirtanen, C. A. 1967, *Publ. Lick. Obs.*, **22**, Part 1.
Shapiro, P. R. 1984, in *Clusters and Groups of Galaxies*, ed. F. Mardirossian, G. Giuricin, and M. Mezzetti (Dordrecht: Reidel), p. 447.
Shapiro, P. R. 1989, in *Fourteenth Texas Symposium on Relativistic Astrophysics*, ed. E. J. Fenyves, *Ann. N. Y. Acad. Sci.*, in press.
Shapiro, P. R., and Struck-Marcell, C. 1985, *Ap. J. Suppl.*, **57**, 205.
Shapiro, P. R., Struck-Marcell, C., and A. L. Melott 1983, *Ap. J.*, **275**, 413.
Shapley, H. 1936, *Harvard Coll. Obs. Bull.*, **903**.
Sharp, N. A. 1981, *M.N.R.A.S.*, **191**, 857.
Sharp, N. A., Bonometto, S. A., and Lucchin, F. 1984, *Astr. Ap.*, **130**, 79.
Shaver, P. A. 1989, in *The quest for the Fundamental Constants in Cosmology*, 24th Rencontres de Moriond, in press.
Shaver, P. A., and Pierre, M. 1988, *Astr. Ap.* submitted.
Shaya, E. J. 1984, *Ap. J.*, **280**, 470.
Shectman, S. A. 1985, *Ap. J. Suppl.*, **57**, 77.
Silk, J. 1974, *Ap. J.*, **193**, 525.
Silk, J. 1977, *Ap. J.*, **211**, 638.
Silk, J. 1982, *Ap. J.*, **256**, 514.
Silk, J. 1983, *M.N.R.A.S.*, **205**, 705.
Silk, J. 1989, *Ap. J. (Letters)*, **345**, L1.
Sodré, L., Capelato, H. V., Steiner, J. E., and Mazure, A., 1989, *A.J.*, **97**, 1279.
Soifer, B. T. *et al.* 1984, *Ap. J. (Letters)*, **278**, L71.
Souriau, J. M. 1974, in *Coll. Internationaux CNRS*, **237**, 59.
Souriau, J. M. 1985, *C.R. Académie des Sciences*, serie gén. **2**, 213.

Starobinsky, A. A. 1986, in *Field Theory, Quantum Gravity and Strings*, ed. H. J. de Vega and N. Sanchez, *Lecture Notes in Physics*, **246**, (Berlin: Springer), p. 107.

Starobinsky, A. 1982, *Phys. Lett.*, **117B**, 175.

Starobinsky, A. 1985, *JETP Lett.* **42**, 152.

Staveley-Smith, L., and Davies, R. D. 1989, *M.N.R.A.S.* **241**, 781.

Stebbins, A., and Turner, M. S. 1989, *Ap. J. (Letters)*, **339**, L13.

Strauss, M. A. 1989, Ph.D. dissertation, University of California, Berkeley.

Strauss, M. A., and Davis, M. 1988, in *IAU Symposium 130, Large-Scale Structures of the Universe*, ed. J. Audouze, M.-C. Pelletan, and A. Szalay (Dordrecht: Reidel), p. 191.

Strauss, M. A., and Davis, M. 1988, in *Large-Scale Motions in the Universe*, ed. V. C. Rubin and G.V. Coyne, S. J., (Princeton: Princeton University), p. 255.

Strauss, M., A., Davis, M., Yahil, A., and Huchra, J. P. 1990, *Ap. J.*, **361**, 49.

Struble, M. F., and Peebles, P. J. E. 1985, *A. J.* **90**, 582.

Struble, M. F., and Rood, H. J. 1987, *Ap. J. Suppl.*, **63**, 543.

Struble, M. F., and Rood, H. J. 1987, *Ap. J. Suppl.*, **63**, 555.

Struck-Marcell, C., and Shapiro, P. R. 1989, in preparation.

Sutherland, W. J. 1988, *M.N.R.A.S.*, **234**, 159.

Sutherland, W. J., and Efstathiou, G., 1989, in preparation.

Suzuki, M. 1981, *Adv. Chem. Phys.*, **46**, 195.

Szalay, A. S., Hollósi, J., and Tóth, G. 1989, *Ap. J. (Letters)*, **339**, L5.

Tammann, G. A., and Binggeli, B. 1988, in *High Redshift and Primeval Galaxies*, ed. J. Bergeron, D. Kunth, B. RoccaVolmerange, and J. Tran Thanh Van, (Paris: Editions Frontieres), p.149.

Tammann, G. A., and Sandage, A. 1985, *Ap. J.*, **294**, 81.

Teague, P. 1988, Ph.D. dissertation, Australian National University, Canberra.

The, L. S., and White, S. D. M. 1986, *A. J.*, **92**, 1248.

Thorstensen, J., Wegner, G., Hamwey, R., Boley, F., Geller, M. J., Huchra, J. P., and Kurtz, M. J. 1989, *A.J.*, **98**, 1143.

Thuan, T. X., Gott, J. R., III, and Schneider, S. E. 1987, *Ap. J. (Letters)*, **315**, L93.

Tóth, G., Hollósi, J., and Szalay, A. S. 1989, *Ap. J.*, **344**, 75.

Triay, R. 1981, Ph.D. dissertation.

Tully, R. B. 1987, *Ap. J.*, **323**, 1.

Tully, R. B. 1988, *A.J.*, **96**, 73.

Tully, R. B. 1988, *Nature*, **334**, 209.

Tully, R. B., and Fisher, J. R. 1977, *Astr. Ap.*, **54**, 661.

Tully, R. B., and Fisher, J. R. 1987, *Nearby Galaxies Atlas*, (Cambridge: Cambridge University).

Tully, R. B., and Shaya, E. J. 1984, *Ap. J*, **281**, 31.

Turok, N. 1989, *Phys. Rev. Letters* **63**, 2625.

Veron-Cetty, M. P., Veron, P. 1987, *Catalog of Quasars and Active Nuclei, 3rd Edition*, (Garching: European Southern Observatory).

Viegas-Aldrovandi, S. M., and Gruenwald, R. B. 1989, in *Proceedings of the II Texas-Mexico Conference on Star Forming Regions*, in press.

Vilenkin, A. 1985, *Phys. Reports*, **121**, 263.

Vilenkin, A., 1987, in *300 Years of Gravitation*, ed. S. W. Hawking and W. Israel, (Cambridge: Cambridge University).

Villumsen, J. J. 1989, *Ap. J. Suppl.*, **71**, 407.

Vishniac, E. 1986, in *Inner Space/Outer Space*, ed. E. W. Kolb, M. S. Turner, Olive, K., Seckel, D., and Lindley, D. (Chicago: University of Chicago), p. 190.

Vittorio, N., Juszkiewicz, R., and Davis, M. 1986, *Nature*, **323**, 132.

Vorontsov-Velyaminov, B. A., and Arhipova, V. P. 1963-1974, *Morphological Catalog of Galaxies*, (Moscow: Moscow University), Parts 2 to 5.

Walker, A. R., 1984, *M.N.R.A.S.*, **209**, 83.

Weinberg, D. H. 1988, *Pub. A.S.P.*, **100**, 1373.

Weinberg, D. H., Gott, J. R., III, and Melott, A. L. 1987, *Ap. J.*, **321**, 1.

Weinberg, D. H. 1989, Ph. D. thesis, Princeton University.

Weinberg, D. H., and Gunn, J. E. 1989, *Ap. J. (Letters)*, **352**, L25.

Weinberg, D. H., Ostriker, J. P., and Dekel, A., 1989, *Ap. J.*, **336**, 9.

Weinberger, R. 1980, *Astr. Ap. Suppl.* **40**, 123.

Weliachev, L., Sancisi, R., Guelin, M. 1978, *Astr. Ap.*, **65**, 37.

West, M. J., and Richstone, D. O. 1988, *Ap. J.*, **335**, 532.

West, M. J., Dekel, A., and Oemler, Jr., A. 1989, *Ap. J.* **336**, 46.

Weyman, R. J., Williams, R. E., Peterson, B. M., and Turnshek, D. A. 1979, *Ap. J.*, **234**, 33.

White, S. D. M. 1979, *M.N.R.A.S.*, **186**, 145.

White, S. D. M., Frenk, C. S., Davis, M., and Efstathiou, G. 1987, *Ap. J.* **313**, 505.

White, S. D. M., and Silk, J. 1979, *Ap. J.*, **231**, 1.

White, S. D. M., Frenk, C. S., Davis, M., and Efstathiou, G., 1987a, *Nature*, **330**, 451.

White, S. D. M., Tully, R. B., and Davis, M. 1988, *Ap. J. (Letters)*, **333**, L45.

Wolstencroft, R. D., Clowes, R. G., Kalafi, M., Leggett, S. K., MacGillivray, H. T., and Savage, A. 1986, in *Light on Dark Matter*, ed. F. P. Israel (Dordrecht: Reidel), p. 425.

Xia, X. Y., Deng, Z. G., and Zhou, Y. Y. 1987, in *IAU Symposium No. 124, Observational Cosmology*, ed. A. Hewitt, G. Burbidge, and L. Z. Fang, (Dordrecht: Reidel), p. 363

Yahil, A. 1988, in *Large-Scale Motions in the Universe: A Vatican Study Week*, ed. V. C. Rubin and G. V. Coyne, S. J., (Princeton: Princeton University), p. 219.

Yahil, A., and Vidal, N. V. 1977, *Ap. J*, **214**, 347.

Yahil, A., Sandage, A., and Tammann, G. A. 1980, *Ap. J.*, **242**, 448.

Yahil, A., Walker, D., and Rowan-Robinson, M. 1986, *Ap. J. (Letters)*, **301**, L1.

Yoneyama, T. 1972, *P.A.S.J.*, **24**, 87.

York, D. G., Caulet, A., Rybskii, P., Gallagher, J., Blades, J. C., Morton, D. C., and Wamsteker, W. 1989, *Ap. J.*, submitted.

York, D. G., Dopita, M., Green, R., Bechtold, J. 1986, *Ap. J.*, **311**, 610.

Yoshioka, S., and Ikeuchi, S. 1989, *Ap. J.*, **341**, 16.

Young, P., Sargent, W. L. W., Boksenberg, A. 1982, *Ap. J. Suppl.*, **48**, 455.

Zel'dovich, Ya. B. 1970, *Astr. Ap.*, **5**, 84.

Zel'dovich, Ya. B., Kobzarev, I. Yu., and Okun, L. B. 1975, *Sov. Phys. JETP*, **40**, 1.

Zel'dovich, Ya. B., and Novikov, I. D. 1983, *The Structure and Evolution of the Universe*, (Chicago: University of Chicago).

Zwicky, F. 1933, *Helv. Phys. Acta*, 6, 110.

Zwicky, F., Herzog, E., Wild. P., Karpowicz, M., and Kowal, C., 1961-78, *Catalog of Galaxies and Clusters of Galaxies*, Vols. 1-6 (Pasadena: California Institute of Technology).